But Is It True?

But Is It True?

A Citizen's Guide to Environmental Health and Safety Issues

Aaron Wildavsky

Harvard University Press

Cambridge, Massachusetts, and London, England

First Harvard University Press paperback edition, 1997

Library of Congress Cataloging-in-Publication Data

Wildavsky, Aaron B.
But is it true? : a citizen's guide to environmental health
and safety issues / Aaron Wildavsky.
    p. cm.
Includes bibliographical references and index.
ISBN 0-674-08922-7 (cloth)
ISBN 0-674-08923-5 (pbk.)
1. Environmental risk assessment—Popular works.
2. Environmental policy—Citizen participation—Popular works.
I. Title.
GE145.W55 1995
363.7'0525—dc20   94-40407
CIP

Preface

Aaron Wildavsky died of lung cancer on September 4, 1993. At the time, the manuscript for this book was almost complete because he had given it a thorough going over while he was on leave at Maastricht. During the last few months of his illness, I started to help Aaron put some finishing touches to the manuscript, particularly to some of the last chapters. After he died I continued so that the book could be published.

In this effort, I am especially indebted to Doris Patton, Aaron's administrative assistant, who stayed on top of things and helped push the project along to completion, no small task considering the number of chapters and authors. The contribution of research assistants David Schleicher and Brendon Swedlow should also be noted. They kept me informed as to Aaron's intentions, particularly when I wanted to cut out a section or chapter. They read the chapters for content and tried where possible to improve the documentation. David Vogel helped edit Chapter 8 on rodent studies, while S. Fred Singer was invaluable for his scientific review of Chapters 9, 10, and 11, on acid rain, CFCs, and global warming. Diana Menkes did a great job tightening and unifying the prose of the various authors. I am sure that if Aaron were alive he would have many others to acknowledge and thank for their contributions. Finally, I want to thank Mary Wildavsky for giving me the opportunity to be of service to Aaron, my mentor, colleague, and friend for over thirty years.

Arnold J. Meltsner
Professor of Public Policy, Emeritus
University of California, Berkeley

Contents

But Is It True?

Introduction: Toward a Citizen's Understanding of Science and Technology

This is a book about citizenship in a scientific and technological age. Its subject is the relationship between knowledge and action in major environmental and safety issues. That relationship involves charges that modern technology can harm human beings, other living creatures, and the natural environment. The evidence cited is based on scientific understandings, which have been modified through political processes and have led to governmental actions. The charges and the countercharges by environmental and industrial groups, and the ensuing governmental decisions, are all bids for popular support. With that support governments enforce major changes in industrial practices; without that support the environmental movement would collapse.

The knowledge I have in mind is publicly available in the form of scientific observations, albeit necessarily incomplete and often written in specialized language. The action I contemplate for myself, my students, and my readers (citizens all) involves understanding the scientific bases for rival claims, engaging in informed discussion, and making reasoned judgments—that the claims are well founded, ill founded, deserving of governmental regulation, hardly worth anyone's attention, or somewhere in between. I do not envisage us as apprentice scientists or scientists at all. Rather, I think of us as reasoned deliberators capable of taking informed action in fields not necessarily determined by but infused with conflicting scientific and technological assertions.

Participation in decisions with scientific content is a tall order. Indeed, my own reaction at first hearing a lecture on ozone depletion and, separately, trying to read a scientific article on the grain fumigant EDB, was a combination of confusion and consternation: I literally did not understand a single thought. And, I realized, if I could not make informed decisions (despite practicing an occupation supposedly dedicated to the acquisition of knowledge), I therefore could not fulfill the requirements of citizenship. If that was so, the notion of contributing to self-government—my rationale for becoming and remaining a political scientist—was also in peril.

Fortunately, through trying to comprehend the scientific bases for action in the issues reported here, from Love Canal and DDT to Agent Orange, acid rain, and global warming, I have come to believe that we the citizenry can achieve an informed understanding. Were the requirement acquisition of substantial scientific knowledge, of course, the cause would be hopeless. No large population has ever achieved substantial knowledge across scientific fields. I doubt that even scientists can claim as much. In this regard, I think that scientific education for the general public and for high school and college students misses the mark. Most of us forget pretty quickly, if we ever understood, the content of physics or chemistry or the sequences of historical geology. Why should we remember? The knowledge is not tied to anything that matters to us. And we would have to know a great deal more than can be expected, that is, to know as much as scientists in those fields, to appreciate knowledge for its own sake. But by altering the criterion for citizen participation from extensive factual knowledge to basic procedural knowledge—to learning how to acquire the information one needs—I hope to demonstrate that all of us can be citizens in a scientific age.

Reading the scientific literature at the core of controversies over technological dangers and benefits is not easy. When I discussed my method for accomplishing this feat at a meeting of the National Nuclear Waste Management Board at Woods Hole, participants referred jocularly to the professor's "200-hour fix." Fair enough. "Difficult but doable" preserves citizenship; "impossible" does not.

Although I believe that the evidence warrants the conclusions this book presents about the risks of new technology, policy persuasion is not my main purpose. After due consideration of the evidence, including the ways in which it has been derived, readers may well rate the risks differently. What I hope most to convince them, whether the subject is ozone depletion or Superfund, is that they can learn to make informed judgments on environmental and safety issues. If more citizens when confronted with claims of harm begin to ask "But is it true?" I shall have achieved my purpose.

This is meant to be a self-exemplifying book, demonstrating the feasibility of citizenship in science. By revealing how I have reached my conclusions, I want to persuade readers that they can decide for themselves not only on the many matters discussed here but also on new issues as they arise. Thus it is important that I describe exactly how this book was researched and written.

Composition

Everything in this book is here at my instigation and every word is in accord with my views. That said, the bulk of the book results from a joint enterprise involving me and my students, graduate and undergraduate. The enterprise has been a collective one partially because of its breadth. In addition, engaging students, most of whom had no scientific background, was a deliberate maneuver to see if the method of learning I advocate for citizens would work. Professors sometimes carry out experiments with students because they are handy, not necessarily because they suit the task. In my case, because the subject is how lay citizens relate to scientific knowledge, I chose students as an appropriate group of collaborators. Members of garden clubs would have been equally good, but I could not offer them course credit.

Recruitment

Students came to the project in diverse ways. I did not ask about their environmental or safety views, let alone their ideological preferences. The closest anyone came to giving me a clue was Maria Merritt, a graduate student in the Philosophy Department, who told me she wanted to split her time between my project and the Environmental Defense Fund. For this characteristically forthright statement I assigned her alar to study.

Leo Levenson, a doctoral candidate at the Graduate School of Public Policy, had been a project director for the Superfund program; he was interested in discovering a better way to determine whether chemicals were harmful at low doses. I am indebted to Leo for introducing me to several other Superfund project directors. Their war stories—a mixture of dismay and astonishment at huge expenditures for minor purposes—informed my understanding. Anyone interested in chemical wastes should seek out present or former Superfund project directors to hear their accounts. When the people on the ground commonly regard what they do as ludicrous, something is wrong.[1]

In his write-up of Superfund, I asked David Schleicher to include a series of case histories so the reader could see, step by step, how risk estimates are made and the results implemented. To Brendon Swedlow, who had worked with me previously on political cultures,[2] I assigned the episodes of Times Beach, Missouri, and Agent Orange in Vietnam, involving the chemicals called dioxins. Robert Rye, a fresh college graduate, had heard about my project and decided it would be a good way station between a scientific career and a vocation of explaining science to the general public. I hired him to do a paper on the science of acid rain and one on ozone depletion. I also set out a series of criteria that scientific studies should meet in order for their results to be useful in protecting public health. He applied these to study after study, revised according to my comments, and persuaded me to alter certain judgments until we achieved a common view.

I gave students in a seminar I was teaching a wide choice of paper subjects, some of which fit into the category of risks and benefits from technology. Included here are Tracey Hessel on saccharin, Guillermo Frias on dieldrin,

Michael Sweeney on nitrites, and Darren Schulte on asbestos. These papers fortified my faith in student abilities. David Brown helped me understand global warming; his historical gloss on the global cooling scare of the 1970s led me to ask why, if warming and cooling were different, the policy prescriptions to ameliorate their effects were so similar?

One fall I received a call from longtime colleague Robert Piron, of the Oberlin College Economics Department, asking whether I would take a student, Jesse Malkin, for a month between semesters. I did. The result was a coauthored journal article and then, during a summer in Berkeley, preparation of the eighth edition of *Presidential Elections*.[3] The following summer, before he went to Oxford to study under a Rhodes scholarship, I asked him to work on DDT. When his friend Michelle Maglalang, an Oberlin undergraduate, discovered that she did not have the job she thought she had, I asked her to do Love Canal.

Another entry point was through a new program enabling undergraduates to volunteer to work with professors on their research. I liked Laura Evans's studies of how media treated various issues so much that I employed her to do more of the same during the summer of 1992. Without her careful and exacting data collection, Brendon Swedlow and I, with Laura, would not have been able to write the chapter on the media.

Reading papers of such high quality and observing the sense of empowerment on the faces of undergraduates who have mastered an issue, perhaps a chemical literature with a forbidding vocabulary, have increased my confidence that citizens can do it. For these students are the America from which citizenship comes.

Instruction

My charge to all these students was the same: read the scientific literature, find as much of the truth as possible, and appraise the relationship between knowledge and action. What were the original claims? What was known about their validity at the time they were made? What did the newspapers and news magazines write? Did they diminish the force of the claims, present them as received, or increase their seriousness? What did the government do? What was its scientific rationale and how well did that rationale hold up in the face of existing knowledge? When new legislation was passed or regulations promulgated, the same questions were to be addressed again, helping reveal the trajectories over time of action and knowledge. Have they come closer together or grown farther apart? And, finally, as knowledge grows, does it point in the direction of greater or lesser harm than originally envisaged?

As we began to understand better the environmentalist paradigm that justifies our contemporary regulatory regime, we discussed deeper questions. Would public health be better served if the standard were possibility of harm, as environmentalists advocate, or probability of harm, as used to be the case? Should we stick with the traditional standard of toxicology, the poison depends on the dose, or with the environmentalist version, any dose is a poison? Should any evidence of harm be enough to ban or restrict a chemical or should society return to "preponderant evidence" as a criterion of regulation?

Finding out "the truth" may seem to be a disingenuous instruction. It may appear that there is no truth to discover, that there are no clear cut facts, or that the evidence supports several sides of the question. Perhaps the feeling of uneasiness engendered by the word "truth" comes from the belief that there is no reconciling the conflicting interpretations of the evidence. Perhaps. But there is no knowing until we try to find the facts that are there.

Actually, we face confusion between two subjects, risk perceptions and risk consequences. People perceive potential dangers vastly differently given the same evidence. And there is no limit to how far apart these perceptions may be. Risk consequences refer to the actual effects of engaging in certain activities or being exposed to certain substances. However much we would like to control these consequences, we cannot do so unless we are in touch with the way the world actually works. This book is about risk consequences—actual effects—not perceptions. There are, to be sure, differences in perception of effects, especially as they concern future events. The acid test, however, remains the ability to cause intended effects. It is one thing for former Secretary of Health, Education, and Welfare Joseph Califano to predict an epidemic of cancers from trace exposures to a chrysotile asbestos and another for it to come about.

Scientific evidence does matter. I notice that no mention is made of witchcraft as a rationale for regulation, but rather obeisance is made to science whether or not it is what matters. Nor does any responsible person get up and say that his ideology or her worldview requires inventing or denying dangers and to hell with the evidence. As long as science is the only publicly acceptable rationale, it matters.

There may be properly suspicious types who think that my student researchers already knew my predilections and sought to satisfy them. Of course they knew. Before beginning the research for this book I had written two others on related subjects. In the first (with anthropologist Mary Douglas), *Risk and Culture,* I argue that people choose what to fear to support their way of life.[4] But this proposition holds true for all sides, those who minimalize as well as those who maximize dangers from technology. *But Is It True?* by contrast contains no attribution of motive, no discussion of why people take the positions they do, only analysis of evidence. In a second book, *Searching for Safety,* I argue that health and safety are the product of a society's standard of living; if it declines so does our health. Thus the benefits of programs designed to secure safety have to be considered in the context of harm done by these measures themselves. Moving dirt and climbing ladders also harm health; people fall off ladders, and moving earth leads to a low but palpable number of accidents.[5] As a result of these publications, it is widely known that I am not a supporter of the current approach to evaluating risks and benefits from technology.

I thus had all the more reason for insisting that my research assistants guard against this bias. For one thing it would do me no good to argue false positions that could be easily found out. For another, I did not know enough to recognize what sort of fabrication would help my stance. My interaction with these student researchers was largely concerned with getting them to bolster their positions with better evidence. From the beginning to the end our understanding has been that I would always yield to the evidence.

To reinforce the notion that I wanted the truth (and nothing but), the group of researchers as a whole met and criticized one another's papers. Long before this I met with each researcher individually and discussed whatever difficulties had been encountered. We discussed every section of every paper as it was being written. In part, we considered exposition, our common objective being to write so as to be understood by lay people like us. It might take the student several revisions to achieve sufficient clarity. As the paper progressed, I would become more and more concerned with evidence, asking in different ways for backup, requesting further reading, probing for gaps. Were there not alternative explanations? Could reasonable people not differ? How should various types of evidence be appraised? How does the evidence stand up to critical review? Similarly, as this book was being written, I circulated sections to other scholars for critical comments.

I also created a scientific advisory group, composed of Robert Budnitz, a consultant and former head of research for the Nuclear Regulatory Commission, to review work on nuclear waste; Bruce Ames and Lois Gold, at Berkeley, and Michael Gough, at the Office of Technology Assessment, all three to review different chapters on chemicals; and Fred Singer, director of the Science and Environment Policy Project, to comment on global environmental issues. Each provided detailed comments. In addition, the authors of each paper provided me with a list of the major figures in their particular controversy. They then phoned or wrote to these individuals with requests for comments, and many responded in detail. The rule was that we accepted criticism if we could not refute it. Often we went back and forth between critics to get a better fix on disputed points. These scientists deserve credit for responding to students and to a professor far out of their own fields. Their names are not listed so as to avoid the slightest imputation that these are their views. I gratefully acknowledge their help while absolving them of all responsibility, which is mine alone.

I advise people acting in their capacity as citizens to read and ponder and discuss *before* calling on experts. A flaw in the approach followed by residents of Love Canal and other neighborhoods under stress was that they called for expertise before they knew enough to know what to ask. I have tried to follow my own advice.

Selection Bias

It could be said that I have selected certain environmental and safety issues to prove a point. Anticipating this charge, we have covered a wide range of topics—all the great global issues (acid rain, ozone depletion, global warming), all the most publicized chemical controversies in the United States (DDT, PCBs, Love Canal, Times Beach, Agent Orange, asbestos removal, nuclear waste, saccharin, Superfund). So many episodes are included—so much in the history of regulation is covered—that my primary purpose of engaging the reader in a citizen-oriented endeavor to make sense of the uses and abuses of science in environmental and safety issues can hardly said to be affected by my own bias in selection.

To enhance coverage and reduce bias still further I have relied on three books that contain studies, usually by scientists but always by professionals, that compare knowledge with action and that maintain high evidential standards. All examine chemicals and radiation. Where our studies overlap, as with dioxins and PCBs, I have checked and supplemented our accounts. Where the books add coverage, such as with alleged cancers from video display terminals and from spermicides, I have summarized their evidence and conclusions.[6]

Nevertheless, there are deliberate exclusions that should be brought to the reader's attention. Most important, I have not addressed continuous occupational exposures to chemicals. The reason is twofold: one is anticipatory exhaustion at the prospect, given an already immense coverage; the other is my desire to focus this book on potential dangers to the general population, not to particular groups of workers. I do wish to emphasize, as I have done before,[7] that any time living creatures face huge occupational exposures they should be very careful. I have also left out problems involving medical drugs, since they require more knowledge and energy than I could muster.

Several well-known episodes involving technological risks have been omitted for special reasons. Problems of safety in nuclear power plants are covered in *Safer Power,* written with Elizabeth Nichols and forthcoming in 1995; we discuss the Chernobyl nuclear explosion there. I do not take up the destructive episode in Bhopal, India, because there is no dispute over the great harm done. Although the initial cause may remain forever obscure, the composition of the chemical cloud and the sickness and death it caused are acknowledged on all sides. The Challenger accident remains a single episode.

As to questions of the loss of species, the release of genetically engineered organisms into the environment, wetlands, destruction of tropical rain forests, and general damage to the world's ecosystems, Julian Simon and I have written elsewhere about species loss.[8] And in a volume edited by Bernard Davis, I combined an evidential, a cultural, and a policy analysis in rebutting claims of harm from biotechnology.[9]

Language and Logic

When reading scientific arguments the citizen must be aware of the ways that language and logical reasoning can be used to mislead. For example, a discussion of global warming that uses the term "CO_2 equivalent" may give the impression that carbon dioxide itself is increasing in the atmosphere at a much faster rate than it is.

The rationale for governmental regulation—say, the banning of a weed killer—may seem to be irrefutable, based on conclusions deduced or induced from a valid sequence of arguments. True, the reasoning may be logically flawless, but what about the major premise in the syllogism? How true is the premise that whatever substance harms rodents harms humans also, no matter what the dose?

Chemical regulation is largely based on animal cancer tests. If they are invalid (not imperfect), as I argue, then different methods of regulation must be used. This is no small matter. If scientific opinion continues to invalidate rodent animal tests as proxies for determining cancer causation, as I expect it will, the entire regulatory apparatus will have to undergo radical change. Rodent studies are currently treated as dispositive, which is why I draw so much attention to them.

The canon of "decide in favor of safety," sometimes known as the "precautionary principle," pervades analysis and action in all risk issues. It is the ubiquitous argument sustaining most regulations from the Clean Air Act to the still-existing Delaney Clause, which bans any food additive that causes cancer in laboratory animals. The power of this premise is undeniable: which legislators would publicly refuse to do all they could to guard the lives of their constituents? Without arguing this important point here—I do in the final chapter—I shall propose another rule to take the place of "safety first."

The Milk Room

Consider the allegory of the milk room. Placed in a sealed room full of milk—a great source of human health—one would die by drowning. If the milk level dropped below head level, drowning would not be inevitable. Here the poison does lie in the dose.

Suppose, however, that the milk level dropped to one foot: could a person drown? It is possible; a weak person might fall into the milk. Suppose the milk level dropped to one inch: could a person drown? A determined person might manage to, making it improbable but not necessarily impossible. How about a milk level of a thousandth of an inch? Could we say that drowning was impossible? Not quite. We could use a locution such as "highly improbable." Perhaps, as my wife suggests, a large room might be tilted so that enough milk accumulated in a corner to drown a person.

If anyone can imagine a level that low, how about a millionth of an inch of milk: could a person conceivably drown at that invisible level? My inclination is to think not, to imagine that we have at long last reached the impossible level. But that might be to limit human ingenuity; perhaps a suicidal genius could manage this feat.

Large, lifetime occupational exposures to chemicals occur in the equivalent of rooms with high milk levels, and people who work in such places should be very careful. But the small, sporadic exposures that we as citizens are subjected to are closer to the millionth of an inch of milk. One cannot quite say that significant harm is impossible, but the evidence suggests that it is extremely improbable. Only by substituting possibilities for probabilities as guides to action does being in the milk room constitute a hazard. The basic reason that virtually all allegations of harm from trace exposures to chemicals turn out to be false is that the amounts involved, the supposed causes, are too small to generate the substantial effects attributed to them. The lesson of the milk room is that while too much of even a good thing can be lethal, minuscule causes have minor effects.

Preferences, Yes; Disregard of Evidence, No

There are those who get upset when scientists disagree. I do not. How can one spend a long time in a field and not develop opinions? A climatologist without views on global warming is like a eunuch in a harem, not immediately dangerous to those around him but not in danger of starting anything new either. What is wanted is not scientific neuters but scientists with differing points of view and similar scientific standards. Society at large is helped not harmed by the formulation and defense of different hypotheses, a state of affairs that Michael Polanyi celebrated in his seminal essay "The Republic of Science."[10] The integrity of science as a process of seeking knowledge does not depend on the honesty or even the capability of individual scientists. History is full of examples of lying, cheating, and stealing scientists, like the great Isaac Newton.[11] Instead, science depends on institutions that maintain competition among scientists and scientific groups who are numerous, dispersed, and independent.

However, should the epistemology of science—the criteria for what counts as knowledge—weaken, competition alone could not informally coordinate scientific activity. Were a scientific hierarchy of values, honoring common standards with agreed instruments, let us say, to be replaced by inchoate feelings, as in what its adherents call the field of "clinical ecology," then science as a search for universal truths would be replaced by personal testimony. "What is true" would be replaced by "what is personally authentic." Science, in a word, would be replaced by sincerity.

If science is to be preserved as a universal search for knowledge, the emphasis must be on the integrity of its processes and procedures. If we accept or reject knowledge on the basis of who proposes it, for example, as Ian Hacking has noted, we will be back to the prescientific age when "probable cause" meant being supported by an eminent authority.[12] Ridiculous today? Then why is it that the work of scientists employed by industry is routinely disregarded? The answer given is that because they work for companies with a financial stake in the matter at issue, they are too biased to be worthy of consideration. If we want a universal science, however, it cannot be the origin of the work but its value as science that matters. Of course, we are not required to be naive. We may conclude that industry-sponsored research is likely to underestimate and environmental groups to overestimate dangers and thus seek to counter these biases, if they exist. Yet either of these putative extremes may be right. The assumption that "the truth is somewhere in between" is not a logical one and could well be false.

Money may be a root of evil, but it is not necessarily the root of all evil. Whatever happened to the seven deadly sins? Why is fame no longer the spur, only money? The sources of bias, the point is, are numerous and the process of expanding our knowledge must guard against all of them, not just one. What we need is not objectivity in the sense of having no views but a willingness to consider different hypotheses and, when considering each, to be guided by evidence.

1

Were the Early Scares Justified by the Evidence? Cranberries, Dieldrin, Saccharin

I. The Cranberry Scare of 1959

with Leo Levenson

Alongside headlines about the TV quiz show scandal and the campaigning of presidential hopefuls Richard Nixon and John Kennedy, cranberries became front-page news in November 1959. Seventeen days before Thanksgiving, Arthur Flemming, Secretary of the Department of Health, Education, and Welfare (HEW), announced that cranberries from Oregon and Washington had been found to contain residues of a weed killer, aminotriazole, that might cause cancer. Flemming advised people not to buy cranberries unless they knew where they were grown, and he detailed one hundred inspectors to test cranberries for residues.

Cranberry sales plummeted as supermarkets took them off their shelves and restaurants removed them from menus. Popular magazines joked about cutting your risk of cancer by quitting smoking—cranberries. The U.S. government seized over three million pounds and growers estimated total losses at $40 million. The U.S. Agriculture Department gave cranberry farmers who did not misuse aminotriazole over $8 million as reimbursement for losses suffered by the drop in sales. Cranberry sales did not fully recover for several years.

The great cranberry scare of 1959 represents the first battle in the long-running crusade of U.S. environmentalists to eliminate all traces of potentially carcinogenic agricultural chemical residues from the nation's food supply. Many of the same scientific and policy issues first publicly raised in the cranberry scare are still being debated today.

Background

The issue of the use of aminotriazole with cranberry cultivation had been simmering for two years before Flemming's announcement. Known also as amitrole and 3-AT, aminotriazole was developed in the early 1950s as a broad-spectrum weed killer that acted by blocking photosynthesis. Because it degraded quickly in the soil, a farmer could clear a field before planting, wait a few weeks, and then grow a crop without harm to the new growth. In feeding 3-AT to animals, scientists observed that high doses blocked normal iodine uptake to the thyroid gland, an effect that if unabated would lead to goiter, an enlargement of the thyroid. Animals fed lower doses of 3-AT or intermittent high doses had no adverse thyroid effects.[1]

In 1956 the U.S. Department of Agriculture (USDA) registered 3-AT for controlling weeds on noncrop land, such as roadsides and forests. Cranberry research stations soon found 3-AT to be effective in getting rid of poison ivy and other weeds that grew in cranberry bogs, and as word spread some cranberry farmers began using 3-AT in 1957 before it had been registered for food crops by the USDA. Meanwhile, the Food and Drug Administration (FDA), housed within HEW, required the manufacturer of 3-AT to submit chronic toxicity tests before any residue would be permitted on foods.

When FDA regulators became aware that some farmers had illegally applied 3-AT, they negotiated an agreement with the national cranberry cooperative, Ocean Spray, to freeze 3.5 million pounds of cranberries from farmers who admitted using 3-AT. The cranberries were to remain frozen until the FDA could develop a sensitive test for residues and decide whether any residues found were safe. At the time Ocean Spray marketed about 75 percent of the nation's cranberries. In January 1958 the USDA registered 3-AT for use on crops under conditions that it believed would leave no detectable residues on food products. Among the permitted uses was application to fields in the fall, after harvest, to keep down weed growth over the winter.[2]

During 1958 the issue of the potential risk of cancer from food additives was debated in Congress. In September Congress passed the "Delaney Clause" amendment to the Federal Food, Drug, and Cosmetics Act, which said, "No additive shall be deemed safe if it is found to induce cancer when ingested by man or animal." In early 1959 the FDA received the results of two-year studies of rats fed 3-AT conducted by a laboratory under contract to the manufacturer, American Cyanamid. Out of twenty-six rats fed 100 parts per million (ppm) of 3-AT over two years, four developed cancerous thyroid tumors (adenocarcinomas). Two of fifteen rats fed 50 ppm over their lifetime had growths that American Cyanamid pathologists thought were not cancerous (adenomas) but that FDA researchers thought could have been cancerous. One out of ten rats fed 10 ppm had a benign thyroid growth (American Cyanamid) or a potential thyroid cancer (FDA). Of the five rats in the control group fed no 3-AT, one had tissue changes that may have been a forerunner of an adenoma.[3]

To put the rat dose in perspective, if a 150-pound man ate a pound of cranberries every day of his life, all contaminated at the maximum residue level of 1 ppm requested by American Cyanamid, he would ingest a dose (scaled by relative body weights) about 1,500 times lower than the 10 ppm dose in rats that caused a growth in one of the ten rats, which may or may not have been cancerous and may or may not have been caused by the 3-AT. In the past such small doses would have been rejected out of hand as candidates for regulation.

Representatives of American Cyanamid identified a mechanism by which 3-AT was likely to stimulate cancer development at high doses. Such doses were known to inhibit iodine uptake to the thyroid. Interference with iodine uptake or an insufficiency of iodine in the diet are well known to cause benign thyroid tumors, perhaps engendered by an increase in the release of thyroid-stimulating hormones trying to force the gland to work overtime to compensate for the iodine deficiency. Prolonged treatment with the antithyroid compound thiourea was known in 1959 to create thyroid cancers in rats. However, rats fed low or high intermittent doses of antithyroid compounds did not have a reduction in their average uptake of iodine to the thyroid and showed no increase in benign or cancerous thyroid growths. Antithyroid compounds had been isolated from many foods, including beans, peas, cabbages, and milk. As early as 1928 scientists had induced thyroid growths in laboratory rabbits by feeding them high-cabbage diets.[4]

American Cyanamid scientists believed the body of evidence indicated that 3-AT was not itself a direct carcinogen; rather, the thyroid cancers in rats were likely to have been caused by the continuous blockage in iodine uptake from constant exposure to a high 3-AT diet. They maintained that residues well below those inhibiting iodine uptake would not pose an added cancer risk to people.[5] In the case of cranberries the risk to humans would be especially low, since most cranberry consumption occurred over a short holiday period and almost no one ate cranberries every day throughout the year. (Marketing of cranberry juice did not begin until the mid-1960s.) Nevertheless, the FDA decided in May 1959 that 3-AT should be considered a potential carcinogen at any dose and that no residue on food products could be considered safe.[6] Over the summer the FDA improved the analytical method for detecting 3-AT residues, allowing inspectors to detect less than 1 ppm for the first time. Meanwhile the FDA denied a request by Ocean Spray to release the 3.5 million pounds of frozen cranberries from bogs treated with 3-AT in 1957. Ocean Spray decided to bury the cranberries to make room in the freezers for the new crop.

The 1959 crop reached a record 1,250,000 barrels, with a projected value of $45–50 million. In the summer of 1959 the FDA discussed the potential carcinogenic hazard of 3-AT with Ocean Spray representatives. On September 18, 1959, Ocean Spray instructed all of its growers to discontinue use of the weed killer and required them to sign affidavits that they had not in the past applied the chemical improperly, that is, at any time other than in the fall after harvest. Of Ocean Spray's 1,079 growers, 10 or 11 refused to sign affidavits and saw their crops impounded, along with other crops on adjoining areas, totaling about three million pounds.[7] The FDA informed Ocean Spray that inspectors would still confiscate all berries with detectable residues of 3-AT, regardless of when the chemical had been applied.[8]

In October 1959 newspapers in the Pacific Northwest began carrying stories about 3-AT and Ocean Spray's burial of the cranberries harvested in 1957. An Oregon congressman called FDA Commissioner George Larrick to request an investigation.[9] FDA scientists began examining the 1959 crop of cranberries using its new sensitive 3-AT test procedure, and on November 6 they informed Secretary Flemming that two of the first seven shipments showed traces of 3-AT below 1 ppm and that preliminary tests on twelve additional lots showed probable presence.[10] Flemming, who held regular news conferences, expected to receive questions at one scheduled for the week of November 9 about reports of cranberry contamination in Oregon. FDA officials told him they could not determine the extent of 3-AT contamination: they did not know how effective Ocean Spray's campaign had been to prevent cranberries from bogs treated with 3-AT from reaching the market. Flemming decided to release the results of the FDA tests himself rather than wait for questions to come up at the press conference.[11]

Thus on November 9 Flemming announced the cranberry seizures and advised consumers not to buy canned or fresh cranberries unless they could tell what state they were grown in. "Many of this year's tests," Flemming said, "revealed residue of the chemical which would have produced cancer in rats." He added, "I don't have any right to sit on information of this kind."[12] Flemming later explained that "while in theory there may be a minute quantity . . . which is safe in foods, in actuality our scientists do not know whether this is true or how to establish a safe tolerance."[13] Flemming also announced that he was ordering the FDA to launch a massive effort to test cranberries throughout the nation to certify safe fruit in time for Thanksgiving. In the context of his remarks "safe" could mean only that there were no detectable residues whatsoever. During the next few days, grocers removed cranberry products from stores and restaurants eliminated cranberry sauces from menus. Several state health officials banned cranberry sales. American soldiers stationed in Japan did not get cranberry sauce with their turkey at Thanksgiving, because the Pentagon could not get "safe" cranberries to them in time.[14] However, the two leading presidential candidates, Richard Nixon and John Kennedy, announced they would eat cranberries at their dinners on the campaign trail.[15] By January 1960 the FDA had cleared 33.6 million pounds of cranberries while seizing only 326,000 pounds, or less than 1 percent of the total tested.[16]

The cranberry industry estimated losses of $40 million in 1959 from the sales bans and negative publicity.[17] Intensive lobbying by industry representatives led to a congressionally mandated plan to compensate cranberry farmers. Growers filed claims with the Agriculture Department for $9,014,825 and by October 1962, $8,500,000 had been paid to 1,215 claimants.[18] Losses to the primary manufacturer of 3-AT, American Cyanamid, were said to be "relatively unimportant," though the company did close a 3-AT production plant in Niagara Falls, Ontario.[19] The federal government made efforts to stimulate demand for cranberries by increasing their presence on the menus of school lunch programs, federal hospitals, and the armed services.

Despite the 3-AT ban, cranberry production continued to increase as new irrigation methods and improved cultivars expanded yields.[20] The 1960 crop set another record, 1,344,000 barrels.[21] In November 1964 the *New York Times* heralded a development that led to boom times for cranberries—the successful marketing by the Ocean Spray cooperative of a variety of new products, including cranberry juice cocktail, cranapple juice, jams, jellies, and breads.[22]

Flemming's Philosophy of Carcinogen Regulation

Secretary Flemming explained the theory underlying his cranberry announcement in testimony before Congress in January 1960. This testimony is particularly interesting because his philosophy of regulation in regard to potential carcinogens became fundamental to the arguments of environmental groups over the ensuing decades. As background Flemming explained his belief that technology, though often beneficial, created hidden hazards that might not emerge for many years after the initial damage was done. (The concept of hidden dangers with long lead times first makes its appearance here.) Controls on technology to promote health, he claimed, could reduce "not only known hazards, but unknown or little understood ones as well."[23] Flemming noted that some occupational cancers had been linked with industrial chemicals, and he explained that although people are inevitably exposed to carcinogens such as sunlight, soot, and dust, "it becomes all the more imperative to protect the public from deliberate introduction of additional carcinogenic materials into the human environment." A responsible government must make public health its top priority, "even though by so doing it may be . . . acting against the economic interests of a segment of our society." (Now he asserts the idea that health and wealth are opposed to each other.)

To respond to the notion that some carcinogens could be safe at low doses Flemming said, "We have no basis for asking Congress to give us discretion to establish a safe tolerance for a substance which definitely has been shown to produce cancer when added to the diet of test animals . . . because no one can tell us with any assurance at all how to establish a safe dose of any cancer-producing substance" (hence the no-safe-dose thesis).

Flemming put in the record a report by Dr. G. Burroughs Mider, associate director for intramural research at the National Cancer Institute, which included a specific discussion of chemicals causing thyroid dysfunction and their association with thyroid tumors. Mider admitted that diets low in iodine could cause thyroid tumors in animals, that many natural antithyroid compounds exist in common foods, and that "one would not be surprised if they too produced cancer if fed in adequate amount for prolonged periods." Toxic effects on tissues could stimulate cancer development; for example, "a chemical compound which produces cirrhosis of the liver might well produce cancer of that organ." Mider concluded, "No one at this time can tell how much or how little of a carcinogen will be required to produce cancer in any human being."[24]

Once a chemical had been associated with cancer in one animal experiment, Mider argued, any other negative experiments should be ignored: "The failure to obtain high tumor yields under any given experimental circumstances should in no way be interpreted as minimizing the potential hazard to man . . . As yet, we know of no tolerable dose of a carcinogen."[25] But why isn't the probable cause, thyroid dysfunction, preferred over the improbable cause, 1 ppm or less of a chemical residue, aminotriazole?

Three important themes in Flemming's testimony and the Mider report still influence our understanding of chemical carcinogenesis today, as they remain elements of the creed of environmental advocacy groups concerned with synthetic chemicals:

1. Suspect chemicals should be considered dangerous at any dose unless scientists can prove that low doses do not cause cancer.

2. "Negative" experiments associated with a chemical that show no increased cancer in animals do not prove anything, because the number of animals used might have been too small to reveal a slightly increased cancer risk. Thus suspect chemicals can never be proven to have a safe dose.

3. Economic interests should not be considered when deciding whether to ban a suspect carcinogen.

Together, these arguments claim a unique place in public policy for the regulation of chemicals that are potentially cancer causing. According to Flemming and Mider, residues of chemicals found to cause cancer in animals should be permitted only when there is absolute proof of their safety—a demand that is manifestly impossible to achieve, since any negative test result can be rejected. No argument, however plausible, that the mechanism by which a chemical may cause cancer in animals is dose-related—and that the cancer-causing effects would disappear at lower doses—should be accepted. Nor should we accept any argument that the negative consequences of a decision to ban all residues of a chemical might outweigh the hypothetical benefits.

In his testimony Flemming never claimed that he thought 3-AT residues on cranberries were likely to cause cancer in American consumers. He admitted the possibility that cancers in rats were caused by the high continuous doses of the chemical. But he maintained that once a single cancer test was positive in a rat species, his responsibility was to assume any residue could be dangerous to people.

Media Reporting of the Cranberry Announcement

In the 1970s and 1980s numerous stories about various "cancer-causing chemicals" in food, drinking water, and household products appeared in the media. Rachel Carson's *Silent Spring,* widely credited with popularizing environmentalists' fears about the potential health effects of synthetic chemicals, was published in 1962. In 1959, however, when Flemming made his cranberry announcement, the idea that tiny residues of farm chemicals on food could cause cancer was new and strange.

Newspapers and magazines gave different versions of the story, varying the emphasis on the risks from 3-AT residues and on the economic plight of cranberry growers and marketers who saw their products pulled from the marketplace. The *New York Times*'s lead paragraph on November 10, 1959, read: "The Federal Government warned the public today that some cranberries grown in Washington State and Oregon had been contaminated by a weed-killer that induces a cancerous growth in the thyroid in rats." The sixth paragraph noted: "It has not been proved that the weed-killer produces cancer in human beings." Later paragraphs quoted "industry experts" saying that human beings would have to consume "carloads" of berries to be affected.[26]

The following day a *Times* editorial mentioned that less than 1 ppm of 3-AT was found on cranberries, while the FDA was citing laboratory tests on rats eating cranberries with 10 ppm of 3-AT.[27] The editorial's tone implied that the government was acting responsibly and that no one should be worried or upset. The piece did not mention that the rats were fed the 3-AT diet daily over two years. Three days later the *Times*'s opinion of Flemming's actions had changed to clear disapproval: "We must say that Secretary Flemming went too far when he urged people not to buy cranberries . . . Apparently, even if humans should be as susceptible to the drug as rats, people would have to consume fantastic quantities of contaminated berries to suffer any such ill effects. Then, too, only a very small proportion of the Western crop—which is only 8 per cent of the total—has so far shown any traces of the poison."[28]

The *New Republic* editorialized that the cranberry scare was overblown, stating in addition that general bans on animal carcinogens would prove "unworkable," because foods contain many natural carcinogens such as arsenic.[29] Two letters printed by the *New York Times* praised Flemming's cranberry announcement but questioned the slow government action on confronting the much greater cancer risk from tobacco.[30]

The *Commonweal* supported Flemming's actions:

Perhaps, as has been charged, Secretary Flemming did act hastily, and perhaps the wording of his warning did tend to punish the innocent berries along with the guilty. But . . . the arguments advanced by the cranberry growers were hardly reassuring . . . There is a familiar ring to the claim that the poisonous effect of the chemical on humans has yet to be proved, although the cancer inducing effect on rats is acknowledged . . . That Secretary Flemming leaned over in the direction of public health and welfare is not to his discredit, we think, in the light of previous experience.[31]

Note the linkage between suspicion of the chemical industry and "leaned over in the direction of public health."

NBC's *Today Show* the day after Flemming's announcement featured an American Cyanamid toxicologist telling the audience that one benign growth in one rat fed 10 ppm of 3-AT per day was probably not caused by the chemical, while FDA Commissioner Larrick countered that his toxicologists said the growth could have been cancerous and caused by 3-AT.[32]

It would have been difficult for anyone to evaluate the risk posed by cranberries from such media reports. The *New York Times* and popular periodicals contained no discussion of the scientific basis for assuming 3-AT was dangerous at the levels found on cranberries. None of the stories discussed the implications of American Cyanamid's argument that the likely cancer mechanism in rats from 3-AT—hormone stimulation due to reduced iodine uptake—was highly unlikely to occur in humans eating a few dishes of cranberries with 3-AT residues. None of the early news stories made the point that analogous chemicals were found in nature or compared natural occurrences to the residue levels. Perhaps that sort of sophistication is a product of our ability in the 1990s to take a retrospective view of what matters more and what matters less for human health.

Aftermath of the Scare

No evidence since 1959 has countered the prevailing assumption among toxicologists that 3-AT is not carcinogenic at doses below those necessary to impair thyroid function. Mice and hamsters fed 3-AT in long-term tests since 1959 failed to show any carcinogenic effect, even when given daily doses of 100 ppm for their lifetime (at least 100 times greater than the residues originally found on cranberries). Tests have also shown no increase in birth defects in the litters of pregnant mice fed 500 ppm.[33]

The Environmental Protection Agency (EPA) canceled all food crop registrations for 3-AT in 1971, based on the original rat-feeding tests from 1957 to 1959. In 1974 a joint panel of Food and Agriculture Organization and World Health Organization experts recommended a conditional tolerance of 0.02 ppm for residues of 3-AT in raw agricultural commodities. Because it is banned for use on foods in the United States, the EPA has not set an agricultural tolerance level for it.

The same scientific issues raised in the cranberry episode are being debated again in the 1990s. Recently, for instance, the FDA has come under pressure from environmental groups and members of Congress to invoke the Delaney

Clause on veterinary use of the drug sulfamethazine, which could leave residues in beef or pork, on the grounds that continued high doses of the drug have caused thyroid cancer in test animals. Like 3-AT, sulfamethazine blocks iodine uptake to the thyroid in test animals when it is administered at concentrations much higher than appear in the meat of slaughtered farm animals. As of this writing, FDA administrators are considering taking a different tack from that taken during the cranberry episode—permitting low residues of these chemicals on the assumption that the carcinogenicity is probably caused only at high levels as a result of inhibited iodine uptake.[34]

To some Americans the cranberry episode was the source of jokes about what happened to people who ate a pound of cranberries a day for a lifetime. To cranberry farmers and marketers it was a major financial blow. To the nascent environmental movement the episode represented the first shot in a campaign to raise public fears about cancer from synthetic chemicals in food and water. The next significant shot, in some ways the most significant, was the publication of Rachel Carson's *Silent Spring* in 1962. It would not be appropriate here to do a full-dress study of this immensely influential book; in keeping with our subject matter, the second part of this chapter compares her statement about a more potent chemical, dieldrin, with the evidence.

II. *Silent Spring* and Dieldrin

with Guillermo Frias

Silent Spring had an enormous impact on the debate about pesticides. The controversy that accompanied its publication and the subsequent influence it exercised led many to compare Rachel Carson to Harriet Beecher Stowe, with Carson as an advocate of liberation from bondage to chemicals. *Silent Spring* like *Uncle Tom's Cabin* achieved instant national notoriety and produced vigorous social discussion. Its impact was not limited to the United States: it was translated into many languages and distributed worldwide. Some claim that its effect was so great that "within a few years of its publication in 1962, every country in which *Silent Spring* was widely read was holding hearings on environmental legislation."[1] There can be no doubt that the message delivered by Carson's book advanced the burgeoning environmental movement and contributed to the banning of DDT in the United States in 1972. What can be questioned, however, is the accuracy of some of its premises, especially about dieldrin, in the light of subsequent evidence.

Trained as a marine biologist, Rachel Carson first wrote eloquently about sea life. Among her books, *Silent Spring* alone calls for immediate environmental action. "The book was activist, not just expository; it was written to reform, not just to have a forum."[2]

With *Silent Spring* Carson hoped to become a representative for the interests of the environment, for she agreed with E. B. White of the *New Yorker,* and one of the finest writers in America at that time, when he pointed out to her that "the whole vast subject of pollution . . . is of the utmost interest and concern to everybody . . . Always some special group or interest is represented, never the earth itself."[3] She assumed the role of environmental crusader because she felt that the environment was gravely threatened by pesticides and that there was "an enormous body of fact waiting to support anyone who will speak out to the public."[4]

In her book Carson expressed her dismay with the liberal use of pesticides:

> These sprays, dusts, and aerosols are now applied universally to farms, gardens, forests and homes—nonselective chemicals that have the power to kill every insect, the "good" and the "bad," to still the songs of birds and the leaping fish in the streams, to coat the leaves with a deadly film, and to linger on in the soil—all this though the intended target may be only a few weeds or insects. Can anyone believe it is possible to lay down such a barrage of poisons on the surface of the earth without making it unfit for all life? They should not be called "insecticides," but biocides.[5]

The threat to the environment becomes more acute, according to Carson, when insects develop resistance to pesticides because the usual response is to spray in greater quantities, without adequately considering the effect on the environment. Carson believed that we should abandon our reliance on technology for the control of insects and adopt alternative methods, such as sterilizing and releasing large numbers of male insects and introducing insect-killing microorganisms, parasites, and predators. This "other road" would lead to a world in which humankind would no longer dominate nature.[6] Thus the ability of insects to develop resistance to pesticides could achieve what the government had been unable or reluctant to—an end to pesticide use and the pesticide industry.[7]

Silent Spring on Dieldrin

The title of Carson's book came from her concern that "over increasingly large areas of the United States, spring now comes unheralded by the return of birds, and the early mornings are strangely silent where once they were filled with the beauty of bird song."[8] Carson's main target in her investigation of the demise of birds—among them robins and bald eagles—was DDT, a chlorinated hydrocarbon whose insecticidal properties were first discovered in 1939. Other compounds in this organochlorine series were developed, with a wide range of potential toxicity, dieldrin being much less safe than DDT, as Carson noted: "As tested on quail and pheasants, [dieldrin] has proved to be about 40 to 50 times as toxic as DDT."[9]

Dieldrin's toxicity, stability, and long residual action made it one of the most widely used pesticides in the world and for the same reasons caused it to be severely restricted or banned in several countries beginning in the early 1970s.[10] Apropos of Carson's concern about dieldrin and birds, studies have found lethal levels of the chemical in the brains of a few falcons and eagles. In a study from 1964 to 1970 scientists concluded that out of the 153 eagles analyzed, 15 of them possibly died from dieldrin poisoning.[11] Another study reported that high doses of dieldrin, found chiefly in the liver of birds analyzed, were responsible for 5 to 20 percent of the peregrine falcons found dead in Britain.[12] On the whole, data from studies analyzing large dieldrin consumption by birds suggest that heavy dosages of dieldrin are lethal to birds. What about smaller doses?

In a study published in 1956, quails fed a dieldrin diet of 0.5 or 1 milligrams per kilogram (mg/kg) did not show any change in egg production, fertility, or hatchability. Pheasants fed 25 mg/kg also showed no significant effects in fertility or hatchability. When the dose was raised to 50 mg/kg, however, there was a clear effect, and it was concluded that heavy doses of dieldrin interfered with the reproductive process of pheasants and quails.[13] The eggs from chickens fed a diet of 1 mg/kg of dieldrin for two years showed no change in fertility or hatchability; this was true even though the yolks had heavy concentrations of dieldrin. When the diet of dieldrin was increased to 10 mg/kg, the fertility and hatchability of the eggs decreased slightly.[14]

Numerous later studies have shown that depending on the duration of exposure, dose levels of 5 to 10 mg/kg reduced the survival of the parent birds.[15] The effect on eggs varied; in some cases egg production increased while in others a decrease was reported. In general, the data show that the fertility rate of the eggs remained unaffected. The hatchability and survival of the chicks, in most cases, also were not affected. The studies report an apparent trend in which "the overall reduction in reproductive success occurs only if the parent birds are showing signs of being affected by dieldrin, e.g., reduced food intake with consequent loss of weight and poor condition."[16]

Similar results were reached in a three-generation study published in 1974 of the effects of dieldrin on pheasants. Hens were given 6 or 10 mg of dieldrin per week, and cocks were given 4 or 6 mg. These doses were sufficient to reduce the life of the breeding birds but did not affect the production, fertility, or hatchability of the eggs. The viability of the chicks at the time of hatching was not affected in a consistent manner in relation to dose or to generation, but the survival rate of the chicks from the hens that were fed a diet of 6 or 10 mg per week was reduced.[17] In another study published the same year, in which seven-week-old Japanese quails were given diets containing 3.1 or 50 mg/kg of dieldrin for twenty-one days, significant reduction in egg production occurred.[18]

The data from these experiments suggest that small doses of dieldrin have a negligible effect on bird reproduction but that large toxic doses can impede bird reproductive processes. It is important to keep in mind, however, that the "birds that were tested are precocial species [covered with down and capable of independent activity when first hatched] which show no parental feeding of the young."[19] A study on homing pigeons reported that the parents, which were free-flying, brooded their eggs and fed their young until they fledged.[20] Parental behavior of these birds as well was not much affected by the intake of comparable amounts of dieldrin. Few birds in the environment are precocial, so reproduction generally involves a period of full dependency of the offspring on parental care.

The relationship between dieldrin and birds is complex. All birds do not have the same reaction to dieldrin. We cannot blame the decline of several bird populations solely on pesticides like dieldrin, because a vast array of environmental factors influences the existence of birds. It is also wrong to say that dieldrin has no effect on the reproductive processes of birds, for the tests show that large doses of dieldrin do interfere with these processes.

Dieldrin and Eggshell Thinning

Ever since 1967, when D. A. Ratcliffe suggested that pesticides were responsible for the thinning of eggshells,[21] a number of studies have been conducted to examine this claim. These studies have not offered conclusive results and in fact conflict with one another. For example, mallard ducks who were fed a diet of 1.6 mg/kg of dieldrin for sixteen months produced eggs 3.4 percent thinner than the control birds. However, when the diet of dieldrin was increased to 4 mg/kg, the difference between the treated birds and the control was actually lower, 2 percent.[22] Table 1.1 summarizes the results of various experimental studies.

In certain species the connection between dieldrin and eggshell thickness may result not from the chemical residue found in the egg but rather as a by-product of dieldrin consumption by the parent. Eggshell thinning after dieldrin consumption may be the product of reduced food consumption. Mallard ducks, which in several studies laid eggs with thinner shells after they had been exposed to dieldrin, also laid eggs with thinner eggshells when they were deprived of food for thirty-six hours.[23]

In sum, the data show that when birds take dieldrin in large doses, reproductive failure and death result; low doses of dieldrin have a negligible effect on their reproductive processes. The data also show that not all birds have the same reaction to the chemical.

Table 1.1 Effects of dieldrin on eggshell thickness

Species	Dose of dieldrin (mg/kg diet)	Duration	Difference between eggshell thickness of treated and control birds (percent)
Mallard duck (*Anas platyrhynchos*)	1.6	16 mos.	−3.4
	4.0	16 mos.	−2.0
	10.0	16 mos.	−4.3
	4.0	90 days	−4.2
Japanese quail (*Coturnix coturnix japonica*)	3.1	21 days with changes in photoperiod	−8.0
	50.0		−8.0
Domestic fowl (*Gallus domesticus*)	10.0	12 weeks	0.0
	20.0	12 weeks	+0.3
	10.0	13 mos.	0.0
	20.0	13 mos.	+9.7
Pheasant (*Phasianus colchicus*)	6 mg/hen/wk	—	0.0
	10 mg/hen/wk	—	0.0
	(6)0 mg/hen/wk	—	+4.1
	(6)6 mg/hen/wk	—	+4.1
(10)0 mg/hen/wk	—	+4.1

Source: World Health Organization, "Aldrin and Dieldrin," in *Environmental Health Criteria*, vol. 91 (Geneva: WHO, 1989), p. 177.

Mice, Rats, and Monkeys

What can we learn from the effects of dieldrin at varying doses on different species? In 1962 K. J. Davis and O. G. Fitzhugh fed groups of 100 male and 100 female mice a diet containing dieldrin at 0 to 10 mg/kg for two years. The lifetime of the mice was shortened by two months, and the researchers also noted an increase in the incidence of benign liver tumors.[24] In another study, in which 100 male and 100 female mice received a similar diet for the same period of time, there was an increase in the number of animals with hepatic hyperplasia and benign liver tumors within the treated groups; there was no increase in malignant liver tumors. The survival rate of the treated group was lower than that of the control.[25] Additional studies on mice showed that the overall no-effect level over a lifetime was 0.5 mg/kg of dieldrin.[26] At a diet of 1 mg/kg or more, dose-related changes in the liver occurred. At levels of 10 mg/kg or more, typical signs of organochlorine toxicity, such as irritability, tremors, and convulsions, appeared. In all of these studies there was no increase in the incidence of tumors in the liver or other organs of the animals.[27] Groups of 5 male monkeys were given diets containing dieldrin levels of 0, 0.01, 0.1, 0.5, 1, or 5 mg/kg for approximately six years. The maximum level of dieldrin was reduced to 2.5 mg/kg after 2 monkeys died. This later was reduced to 1.75 mg/kg. The study reported that the liver/body weight ratios and the liver DNA and RNA of the test animals were not different from those of the control animals. No subcellular changes were seen. Especially important, the dieldrin concentrations in the livers of the monkeys were approximately 200 times higher than those in the male rats receiving a dieldrin diet 3 times higher than the monkeys, and they were similar to the concentrations in the livers of male mice consuming dieldrin at a level 50 times higher.[28] Different creatures must process dieldrin differently.

Dieldrin Poisoning in Humans

Most cases of dieldrin poisoning have been either accidents involving children who have consumed the chemical by mistake or suicides. In 1975 there were twelve cases of neurotoxicity in India that resulted from the consumption of wheat into which dieldrin had been mixed accidentally. The victims consumed this wheat for six to twelve months before they began to show the symptoms of dieldrin poisoning, the most dramatic of which were severe convulsions. The electroencephalographic tracings were consistent with the diagnoses of insecticide poisoning. Treated with phenobarbital and diazepam, all of the patients recovered.[29]

The threshold dieldrin concentration below which no adverse effects have been observed and at which none should be expected is 105 micrograms per liter (μg/liter). The World Health Organization (WHO) reports that for the most part "the blood levels in the general population of countries where this has been investigated is well below the threshold level." However, the same report also notes rare instances where low concentration of dieldrin has induced intoxication effects.[30] If prompt and adequate treatment for dieldrin poisoning is given, the patient recovers.

A popular explanation for the persistence of dieldrin poisoning is that during exposure dieldrin circulates through the blood and becomes stored in body fat. Rachel Carson was greatly alarmed by the possibility that pesticide residue may collect in the body and may one day, unexpectedly, cause poisoning. "Because these small amounts of pesticides are cumulatively stored and only slowly excreted, the threat of chronic poisoning and degenerative changes on the liver and other organs is very present."[31] Is it?

A study conducted on four females and fifteen males who underwent surgery showed that the concentration of dieldrin in the blood was unaffected by the catabolic response to surgery. These patients did not experience any sign of intoxication. In another study, concentrations of dieldrin in four women who underwent voluntary near-starvation diets, which resulted in weight losses of up to 7.5 kg/week, remained constant and the women did not experience dieldrin intoxication. Another study reported no significant difference in the dieldrin concentrations of dieting and nondieting mothers before and after delivery.[32]

Carson was especially concerned with the possibility that the residues found in mothers might be transferred to their children: "Insecticide residues have been recovered from human milk in samples tested by Food and Drug Administration scientists. This means that the breast-fed human infant is receiving small but regular additions to the load of toxic chemicals building up in his body."[33] On the health effects of those residues, the WHO concludes: "[Dieldrin] concentration in the blood and adipose tissue of suckling infants does not increase with age during the first six months, nor is their blood dieldrin level higher than that of bottle-fed babies. Under these circumstances, the benefits of natural breast feeding still make it the preferred method of infant feeding, in spite of the dieldrin residues."[34]

Controlled Human Studies

In a pharmacodynamic study thirteen volunteer male college students were given oral doses of dieldrin of 0, 10, 50, or 211 µg for eighteen months. For six more months the volunteers who were given 50 µg continued to receive dieldrin at this level while all the others, including four in a control group, received 211 µg/day.[35] This study had three objectives. The first was to establish the relationship between the daily intake of dieldrin and its concentration in human blood and tissue. The second was to establish the blood/fat ratio in human beings. The last objective was to establish the relationship between the concentrations of dieldrin in blood and fat and the length of exposure. Clinical, physiological, and laboratory findings remained constant throughout the whole experiment period of eighteen months, and the subjects remained in excellent health. The conclusion was that a total daily intake of 211 μg of dieldrin per person for two years had no effect on human health.[36]

Perhaps the most important finding from studies of dieldrin is that, as the WHO reported, "the body burden resulting from the present level of exposure constitutes no health risk to the general population."[37] The concentration of dieldrin found in those who work with the chemical caused no health problems, with the important exception of those who spray it; this finding strongly suggests that the concentrations to which the general public is exposed pose no risk to human health. Dieldrin is certainly not a harmless chemical; it is highly toxic and dangerous. But used properly, it is safe for human health. As the World Health Organization concludes, "All the available information on dieldrin, including studies on human beings, supports the view that for practical purposes, [dieldrin] makes very little contribution, if any, to the incidence of cancer in human beings."[38] The same could be said for saccharin, our next topic.

III. The Saccharin Debate

with Tracey Hessel

As early as 1900 saccharin ($C_7H_5NO_3S$) was being utilized as a food preservative and antiseptic as well as a drug for the treatment of urinary tract infections. Soon after, it became most widely used by canners who recognized that this calorie-free substance provided an inexpensive sweetener for fruits and vegetables. Depending on its form, it tastes 350–500 times sweeter than cane sugar. Then regulatory battles began. There was concern that high levels of saccharin might cause digestive disturbances.[1] By 1902 French, German, Portuguese, and Hungarian governments had banned its use in food and beverages for this reason.[2]

Why did the United States not join in the banning of saccharin? In 1907 Dr. Harvey Wiley, secretary of the Department of Agriculture's Bureau of Chemistry, was summoned to advise President Theodore Roosevelt on food additives. A congressman had inquired about saccharin, saying his company used it as a substitute for sugar in canned corn. Wiley stated that everybody who ate that corn was deceived, thinking they were eating sugar rather than a nonnutritive substance that was "highly injurious to health." The president ended the discussion and temporarily stalled attempts to ban the sweetener with his forceful yet subjective response: "Why, Dr. Rixey gives it to me every day. Anybody who says saccharin is injurious to health is an idiot."[3]

Scientific support for the president's declaration soon followed. A panel he created to investigate potential health threats, the Referee Board of Consulting Scientific Experts, concluded that only "chronic saccharin use at levels exceeding 0.3 grams per day was liable to impair digestion," acknowledging that substituting it for sugar "reduces the value of the sweetened product and hence lowers its quality." Based on the 2–3 grams (g) of sugar generally added to a ten-ounce (oz) can of vegetables today, the equivalent saccharin substitution would probably not surpass a level of 0.02 g. Even Wiley conceded that saccharin would not be harmful, for at that time it was unimaginable that saccharin consumption would ever exceed the amount deemed safe by the panel.[4] Studies on both the health benefits and dangers continued but did not attract public attention again until the 1970s, when the questions most of us are familiar with surfaced.

The beginnings of the recent controversy can be traced to the 1950s and 1960s, when Americans became calorie conscious and the use of cyclamates and saccharin as artificial sweeteners increased substantially, no longer limited to people suffering from diabetes or heart disease. With the ban on cyclamates by the FDA in 1969—when ingestion of large quantities appeared to cause cancer in some animals—saccharin consumption escalated. By 1977 bottlers were selling $1.5 billion in diet soft drinks annually, representing 15 percent of the soft drink market.[5]

Increased consumption led to a renewed interest in saccharin on the part of the FDA and many research groups. In 1967 the Food and Agriculture Organization/World Health Organization (FAO/WHO) Joint Expert Committee on Food Additives established an unconditional acceptable daily intake (ADI) level for saccharin of 5 mg/kg of body weight. It also adopted a conditional ADI of 15 mg/kg that was to be applied only to diabetic foods. These levels were based on the results of past experiments that had indicated there was no effect when rats were fed saccharin as 1 percent of their diet, which is equivalent to about 500 mg/kg per day in humans. The committee divided this no-effect dosage by arbitrary safety factors of 100:1 and 30:1 to arrive at the unconditional and conditional ADIs respectively.[6]

In 1970 the National Academy of Sciences (NAS), while adopting the FAO/WHO committee's no-effect level of 1 percent, concluded that "the application of a 100:1 safety factor was unduly conservative" and instead recommended the 30:1 safety factor dosage of 15 mg/kg per day, or about 1 g per day for the average adult.[7]

While anticipating increased saccharin consumption as a result of the cyclamate ban, the NAS committee predicted that reaching an intake level of 1 g per day was highly improbable.[8] Following this reasoning, it concluded that "the present and projected use of saccharin in the US does not pose a hazard."[9]

Using the same method, the average soft drink consumption level (16.6 oz per day) over all age groups yielded an average exposure estimate of 134 mg/day (0.005 oz). Since most sodas are sold in 12 oz cans, the average consumption level was less than 1.4 cans a day, well below the level of 1 g (1,000 mg) per day recognized as safe by the National Academy of Sciences.

As for the NAS recommendation for further studies, three were released within the following seven years, and some believed these studies addressed the remaining question of carcinogenicity. The FDA, the Wisconsin Alumni Research Foundation, and the Canadian government released concurrent findings demonstrating that a diet containing saccharin of at least 5 percent caused bladder tumors to develop in male rats. The studies immediately received criticism from many in the scientific community because of the high doses administered and the possibility that positive findings were due to impurities discovered in the samples. FDA Commissioner Sherwin Gardner, however, stated that the studies provided "clear evidence that the safety of saccharin does not meet the standards for additives established by Congress."[10] In March 1977 Gardner announced that initial steps were being taken toward withdrawing approval of saccharin. In the following months the FDA received over 100,000 letters, almost all opposing the proposed legislation.[11]

As it became evident that there was broad support for Congress to take action to prevent a ban on saccharin, Senator Edward Kennedy (D-Mass.) and Representative Paul G. Rogers (D-Fla.) initiated a series of congressional hearings to evaluate the current body of knowledge. The hearings resulted in a moratorium on saccharin, which President Jimmy Carter signed into law on November 23, 1977. The legislation postponed FDA action for eighteen months. Arrangements were also made for two studies by the NAS on the dangers and benefits of saccharin as well as the policy issues involved in regulating a potential carcinogen. Also included was the requirement that all foods containing saccharin be labeled with the warning statement with which we have become so familiar.[12]

Since 1977 the moratorium has been extended five times. The studies continue, yet interpretations differ. How seriously should we take these warnings? How likely is it that products containing saccharin "may be hazardous to your health"? To cast light on the issue let us look at some of the studies that influenced the FDA.

The FDA and Its Evidence

In June 1971 the FDA published in the *Federal Register* what it said was its response to NAS reports: that current and projected use of saccharin did not appear hazardous. The FDA called for the removal of saccharin from the Generally Recognized as Safe (GRAS) list, while temporarily permitting its use under an interim food additive regulation. A better understanding of this unexpected reaction is provided by a press release issued the same day, in which the results of a recent study were mentioned as having contributed to the FDA's action.[13]

The study, from 1970, had reported that pellets of cholesterol plus saccharin at a dose of 500 mg/kg of body weight produced tumors when implanted into the urinary bladders of mice. Of 66 mice implanted with saccharin, 31 (47 percent) were found to possess tumors compared with 8 of the 32 rats receiving cholesterol alone (25 percent).[14] While quantitatively impressive, the results held little significance to many in the scientific community. The chair of the NAS Food Protection Committee, Dr. Julius M. Coon, said that his panel deemed implantation an unreliable method of gauging carcinogenic properties compared with the more natural oral administration.[15] In a *Science* review (on sodium cyclamate), the implantation method was called into question: "If we ignore, as the authors did, the entire problem of which molecules would normally end up in urinary bladders in amounts similar to . . . concentrations introduced by their technique, then it is interesting to speculate that sucrose (i.e. sugar), for example, could also have a carcinogenic effect."[16]

In other words, implantation bypasses the usual alteration, through metabolism, of the concentration, molecular form, and pathway of a given substance. With this method there is no way to ascertain whether such conditions might occur in the body or whether results reflect other causes. For instance, the results of the study could be caused by a reaction to unrealistically high and toxic levels of saccharin or simply by the bladder's normal response to abrasion caused by salt concentrations elevated to the point that they precipitate out of solution. Use of the implantation method could yield similar results for many other substances. It has been demonstrated that calculi, or hardened aggregations of mineral salts, are themselves carcinogenic in the rat bladder. Foreign material introduced into the bladder induces the formation of tumors.[17] The effects that caused the FDA to remove saccharin from the GRAS list could well have resulted from the method of administrating the sweetener rather than from the saccharin itself.

While saccharin remained under interim regulation, the FDA hoped that the results of animal studies under way at the time as well as yet another NAS review would settle any remaining questions about saccharin's carcinogenicity. But that was not to be.

Two studies were released within a year that supported the postulated relationship between saccharin and bladder tumors. One had been conducted by the FDA itself, the other by the Wisconsin Alumni Research Foundation (WARF), the group that contributed the research responsible for the eventual banning of cyclamates. The WARF studies were commissioned by the sugar industry. The only two-generation studies undertaken, the WARF and FDA studies were similar in design. The parent generation of rats, F_0, was fed a given amount of saccharin from weaning until reproduction with a mate from the same dosage group. The second rat generation, F_1, from birth received the same dosage of saccharin as its parents until death, upon which it was examined for tumors.

In the FDA study rats were divided into seven groups. Five were fed sodium saccharin at levels of 0.01, 0.1, 1, 5, and 7.5 percent of the diet. A control group was fed 1.51 percent sodium carbonate, the amount determined to provide the same sodium ion concentration as would 5 percent sodium saccharin. The seventh group received a diet consisting of 5 percent calcium cyclamate. This last group, it was anticipated, would serve as a control "based on the positive findings of urinary bladder neoplasms by Price et al. (1970) when a combination of saccharin and cyclamate was fed."[18]

The expected bladder tumors were not found in the sodium cyclamate group, however. Of the total 406 F_1 rats examined, 11 developed carcinomas of the bladder: 7 of 23 (30 percent) male and 2 of 31 (6 percent) female rats fed the highest dose of saccharin (7.5 percent of the diet) developed bladder tumors, as did 1 each of the 29 male controls and the 24 male rats fed 5 percent saccharin. None of the remaining 299, the rats fed lower levels (5, 1, 0.1, and 0.01 percent) and the female controls, developed urinary bladder abnormalities.[19] It also should be noted that "in litters from parents receiving 5.0 and 7.5 percent sodium saccharin and 5.0 percent calcium cyclamate . . . [rats] had lower average body weights throughout the study."[20] We will discuss this fact later, when considering using high doses of a substance to determine possible carcinogenicity in animals.

In the WARF study, offspring of F_0 rats were raised for 100 weeks on the same diets their parents had received. Twenty male and 20 female weanlings made up each test diet group, receiving 0, 0.05, 0.5, and 5 percent sodium saccharin in a basal diet of Purina Lab Chow. Bladder tumors were discovered only in the group of 20 male rats that received saccharin at the highest dosage. Of those, 6 rats were found to have bladder carcinomas. Rats from the high-dosage group were observed to have below normal weights during weaning and at nine weeks. By the thirteenth test week, growth curves for the rats were similar, "though body weights of the high-level groups remained slightly smaller."[21]

The NAS committee considered these studies in its 1974 report, which determined that on the basis of the new evidence, saccharin could be declared neither hazardous nor safe. The NAS had reviewed eleven feeding studies and dismissed all except those of the WARF and FDA as failing to demonstrate an association between saccharin and increased tumors. The FDA experiment was the only one to yield statistically significant results, demonstrating at least a 95 percent likelihood that the bladder cancer in male rats was not due to chance. But even those findings, while suggesting that under test conditions "the bladder tumors observed were related to the consumption of the saccharin samples used . . . cannot be interpreted as showing that saccharin itself was the cause of tumors."[22] Neither could the many reported negative tests be interpreted as showing saccharin not to be a bladder tumorigen because they "were at fault in not involving in utero exposure of the test animals."[23]

NAS was concerned about whether the tumors were caused by saccharin or by the impurities that were discovered in the samples. The major impurity in the saccharin used in feeding studies was ortho-toluenesulfonamide (O-TS), known to inhibit carbonic anhydrase, an enzyme involved in the acidification of urine. Decreased activity of this enzyme could produce alkaline urine and bladder stones, irritation from which is known to cause the development of tumors.[24]

Also questioned by NAS was the use of dosages as high as 5 and 7.5 percent. It noted that rats fed saccharin at those levels had tended to grow less than did controls and rats fed lower saccharin levels. High sodium levels found in the animals were also identified as possibly being responsible for altering the metabolism of saccharin and precipitating the formation of bladder calculi.[25] The FDA referred to the need to investigate such "toxicity factors" when it decided to extend the interim regulation. By recognizing these uncertainties, the agency was in essence acknowledging the possibility of a threshold for saccharin's carcinogenicity.[26]

Consequently saccharin's fate remained in limbo while the FDA worked closely with its counterpart in the Canadian government, the Health Protection Branch of the Department of Health and Public Welfare, to design another two-generation study that would address the remaining concerns. The Canadian study was released in March of 1977. It involved six groups of 100 rats, 50 of each sex and a random selection of their offspring to obtain the same number and proportions of each. One group served as the control. Others were fed the same basal diet of Master Laboratory Cubes with 2.5, 25, and 250 mg O-TS/kg/day. A fifth group received 250 mg O-TS/kg/day along with 1 percent ammonium chloride. The sixth was fed 5 percent sodium saccharin that contained no O-TS.[27]

The pure saccharin and the O-TS were administered to determine whether the results of the FDA and WARF studies were due to saccharin or O-TS. Ammonium chloride has a tendency to produce alkaline urine and therefore was used to test whether a pH change could be responsible for bladder stones and, thus, tumor formation. In addition, the highest dosage level was kept to 5 percent to address the concern of toxicity at the 7.5 percent level.

In this study the parent (F_0) and second (F_1) generations were examined for bladder tumor development. Of the rats fed 5 percent saccharin, 3 F_0 and 8 F_1 male rats developed tumors believed to be malignant. Tumors were also found in the bladders of 2 of the F_1 females with the same diet. None of the other 587 rats exhibited signs of neoplasia.[28]

How should these results be interpreted? Should they be taken as an indication that saccharin poses a minimal threat, if any, to those who consume moderate amounts of foods and beverages containing the sweetener? The Office of Technology Assessment in 1977 acknowledged saccharin to be "among the weakest of carcinogens ever detected in rats."[29]

Bruce Ames and Lois Gold have compiled a scale (Human Exposure Dose/ Rodent Potency Dose: HERP) that ranks potential hazards, considering how much of a given material elicits a carcinogenic response in laboratory animals and the amount to which the average person might be exposed. HERP does not predict the actual chances of developing cancer, but it helps to put data, which mean little to those of us unfamiliar with animal carcinogenicity research, into perspective, enabling us to evaluate possible carcinogens in the context of potential risks to which we daily expose ourselves. According to Ames and Gold and their colleagues, the risk of contracting cancer from eating each day one raw mushroom, which contains hydrazines (natural pesticides), is 1.7 times greater than drinking 12 oz of diet cola containing saccharin. A daily drink of 12 oz of beer or 8 oz of wine, both of which contain ethyl alcohol, represents a carcinogenic threat greater than diet soda's by a factor of 47 and 78, respectively.[30]

Should we then, given the knowledge that saccharin is a relatively weak carcinogen, be informed of the potential hazard and make an independent decision about what risks we chose to accept? Or should any negative evidence, regardless of the strength of that finding or the potential benefits of a substance, be enough to warrant its elimination from the market? There was no available substitute for saccharin during the time when its fate was being determined; thus its ban would have meant the discontinuation of dietetic products for those suffering from diabetes and obesity.

The results of the Canadian study were, if not conclusive, highly influential. In Canada soft drinks containing saccharin were prohibited as of July 1977. Such immediate action was legally impossible in the United States, but the day after the findings were released FDA Commissioner Gardener announced plans to begin the process toward a saccharin ban: "Science and law dictate that saccharin be removed from our food supply."[31] To avoid the anticipated panic that such an announcement could generate, Gardner ended up providing the press and opponents with the material they would later use to gain much of their public support for eliminating saccharin restrictions. He reminded the public that saccharin had been in use for over eighty years and had never been shown to harm people and that the Canadian data did not indicate an immediate hazard to public health. Ending on an even lighter note, he said that a person would have to drink 800 diet sodas per day to receive the equivalent to the dosage of saccharin fed to the Canadian rats.

The Reaction

The press immediately picked up on the 800 cans per day. During the year following Gardner's announcement several articles covering the controversy expressed skepticism, portraying an FDA that was either given to overreaction or too rigid to make a decision that would take into account the harm and benefits of a potential carcinogen when determining policy. With titles such as "Reappraising Saccharin—and the FDA,"[32] "The Sour Taste of the Saccharin Ban,"[33] and "Of Rats and Men,"[34] many of the articles were accompanied by satirical cartoons that focused on the questionable applicability of animal testing with such huge doses and the legislation that limited the FDA's legal alternatives when dealing with a known animal carcinogen.

A few articles presented the other side of the debate and lent support to the FDA and its actions. In the *New York Times* Charles Wurster, associate professor of environmental sciences at the State University of New York at Stony Brook, argued that "cancer causation by a chemical at any dosage in laboratory animals is a warning of hazard to man." He went on to defend the rigidity of the Delaney Clause, saying, "It wisely allows no human discretion based on dosage . . . since there is no valid scientific basis for such discretion."[35] Neither in this nor in any similar statement is there an indication of what a "scientific basis" might be, other than proof of a negative—that cancer could never be caused. While those in the press who supported the FDA were outnumbered by its critics, the vast majority of articles on saccharin were neutral. Reflecting the nature of the debate, journalists often presented the two opposing views without drawing a conclusion. A commentary by John P. Wiley, Jr., for example, commended the early crusade fought by his grandfather, Secretary of Agriculture Harvey Wiley, against "deleterious additives," including saccharin: "He was on the right side then . . . when you could buy soothing syrups for crying babies that contained morphine and opium." However, recognizing the additional complexity of the issue over seventy years later, Wiley asked, "Should the government ban [saccharin] or simply inform us and let us decide what we want to do about it?"[36] A *U.S. News and World Report* article presented conflicting opinions as well, placing an interview with Senator Gaylord Nelson (D-Wisc.), who favored the ban, alongside that of Representative James G. Martin (R-N.C.), author of legislation that would allow exceptions to the Delaney Clause.[37] Although we do not know how much the coverage of saccharin influenced public opinion, readers appear to have had access to the varied arguments surrounding the debate.

The FDA commissioner's combined statements about the lack of evidence demonstrating saccharin to be a carcinogen in humans and the Canadian data not representing an "immediate hazard" to humans caused many to be dubious about the relevance of the rat data. Representative Andrew Jacobs, Jr., an Indiana Democrat, probably best captured the sarcasm with which many viewed the FDA's findings by introducing a bill that would require saccharin-sweetened products to carry the label, "Warning: the Canadians have determined that saccharin is dangerous to your rat's health."[38]

The director of the FDA's National Center for Toxicology Research, Morris Cramer, was one of many to criticize the inflexibility imposed by the Delaney

Clause. In a report to the commissioner, Cramer argued that the law failed to recognize that the potential risk of cancer from saccharin might be outweighed by the benefits to diabetics and the obese.[39] The American Diabetes Association (ADA) declared the ban "premature," since saccharin had not been shown to cause cancer in humans.[40] The head of the ADA, Dr. Donnell Etzwiler, scheduled an emergency meeting to deal with the saccharin "crisis," at which he said, "Taking away low-calorie sweeteners may well be a more serious health threat than this cancer 'threat.' "[41] Arguing the need to reevaluate food and drug laws, former FDA Commissioner Alexander Schmidt said, "Our scientific capacities to detect chemical residues have in many cases outstripped our scientific ability to interpret their meaning."[42]

To evaluate the potential human threat, some have looked to the scientific evidence that was not addressed by the FDA. By the spring of 1977 six epidemiological studies involving saccharin had been completed, only one of which (done in Canada) supported a positive relationship between saccharin use in males and bladder hypoxia. This 1977 Canadian study immediately received criticism from the scientific community. It identified all newly diagnosed bladder cancer patients in three provinces of Canada during a twenty-two-month period. For each of the 632 cases a control was matched for sex, age within five years, and neighborhood. Both groups completed a questionnaire that attempted to characterize the composition of their diets. "To test the hypothesis . . . that use of both saccharin and cyclamate data might increase the risk of bladder cancer, three questions were asked. The first was: 'Do you now, or have you ever used sugar substitutes?' If 'yes,' the number of tablets or drops usually used and the frequency and duration of use were determined."[43] The other two questions concerned the use of dietetic foods and soft drinks.

In females there was no correlation between saccharin exposure and outcome. In males there was a relationship between sweetener consumption and bladder cancer with a risk ratio of 1:6. In other words, 16 percent of the men interviewed who had consumed artificial sweeteners at least once in their lives had bladder cancer. A significant risk ratio of 1:8 (12.5 percent) was observed as well between bladder cancer and both coffee intake and smoking in males.[44] Critics attacked the methods of the research, charging that results from recall surveys tend to be inherently biased.[45] Another important weakness was that the questionnaire asked patients about past use of "sweeteners," without differentiating between cyclamates and saccharin, making it impossible to determine whether the questionable positive relationship would be between bladder cancer and smoking, coffee, saccharin, or cyclamate consumption.[46]

Human studies were not alone in failing to demonstrate saccharin to be a carcinogen. Laboratories had investigated saccharin for its effects on the bladders of female rats, mice, hamsters, rabbits, and monkeys and were unable to find evidence in any of these other animals.[47] Doubts still remain about the significance of the findings of male rats. Scientists, such as Duke University biochemist Henry Kamin, contend, "The dosages are so large that the result means nothing."[48] Even assuming for a moment the findings to have been meaningful, what does a positive finding in male rats at huge doses mean in the context of negative results for all of these other species?

Why were bladder tumors seen only in male rats? To answer this question, and to determine whether there was a dose-response relationship to saccharin, another two-generation study was undertaken, this time by the International Research and Development Corporation. As usual, the most sensitive species of rats were used, but this time a very large number, 2,500, were included, so that a variety of conditions could be tested. The study demonstrated that bladder tumors were not proportional to the administered dose—the linear assumption—but that the number declined rapidly with the dose. Extrapolated to human beings, if one believes that rats and people are alike in their reactions to saccharin, then over a lifetime the risk of getting cancer from drinking little more than two 12 oz cans of saccharin-sweetened drink a day would be less than one in a million.[49]

Why were the results of the saccharin tests so difficult to interpret? The general answer relates to the extremely low level of cancer correlation in especially sensitive strains of rats. By the time the National Academy of Sciences conducted its study, there had been twenty-one short-term bioassays of which sixteen were negative and five were positive. Why the differences in outcomes? One reason, biologist Clifford Grobstin notes, is that in evaluating the causes of bladder cancer among caffeine drinkers and smokers who also ingested saccharin, one would not be able to separate the effects. It is true that studies of diabetics, who consume a great deal more saccharin than most other people, "show no positive correlation between saccharin consumption and bladder cancer. However, the negative finding is not regarded as persuasive for the general population because diabetics are not typical in genetic background or metabolistic characteristics." And when studies conflict, Grobstin continues, "given the small difference in the actual results, the methodological differences between two studies make a decision between the conflicting possible interpretations impossible."[50]

Saccharin differs from other issues in that it has a powerful constituency of diabetics and weight-conscious consumers; otherwise, it probably would have been banned on the grounds of "causing cancer" in laboratory animals. We have presented the bare outlines of the history and science of saccharin in order to focus attention on the criteria of choice involving potential dangers from technology. Following the Delaney Clause, Commissioner Rowland was undoubtedly right in invoking a ban. A more relaxed version, not in current law, requiring that the doses fed to rodents approximate human doses, with a safety factor of 100 to 1,000 times less, would have rescued saccharin. But a similar rule would also rescue the vast majority of chemicals that have been fed to laboratory animals in quantities thousands and tens of thousands of times greater than humans would imbibe. If rodent cancer tests were invalidated as unrelated to human cancers, then only epidemiological studies (those involving humans) would count, and saccharin would be exonerated. If we considered any adverse finding decisive, then saccharin would be banned or severely limited. If regulation required a preponderance of studies, then saccharin would be acquitted.

Given its exceedingly widespread use, saccharin should be evaluated cautiously. The animal evidence shows it to be a very weak carcinogen in only one species at huge doses; human evidence is negative except for a single study based on recall evidence. Overall, we would judge that the evidence does not justify regulation. Readers might come to a different conclusion, but they should now be aware that it is not only the studies themselves but the criteria of interpretation and the philosophies behind them that matter.

Evaluating the Risks of the Three Substances

Viewed as studies of politics, not of knowledge, the different fates of the artificial sweetener saccharin, of dieldrin, the pesticide, and of aminotriazole, the weed killer used in cranberry bogs, are instructive. The chemical with the largest constituency won, while those used in specialized industry or agriculture lost. True, saccharin, if a carcinogen in rodents at all, is an awfully weak one, but the same could be said of the cranberry weed killer. We are being inducted into the strange world of weak causes and infinitesimal effects. If it is inaccurate to say that the substances involved cause cancer in human beings at doses comparable to those tried on animals, neither would it be entirely accurate to say that they could not, now or ever, cause cancer. Proving a negative, absent total knowledge, is not possible. The research results bandied about are often bewildering, not because the subject matter is unfathomable to ordinary mortals but because infinitesimally small effects are combined with a demand for conclusive negative evidence. Usually in our lives we ask for positive or probable proof that an effect will occur or we dismiss it. If cars are not in the habit of jumping curbs, we are not likely to walk defensively, though we know that automobiles have in rare cases run up onto sidewalks and maimed or killed pedestrians. Concrete or steel fences lining every curb could be justified on this basis; but we do not entertain such stultifying ideas because ordinarily we do not take seriously the demand for negative proof under possible but improbable circumstances. Now, in regard to substances accused of causing cancer in laboratory animals, we do.

Dieldrin is different in that in substantial doses it has proven toxic to mammals in general and humans in particular. But it also has beneficial uses. A grave fault of the antipesticide literature, especially *Silent Spring,* is that it makes virtually no mention of pesticides' benefits to health, not only for control of noxious insects but also in terms of lower costs of foods to consumers and higher income to farmers. Instead of balancing health benefits, we end up concentrating totally on harms. Insistence that possible minuscule dangers cannot be ignored or outweighed by proven benefits has produced the "cancer-of-the-week" syndrome, which has occasioned some ridicule but is, nevertheless, official government policy.

2

PCBs and DDT: Too Much of a Good Thing?

I. Which Regulations Governing PCB Residues Are Justified?

with Leo Levenson

Polychlorinated biphenyls, or PCBs as they are popularly called, were once regarded as valuable industrial chemicals. First marketed in 1929, they were used as fire-resistant insulators in electrical transformers, capacitors, and industrial machinery and as components of "carbonless" copy paper, paints, and pesticides. PCBs were little known to the general public until the 1970s, when reports spread that PCB residues in the environment might raise cancer rates and harm wild animal populations.

In 1973 the Food and Drug Administration (FDA) set strict limits on allowable levels of PCB residues in food. In 1976 Congress banned the manufacture of PCBs, setting in motion a series of regulatory activities by the Environmental Protection Agency (EPA) regarding the cleanup of PCBs in soil and water and ordering the removal of PCBs from electrical equipment.

Until the 1960s concern about PCBs was restricted to protecting the health of workers handling the chemicals. In the 1930s several reports were published about widespread acnelike skin problems and occasional liver damage among workers exposed to PCBs along with other chemicals in poorly ventilated environments. One fatal case of jaundice was reported. With improved ventilation, the symptoms usually declined or disappeared.[1]

In 1966 a Swedish biologist, Soren Jensen, looking for residues of DDT and other pesticides in wildlife, used newly refined scientific instruments to discover PCBs stored in the fatty tissues of fish and fish-eating birds. Jensen sounded the alarm that PCBs might interfere with animal reproduction and might be carried far from their industrial sources in water or air without breaking down. He feared PCB mixtures would prove a more important threat to wild species than DDT, since they had been manufactured and released into the environment in much greater quantities.[2]

In 1968 over 1,800 Japanese became sick after consuming rice oil contaminated with an industrial coolant containing PCBs. Victims suffered from skin eruptions called chloracne, gastrointestinal problems, and nervous disorders. Symptoms continued for five years or longer for some patients.[3] Some babies born to victims had unusual dark brown skin pigmentation, though their growth patterns appeared normal. The event became known as the Yusho oil disease episode. A similar mass poisoning involving PCB-tainted rice oil occurred in Taiwan in 1979.[4]

The rice-oil poisonings appeared to show that PCB mixtures could be exceedingly toxic to humans. But by 1975 scientists determined that PCBs alone were almost certainly not the primary cause of the Yusho illness, for Yusho patients had smaller concentrations of PCBs in their blood and fat than healthy workers who handled PCBs. Instead, potent compounds called polychlorinated dibenzofurans (PCDFs) found both in the Yusho and Taiwan episodes were eventually implicated as causing the illnesses.[5]

After Jensen's alarm and the Yusho episode scientists began looking for PCB residues in the environment and found them widely distributed in stream and lake sediments and in the fatty tissues of fish, fish-eating animals, and people.[6] In the Hudson River, where two General Electric plants were estimated to have discharged at least thirty pounds per day of PCBs into the river for decades, fatty fish, such as salmon and trout, were routinely found to have 10–100 parts per million (ppm) of PCBs in their flesh.[7] There was no evidence that this level of PCBs was hurting the fish, but scientists expressed concern about the potential health effects on wildlife and people who regularly ate the fish.[8]

PCBs have low solubility in water and bind tightly to most types of soil and sediment.[9] Although sediments in industrial waterways near factories using PCBs had been found to contain up to 1 percent PCBs, the maximum concentrations found in water were about 3 parts per billion (ppb), or 0.0000003 percent.[10] Because of the tendency of PCBs to bind to soil, most PCBs spilled on the ground or disposed of in landfills could be expected to remain in place, without entering waterways or wildlife.[11]

In 1970, after early studies showed animal reproductive problems associated with PCBs, Monsanto Corporation (the one major U.S. producer) announced that it would sell the chemicals for use only in "closed systems," primarily capacitors and transformers, for which no equivalent fire-resistant replacements existed at the time. PCBs would disappear from pesticides, copying paper, most hydraulic fluids, inks, and paints, as old stocks of these materials were used up or discarded.[12]

In 1971 an animal counterpart to the Yusho incident was discovered after reproduction among breeder hens suddenly declined in a Holly Farms chicken facility. Holly Farms analyzed the feed and discovered about 150 ppm of

PCBs.[13] The Department of Agriculture traced the contamination to leaking machinery in a North Carolina fish meal plant that had distributed at least sixteen tons of feed containing PCBs. Holly Farms voluntarily destroyed 77,000 contaminated chickens.

As far as we can tell, the 1968 Yusho poisonings in Japan went unreported at the time by major U.S. news media.[14] Three years later, after the Holly Farms affair, the leading weekly news magazines, *Time* and *Newsweek,* carried their first articles on the potential dangers of PCBs. *Time*'s story, "The Menace of PCB," characterized PCBs as a "potent new threat to the environment." The story mentioned that PCBs had been found to cause liver damage in mice and fragile eggshells in birds. Only in the final paragraph did *Time*'s story say that there was no evidence that existing levels of PCBs in the environment or food supply were dangerous to people.[15]

The *Newsweek* piece, "The PCB Crisis," contained more reports of potentially harmful effects on humans from PCBs, including what was apparently the first major published reference to the Yusho episode three years earlier. The article related that large doses of PCBs could be fatal and that lesser amounts were known to result in "acne, impaired vision, abdominal pains and liver ailments," along with "suspicions of genetic damage." *Newsweek*'s writers did not compare these "large doses" and "lesser amounts" with the amounts generally consumed in food;[16] if they had, the situation might have seemed less of a "crisis."

These articles were early examples of what would become a characteristic pattern for stories on PCBs: descriptions of the worst consequences of high-dose exposure, with little or no comparisons offered between toxic levels and the environmental concentrations people and animals were likely to encounter.

By the early 1970s people knew that relatively small leaks of PCB mixtures from machinery directly into food and animal feed could harm humans and animals.[17] PCB residues were widely found in the fatty tissues of people and animals and in industrial riverbeds, but no one knew whether these residues would cause toxic effects. No epidemiological studies were available to provide evidence as to whether people exposed to PCBs (other than in the Yusho episode) had suffered long-term health problems. Numerous animal research projects were then launched to try to determine whether PCB levels found in the environment were dangerous.

FDA Regulation of PCBs in Fish in 1973

After conducting a series of studies from 1970 to 1972, FDA scientists found traces of PCBs in many foods, with the largest concentrations in fish. The FDA estimated that American adults ate on average 4 millionths of a gram (micrograms, µg) of PCBs per day,[18] equivalent in weight to about 1 millionth of a teaspoon of salt. With the discovery that most Americans were regularly ingesting PCBs, the FDA came under increasing pressure to write regulations defining what level of PCB residues, if any, should be permissible in foods.

After receiving comments on proposed regulations, the FDA enacted a regulation in 1973 defining a tolerable daily intake (TDI) of 1 μg per kilogram of body weight per day (1 μg/kg/day), or about sixteen times the national average intake. This meant that FDA scientists believed that a 150-pound (70-kg) person could eat as much as 70 μg of PCBs every day without suffering ill effects. Of course someone weighing 100 pounds (45 kg) would reach the standard after eating 45 μg per day. To reduce the probability that people would exceed the TDI, the FDA set maximum allowable PCB levels in fish of 5 ppm.[19]

Why did the FDA allow any residues, instead of banning the sale of fish with detectable PCBs, as some environmental groups had suggested? The FDA concluded that a ban would not be in the best interests of the consumer: because small PCB residues were widely found in major waterways, such an action would have effectively banned commercial fishing in many areas, devastating individual fishermen and local economies, and raising some fish prices. Meanwhile, there was no evidence of a public health problem from occasional ingestion of very small amounts of PCBs. As the FDA's standard-setting document declared, a complete ban on PCB residues would "unnecessarily deprive the consumer of a portion of his food supply and disrupt the Nation's food distribution system."[20]

It is worth noting that a 150-pound person who regularly ate fish contaminated with PCBs at the legal limit of 5 ppm would probably exceed the FDA's TDI of 1 μg/kg/day. In 1984 the FDA decided that a stricter residue standard for fish could be met without undue hardship for the fishing industry, and the maximum tolerance for PCBs in fish was lowered to 2 ppm.[21]

How did the FDA conclude that allowing some traces of PCBs at the level of parts per million was unlikely to be a public health problem? FDA researchers used both animal data and early analysis of the Yusho episode. Animal studies reported that dogs and rats appeared to suffer no toxic effects when fed PCBs at a rate of 250–300 μg/kg/day, which is equivalent to about 250–300 times the standard the FDA developed for humans.[22] The FDA also cited rough estimates that the smallest known PCB dose associated with symptoms of poisoning in the Yusho incident was about 200 μg/kg/day over fifty days, for a total PCB intake of 10,000 μg/kg body weight over the period of exposure.[23] If a person ate PCBs in food for a much longer period, say three years or roughly a thousand days, and the PCBs accumulated in the body, a diet of 10 μg/kg/day would result in the same total intake. If FDA researchers had known about the probable role of PCDFs in enhancing the toxicity of the Yusho PCBs, they might have concluded that even higher amounts of PCBs would be safe as long as PCDFs were not also present. But convincing evidence of the PCDF role was not available until 1981.[24]

The evidence used by the FDA in 1973 to establish a basis for regulating

PCBs was relatively straightforward. It did not rely on the elaborate and controversial statistical models of cancer risk later used by the EPA in determining PCB regulations. Lacking human evidence other than the Yusho episode, the FDA made an explicit attempt to set an allowable PCB intake that would not be too expensive for consumers and food producers and that would represent a limit at least a hundredfold less than what was then estimated to cause illness in animals.[25]

From our reading of the story, the FDA's actions were reasonable, given evidence that PCBs could pose a hazard in food if present in large enough quantities. The residue limits set by the FDA were achievable at a reasonable cost and were well below levels that would be expected to cause a significant health hazard, based on the evidence available.

Media attention to PCBs died down for a few years after the animal feed contamination episodes in 1971 but escalated in 1975 when state officials announced that salmon and bass containing more than the FDA tolerance level of 5 ppm were prevalent in industrial areas of the Hudson River and the Great Lakes. *New York Times* stories referred to PCBs in the river as a "toxic peril."[26] Scientists argued that since they took so long to break down in the environment, all the PCBs produced would continue to accumulate and might cause environmental damage. Concerned groups began lobbying for a complete ban on PCB manufacturing as soon as a fire-resistant replacement was available for capacitors and transformers.[27]

When companies reported in 1975 that they had developed effective, fire-resistant, and less toxic replacement fluids for capacitors and transformers,[28] the stage was set for ceasing all manufacture of PCBs. In October 1976, three weeks before losing his office to Jimmy Carter, President Gerald R. Ford signed the Toxic Substances Control Act (TSCA), which included the PCB ban.

Section 6(e) of the act required the EPA to promulgate regulations that would phase out the use of PCBs except under "totally enclosed" conditions that "will not present an unreasonable risk of injury to health or the environment."[29] It was up to the EPA to decide what disposal methods for PCBs in equipment and soil and what continuing uses of old equipment containing PCBs could be allowed under such legislative language. PCBs were now largely illegal. But in the doses and exposures involved were they dangerous to animals or people?

Animal Studies

New animal studies presented mixed findings. During the 1970s animal tests still showed PCBs to have a low short-term toxicity. Thanks to their low solubility in water, environmental concentrations in waterways were not sufficient to kill fish.[30] Rats and mice could survive one-time doses of several grams of commercial PCB mixtures.[31] But doses thousands of times smaller—in the range of milligrams per day—if fed to animals over an extended period of weeks or months, were found to cause liver damage and reproductive problems.[32] The animal experiments that became most prominent during debates over regulatory standards for PCB residues were those measuring cancer in rats and reproductive success in rhesus monkeys.

In 1975 a study led by Public Health Service scientist Renate Kimbrough found that female rats fed 100 ppm Aroclor 1260 (a commercial PCB mixture containing 60 percent chlorine) every day for twenty-one months developed an increased number of liver tumors late in their lives compared with untreated animals.[33] The liver tumors did not spread to other tissues nor shorten the lifespan of the treated rats. The 100 ppm level was equivalent to about 5,000 µg/kg/day, or about 5,000 times the tolerable daily intake of PCBs set by the FDA in 1973.

The Kimbrough study would be frequently cited during the development of PCB policy as evidence that the chemicals posed a serious cancer risk. But little attention was paid to another interesting finding of the study: the treated rats actually had fewer reproductive-system cancers, making their overall cancer rate no higher than that of untreated rats.[34] Could PCB exposure actually prevent some types of cancer?

In 1984 a German study led by Ekkehard Shaeffer echoed Kimbrough's mixed findings about PCBs and cancer: rats fed diets containing 100 ppm of commercial PCB over most of their lives had more liver tumors and observable liver damage at time of death than control rats, but again they had significantly fewer tumors of other tissues.[35] In fact, rats fed two different types of commercial mixtures had significantly lower overall cancer rates and better survival rates than the control rats on a diet without PCBs. The authors theorized that the reduced mortality and cancer rate might have been caused by "PCB-induced alterations in the immune system" (presumably one stimulating cancer-preventing or cancer-fighting capabilities).[36]

PCBs and Humans

The most serious evidence of potential reproductive problems to humans from PCBs came from tests on rhesus monkeys. Eight females fed approximately 2.5 ppm of the 48-percent-chlorine Aroclor 1248 (equivalent to about 100 µg/kg/day) developed facial swelling and acne within two months and had irregular menstrual cycles by four months. All eight monkeys conceived, but three had spontaneous abortions. The untreated animals had no spontaneous abortions or difficulty conceiving. Female monkeys on a diet of 5 ppm (200 µg/kg/day) Aroclor 1248 did worse, with two out of eight unable to conceive, and four out of the remaining six experiencing spontaneous abortions. Male rhesus monkeys on the diet of 200 µg/kg/day suffered no decline in fertility.[37]

The effects of PCBs on reproduction were not consistent across different monkey species. In a later anecdotal report squirrel monkeys appeared to reproduce successfully in a PCB-contaminated environment where female rhesus monkeys had severe reproductive difficulties and other signs of PCB toxicity.[38] The diet of 100 µg/kg/day PCBs that caused harmful effects in the rhesus monkeys was about 100 times greater than the 1973 FDA TDI and about 1,600 times more than the FDA's estimate in 1973 of the average U.S. intake of PCBs.

The monkey studies raised concern about pregnant women who regularly ate fish from industrial waters. A 150-pound pregnant woman would have to eat every day about 3 pounds of fish containing the FDA limit of 5 ppm PCBs to match the diet of 0.1 mg/kg/day PCBs shown to cause reproductive harm in the rhesus monkeys. There may be no one who actually eats that much fish, but our judgment is that the monkey diets were close enough to possible human intakes of PCBs, and monkeys close enough to humans, to give pregnant women reason to avoid a daily diet of fish containing significant amounts of PCB residues.

By the mid-1970s studies had generated information that various animal species could suffer liver toxicity, apparent hormone imbalances, liver cancer, and spontaneous abortions when fed high enough amounts of commercial PCB mixtures. Not enough epidemiological evidence in humans was yet available to evaluate whether people might be more or less susceptible than laboratory animals to the toxic effects of prolonged exposures to PCBs.

The best evidence of PCB toxicity in humans appeared to be the Yusho episode. But, as we noted earlier, Japanese researchers had made a puzzling discovery: Yusho victims with serious poisoning symptoms had lower amounts of PCBs in their blood than relatively healthy workers occupationally exposed to PCBs (1–12 ppb PCBs for a sample of twenty-five Yusho patients versus 60–920 ppb PCBs for a sample of twenty-three Japanese PCB production workers; two of the workers showed signs of skin problems).[39] A plausible explanation for this anomaly was provided by the discovery, reported in 1975, that the rice oil contained not just PCBs but also unexpected amounts of much more toxic PCDFs, probably formed when PCBs were heated during the oil processing or later cooking. Sampling techniques for PCDFs in oil were not precise, but two different researchers estimated the concentration of total PCDFs in preserved samples of the original rice oil at roughly similar levels: 2 and 5 ppm. They could not estimate how many more PCDFs might have been formed during cooking with the rice oil.[40]

American commercial PCB mixtures were soon found to contain trace quantities of PCDFs—but at much lower relative concentrations than in Yusho oil. Whereas the ratio of total PCBs to total PCDFs in Yusho oil was estimated to be between 100:1 and 200:1, the equivalent ratio in commercial American PCB mixtures ranged from 500,000:1 to 1,000,000:1.[41] The new information on Yusho had two major implications. Exposure to commercial PCB mixtures not containing concentrated levels of PCDFs was much less dangerous than originally thought from the Yusho incident. But heated PCB mixtures, such as smoke from transformers caught in a fire, might generate elevated levels of PCDFs and be more toxic than would be expected based on animal tests of unheated commercial PCB mixtures.

After passage of the PCB ban by Congress in 1976, the EPA wrote a number of rules regulating the allowable uses and disposal of remaining equipment containing PCBs and the cleanup standards for PCB spills. The total cost of these rules, compared with replacing PCB electrical equipment gradually as it wore out, was estimated by the EPA to be over $700 million (in 1982 and 1985 dollars).[42] In each of these instances there was little disagreement that some level of regulation was appropriate to prevent people and animals from coming into contact with large amounts of PCBs. The highest financial stakes and most bitter controversies came over the regulation of electrical equipment and soil containing small residues of PCBs.

EPA Regulation of Capacitors and Transformers

In 1979 the EPA issued a regulation designating all nonleaking capacitors and transformers other than railroad transformers as "totally enclosed," permitting utilities to continue to use them without major restrictions. This enclosure determination would have reduced the economic cost and disruption of the PCB ban, allowing utilities gradually to replace and dispose of their PCB equipment as it wore out.[43] But this EPA regulation was successfully challenged in court by the Environmental Defense Fund and the Natural Resources Defense Council, which argued that electrical equipment can leak and therefore cannot be considered "totally enclosed." In 1982 the EPA issued new regulations requiring the removal of large PCB-containing capacitors in public areas and all PCB equipment that could potentially leak into food storage and processing areas.[44]

The limited removal of certain PCB capacitors and transformers was estimated by the EPA to cost $16 million and was relatively uncontroversial. The liability risk alone of one major PCB leak into food could easily be more expensive than the entire cost of the regulation. But the provision calling for the replacement of all large PCB capacitors from unfenced outdoor areas, carrying a cost estimated by the EPA at $140 million, caused greater protest from utility companies.[45]

Large capacitors are present on electricity distribution lines to regulate voltages. Almost all of them manufactured before 1978 used an average of 17 pounds of PCB mixtures as a fire-retardant insulator.[46] Periodically, as a result of equipment failure, lightning, or sudden voltage surges, capacitors would explode, spraying their insulating fluid 10 feet or more.

What were the reasons for requiring the early removal of these capacitors? In introducing the new regulations the EPA claimed that the rule would benefit the environment by preventing the release of 572,000 pounds of PCB.[47] That sounds like a large amount, yet according to EPA estimates the releases would come from some 34,000 different capacitors. In cases where the PCBs fell onto soil or other porous material, most would be absorbed where they fell. According to estimates used by the EPA, the likely evaporation of PCBs would be 0.2 percent per day, and movement into water would be at least 100 times less than that, so a 17-pound PCB spill would result in about 0.03 pounds (0.5 oz) of PCBs being volatilized per day, assuming no cleanup of the spill.[48]

How dangerous would it be if 0.5 oz of PCBs were to enter the atmosphere? Only a tiny fraction of the volatilized PCBs would actually be breathed in by people. In most cases PCBs from these spills would be outdoors, where diffusion would rapidly dilute the PCBs to nondetectable levels anywhere other than directly above the spill. In the rare cases when a capacitor in a residential area might explode, sending PCBs through the open window of a house or car, the owners would probably clean up the oil and wash their hands. There would be human exposure to only a tiny fraction of the PCBs in the capacitors.

If high enough heat occurs during capacitor explosions to form large quantities of highly toxic PCDFs, then concern about incidental human exposure might be warranted. But even in this case it seems unlikely to us that the amount of PCDFs that would actually be breathed in or absorbed through the skin would be significant. The EPA did not report evidence about the composition of exploded capacitor fluid in the preamble to the regulations. Reviews of PCB toxicity lack mention of any observed skin problems or other toxic effects on passers-by exposed to capacitor explosions.[49]

In summary, the projected $140 million expenditure for early removal of capacitors seems to have had the intended goal of preventing about 34,000 spills of PCB-containing fluids averaging 17 pounds of PCBs each, without supporting evidence that these small spills would result in human exposures to toxic concentrations of the chemicals.

Two crucial decisions by the EPA concerning PCB concentrations appeared first in the 1979 draft regulations and survived to reappear in the 1982 rules. The EPA defined "PCB-contaminated" transformers as containing oil with between 50 and 500 ppm (0.005–0.05 percent) of the substance. "PCB-transformers" were defined as any containing more than 500 ppm (0.05 percent). Both categories of transformers required expensive special handling and disposal.

According to EPA estimates, about 130,000, or 0.5 percent, of the 25 million or so transformers in service by the end of 1981 were filled with commercial PCB mixtures (containing about 50 percent PCBs).[50] The other 99.5 percent were originally designed to contain cheaper (and less fire-resistant) mineral oil. About 10 percent, or 2.5 million, of the mineral oil transformers were estimated to contain small residues of PCBs of between 50 and 500 ppm (the "PCB-contaminated" category). These residues were introduced when the transformers were drained and refilled by means of the same equipment used on PCB transformers. Utilities could not be sure which mineral oil transformers contained 50 ppm or more without testing fluid samples from each one.

The 50 and 500 ppm cutoff levels chosen by the EPA persisted as stricter regulations on PCB-containing electrical equipment were promulgated during the 1980s. The EPA rejected arguments from the utility industry that risks from dilute concentrations of PCBs would be negligible and that regulations should be limited to mixtures containing at least 5,000 ppm (0.5 percent) PCBs.[51]

Choosing the 50 and 500 ppm levels for regulated transformers, rather than limiting the scope to equipment originally filled with PCB mixtures, had expensive consequences. Reclassifying PCB-filled transformers as non-PCB transformers by replacing the PCBs with alternative material required repeated flushing to eliminate the sticky residues. The EPA required that refilled transformers be put into service for three months and then resampled before they could be reclassified. Under EPA rules, malfunctioning PCB transformers could not be serviced, so many potentially repairable pieces of equipment had to be scrapped.[52]

Why did the EPA decide to treat mineral oil transformers containing a fraction of a percent of PCBs like PCB-filled transformers? The EPA argued that dilute PCBs in equipment could be dangerous. As a 1979 EPA document stated, "The Agency disagrees there are insufficient adverse health effects data to warrant regulations [on electrical equipment] below 500 ppm. PCBs at levels below 500 ppm have been shown to cause a variety of adverse health effects in animals including malignant and benign tumors."[53]

The EPA's comparison of the concentrations of PCBs in electrical equipment to PCB doses fed to animals in laboratory experiments was completely inappropriate. Yes, animals get sick when fed 500 ppm of PCBs over an extended time. But people would not be eating the PCBs in electrical equipment; they would be exposed indirectly—through breathing vapors or absorbing fluids on their skin from maintenance, disposal activities, spills, or fires. These routes of exposure would mean inhaling or ingesting only a tiny fraction of the PCBs present in the equipment.

The EPA argued that the total amount of PCBs in dilute mixtures spread among millions of pieces of equipment together represented enough chemicals to cause damage to "the environment," presumably by harming susceptible species of animals, plants, and microorganisms. The preamble to the 1979 regulations stated that EPA set the lower cutoff level for "PCB-contaminated" equipment at 50 ppm rather than 500 ppm because 1 million additional pounds of PCBs would thus be controlled.[54]

This "cumulative weight" argument implies that 1 pound of PCBs is as dangerous in a concentrated form as it is in dilute mixtures spread out among many pieces of equipment. Eventually, the argument goes, all of the PCBs will enter air or waterways, build up in animal tissues, and be a potential cause of disease. What this argument ignores is that most dilute PCBs spilled on soil or disposed of in landfills will remain attached to the soil, never being absorbed, inhaled, or ingested by any animal or plant. The less concentrated a PCB spill, the less likely that a significant amount will volatilize into air or be transported into groundwater or surface water.[55] PCBs buried in landfills are not suspected to be a significant source of contamination of waterways, fish, or birds.[56] It is unlikely that special handling and disposal of the dilute PCBs in mineral oil transformers would make a discernible or significant difference in environmental PCB levels.

Another argument used by the EPA to maintain regulations on dilute PCBs was that industry could afford the costs. The 1979 regulatory preamble stated, "No evidence was presented that indicated that industry is technologically or economically unable to comply with the more stringent standard."[57] In 1982, when the agency refused industry requests to change the definition of "PCB-contaminated" transformers from 500 to 5,000 ppm, it explained: "Such an approach would result in unnecessary and avoidable exposure to PCBs since there is information in the rulemaking record that indicates that technology is available at reasonable cost to reduce the PCB concentration in transformers to below 500 ppm."[58] The "information" was a background document predicting that flushing transformers to reduce PCB concentrations below 500 ppm would cost about $25 per gallon of PCBs.[59] In other words, to the EPA it seemed as though the PCB definition did not matter much: if it was inexpensive to reach the lower-level definition, why not do it?

In fact, getting transformers to below 500 ppm turned out to be very expensive. A 1988 industry paper reported that retrofilling a 1,000-kilowatt industrial PCB transformer containing 300 gallons of PCB fluid to meet the EPA's 500 ppm limit required at least two power outages and cost $55,000, or about $180 a gallon.[60] This calculation represents about six times the EPA's 1982 estimate (adjusted for inflation).[61]

Treating transformers with 0.05 percent PCBs like pure PCB transformers may have prevented occasional exposures to dilute PCBs, but it added greatly to the volume of waste fluid that would require special, and expensive, handling and disposal. Meanwhile, no human or animal evidence suggested that transitory exposures to dilute PCBs were likely to be dangerous.

The PCB regulations on electrical equipment were designed to protect against potential environmental and health risks, but little attention was given to the possibility that the regulations could themselves create risks, primarily to utility workers who would have to work near high-voltage equipment and high above ground to remove or test equipment. One early reader of this chapter recalled that a utility worker was killed by accidental exposure to high voltages while sampling a piece of equipment for PCBs to comply with EPA regulations.[62]

Scientists were originally concerned about the dangers of PCBs accumulating and causing harm to wildlife and entering human food supplies through fish products from industrial waterways or from food-processing machinery. After PCDFs were implicated in the Yusho and Taiwan rice-oil poisonings, another threat was identified. What if PCBs in electrical equipment were heated in a structural fire and formed smoke containing PCDFs? Would the smoke be exceptionally toxic? Should special measures be taken to remove PCBs from inside buildings where fires might occur and people could be exposed to the smoke? The issue became more than hypothetical when on February 5, 1981, an electrical fire broke out in a state office building in Binghamton, New York, leading to what has probably been the most expensive cleanup operation ever for a single office building.

The Binghamton fire caused the rupture of a transformer containing 180 gallons of concentrated PCBs. The heat of the fire vaporized the transformer fluid, releasing PCB-laden soot. When the soot was later analyzed it was found to contain about 2,000 ppm of PCDFs and 20 ppm of equivalently toxic poly-chlorinated dibenzo-dioxins (PCDDs), formed when PCBs and solvents in the transformer were heated by the fire. Luckily, the fire occurred at 5:30 A.M. when the building was unoccupied. Unluckily, the soot entered a ventilation system and was spread into every corner of every floor of the building.

State janitors immediately began cleaning up the building before administrators learned that PCDDs and PCDFs were present in the soot. Initially the janitors did not wear protective clothing and tracked soot from the state building into the adjacent city government building used as a staging area for the operation. One janitor was sent to the hospital with a rash on his face after working on the cleanup.[63]

The New York State Department of Health (DOH) faced the task of deciding when the building was "safe" to reoccupy. At the time there were no formal safety standards for dioxins or furans in air or on surfaces, and developing a standard for the complex mixture present at the Binghamton building was a daunting task. DOH scientists decided to set cleanup standards by comparing the soot's toxic effects in animals with the toxic effects of a relatively well-studied type of dioxin, 2,3,7,8-tetrachlorodibenzo-*p*-dioxin, or TCDD. Subsequently all concentrations of chemical mixtures in the Binghamton building were converted to TCDD equivalents.[64]

By 1983 DOH staff had measured the average toxicity of the air inside the building at about 14 picograms (trillionths of a gram, pg) of TCDD equivalents per cubic meter (m^3), with considerable variation in different parts of the building.[65] An advisory panel accepted a DOH recommendation that the building not be reopened until the air in all areas contained less than 10 pg/m^3 of TCDD equivalents and all surfaces contained no more than 25 nanograms (billionths of a gram, ng) per square meter (m^2). Using various assumptions about air intake and absorption of chemicals, the DOH estimated that this would restrict exposure of building occupants to TCDD equivalents about 500 times less than the maximum amount of TCDD that rats had been observed to ingest without suffering harmful effects (adjusting for the difference in weight between rats and people).[66]

After the initial cleanup tiny amounts of chemicals continued to diffuse slowly from the building, keeping the internal air contamination above the DOH cleanup levels. The quickest way to reduce air contaminant levels would have been to pump clean air aggressively through the building, venting chemical residues to the outside, where they would have been immediately diluted to insignificant concentrations.[67] Nevertheless, perhaps fearing local protest or potential liability, authorities decided to seal the building and place carbon filters on ventilation ducts, greatly slowing the process of removing contaminated air. By 1988, seven years after the fire, all of the building had cleared reentry standards except the equipment room where the fire started, which still had chemical levels in air at about double the standard. Even though maintenance workers would not be working continuously in the equipment room, the DOH decided that the same standard should apply to all parts of the building.[68]

As of December 1991, three years of extensive work on the equipment room had still not brought the concentrations down to the DOH standard. More than ten years and $40 million in cleanup costs after the original fire, the building had not reopened.[69] Numerous multimillion-dollar lawsuits against the state by firefighters and cleanup workers exposed to the soot are still pending.

Did the cleanup need to be so expensive and prolonged? Our reading of the story is that the overall goals were reasonable, based on the animal evidence available at the time. However, it appears that the job could have been completed more quickly and millions of taxpayer dollars saved without compromising public health if state officials had been allowed to vent the building air directly to the outside during the cleanup and to leave higher residues in the equipment room than in the continuously occupied areas.

Responding to the Binghamton fire and other PCB episodes involving office buildings, the EPA published a regulation in 1985 that required replacing about 7,400 high-voltage PCB transformers in commercial buildings and adding new "electrical protection" adjustments and labels to the other 57,000 or so commercial PCB transformers. The EPA estimated the regulations were likely to prevent PCB transformers from being involved in about thirty-five commercial fires, at a cost of roughly $600 million.[70]

Many utilities agreed that removing concentrated PCBs from transformers inside commercial buildings made sense and had already began doing so before the EPA rule was finalized. But was it also worth worrying about tiny concentrations of PCB residues, as the EPA regulations had demanded since 1979? While developing the "fire rule," EPA regulators never publicly considered one key change that could have made the rule much less expensive: relaxing the arbitrary definition that any transformer with more than 500 ppm PCBs represented a "PCB-transformer." Transformers originally filled with commercial PCB mixtures, such as those in the Binghamton fire, contained about 50 percent, or 500,000 ppm PCBs. Simply draining and refilling a pure PCB transformer with replacement fire-resistant fluid could remove 95 percent of the PCBs, bringing the concentration down to about 2.5 percent or 25,000 ppm PCBs.[71]

Had the EPA allowed this level, what consequences would have resulted? Any fires that still occurred would contain at least 95 percent less PCBs and breakdown products than the pure PCB fires. How much money would have been saved by allowing dilute concentrations of PCBs to remain in commercial transformers? How much more quickly would utilities have been able to comply with the rule, perhaps avoiding the involvement of concentrated PCBs in new fires? In the EPA's *Federal Register* article on the regulation, the agency gave no indication that it had analyzed the risks and benefits of leaving dilute residues of PCBs in place.[72]

Just recently these regulations came close to home when the University of California at Berkeley was caught having not complied with all EPA rules regarding PCB transformers. The university agreed to pay a fine of $150,000 and to spend approximately $15 million to replace PCB transformers.[73] That money, which will result in no benefits to health, could have paid for quite a few professors' salaries and even more graduate students' stipends.

The most expensive PCB regulations governed how to deal with soil with PCB residues left over from leaking or discarded equipment. After 1980, EPA regional decisions and state regulations created widely varying procedures for cleaning up PCBs in soil. In some parts of the country soil removal was required only for PCB residues containing more than 50 ppm PCBs.[74] In other areas, EPA regional offices or states required all detectable PCB residues to be removed. The stakes were enormous. In 1987 the EPA estimated that if all capacitor spills in the country had to be cleaned up to nondetectable levels, the cost would be anywhere from $100 million to over $2 billion *annually;* the estimate for a 50 ppm level was $42–80 million annually.[75]

In 1987 the EPA ratified an agreement negotiated by industry groups, the Environmental Defense Fund, and the Natural Resources Defense Council to set a nationwide standard allowing 50 ppm residues in fenced electrical substations and 10 ppm residues in residential areas (covered with 10 inches of clean soil). Industry had argued for higher residue limits but accepted the agreement as better than the status quo. The "compromise" was still very expensive, as it left unchanged the hazardous waste disposal laws that require all soil removed from PCB spills to be disposed of in special chemical waste landfills or hazardous waste incinerators. In addition, as of 1991, twenty-seven states continued to regulate PCBs under their own statutes, often with more stringent guidelines than the EPA compromise.[76]

The Dangers of PCB Residues in Soil

How dangerous are parts-per-million concentrations of PCBs in soil? People can be exposed to PCBs in soil primarily by breathing in dust, getting dirt on food materials, and eating with dirty hands. Children who play outside are likely to absorb far more soil than adults relative to their body weight.

According to recent estimates, an average 33-pound (15 kg) child playing all day in soil containing 10 ppm PCBs would end up absorbing about 0.6 μg of PCBs, for a total exposure of 0.00004 mg/kg/day.[77] This is 2,500 times less than the daily diet that appeared to cause hormonal problems in pregnant rhesus monkeys (0.1 mg/kg/day) and 125,000 times less than the daily diet associated with increased cancer incidence in two strains of rats (5 mg/kg/day). In reality, children would not play all day, every day in contaminated soil, so they would have even smaller exposures.

Under current regulations the cost of removing PCB residues, even from small sites, is enormous. To take just one example among many, Honolulu taxpayers faced a cost of over $1 million to clean up PCB residues in concrete at the construction site of a new police station.[78] The highest concentration of PCBs found in the concrete was 33 ppm, or one half of one thousandth of a gram per pound of concrete.[79] Setting a 10 ppm versus a 100 ppm standard for residential areas diverts many tens of millions, if not hundreds of millions, of dollars of private and public money to cleanups that produce no likely health or environmental benefits.

Banning the disposal in municipal landfills of soil containing more than 50 ppm PCBs seems equally overprotective. No one would ingest PCB soil that is inside a landfill, and once in the landfill the soil would quickly be mixed with thousands of tons of other material, resulting in a tremendous dilution of the original PCBs. Modern landfills are sited and constructed to avoid rapid migration of the contents into ground or surface water. Most of the PCBs would remain tightly bound to soil particles. Thus we find no evidence to suggest that PCBs greatly diluted inside a modern landfill pose a health hazard to anything, human, animal, or plant.

The Epidemiological Studies of the 1980s

Those who developed PCB regulations imagined that small concentrations of PCBs posed a health threat, particularly with regard to increasing rates of cancer and reproductive problems. If that were true, we would expect to see proportionately greater health problems among people exposed to higher amounts of PCBs on a daily basis. Epidemiological studies, however, have not provided a consistent picture of significant increases in illnesses among workers occupationally exposed to PCBs or of women who ate fish from waterways containing elevated levels of PCBs.

Two major epidemiological studies done in the 1980s have assessed cancer rates among former capacitor workers, most of whom were exposed to PCBs on a daily basis for months or years. An Italian study of 2,000 workers exposed to PCBs between 1946 and 1970 reported significantly more cancers among males than expected (14 observed versus 2 expected), of which the largest increase was in gastrointestinal cancers.[80] Yet a government study of 2,500 workers employed in U.S. capacitor manufacturing plants since the 1940s found no increase in gastrointestinal cancers. In fact, this study showed a slightly lower number of total deaths and deaths specifically from cancer than would be expected on the basis of national statistics (295 deaths versus 318 expected; 62 deaths from cancer versus 80 expected).[81] Both the Italian and American workers were exposed to other chemicals in addition to PCBs. Taken together, the Italian and American epidemiological studies do not provide conclusive evidence about the effects of elevated human exposures to PCBs.[82]

In the 1980s three major studies examined the pregnancies and births of PCB-exposed women. In one study 242 Michigan women who regularly ate PCB-containing fish from Lake Michigan were compared with a control group of 71 women who never ate Lake Michigan fish.[83] A second study compared a group of 200 women employed in capacitor manufacturing plants who worked with PCBs with 205 women who never had a job with direct exposure to PCBs.[84] Both studies reported heavier PCB exposure to be associated with slightly shorter gestation periods and slightly smaller infants (though their weight was still above the national average). Reviewers of the studies noted that other possible factors affecting birth weight were not adequately accounted for to ensure that differences were caused by PCB exposure.[85] A third study of women in North Carolina found no association between maternal PCB levels and birth weight.[86] As in Lake Wobegon, everyone (exposed to PCBs or not) was above average.

In follow-up reports neither the Michigan nor the North Carolina studies showed a correlation between length of breast-feeding by the higher PCB-exposed mothers and adverse outcomes in the children. The Michigan study found slightly lower scores on a short-term memory test among four-year-olds born to mothers with the highest PCB levels in breast milk. The North Carolina researchers did not report results related to memory tests but found slightly lower scores on motor-skills tests among children whose mothers had the highest PCB levels, though all differences disappeared by the time the children turned three or four years old.[87]

Reviewing these studies together, scientists have been reluctant to state firm conclusions about PCB toxicity to fetuses from maternal occupational, fish eating, or incidental PCB exposure.[88] It still appears prudent for pregnant women to avoid regularly eating fish caught in waters known to contain high amounts of PCBs.

Between 1968 and 1990 not a single article in the *New York Times,* the

Washington Post, or major news magazines carried an analysis of whether EPA regulations governing dilute PCBs were likely to provide public health or environmental benefits to counterbalance the enormous costs of complying with the regulations. Occasionally newspapers and magazines did report on PCB exposure issues in greater detail, usually at the end of articles on contamination scares or on inside pages. An example of a special effort by a major newspaper to place PCB-related cancer evidence in perspective was provided by *San Francisco Chronicle* science editor David Perlman after a fire in a city transformer vault. Titled "PCB-Cancer Controversy," the article states: "There is little clear evidence that the toxic chemical known as PCB can cause long-term serious illness to humans, although conflicting studies over many years have indicated that at least in laboratory rats the chemicals induce liver damage and perhaps liver cancers. Many human studies—some continuing for more than 20 years—have examined the effects of the compound on workers . . . the most significant human effect positively identified has been a serious but usually temporary skin ailment."[89]

Two days after this page-six article appeared, PCDFs were found averaging 3 ppm in ash on the sidewalk grate above the transformer fire. The next day the big, bold front-page headline in the *Chronicle* read: "New Chemical Peril from Highrise Fire."[90] The paper had returned to the usual media practice of emphasizing peril over perspective.

We are willing to believe that the FDA's early restrictions on PCBs in food were a prudent attempt to prevent potentially significant public exposures. Likewise, once PCB substitutes were available, the ban on PCB manufacturing may have been justified to cut off major new releases of PCBs and limit the production of highly toxic PCB by-products. But the subsequent regulations on dilute PCBs in soil and equipment seem to us far more stringent and expensive than necessary.

Proponents of strict regulations used evidence about concentrated PCBs to condemn tiny residues. As a result, Americans had to pay hundreds of millions of dollars extra on utility bills and public and private cleanups to remove trace amounts of chemicals that could not have resulted in significant contamination of water, air, or food.

In addition to their financial costs, the stringent regulations also created their own potential health risks—mental stress engendered in people warned about trace chemicals that were probably harmless, and potential accidents that could occur from the additional cleanups and testing required. Research should be undertaken not only on possible harm from chemicals but also on possible harm from regulation and cleanup. The environmental safety field needs to be made more symmetrical; there should be risk-risk (or double-entry) analysis: the harms reduced versus the harms created by a regulation.

Already the alert citizen has important clues in distinguishing worse from better evidence. Look for other causes. If that advice seems a bit vague, try this:

if X chemical is a reputed cause, compare levels in sick and healthy exposed individuals. If the healthy have equal amounts or more of the stuff in their bodies than the sick, look elsewhere. Look closely at animal evidence; if the doses used are much higher than human exposures, be skeptical. Use industrial (large, continuous) exposures as a guide; for if workers receiving huge doses over years remain healthy, the chances of tiny, fleeting exposures doing harm must be minuscule.

If there ever was a chemical with a bad reputation, it is DDT, first made infamous by Rachel Carson. It may come as a surprise to all but the older generations to realize that DDT was once considered a savior. Truth and legend about DDT have figured so prominently in stories about harm from chemicals that the citizen risk detective must ask whether DDT deserves its reputation.

II. Is DDT a Chemical of Ill Repute?

with Jesse Malkin

When the insecticide DDT was introduced during World War II it was hailed as a miracle. It saved millions of lives by stopping the spread of insect-carried diseases; it increased crop yields, making produce more affordable; and it was an effective agent against pests that defoliated trees. After the publication of Rachel Carson's *Silent Spring* in 1962, however, attention began to shift to DDT's potential for harm. It was learned that the chemical had spread to the corners of the globe and persisted long after application in soil. Worse, it was detected in the tissues of living creatures, where it seemed to pass up each link of the food chain in successively higher concentrations. Environmentalists charged that the chemical damaged bird reproduction and posed a risk to the human race itself. The government reacted with panic, ordering a near-complete ban on DDT in 1972. Introduced as a symbol of life, DDT went down in the history books as a symbol of death.

The allegations against DDT were repeated so often and stated with such passion that the public remains convinced of their validity. But once the unsubstantiated charges are separated from known facts, it becomes clear that the hazards posed by DDT were exaggerated. Moreover, people lost sight of the many benefits that DDT offered humanity, benefits that could not be produced by other, less maligned substances.

Origins

Dichlorodiphenyltrichloroethane (DDT) was first synthesized in Strasbourg in 1874 by a young chemistry student and again in 1939 by Paul Müller, a chemist at the J. R. Geigy laboratories in Basel, who discovered its value as an insecticide. He found that tiny amounts of DDT could kill flies, aphids, Colorado beetles, and the body lice that spread typhus fever. DDT was later shown to kill mosquitoes that transmit malaria and was used extensively in the Pacific during World War II. By 1959 malaria had been nearly eradicated in the United States, Europe, portions of the Soviet Union, Chile, and several Caribbean islands through the use of DDT. Müller was awarded the Nobel Prize in physiology medicine in 1948 for his life-saving discovery.

Immediately after the war ended, DDT became available for general civilian use. Farmers had earlier used arsenic, fluorine compounds, lime, sulfur, and other inorganic chemicals as insecticides, some of which were potent carcinogens. DDT, by contrast, while highly toxic to a wide range of insects, was very low in toxicity to mammals. Its persistence often made repeated applications unnecessary. An outstanding feature of DDT was its low cost: in 1968 one pound cost $17\frac{1}{2}$ cents, making it "the most economical insecticide ever sold."[1] Millions of tons were produced, the largest amounts being used to combat cotton pests, with major increases in cotton yield. To agriculturalists the chemical appeared as "a gift from heaven."[2]

Soon after its introduction indications of DDT's dangers became apparent. As early as 1944 scientists noted that DDT was toxic to fish and frogs.[3] By 1945 it was known that DDT accumulated in the body fat and milk of laboratory animals.[4] Other limitations became evident in 1947 when it was learned that houseflies in Sweden had developed resistance to DDT. As time went on, reports began to accumulate from throughout the world: flies, mosquitoes, lice, and many other insects were not succumbing. A rapid evolution was occurring: those few insects with genes that made them resistant to DDT survived and reproduced to form largely resistant populations. In response, heavier and heavier doses of DDT were used, but this tactic was not usually effective—the pests simply became resistant to high doses. Ironically, in these instances, it was discovered that "less is more": when DDT was applied in smaller doses, resistance was less likely, as "non-resistant individuals swamped the resistant and control was still possible."[5]

In 1949 New York physician Morton Biskind published a series of articles claiming that DDT caused the mysterious "virus X" in humans—a disease that had been linked with polio.[6] His claims were refuted a few months later,[7] but the scare nevertheless increased public concern about DDT.

Research at Clear Lake, California, in 1960 led to the first of a stream of articles on "biomagnification," the idea that pesticides are passed up each step of the food chain in higher and higher concentrations.[8] This theory had far-reaching implications: it laid the groundwork for future suggestions that even in small doses DDT posed a threat to all living creatures. Anesthetized by the "virus X" episode, the pesticide industry tended to dismiss the biomagnification theory as scare mongering and did not look closely enough at the real dangers that could exist with heavy spraying.[9]

In 1962 Rachel Carson's shocker, *Silent Spring,* riveted attention on the dangers of pesticide usage. Pesticide control became a matter of urgent public debate and governmental concern. Congress held hearings, and President John F. Kennedy appointed a Science Advisory Committee to investigate the hazards of pesticides. In 1963 the panel released a report, *Use of Pesticides,* recognizing both benefits and harms, while recommending the long-term goal of eliminating persistent pesticides.

Perhaps most important, *Silent Spring* mobilized environmental scientists. Almost immediately they began publishing studies indicting DDT. Residues were observed in samples of soil, rivers, and air;[10] even Antarctica's penguins, seals, and snow contained traces of DDT.[11] Birds' eggs were cracking allegedly because DDT had thinned the shells.[12] Feeding DDT to mice at high doses induced tumors.[13] The main cause for alarm was the revelation that DDT persisted everywhere. Still, there was no evidence of harm to humans and little firm evidence of harm to wildlife.

In January 1968 Arizona imposed an experimental one-year ban on DDT. In New York and Michigan the Environmental Defense Fund succeeded in halting the use of DDT in mosquito and Dutch elm sprayings. The rapid spread of opposition to DDT across the country was followed by a U.S. Department of Agriculture (USDA) decision on July 10, 1969, to suspend the use of DDT in its spraying programs. Later that year California and Wisconsin announced bans. With a "better safe than sorry" attitude the federal government began phasing out DDT in November 1969, the same year that Sweden imposed a two-year ban, the first time a country had done so. A virtually complete ban on DDT went into effect on January 1, 1973.[14]

The Media

When DDT was introduced during World War II the media trumpeted DDT's potential to help humanity. *Newsweek* stated: "One of the three greatest medical discoveries to come out of the war (plasma and penicillin are the others) DDT has enormous peacetime possibilities as an insecticide. A representative of the Surgeon General's office said last week: 'DDT will be to preventive medicine what Lister's discovery of antiseptic was to surgery.' "[15] By 1946, however, DDT's limitations began to become evident. Reports of DDT's dangers, published in both scientific journals and the popular press,[16] did little to temper the public's initial unabashed enthusiasm for DDT. Over the course of the next fifteen years, two unrelated events promoted public uneasiness over chemicals in general. One was Europe's thalidomide tragedy in 1961, and the other was the discovery that fallout from atomic bomb testings had contaminated milk.[17]

The press misled the public in numerous ways, one method being the damning of DDT by association with more toxic pesticides. An article in *Time* magazine read: "The farm workers may have lost one round in the case, but the hearings gave them ammunition for a larger suit to ban the use of DDT in California. Dr. Irma West of the state department of public health . . . testified that in 1965, one California farm worker died of pesticide poisoning, and between 200 and 300 had been nonfatally poisoned. In addition, some 1,000 workers had experienced 'dermatitis, chemical burns of the skin and eyes, and other miscellaneous conditions, resulting from contact with pesticides.' "[18] The clear implication is that DDT was among the pesticides that poisoned farm workers. In fact, this was not the case. Both critics and advocates of DDT agree that it is remarkably low in toxicity to humans. Indeed, despite millions of incidences of intensive exposure, the only injuries resulting from DDT have been caused by massive accidental or suicidal ingestion.[19] Farm workers were poisoned not by DDT but by organophosphate insecticides, such as the parathions, which are hundreds of times more toxic to man than DDT and which were touted as superior substitutes to DDT.

Another distortion was caused by omission. Shortly after DDT was banned in 1972, a tussock moth infestation of Douglas firs in the Northwest caused millions of dollars worth of damage.[20] Various pesticides were used, but none were effective. After over two years of damage and pleading by forestry officials, in 1974 the EPA temporarily permitted the use of DDT on infested forests in Washington, Oregon, Idaho, and Montana. The DDT promptly halted the moth and caused no observable environmental damage. One would think that this episode would have been reported, yet it received no mention in the national weekly news magazines.[21]

Cancer

Perhaps no statement was more damning to DDT than the assertion that it might cause cancer. Prior to DDT's ban in 1972 the EPA carried out an eighty-day hearing to determine the risks and benefits of DDT use. The hearing examiner concluded that "DDT is not a carcinogenic hazard to man."[22] Less than two months later, EPA administrator William Ruckelshaus banned DDT on the grounds that "DDT poses a carcinogenic risk" to humans.[23] Ruckelshaus cited three reasons for his finding: (1) experiments demonstrate that DDT causes tumors in laboratory animals; (2) responsible scientists believe tumor induction in mice is a valid warning of possible carcinogenic properties; (3) there are no adequate human epidemiological data on the carcinogenicity of DDT, nor is it likely that they can be obtained. These arguments will be addressed in turn.

LABORATORY EXPERIMENTS

Two studies constituted the bulk of the evidence of tumors in laboratory animals supporting Ruckelshaus's claim.[24] The first was a multigenerational study on mice in 1969 by two Hungarian scientists.[25] The researchers fed small doses of DDT—about 3 ppm in food per day—to 683 mice spanning five generations. (This dosage was roughly ninety times greater than the average exposure of the public at that time.) A normal diet was fed to a control group of 406 mice. The incidence of lung carcinomas was 16.9 percent in the treated mice and 1.2 percent in the controls. The authors also noted higher rates of lymphomas, leukemias, and other tumors in the experimental mice as compared to the controls.[26]

This study was challenged for two main reasons. First, although other researchers had examined mice, nobody else working with similar dosage levels had discovered cancer. Second, the control group of mice showed leukemia, despite the authors' claim that leukemia was unknown in the strain of mice they used. In addition, an investigation by the World Health Organization demonstrated that the findings might have been caused by spoiled food contaminated with aflatoxin, a cancer-causing substance produced by mold.[27]

The second experiment cited by the EPA was the Bionetics Report of 1969, also called the Innes paper.[28] Two hybrid strains of mice were given single doses of 46.4 milligrams per kilogram (mg/kg) of DDT by stomach tube from seven to twenty-eight days of age. Then for about eighteen months they were fed the maximum tolerated dose (MTD, the highest an animal can take without major weight loss or shortened lifespan due to toxic effects other than cancer). This dosage was equivalent to 853 times the average amount found in the human diet.[29] Among the 72 experimental mice, 23 with hepatomas (liver cell tumors) were found, as compared with only 14 among 338 controls.

The study created the impression that DDT caused cancer in the mice, but this conclusion had no firm basis. There was no evidence that the hepatomas were cancerous and not merely benign nodules.[30] In addition, the experimental design of the study was criticized. C. S. Weil, for example, argued that the statistical analyses applied were meaningless, because the researchers did not ensure random distribution of the litters.[31]

Several studies conducted in subsequent years, however, confirmed that high doses of DDT can cause cancer in mice. In 1972 and 1974 L. Tomatis and colleagues released studies showing that DDT had produced liver tumors in mice, but the second study also noted that hepatomas "rarely metastasize and in many instances do not show obvious signs of invasiveness." The researchers also discovered that DDT had no effect on hamsters.[32] Other studies showed that mice fed 250 ppm[33] or 100 ppm[34] of DDT were far more susceptible to liver tumors than controls. Yet in several studies on rats, DDT was not shown to result in an increased incidence of tumors.[35] Indeed, at least one study found

DDT to be an anticarcinogen: "Pretreatment of female Sprague-Dawley rats with p,p'-DDT significantly reduces their subsequent liability to mammary tumor induction by [the carcinogen] dimethylbenzanthracene (DMBA)."[36]

MICE AND MEN

Many observers believe mice experiments were among the chief factors that led to banning DDT. Thus, it is essential to examine whether substances that induce liver tumors in mice predict cancer in humans. In the 1972 hearings on DDT, EPA administrator Ruckelshaus asserted, "scientists agree that tumor induction in mice is a valid warning of possible carcinogenic properties."[37] In its 1975 review of the DDT ban the EPA cited only one paper, a review of the literature, as supporting Ruckelshaus' assertion.[38] EPA's conclusion, in its entirety, read: "Although the target tissue may be different, the mouse can, in specific cases, serve as a reliable and proven indicator of the carcinogenicity of a chemical in other species, including man. However, although carcinogenic effects in mice are valid when dealing with certain chemicals, the results can vary greatly depending on the compound tested and may not always be a reliable basis for extrapolation to other species."[39] The EPA was right to add the disclaimer, for there is no reason to expect that what causes cancer in mice will cause cancer in humans. Many natural substances—chemicals that most of us would not think of calling "potential human carcinogens" (such as vitamin A and salt)—have been shown to cause cancer in animals when administered experimentally at high levels.[40] Today, scientists know that rodent carcinogens are virtually everywhere. They are present in almost all fruits and vegetables, including apples, bananas, broccoli, brussels sprouts, cabbage, mushrooms, and oranges.[41]

As early as 1972 Ruckelshaus was aware of such arguments. He responded that DDT advocates "take refuge under a broad canopy of data . . . and support it with the increasingly familiar argument that exposure to any substance in sufficient quantities may cause cancer." He summarily dismissed this "familiar argument" in one sentence: "The 'everything is cancerous argument' fails because it ignores the fact that not all chemicals fed to animals in equally concentrated doses have produced the same tumorigenic results."[42] As Max Sobelman noted, "This argument was made nowhere in the proceeding, and significantly the Administrator cited no reference for it."[43]

STUDIES ON HUMANS

From 1956 to 1958 DDT was fed to prison volunteers for prolonged periods. Some men received as much as 35 mg/day for 21.5 months, approximately 1,000 times as much as the average American adult ingested during the heaviest years of DDT usage (1956–1966). The subjects reported no complaints, and no adverse effect was observed. Five years after completion of the experiment, some of the subjects were reexamined, and, again, no adverse effects attributable to DDT were observed.[44]

Two occupational studies were conducted prior to the DDT ban. The first, by Dr. Mark Ortelee of the U.S. Public Health Service, examined forty workers employed by firms that manufactured DDT. Ortelee concluded: "With the possible exception of hypersensitivity reactions, it is considered unlikely that any illness or symptom complex identifiable as chronic DDT poisoning exists in people exposed to DDT at the current dietary level, because no such effects were found in men exposed for as long as 6.5 years in such a way that they absorbed 200 times as much DDT as that absorbed by the general population from their diet."[45]

The second study involved workers at the Montrose Chemical Corporation in Torrance, California, which from 1947 to 1982 produced nothing but DDT. (After DDT was banned in the United States in 1972, it continued to be manufactured for exports to countries with malaria problems.) In 1967 Dr. David Laws examined thirty-five men who had worked at Montrose for at least eleven years, the mean being fifteen years. The men were categorized according to level of exposure (twenty, twelve, and three men in the high, medium, and low exposure groups, respectively). Analysis of body fat showed that the men carried DDT levels of up to 647 ppm as compared with an average of 8 ppm for the general population at the time. These men absorbed about 18 mg/day, nearly 300 times more than the intake of the general population, yet the researchers found no ill effects attributable to DDT.[46]

Millions of people have been exposed to DDT. Despite this intensive exposure, "the only confirmed cases of injury [from DDT]," the World Health Organization stated, "have been the result of massive accidental or suicidal ingestion."[47]

Only one study on humans has been interpreted as indicating that DDT is a carcinogen. In 1968 J. L. Radomski and colleagues reported that the concentration of DDT, DDE (DDT's main metabolite), or both was elevated in the body fat of people who had died of liver cancer, leukemia, and other serious diseases, as compared to the level in those who had died in automobile, gunshot, or other accidents. The researchers suggested that the increased DDT concentration caused the disease or that the disease caused the increased concentration. Wayland Hayes, Public Health Service scientist, concluded that the latter was the correct explanation: "What Radomski et al. observed is readily explained by the debilitating nature of the diseases they studied and by the fact that loss of body fat leads to an increase in the concentration of DDT and DDE that remains."[48] Ruckelshaus rejected all of the studies on humans, asserting that their sample sizes were too small, their duration was too short, and no completely unexposed human control group existed, given DDT's ubiquitousness. These are valid points; the experiments were not ideal. The weaknesses of the epidemiological studies, however, paled in comparison to the weaknesses of Ruckelshaus's interpretation of the mouse experiments. In spite of their imperfections, the human studies were far more relevant than experiments in which mice received DDT at the maximum tolerated dose.

A final comment from 1970 is informative: "DDT is not endangering the public health and has an amazing and exemplary record of safe use. DDT, when properly used at recommended concentrations, does not cause a toxic response in man or other mammals and is not harmful. The carcinogenic claims regarding DDT are unproved speculation."[49] This claim was made not by an insecticide manufacturer or an agribusiness spokesman but by William Ruckelshaus, then Assistant Attorney General of the United States.

RECENT STUDIES

Recently, two studies have linked DDT to cancer in humans. One purported to show an association between DDT exposure and pancreatic cancer: "among subjects who had a mean exposure to DDT of 47 months, the risk was 7.4 times that among subjects with no exposure."[50] But this conclusion was based on shaky methodology. The subjects' lifestyle histories, including tobacco and alcohol consumption, were developed from interviews with next of kin, a notoriously unreliable method. Exposure histories were based on evidence not from blood or fat samples, but from written records compiled by the chemical manufacturing firm. The results of this study are best considered with a heavy dose of skepticism.

The second study showed that women who develop breast cancer tend to carry higher residues of DDE in their blood than do women who are free of the disease. Study subjects were a subset of 14,290 New York women enrolled in a breast cancer study program. The study group consisted of 58 women diagnosed with breast cancer within six months of entering the program. They were matched with a control group of 171 cancer-free participants on the basis of such factors as menopausal status, age, and socioeconomic status. The researchers found that concentrations of DDE were roughly 35 percent higher in women with cancer than in the controls. Cancer patients also tended to have higher PCB concentrations, although the difference was not statistically significant.[51]

The results of this study have been widely interpreted as showing that DDE increases the risk of breast cancer in women,[52] but this presumption is not warranted. It may eventually be shown that high blood levels of DDE lead to breast cancer, but it seems much more likely that the causality runs the other way around, that breast cancer leads to increased levels of DDE in the blood. Many debilitating diseases mobilize fat from fat storage depots, which contain fat soluble compounds such as DDE. When this occurs, blood concentrations of those fat soluble compounds increase. Certain drugs will also increase blood levels of DDE. In short, the public should be wary of claims that any study proves a cause-effect relationship on the basis on statistics alone.

Some environmental groups have suggested that DDE may cause breast cancer by imitating the hormone estrogen. But DDE is dwarfed by naturally occurring chemicals that also mimic estrogen. In addition DDE is hundreds of times less potent than estrogens in birth control pills, and there is no conclusive evidence that even those large doses cause breast cancer.

If DDE causes breast cancer, we would expect a sharp increase in the number of breast cancer cases, since every woman in America was exposed to the chemical during the 1950s and 1960s. Yet, media hoopla to the contrary, most specialists suspect that the purported epidemic of breast cancer is actually a statistical illusion resulting from improved detection methods. The actual incidence of breast cancer cases—as opposed to the reported incidence of cases—has probably remained fairly steady in recent years.[53]

Photosynthesis

One of the more melodramatic claims about DDT was that it impaired photosynthesis in the oceans. If true, this would result in a reduction in the earth's oxygen supply and, possibly, the end of all human life. Biologist Paul Ehrlich of Stanford University was one of the authorities urging attention to this matter. He wrote in 1969:

> The end of the ocean came late in the summer of 1979, and it came even more rapidly than the biologists had expected. There had been signs for more than a decade, commencing with the discovery of 1968 that DDT slows down photosynthesis in marine plant life. It was announced in a short paper in the technical journal, *Science,* but to ecologists it smacked of doomsday. They knew that all life in the sea depends on photosynthesis, the chemical process by which green plants bind the sun's energy and make it available to living things. And they knew that DDT and similar chlorinated hydrocarbons had polluted the entire surface of the earth, including the sea.[54]

The study reported in *Science* in 1968 from which these claims originated was done by Charles Wurster, who tested the effects of alcoholic solutions of various concentrations of DDT on five species of marine algae.[55] The DDT was added to produce concentrations of up to 500 ppb. At concentrations as low as 10 ppb of DDT, photosynthetic activity in experimental algae was reduced by about 20 percent. Since the maximum solubility of DDT in water is only 1.2 ppb, a concentration of this level would occur only in the case of complete DDT saturation of the ocean. Yet the levels of DDT in the ocean in 1968 were nowhere near saturation; the amount actually present in the surface water of the North Atlantic Ocean was less than 1 part per trillion, or one thousand times less than the lowest concentration of DDT that Wurster tested.[56]

Subsequent studies published in *Science* showed that the supply of oxygen, a product of photosynthesis, had not changed in the previous sixty years and was not in danger of depletion.[57] The photosynthesis claim was disproved and after a few years was no longer raised by the anti-DDT campaign.

Biomagnification

Biomagnification has often been interpreted as meaning that even small levels of DDT in oceans, ponds, and rivers pose a threat to human beings because magnification up the food chain will culminate in lethal levels. One of the earliest studies of biomagnification was conducted by Roy Barker, an entomologist at the University of Illinois. From 1949 to 1953, 1,400 elm trees on the 430-acre Urbana campus were sprayed twice a year with a 6-percent-DDT water emulsion. Barker said that "very few dead birds were observed in 1949 when DDT was first sprayed. However, before spraying began in the spring of 1950 the numbers of dying robins . . . had attracted unsolicited attention."[58] Analysis of earthworms from the area revealed that they contained DDT, apparently by feeding on leaf litter from sprayed trees.[59] According to Barker, when robins ate the earthworms, they accumulated deposits of DDT. Over the course of two years Barker analyzed 40 dead and dying adult robins' brains and found concentrations of DDT ranging from 0 to 139 ppm, with a median lethal concentration of approximately 60 ppm; DDE concentrations ranged from 0 to 252 ppm.[60] This finding suggested that DDT could be passed up the food chain and that both indirect and direct poisoning of the robins might be occurring.

The most famous example of biomagnification occurred at Clear Lake, California, home to approximately 1,000 pairs of western grebes. Throughout the early 1950s, DDD, a pesticide closely related to DDT, was sprayed on the lake to control gnats. In 1954 DDD was applied at a rate of 0.02 ppm. An examination of the lake revealed that plankton contained an average of 5.3 ppm of DDD—265 times as much as the concentration of the lake itself. Fish that fed on the plankton contained as much as 125,000 times the lake level of DDD. The fish apparently were not greatly affected by DDD residues, but grebes, which ate the contaminated fish, died in large numbers. Analysis of the fat of these birds revealed concentrations of up to 1,600 ppm.[61]

Biomagnification implies a steady increase in the concentration of a contaminant: "Once introduced into the food web," read a typical description, "chlorinated hydrocarbons [such as DDT] are able to move up from phytoplankton to zooplankton to fish or bird or mammal, and they become more and more concentrated as they are retained by a higher level of animals."[62] But whether this in fact occurs depends on the species involved. Some are able to excrete large amounts of DDT; others are not so fortunate. At Clear Lake, for example, grebes contained less DDD than the fish they fed upon. Tuna and barracuda accumulate very little DDT, whereas sharks carry relatively high levels. The cautions urged by Wayland Hayes about biomagnification are apropos:

Actually, the phenomenon may be little more than a rare curiosity requiring a whole series of species that are inefficient at degrading and excreting the compound in question. Since food chains are of finite length, it is only necessary that one or at most a few species in a series be really effective in eliminating the compound, or that several species be moderately effective in eliminating it, in order to limit accumulation to a point that nothing really happens. Furthermore, even in susceptible food chains, biological magnification may be expected in connection with only a few compounds. Most pesticides are degraded readily by living organisms.[63]

The point should be stressed that biomagnification is like a chain letter in one respect: it takes only a single missing link to render it inoperative.

DDT and Birds

Environmentalists based a cornerstone of the anti-DDT case on claims the chemical has caused dramatic population declines of several species of birds, especially birds of prey. Though early claims were exaggerated, a large body of evidence now confirms that DDT can be toxic and cause reproduction problems in a number of bird species.

The robin.   In the early 1960s some naturalists asserted that DDT had brought robins to the brink of extinction.[64] But according to the Audubon Society's annual Christmas bird counts, between 1941 and 1961 (a span that includes years of heavy DDT usage) the number of robins per observer increased twelve times.[65] The counts, however, are disputed. Joseph Hickey said in congressional testimony that the counts were taken by amateur ornithologists and were not "worth a damn as far as scientific data are concerned."[66] Roger Tory Peterson, a respected DDT opponent, wrote in *The Birds* that the robin became the most abundant bird in North America during years that DDT was used.[67]

The Audubon Society counts suggest that many other bird populations increased from 1941 to 1961: 21 times more cowbirds, 8 times more blackbirds, 131 times more grackles.[68] In this period herring gulls became so abundant that the society obtained permission from the Massachusetts Department of Natural Resources to exterminate 30,000 of them on Tern Island in 1971.[69]

Today, few knowledgeable people would argue that the existence of small birds, such as robins or starlings, was ever threatened. But they contend that a few species of carnivorous birds—in particular, the osprey, bald eagle, and peregrine falcon—were imperiled.

The osprey.   Declines in osprey populations in the New England area, first noted in 1890,[70] were apparently caused by a lack of available food supply and by trapping.[71] During the mid-1960s concern grew intense about the osprey's dwindling numbers.[72] Roger Peterson wrote of a "catastrophic decline of the nesting osprey population on the Atlantic coast."[73] In contrast, William Stickel found ospreys nesting successfully in the Chesapeake Bay area: "They are relatively abundant on the east side of the Bay in Talbot County."[74] According to Robert White-Stevens, a biologist at Rutgers, the osprey was driven from its shoreline haunts on the East Coast but "reinforced its numbers inland."[75] The migration count recorded by the Hawk Mountain Sanctuary Observation Post on the Great Appalachian Fly Way confirms White-Stevens's comment, showing a steady and significant increase through the years that DDT was used.[76]

The accuracy of the Hawk Mountain data has been challenged by the Environmental Defense Fund and other opponents of DDT, who point out that it shows some of the same weakness as the Audubon Christmas counts. But Robert Ackerly, a lawyer for the DDT industry, noted, "they do represent the best, perhaps the only, data which can show population trends for a significant period."[77] The addition of another observation site in 1967 might partially account for the apparent increase in ospreys, Ackerly conceded, but the increase began well before 1967.[78]

In sum, osprey declines are well documented, but (1) they started before DDT usage; (2) they occurred only in certain areas on the East Coast, whereas DDT was used everywhere; and (3) some osprey populations increased during the DDT era.

The bald eagle.   In 1958 Charles Broley, a retired banker who had been banding and studying bald eagles since 1940, reported an alarming increase in nesting failures in Florida. Instead of finding his usual 125 nests, that year he found only 3. "I am firmly convinced that about 80 percent of the Florida bald eagles are sterile," Broley said, adding that DDT was responsible.[79]

Advocates of DDT noted that in some regions the bald eagle apparently fared well during years of heavy DDT use. In 1964 the Bureau of Sports Fisheries and Wildlife published an article stating that in Everglades National Park "the number of nesting eagles seems to be remaining constant."[80] In November 1969 in Montana a record number of 373 bald eagles were sighted on a single day, according to San Jose State entomologist J. Gordon Edwards. Significantly, 120 of those were young, indicating successful mating and nesting during years when DDT was widely used.[81]

Taken as a whole, the evidence supports the position of Alexander Sprunt, a specialist in bald eagles, who wrote that "the overall population of the bald eagle has declined slowly over a long period of time. Since World War II a decline has taken place at a considerably accelerated rate in some portions of the eagles' range."[82] To some degree—how much we do not know—this decline may have been due to DDT. Like ospreys, bald eagles have resurged in numerous areas since the DDT ban.

The peregrine falcon.   Around 1960 rumors began to circulate that peregrine falcons were experiencing a massive reproductive failure in the United Kingdom.[83] In 1963 Derek Ratcliffe, a British peregrine falcon expert, confirmed the rumors: "Decline on this scale and at this rate is quite unprecedented. An unprecedented cause must therefore be sought."[84] In the spring of 1964 Joseph Hickey set out to see if a similar population crash was occurring in America. He sent two research assistants on a 14,000 mile trek from Georgia to Nova Scotia. They did not find a single fledgling. Shocked, Hickey organized an international conference on peregrines in 1965, and the sixty biologists, ecologists, and naturalists concluded that a worldwide population crash was occurring.[85]

The peregrine's troubles began before DDT materialized. Hickey wrote in 1942 that the peregrine had been declining since 1890 and that by 1940 there were only about 170 pairs in the eastern United States.[86] He attributed the decline to environmental factors such as habitat loss and timber clearance. A Canadian expert, Frank Beebe, speculated that the trend might have begun with the extinction of the passenger pigeon, a principal prey for the falcon.[87]

A supporter of DDT, Gordon Edwards, has recently argued that peregrine falcons have been persecuted, not by those applying DDT but by environmentalists:

> As soon as U.S. peregrines were officially declared to be "endangered," environmental activists were paid by the government to "study" them. They trapped brooding birds on their nests, removed fat samples from live birds for analysis and operated time-lapse cameras beside the nests night and day. They also collected a large portion of the eggs annually and killed dozens of nestlings by suffocation (for later analysis). They then pointed with alarm at "declining numbers" of the persecuted peregrines.[88]

We conclude that the population of the peregrine falcon has declined over the last century in populated areas. After the introduction of DDT, this decline became more rapid.

DID DDT CAUSE STEEP DECLINES IN BIRD POPULATIONS?

Direct DDT poisoning.   Claims that DDT can kill bald eagles first emerged in 1962. That year James DeWitt and his colleagues at the Bureau of Sport Fisheries reported that captive bald eagles developed severe tumors and died after being fed DDT. The eagles were fed diets containing up to 4,000 ppm,[89] an amount over 120,000 times more than the estimated average daily intake by a human being in the United States at that time.

Since 1960 dead bald eagles have been packed in dry ice by federal, state, and private researchers from all over the United States and sent to the Patuxent Wildlife Research Center in Laurel, Maryland, where autopsies are performed.

From 1960 to 1977, 475 eagles were analyzed. Illegal shooting and trapping were the most frequent causes of mortality, accounting for about one-third (155) of the deaths. Other causes included impact injuries, electrocution, disease, lead poisoning (10), and thallium poisoning (9). One pesticide accounted for 27 deaths, but it was dieldrin; DDE poisoning accounted for 2 deaths.[90] In discussing bird populations in 1986, Robert Riseborough stated: "Although the biocide-induced mortalities are frequently significant local problems, such mortalities have apparently not had an impact upon the population status of any North American species."[91]

Such research does not bear on another argument—that DDE, the persistent metabolite of DDT, caused eggshell thinning. Mark Stalmaster, for example, conceded that levels of DDE were rarely high enough to kill eagles but generally sufficient to impair reproduction. "Levels of DDT or DDE as low as a few parts per million may cause thinning," he wrote. "Sometimes, if contamination is severe, no eggshells are formed at all. Instead, the pliable shell membrane is all that holds the inner contents in place."[92] Eggshell thinning was a major justification for banning DDT.

In 1967 Derek Ratcliffe analyzed eggs from private (and largely illegal) collections and museums throughout Britain and discovered a dramatic decrease in eggshell weight beginning in shells from 1946. Eggs from 1900 to 1946 had a mean shell weight of 3.81, while those from 1947 to 1967 had a mean weight of 3.09, a 18.9 percent decrease.[93] He found a similar decrease in shell thickness among British sparrow hawks and golden eagles, a critically important finding. If overly thin, eggshells could be crushed by the nesting mother bird, which would explain perceived reproductive failures of raptors.

Ratcliffe then noted an important correlation: the drop in eggshell thickness occurred at about the same time that DDT became available for civilian use. Wayland Hayes, who for many years testified that DDT was not carcinogenic, has stated that "no other compound is known to have become generally available about the time when eggshell thinning first occurred, soon after 1945. Polychlorinated biphenyls did not come on suddenly and dieldrin was not available."[94]

Joseph Hickey and Daniel Anderson were the first scientists in North America to follow up on Ratcliffe's study, and they too found an abrupt decline in shell thickness around 1947. Unlike Ratcliffe, who only hypothesized a causal link between pesticides and eggshell thinning, Hickey and Anderson stated the cause-effect relationship as fact: "these persisting compounds are having a serious insidious effect on certain species of birds at the tops of contaminated ecosystems."[95] In the years that followed several other studies documented shell thinning in a variety of bird species, including bald eagles, brown pelicans, and ospreys.[96]

Upon close examination, some observers noted that the sequence of events in Ratcliffe's pathbreaking study was questionable. Ratcliffe had stated that

DDT was first introduced to Britain in 1945–1946 and that the shell thickness of peregrine eggs decreased abruptly in the spring of 1946. But it was not until 1948 that a reduction in the price of DDT allowed the insecticide to be used widely for agriculture in Britain. As Hayes noted in *Toxicology of Pesticides* in 1974, "One must question whether there was enough DDT available in 1946 in the remote areas inhabited by falcons and, if so, whether there was enough time for this chemical to be metabolized to DDE, and for that to work its way through the food chain to produce the thinning of eggshells."[97] In 1970 Ratcliffe speculated that DDT could have been used during the war to dust homing pigeons on which the peregrines fed.[98] Hayes responded: "If one accepts, for point of argument, that there was enough DDE available to the hawks of the United Kingdom in 1946 to explain the observed change in eggshell thickness, then it becomes difficult to explain why there was no progression of thinning after 1948, when use of DDT began in earnest, or how the slight recovery of the population in 1964 could occur; the amount of DDT certainly had increased by that time."[99]

In sum, DDT's introduction roughly correlated with raptor declines, but this correlation was imperfect and did not prove a cause and effect relationship. Even anti-DDT scientists have recently described the evidence as "weak correlations between the timing of the population declines and the introduction of the chlorinated hydrocarbon pesticides and between the areas of decline and areas of pesticide use."[100] They go on to say, however, that since then convincing data have surfaced.

Correlation of eggshell thinning with DDE content.   A strong negative correlation between DDE residues and shell thickness was first documented in 1968 by Hickey and Anderson, who analyzed herring gull eggs.[101] Over the next three years this relationship was reported for populations of peregrine falcons, prairie falcons, brown and white pelicans, great blue herons, and double-crested cormorants,[102] but not for the common tern.[103] When a wider range of species was studied, dieldrin and PCBs as well as DDE appeared important for one species or another.[104]

James Enderson and Daniel Berger noted with astonishment that Canadian falcons were thriving in spite of high DDT residues:

> Total residues in seven whole Peregrine eggs averaged 27.1 ppm, about twice that found in Peregrine eggs in Britain. A seemingly normal average of 2.3 viable eggs or young, or both, was found near the time of hatching in the 15 sites that we observed. All these data suggest that adult Peregrines in northern Canada carry high levels of organochlorine residues acquired over a period of many months, that their eggs bear about twice the levels found in eggs from the stricken British Peregrine population, and that even with these precariously high levels the Canadian Peregrines appear to be reproducing normally.[105]

Later evidence indicated that Canadian populations of peregrines had experienced reproductive failure by 1970,[106] but Hayes noted, "this later result does not negate or explain the earlier observation."[107] Hickey explained the 1968 finding by noting that in Britain, unlike America, "the contamination of peregrine population was due not only to DDE but also to dieldrin."[108] But the peregrine was also reported to be having problems reproducing in Wisconsin, where the average falcon (of ten sampled) carried total DDT residues of 18.8 ppm in its fatty tissue.[109] Why were the Canadian falcons reproducing better than the Wisconsin falcons despite higher concentrations of DDT?

Although correlation between two variables does not prove a cause-and-effect relationship, in the case of DDE and eggshell thickness there was indeed a striking correlation for many species of birds, a correlation undermined by the existence of several exceptions.

LABORATORY STUDIES

After the field studies, attention turned to laboratory research. If DDT-fed birds laid abnormally thin-shelled eggs in controlled experiments, then the case against DDT would be considerably strengthened. At this time, it was believed that the osprey, bald eagle, and peregrine falcon could not be bred in captivity, so scientists utilized other birds. Obviously, this was not ideal, for birds' susceptibility to DDT varies from species to species; but it was the only option.

Some birds, such as chickens and quail, exhibited little or no eggshell thinning, even at high doses. The Bengalese finch laid eggs with shells 7 percent thicker than the control birds.[110] The two species that appeared most susceptible to shell thinning were the duck and the American kestrel.

The hallmark study on ducks was conducted by Robert G. Heath, a researcher from the Department of the Interior. Heath fed mallard ducks 40 ppm of DDE every day for one to two years, and these ducks laid eggs with shells 13 percent thinner than normal.[111] Other researchers who tried variations of Heath's experiment came up with similar results.[112] There is, however, an important weakness in the studies. Heath (and those who followed him) fed the experimental ducks far more than wild ducks could be expected to encounter in the environment.

An experiment on the American kestrel (also called sparrow hawk), by contrast, utilized a dosage thought to be representative of levels this bird would encounter in the wild—10 ppm of DDE every day for one year. The study is of particular interest because the sparrow hawk is of the same species as the peregrine falcon. The birds laid eggs that were 9.7 percent thinner than normal.[113]

At the time that DDT was banned the hypothesis that it caused declines of certain predator bird populations was supported by the following evidence: shell thinning was not noticeable before the introduction of DDT; in many species there existed a strong negative correlation between DDE residues and eggshell thickness; controlled experiments on American kestrels showed that eggshell thinning could be induced by DDE at levels encountered in the wild; and no other chemical had been shown to produce the degree of shell thinning induced by DDT.

The evidence implicating DDT had to be weighed against the following facts: raptor populations began declining decades before the introduction of DDT; decreases in eggshell thickness in Britain's peregrine falcon apparently occurred before DDT was prevalent in the environment; examination of a few species of birds indicated no correlation between DDE and shell thickness; in laboratory studies eggshell thinning was not induced in several species of birds; and in those birds in which thinning was induced, either the dosage was higher or the degree of thinning was lower than that found in nature. While there was reason for concern and for further investigation, the evidence at this point was marginal.

SINCE THE DDT BAN

Since DDT was banned, a variety of evidence has surfaced, most of which supports the claim that DDT was responsible for eggshell thinning.

Several studies have examined the effect of other pollutants at environmental levels on eggshell thickness. Polychlorinated biphenyls—PCBs—which were considered a possible culprit, were found not to induce shell thinning in mallards, American kestrels, and screech owls; they did, however, reduce eggshell strength. Salts of mercury and lead were shown to induce temporary shell thinning in mallards, but in the long term this effect abated.[114] Another study found that while DDT and PCBs had no effect on shell thickness of chickens, mercury caused dramatic thinning. It is now known that mercury levels in the peregrine falcon increased sixteen times from the mid-1800s to the mid-1900s.[115]

It has been some two decades since DDT was banned, providing ample time to examine its impact on raptor populations. All of the species said to be harmed by DDT have at least partially recovered. In 1982 the *New York Times* reported "a tremendous comeback" of the osprey in New York and it was removed from the state's endangered species list.[116] A sixteen-year study on bald eagles concluded that reproduction in northwestern Ontario "declined from 1.26 young per breeding area in 1966 to a low of 0.46 in 1974 and then increased to 1.12 in 1981."[117] Peregrines bred in captivity are currently being reintroduced into the wild and into cities; scientists have confirmed over 1,200 pairs living in the United States.[118] In Alaska, after 1978, productivity began to increase and DDE levels in unhatched eggs began to decline. The decrease in levels has been attributed to restrictions on DDT use in Argentina and southern Brazil, places where peregrines are known to migrate during the winter.[119]

The resurgence of these species is not conclusive proof that DDT caused their decline, but the strong chronological correlation is suggestive. Taken as a whole, the evidence has convinced a large majority of scientists that DDT contributed to eggshell thinning and consequent raptor and pelican declines during the 1950s and 1960s. We agree.[120] But we do not agree that as a consequence the total ban of DDT at that time was justified.

Replacing DDT

In August 1970 Clarence Lee Boyette, a North Carolina tobacco farmer, went to buy a pesticide to control an infestation of worms. The shopkeeper suggested Big Bad John, the local brand name for parathion, because DDT could not be used on tobacco if a farmer wanted to qualify for government price supports. Big Bad John killed the worms, but it also killed Boyette's youngest son; another son barely escaped death.[121]

Environmentalists claimed that DDT could be abruptly abolished without harmful effects on agriculture because suitable alternatives existed. The evidence did not warrant such optimism. At the time DDT was banned the main substitute pesticides were highly toxic, susceptible to resistance by insects, or expensive. For certain purposes alternative insecticides did not exist. Nonchemical forms of pest management, while promising, did not offer immediately viable options.

ALTERNATIVE INSECTICIDES

Chlorinated hydrocarbons.   DDT belongs to a class of insecticides called chlorinated hydrocarbons, or organochlorines, a few of which—toxaphene, lindane, and methoxyclor, for example—are still in use. Most—aldrin, dieldrin, endrin, mirex, and heptachlor—were banned shortly after the DDT decision. Generally speaking, the chlorinated hydrocarbons are powerfully toxic to insects but low in toxicity to mammals. Aldrin was favored for soil application and was also used on a broad spectrum of pests of fruits and vegetables. Dieldrin was used on livestock against lice, ticks, and blowflies. Endrin provided powerful control of caterpillars attacking field and vegetable crops and was one of the few insecticides effective against mites.[122]

The ban on chlorinated hydrocarbons was not implemented because these chemicals were superseded by cheaper insecticides; on the contrary, their cost-effectiveness was unrivaled. Nor was it because the amount of the insecticide required to control pests was very great. Indeed, the hallmark of chlorinated hydrocarbons was the minute amounts needed to ward off pests for long periods of time. Ironically, it was this very quality—persistence in the environment—that led to their downfall.

Organophosphates.   The organophosphates parathion and malathion served as the major substitutes for DDT.[123] Parathion, even more dangerous than malathion, is readily absorbed through the skin, from the stomach and the lung. The lethal dose for man is less than a gram.[124]

As DDT was phased out and organophosphates were used more, mortalities increased sharply. In 1972 estimates of the number of deaths jumped to 118 for the first half of the year alone.[125] The parathions are also toxic in minute amounts to fish, birds, and beneficial insects such as bees and wasps. Farmers who switched from DDT to the parathions were hurt not only physically but also economically, for these chemicals were more expensive than DDT and had to be more frequently applied.

Carbamates. Carbaryl, also known by its brand name, Sevin, is the most widely used carbamate. It is employed on cotton, fruits, and vegetable crops. Low in toxicity to humans, it is highly toxic to beneficial insects such as honeybees. "Although carbaryl is not toxic to birds," Frank Graham, Jr., noted, "its use often coincides with the birds' nesting season, thus killing off the variety of insects they require to feed their young."[126]

Synthetic pyrethroids. Since the DDT ban, synthetic pyrethroids (related to the natural pesticides, pyrethrins, extracted from the pyrethrum flower) have emerged as remarkably effective and safe insecticides.[127] They are expensive, but because only low dosages are required, they are as cost-effective as DDT. They kill almost all of the same pests that DDT did and are also extremely low in toxicity to mammals. They were not commercially available for agricultural use in the United States until 1978, five years after the DDT ban, by which time DDT substitutes had already done substantial damage.

ECONOMIC SHORTCOMINGS OF ALTERNATIVES

Cotton. Few pests have caused farmers as much despair as the boll weevil. At first DDT seemed a godsend to cotton growers: "We had defeated the Axis powers and now we had DDT," entomologist Knox Walker said, "and we were on top of the world."[128] After a few years, however, the persistent boll weevil was back, having developed a resistance to DDT. Agriculturalists then switched to a DDT-toxaphene mixture. Again, the treatment worked for a time, but the boll weevil adapted. One by one, chlorinated hydrocarbons were tried and discarded. Next, parathions were used. They managed to curb the boll weevil but also killed the insects that normally control the cotton bollworm and the tobacco budworm, which "came on like tigers."[129] These two pests had developed resistance to low doses of DDT in 1961, but high doses of DDT, or DDT combined with toxaphene, were effective.[130]

Following the DDT ban the tobacco budworm developed resistance to parathion and rapidly reasserted itself. Cotton yields abruptly plummeted. According to the EPA, yields in 1974, compared with those from 1969 to 1973, decreased 28–29 percent in Arkansas, Mississippi, and Tennessee; 36 percent in Missouri; and 7.5 percent in Louisiana.[131] National cotton prices hit a record high in 1973, nearly tripling within one year.[132] The fact that the DDT ban coincided with low yields and price increases does not necessarily prove a cause-effect relationship. The EPA blamed the low yield not on the budworm but on floods and adverse weather conditions and said that competition from synthetic fibers and declining demand accounted for the price hike. Some farmers thought differently, and by January of 1975 the EPA had received forty-five applications requesting special DDT use to control cotton pests. One group of applicants, Louisiana's farmers, believed DDT was the only insecticide that could control the tobacco budworm. The EPA denied all the requests.[133]

Whatever its causes, the cotton crash devastated thousands of Americans. Cotton, the nation's fifth most valuable domestic crop, contributed over $3 billion per year to cash receipts of farmers.[134] In the Southeast about 500,000 people were directly involved in growing cotton, and an additional 5 million were either directly or indirectly dependent on cotton for their living. In Louisiana alone the budworm infestation "specifically caused the loss of approximately 50–60 million dollars in 1974 in direct and indirect loss to the cotton industry."[135] In California, where in 1969 some 412,000 people earned all or part of their income either directly or indirectly from cotton, producers "were surviving only because they were receiving subsidy payments under federal government programs."[136]

Honeybees.   An unfortunate side effect of malathion, parathion, and especially carbaryl is their high toxicity to honeybees; by contrast, DDT is only mildly toxic. After Arizona banned DDT in 1969 farmers switched to parathion, which caused a decline in the honeybees needed to pollinate Arizona crops. In California, too, DDT restrictions and the consequent increased use of organophosphates led to dramatic declines in bee colonies. John E. Swift, chemical coordinator for the Extension Service at the University of California, reckoned 83,000 bee hives in California were killed by DDT substitutes.[137] For the first time, almond growers experienced difficulty in obtaining an adequate supply of bees for pollination.[138] California crops valued at over $300 million (in 1969 dollars)—almonds, alfalfa seed, apples, cherries, clover seed, cucumbers, melons, pears, plums, prunes, and vegetable and flower seeds—depended almost completely on honeybees for pollination.[139]

Tussock moth.   Use of DDT to control the tussock moth was registered in 1947 by the U.S. Forest Service but was abandoned by the USDA in 1969. Beginning in 1971 the moth population exploded in the Northwest, destroying thousands of Douglas fir trees. Horticulturalist H. J. Conklin of Cashmere, Washington, commented, "The once beautiful forests within five miles of my house are dead. Last summer [1971] it looked like a giant fire had swept the area."[140]

Anticipating another year of moth infestations, in early 1973 the Forest Service asked the EPA for a waiver to spray DDT. Environmental groups maintained that the moth could be controlled without DDT. A member of the Oregon Environmental Council said that pyrethrins and Zectran, a new organic carbamate compound, could safely bring under control the moth infestation. But pyrethrins and Zectran were not yet ready for operational use. Although of all insecticides only DDT had proven its effectiveness against the tussock moth, the EPA denied the Forest Service's request. William Ruckelshaus stated that the benefits of using the pesticide did not outweigh the hazards, that most of the projected damages to Douglas fir trees had already occurred, and that the reintroduction of DDT would have presented a danger to the area's ecosystem.[141]

Contrary to Ruckelshaus's prediction, the moth infestation worsened in 1973, damaging an estimated 800,000 acres of timber in Washington, Oregon, and Idaho, with estimated costs by December 1973 of more than $300 million. Moreover, the defoliation presented serious fire hazards.[142] After over two years of local efforts, the EPA lifted its ban in 1974. The February 1976 issue of the *Journal of Forestry* reported that "the effect of DDT on Douglas-fir tussock moth larvae was dramatic . . . There is little doubt that much of the resource loss occurring in late 1973 and early 1974 could have been avoided if the EPA decision to permit DDT use had been made a year earlier."[143]

And what of environmental degradation? According to entomologist Ralph Sherman, over 4,000 samples of water, air, vegetation, litter, stream sediment, benthic invertebrates, fish, birds, deer, elk, sheep, coyotes, chipmunks, mice, shrews, human blood, milk, and livestock were collected from locations throughout the treated region. "A total of 19 dead birds were found," he stated, "three of which may or may not have been killed by DDT." Although DDT had a short-term effect on aquatic and terrestrial insects, late fall surveys indicated that the insect populations were recovering and no fish mortality was observed.[144]

Sweden, the first country to ban DDT in 1969, quickly lifted its ban after it found that alternative pesticides could not control the spruce budworm and large pine weevil, which defoliate trees. H. H. Eidmann, professor of forest entomology at the Royal College of Forestry in Stockholm, said that he "explained to the Poisons Board that if DDT could not be used on the young stock we must be prepared to lose at least $15 million or better a year due to crop destruction caused by the pine weevil."[145] Similarly, although authorities in Ontario, Canada, had placed tight restrictions on the use of DDT effective January 1, 1970, six months later, in the face of a severe cutworm infestation, the minister of agriculture agreed to rescind the ban temporarily.[146]

BIOLOGICAL CONTROL

Biological control, which manages pests without using chemicals, was not a well-known concept until Rachel Carson popularized it in 1962, but it has been practiced throughout the ages. The first major victory in the United States was the importation of vedalia beetles from Australia to control successfully the cottony-cushion scale, which had multiplied to such numbers in California during the 1880s that orchardists were going out of business.[147]

When it works, classical biological control offers many advantages over chemical control. First, it is cheap. Control of the coconut scale, which threatened the entire economy of Principe off the coast of West Africa, cost only $10,000.[148] Second, it produces no residues and, provided the predator sticks to its host, risks to mammals are nil. Third, when a natural enemy is successfully introduced into an environment, control is permanent. As we have seen with chemical controls, resistance by insects frequently renders insecticides irrelevant. At the time that DDT was banned, Paul DeBach estimated that worldwide there had been 235 successful importations of natural enemies resulting in permanent control of a serious pest. As George Claus and Karen Bolander observed, this is quite an achievement. But since some 3,000 species of insects are considered pests, only a small percentage could be controlled in this fashion.[149] The world in which DDT was banned was not a world in which classical biological control alone could serve as a major weapon against pests.

Great hope was held out for the sterile male technique, a form of control that introduces large numbers of sterilized male insects into the environment in hopes that in mating they will successfully compete with fertile males of the species, thus preventing fertilization. In 1955, in what has been hailed as one of the greatest victories of biological control, this method eradicated the screwworm fly from Florida.[150] But the technique applies to only a narrow range of cases. An advocate of biological control commented that "the technique is expensive and is suited to only a small percentage of pest insects, mainly those that are present in comparatively small numbers."[151]

Another method of biological control uses insect-killing microorganisms or their products, "microbial pesticides"—bacteria, viruses, rickettsia, fungi, or protozoa. *Bacillus popillae,* a bacterial disease that kills Japanese beetles, was disseminated at more than 90,000 sites in thirteen eastern states, and within three years the Japanese beetle was reduced from a major to a minor pest.[152] *Bacillus thuringiensis,* a bacterial pathogen infecting a broad range of pests, is the most common microbial pesticide in use today. The main drawback of this method is the difficulty of finding pathogens that kill pests but not beneficial insects and mammals. The virus *Saproleginia,* for example, causes a fungal disease that kills fish; *Microsporidium* attacks any type of insects but can also harm domestic animals and humans.

Most farmers now use integrated pest management, which combines relatively low doses of chemicals with biological control. Our conclusion is that biological control alone could not serve as an effective replacement for chemical pesticides like DDT.

Government Regulatory Action

Federal governmental proceedings to ban DDT stemmed from a petition filed on October 31, 1969, by the Environmental Defense Fund (EDF) requesting immediate suspension of all uses of DDT. Under the Federal Insecticide, Fungicide, and Rodenticide Act (FIFRA), the Department of Agriculture could ban a pesticide if it presented an unreasonable risk to the environment. Secretary of Agriculture Clifford Hardin canceled four registered uses of DDT (on shade trees, on tobacco, around the home, and in aquatic areas) but refused complete suspension, concluding that DDT was in compliance with FIFRA. Unsatisfied, the EDF filed a petition with the U.S. Court of Appeals of the District of Columbia, seeking review of Hardin's failure to comply with its requests. The court ordered Hardin either to issue a Notice of Cancellation for all remaining uses of DDT or explain in writing why DDT should not be banned. On June 29, 1970, Hardin released a Statement of Reasons, but on subsequent review of those reasons the court determined that there remained sufficiently significant questions concerning the safety of DDT to justify an administrative hearing. It asked that suspension be reconsidered by the newly created EPA, which had been given the authority to administer FIFRA.[153]

Formal public hearings began on August 17, 1971, and proceeded for several months. Over 100 witnesses representing the DDT industry (Group Petitioners), USDA, EPA, EDF, Sierra Club, National Audubon Society, and West Michigan Environmental Council testified, producing over 8,300 pages of testimony. Edmund Sweeney, the examiner who presided over the hearings, emphasized their depth and breadth, and on April 25, 1972, he submitted to the EPA his *Recommended Findings, Conclusions, and Orders:*

1. DDT is not a carcinogenic hazard to man.
2. The uses of DDT under the registrations involved here do not have a deleterious effect on freshwater fish, estuarine organisms, wild birds, or other wildlife.
3. The adverse effect on beneficial animals from the use of DDT under the registrations involved here is not unreasonable on balance with its benefit.
4. The Petitioners have met fully their burden of proof.
5. There is a present need for the continued use of DDT for the essential uses defined in this case.[154]

He recommended no suspension of the remaining DDT registrations under FIFRA.

The final decision, however, lay with the administrator of the EPA, William Ruckelshaus. This Final Order, released on June 14, 1972, overturned Sweeney's findings. "The evidence of record showing storage [of DDT] in man and magnification in the food chain, is a warning to the prudent that man may be exposing himself to a substance that may ultimately have a serious effect on his health."[155] Ruckelshaus ordered a virtual ban on all uses of DDT effective January 1, 1973. Subsequently, it was learned that Ruckelshaus had not read the transcript of the hearings. His decision to ban DDT had been based primarily on summaries written by two EPA staff members, neither of whom had attended the hearings.

Ruckelshaus committed a number of factual errors, such as suggesting the use of carbaryl as an organophosphate, when it is a carbamate, an entirely different insecticide.[156] But the fundamental weakness of the Ruckelshaus order lies with what it left out. As Max Sobelman has written, "Nowhere in the administrator's *Order* does he quantify. Nowhere does he relate DDT levels in the environment with specific injury to bird, fish, animal, or man as determined in the laboratory at those same DDT levels."[157] Instead Ruckelshaus relied heavily on potential detrimental effects: "DDT *can* persist in soils for years and even decades," "DDT *can* persist in aquatic ecosystems," "DDT *can* be transported by drift during aerial applications," "DDT *can* have lethal and sublethal effects on useful aquatic freshwater invertebrates," "DDT *can* affect the reproductive success of fish," "Birds *can* mobilize lethal amounts of DDT residue," "DDT *can* cause thinning of bird eggshells and thus impair reproductive success," "DDT is a *potential* human carcinogen."[158]

All pesticides are poisons; otherwise they would not be effective. Assuming that no pesticide is perfectly selective in its impact on pests, all present some risk to nontarget organisms. That is why Congress, when writing FIFRA, set the criterion for banning a pesticide as "unreasonable risk" and not "potential risk." As Robert Ackerly notes, Ruckelshaus's assertions about the potential harm of DDT do not meet the statutory standard.[159]

The most enduring effect of the DDT saga is the ill-advised pesticide policy it created. Although DDT harmed birds, it was generally acknowledged to be the least damaging of the chlorinated hydrocarbon insecticides to mammals. The irony is that distrust of pesticides has grown so widespread that public pressure now exists to discourage use of insecticides like malathion (one of the safest and least toxic insecticides ever known), despite the significant benefit to humanity of judicious use of such pesticides.

Doing More Harm Than Good

The most important understanding to be gained from the history of DDT is that there are few unalloyed good things in the world. Rarely does one find a substance that has benefits but not costs.[1] Why, then, do we expect technologies, including their chemical components, to be entirely benign, unlike anything else in our lives—our friendships, marriages, children, parents, colleagues? It is the balance of gains and losses, not the absolute amount involved, that matters. Often there are unanticipated consequences, favorable or adverse, that come to be recognized only over time. If the mere possibility of adversity were considered decisive, no action—at least no new action—would be undertaken.

The DDT story contains many gains and losses: great good, near euphoria, disturbing but isolated reports, and discovery of unanticipated adverse consequences, including threats to bird populations. Concern rises along with resistance in certain insect populations. Perhaps human life is threatened. Perhaps biomagnification leads us to predict the unacceptable loss of entire raptor populations. For a while confusion reigns. Over time, however, scientific study goes on and knowledge grows, even amidst acrimony and conflicting findings. The least defensible hypotheses—such as that DDT causes the destruction of ocean oxygen and thus the loss of all life—drop out as insupportable. Fear of harm to humans, though still bruited about in public, diminishes among scientists. DDT use is regulated to reduce accidental overexposures, to make more effective use of smaller quantities, and to protect raptor populations. Use continues at a reduced rate.

The warning signs are there. Incentives exist to develop alternatives with similar effectiveness and low cost but without the same sort of detriments. Before this somewhat disorderly but progressive process can complete its course—with DDT being gradually replaced by other pesticides—it is abruptly aborted. Consequently, though the harm from DDT was reduced faster than it otherwise would have been, the good it was doing and would have continued to do was halted. The health loss from the ban has been much greater than the health gain; the DDT ban flunks a risk-risk analysis.

There is no getting away from it: completely banning DDT did more harm than good. What went wrong? Regulation of DDT parallels that of PCBs—at first desirable but unchecked by subsequent evaluations of the true costs and gains of the chemical. Perhaps this was inevitable. Questions of nuance—how much, how often, how long, under which conditions, compared to what alternatives—went unheard. Perhaps the public campaigns necessary to enact regulation require such denunciation of the substance in question that its rejection becomes the only moral and hence politically feasible act. Should that be so, however, there would be no place for knowledge.[2]

By this time our citizen risk detective has learned to look for high versus low exposure groups: if the highs don't get sick, why should the lows? Deviant studies are suspect. With many pointing in one direction and only one or two in another, the deviants should be rejected. Of course, deviant studies could be right, but the chances are they are wrong. Theories often have weak spots: biomagnification could be real, but if so every link in a long chain must be in place. Timing matters. You cannot get sick from a chemical if it was not around. Going with the best available evidence may mean compromise. Though evidence on DDT's harmful effects on birds is spotty, the resurgence of populations after the ban is suggestive. Always the question of alternatives is worth asking: how healthy is the replacement likely to be? Failure to ask that cost several hundred lives.

Almost uniformly, research findings are variegated, possibly conflicting, with diverse consequences under different conditions. The best use of science, therefore, is conditional. If there is truth in these mobilizing allegations, the results will be more protective public policy, a more enlightened public, and an enhanced respect for the public authorities that take knowledge into account in making decisions. But if there is little truth in the charges, and yet government acts on them, numerous harms may follow.

3

Dioxin, Agent Orange, and Times Beach

with Brendon Swedlow

It is a long way from Vietnam to Missouri, but mention Agent Orange and Times Beach to a couple of friends and the distance disappears.

"Wasn't that the stuff they sprayed on our troops?"

"Yeah. That's it."

"Isn't that the town the government bought because it was contaminated by dioxin?"

"Yeah, dioxin."

"Isn't that the stuff that was in Agent Orange?"

It seems to go without saying that dioxin is deadly. This widely held belief, however, is flatly false unless carefully qualified. Dioxin does cause a variety of diseases and death in some laboratory animals at very small doses. No human, however, has been killed outright by dioxin poisoning. As for disease, some studies have found increased illness, including certain cancers, among people exposed to dioxin. But the more powerful studies have found serious health consequences only among the most highly exposed. Of Vietnam veterans, just the 1,256 U.S. Air Force herbicide sprayers known as Ranch Hands have something to worry about, and even for them the effects are relatively limited. As for Missourians, former residents of Times Beach should not be concerned; tenants of Missouri's Quail Run mobile home park who were potentially exposed to much more dioxin experienced few adverse health consequences. The only Missourians severely affected were four children who played in two horse arenas saturated with dioxin-contaminated waste oil.

What scientists knew about herbicide health effects and what government officials did with this knowledge is our central concern. But we also want to chart science's evolving understanding of the effects of dioxin, as well as indicate how completely this knowledge was shared by others, including military personnel, journalists, and government officials.

The two types of phenoxy acids, 2,4-D and 2,4,5-T, used in the herbicide

Agent Orange were developed by the Department of Defense during World War II and were first used after the war by farmers and foresters. By the time President John Kennedy authorized the use of Agent Orange in Vietnam in November 1961, similar herbicides had been applied in the United States and other countries for two decades. By 1950, 4.5 million kilograms (kg) of 2,4-D and 2,4,5-T were being sprayed in the United States annually. Ten years later U.S. yearly consumption had more than tripled to over 16 million kg.[1] Farmers used these chemicals to control the growth of broadleaf weeds among crops like rice. Foresters used them to keep faster growing deciduous trees from overshadowing slower growing but more valuable evergreens. Ranchers found herbicides effective in keeping grazing ranges free of brush and prickly pear cactus. Utility companies relied on herbicides to clear the areas around power lines, and state and local governments discovered that herbicides were a less labor intensive and therefore less costly way to keep roadways and shoulders free of weeds and brush.

Production versus Spraying

These two decades of domestic application of phenoxy acids provided experience about the toxic dose for hundreds of plant species and the longer-term effects on the environment and humans. There did not seem to be any such effects. In this period scientists were also able to study the health consequences of exposure to these chemicals during their production. Beginning with the 1949 explosion of what has been described as a 500-gallon pressure cooker,[2] or autoclave, in Monsanto's Nitro, West Virginia, plant, there were a dozen incidents of accidental exposure by 1961, mostly in plants in the United States and West Germany, but also in France and Italy.[3]

Because larger numbers of Monsanto workers were exposed in 1949 than in subsequent accidents, and because their exposure occurred far enough back in time for longer-term effects to register, when in 1961 army scientists at Fort Detrick, Maryland, began testing herbicides for use in Vietnam, the workers' histories provided the best information available about the health consequences of exposure to phenoxy acids, here trichlorophenoxy acetic acids. Workers who reentered the building where the autoclave explosion occurred soon reported skin and eye irritation, breathing problems, headaches, dizziness, and nausea. One to two weeks later they began breaking out in blackheads and pale yellow cysts under their eyes and behind their ears, a condition known as chloracne. Three to four weeks later the earliest symptoms, like nausea, subsided only to be supplanted by severe muscle pains in the legs, shoulders, neck, and chest, as well as reports of fatigue, nervousness, irritability, insomnia, sensitivity to cold, decreased sex drive, and even impotence.[4]

Four years later most of these exposed workers appeared to have regained their health, but some continued to have problems. Of the 117 who developed chloracne, 27 still had mild to severe cases. Of 4 severely injured men, 2 had not returned to work because of muscle pain, and 1 man lacked feeling in his legs and feet.[5] During the same round of post-accident medical exams doctors found chloracne among 97 other Nitro Monsanto workers; 25 workers had other symptoms of exposure to phenoxy acids, like muscle pain, eye irritation, and decreased sex drive, only less severe. All these men had been exposed to trichlorophenols during the normal production process, not during the accident or its cleanup.

In 1957 another accident happened, but this one ultimately led to safer conditions for workers producing trichlorophenols and trichlorophenoxy acetic acids. A German laboratory technician at the Institute of Wood Chemistry in Hamburg got a faceful of dioxin while attempting the first synthetic production of this chemical. He developed a severe case of chloracne and went to a University of Hamburg dermatologist, Karl Schulz, who had experience treating chloracne in workers producing trichlorophenol for the Boehringer Sohn chemical company. Schulz and the chemist in whose laboratory the accident occurred proved that dioxin was a contaminant of trichlorophenol production and the cause of chloracne.

This discovery made possible the application of new technologies to detect dioxin in trichlorophenols. By 1965 gas chromatography allowed chemists to detect dioxin concentrations of 10 to 20 parts per million (ppm) in technical-grade trichlorophenol, and 1 ppm in laboratory-grade trichlorophenol.[6] Other advances in technology allowed producers to purify trichlorophenols of dioxin down to these new low levels of detection. Chloracne among workers became milder and less frequent. Dow Chemical Company adopted the 1 ppm dioxin standard in 1965 and later that year supplied the army with its first shipment of Agent Orange.[7]

There is not in fact a single chemical known as dioxin; that name refers to a family of 75 compounds, with 135 chemical cousins called furans.[8] Formed in tiny amounts as by-products of certain manufacturing processes, dioxins and furans have been patented for use in electric insulation and as flame retardants and agents against bacteria, insects, and fungi. Dioxin also results from all but the most high-temperature combustion, which destroys it. Hence forest fires are natural sources of dioxin in the environment. Most human beings have accumulated some dioxin in their fatty tissues, at trace levels of about 6 parts per trillion (ppt).[9]

The most deadly of the dioxins, 2,3,7,8-tetrachlorodibenzo-*p*-dioxin, or TCDD, is the one that contaminates the trichlorophenols and bactericides at the root of the Agent Orange and Times Beach stories. Dichlorophenols (2,4-D) are not contaminated by dioxin; consequently, the dichlorophenoxy acetic acid herbicides made from these chlorophenols are also dioxin free. Not so for the trichlorophenoxy acetic acid herbicides: they are made from trichlorophenols, which are contaminated by this deadly dioxin. TCDD is considered to be the most toxic synthetic substance ever, and for certain rodents it is. It took 64,000 times more sodium cyanide than TCDD to kill half the guinea pigs in a lab test.[10] At the same time, acute, or short-term, toxicity among rodents varies greatly. The dose of TCDD needed to kill hamsters, to give the most extreme example, is 5,000 times greater than the lethal dose for guinea pigs.[11]

It is not clear how much Fort Detrick army scientists knew about the health histories of Monsanto or other workers exposed to trichlorophenols, but it is obvious that they took toxicity to humans and animals into account when testing and selecting from among commercially available herbicides.[12] They knew that it was not trichlorophenols but dioxin that caused chloracne.[13] They even gave brief consideration to dioxin as a chemical warfare weapon. If we assume that these scientists were familiar with the recorded health consequences of industrial exposure available in the early 1960s, their recommendation to use these herbicides in Vietnam is understandable. Worldwide application had not produced adverse health consequences. Only the heavy exposures occurring in production seemed to have unhealthy effects, which—as nasty as these could be—did not appear life threatening. At the same time Vietnamese forests, providing cover to the enemy, seemed deadly to American troops.

In Vietnam Agent Orange was used to suppress vegetation around base camps, so that American troops would be less vulnerable to surprise attack, and also to defoliate river embankments to foil the ambush of patrol boats. Offensively it was used to uncover North Vietnamese bases and infiltration routes. Finally, it and other herbicides were used to destroy crops in the hope of eliminating a significant portion of the enemy's food supply.[14]

Admiral Elmo Zumwalt, who was promoted to chief of naval operations during the war, ordered the use of defoliants along the rivers patrolled by his son, who was also in the navy. After the war the son developed two cancers, Hodgkin's disease and non-Hodgkin's lymphoma, which both father and son attribute to Agent Orange. Despite the admiral's visible role in trying to secure government compensation for his son and other Vietnam veterans, both men believe that using Agent Orange was the right thing to do. Before Admiral Zumwalt ordered the use of defoliants along inland waterways, patrol boat crews had a 70–75 percent chance of being killed or wounded in the course of a year's service.[15] "Certainly thousands, including me, are alive today because of his decision to use Agent Orange," the younger Zumwalt said in 1986.[16] The admiral sorrowfully agreed: "Had I not used Agent Orange, many more lives would have been lost in combat, perhaps even Elmo's. And knowing what I now know, I still would have ordered the defoliation to achieve the objectives it did."[17]

Agents Orange, White, Blue, Purple, Pink, and Green

Some aspects of the course of herbicide use in Vietnam are certain; others are less clear. We know how many different kinds of herbicides and what quantity of each were used, when and on what types of vegetation and in support of what kinds of objectives. We know the average dioxin content of these herbicides. We know where air force pilots sprayed and how much they sprayed. We know less about where the army and the South Vietnamese Air Force used them. We can make some intelligent guesses about what kinds of servicemen were likely to have been more exposed and which less. We know the present dioxin content of a number of veterans' blood lipids based on samples taken during the mid-1980s.

The first shipment of herbicides received in Vietnam on January 9, 1962, consisted of Agents Purple and Blue, two of the six different mixtures, including Orange, that were used in Vietnam over the next ten years.[18] Agent Blue was the only herbicide used throughout that period—to destroy cereal and grain crops and to control grasses around base-camp perimeters.[19] It contained no dioxin because it contained no trichlorophenoxy acetic acids. Agents Purple, Pink, and Green—used until July 1965, when they were replaced by Orange and White—had the highest dioxin content of any herbicides used in the war, at 32 ppm.[20] They contained 2,4,5-T and were manufactured prior to the point in 1965 when gas spectrometry could detect 1 ppm dioxin in these phenoxy acids. The average dioxin content of Agent Orange was consequently only 2 ppm.[21] Agent White, like Blue, contained no dioxin, because it contained no 2,4,5-T.

Although the average dioxin content of Agent Orange was low, the dioxin concentrations of 2,4,5-T made by different manufacturers varied considerably. Dow produced Agent Orange that had only 0.5 ppm dioxin; its 2,4,5-T contained 1 ppm dioxin. When the 2,4,5-T was mixed 50:50 with 2,4-D, the dioxin concentration in the resulting Agent Orange was only 0.5 ppm.[22] Other manufacturers', in contrast, had as much as 47 ppm dioxin; Dow's and Hercules' Agent Orange levels were on the low side compared with Monsanto's and Diamond Shamrock's.[23]

In March 1965, before contracting with the government to produce Agent Orange, Dow hosted a meeting with other producers of 2,4,5-T such as Hercules and Diamond Shamrock (but not Monsanto). Dow was concerned about "surprisingly high levels" of dioxin in its competitors' products and hoped to encourage them to lower those levels to avoid investigation and regulation. Dow cautioned that repeated exposures to even 1 ppm of dioxin could produce chloracne in test animals and informed its competitors about the German firm from which Dow had obtained its testing technology for only $35,000.[24]

By 1965 Agent Orange had replaced Purple, Green, and Pink and had become the most widely used herbicide of the war, at over 42 million liters. Of the total amount of these four herbicides used in Vietnam, 85 percent went to defoliate trees, particularly mangroves along waterways; 8 percent to destroy crops like beans, peanuts, roots, and tubers; and 7 percent to secure base perimeters, cache sites, waterways, and communication lines.[25] It is estimated that the air force sprayed herbicides on 10 percent of the land mass of South Vietnam, an area about the size of Rhode Island and Connecticut combined.[26] All six herbicides were delivered in 55-gallon drums differentiated by colored bands, which is how they got their names.

That Agent White did not contain dioxin is particularly significant for it was the most widely used herbicide in Vietnam after Agent Orange, at over 21 million liters.[27] For every two Ranch Hand C-123 cargo planes spraying Agent Orange there must have been one spraying Agent White. The two herbicides were also similar in appearance: Orange was a reddish brown and White was a dark brown. Moreover, they were directed at similar targets, though Agent White was more likely to be used for inland forests, while Orange was used on coastal mangrove forests. For the most part, then, a casual observer could not tell an Agent Orange spraying from an Agent White spraying. This may be insignificant, however, since both herbicides were usually sprayed well in advance of troop movements to permit the three months needed for complete defoliation.[28] Troops were more likely to have mistaken insecticide for herbicide spraying than Agent White for Agent Orange spraying, since the malathion mosquito-abatement program was carried out close to troops by the same C-123 cargo planes. The only difference was that the planes spraying malathion had no camouflage colors.[29]

Scientists' Reactions to the Spraying Program

American scientists objected to the use of herbicides in Vietnam even before Agent Orange was deployed. In March 1964 the Federation of American Scientists protested herbicide use because it did not discriminate between fighting forces and civilians and because such use constituted chemical and biological warfare.[30] In January 1966 a group of Harvard scientists urged President Lyndon Johnson to end the defoliation campaign. Later that year, in June, a member of the American Association for the Advancement of Science (AAAS), E. W. Pfeiffer, a University of Montana zoology professor, proposed that the AAAS establish a committee to study the biological effects of the chemicals used in Vietnam.[31] That same month Bionetics Research Laboratories in Bethesda, Maryland, informed the National Cancer Institute (NCI) that the herbicide 2,4,5-T (found in Agents Orange, Purple, Green, and Pink) produced great increases in birth defects when injected into pregnant mice.[32]

The NCI asked Bionetics to administer 2,4,5-T to pregnant mice orally and report on the results, which they did—two years later. Meanwhile, in December 1966, the AAAS established the committee Pfeiffer requested and wrote Secretary of Defense Robert McNamara urging studies of the short- and long-term consequences of herbicide use in Vietnam.[33] In February 1967 President Johnson received a second petition calling on him to end herbicide use in Vietnam; it was signed by over 5,000 scientists, including 17 Nobel laureates and 129 members of the National Academy of Sciences. Another AAAS letter to McNamara followed, suggesting that the NAS, the National Research Council, a panel of the president's Science Advisory Committee, or an independent commission study the consequences of the defoliation program. John S. Foster, Jr., director of defense research and engineering, responded by writing that "qualified scientists, both inside and outside our Government, and in the governments of other nations, have judged that seriously adverse consequences will not occur. Unless we had confidence in these judgements, we would not continue to employ these materials."[34]

The AAAS was not satisfied with this response; its members wanted to know what research supported these judgments. Foster referred to a "consensus of informed opinion" among 50 to 70 individuals but conceded that they had no hard data. He assured the AAAS that the Department of Defense was taking steps to back up its claims by commissioning the Midwest Research Institute (MRI), a private nonprofit organization, to review all published, unclassified literature on the ecological consequences of herbicide use.[35] The institute's study was to be reviewed by a panel of the NAS, constituted especially for this purpose at the Department of Defense's request. Both reviews were to be made available to the AAAS for further review.

By December 1967, three and a half months after contracting for the job, the MRI had evaluated over 1,500 reports and interviewed more than 140 experts.[36] At some point during 1967 another research institute, the Rand Corporation, reported to McNamara that infants could potentially receive lethal doses of herbicide in Vietnam from the crop destruction program. McNamara passed this information to the Joint Chiefs of Staff, who recommended that the program continue.[37] The NAS panel completed its review of the MRI report in January 1968, finding that the MRI had done "a creditable job."[38] But the basic question AAAS had been asking could not be answered in the library. Much of the research on herbicides at that point dealt with vegetation management—the effects of civilian applications in noncrop-land settings like roadsides. As the NAS panel noted, "the scientific literature provides markedly less factual information on the ecological consequences of herbicide use and particularly of repeated or heavy herbicide applications."[39] Still, the available evidence allowed the MRI to conclude, in the paraphrase of a Ralph Nader–sponsored critique of the NAS reviewers, that "lethal toxicity to humans, domestic animals, or wildlife was highly unlikely."[40]

Scientists' Efforts in the Field

The AAAS remained unsatisfied. According to a retrospective piece in a 1989 AAAS newsletter, "There was much agreement that the confidence expressed in John Foster's letter that 'seriously adverse consequences will not occur' was quite unjustified." As an AAAS ecologist reviewing the MRI and NAS work put it: "The outstanding fact that emerges is that we don't know."[41] Professor

Pfeiffer kept pushing the AAAS to sponsor a study, and the AAAS responded by publicly asking the United Nations to do the fieldwork. The United Nations (UN) did pass a resolution asking the secretary general to prepare a report on chemical and biological weapons and to endorse the 1925 Geneva Protocol, an international treaty banning the use of these weapons.[42] The Department of State determined that the study recommended by the AAAS could not be undertaken in combat areas but promised cooperation "in responsible long-term investigations of this type as soon as practicable."[43] The Defense Department's position, as expressed by John Foster, was that "while there are a number of scientific questions left unanswered by available studies, these questions would not be answered by additional, short-term investigations. On balance, we continue to be confident that the controlled use of herbicides will have no long-term ecological impacts inimical to the people and interests of South Vietnam."[44]

The pressure to "know" something more about the effects of herbicide use in Vietnam was unbearable, however, so scientists began visiting the country; the State Department sponsored the first inspection. In September 1968, the U.S. ambassador to South Vietnam, Ellsworth Bunker, released the findings of an interagency committee that had evaluated the defoliation program. An Agriculture Department botanist, Fred S. Tschirley, toured Vietnam in early 1968 for this review and reported that coastal mangrove forests had suffered more damage—total denuding and near-total killing—than inland forests, which were substantially defoliated but had few dead trees.[45] In the final words of the interagency report, however, "in weighing the overall costs, problems, and unknowns of the herbicide program against the benefits, the committee concluded that the latter outweigh the former and that the programs should be continued."[46]

Unsatisfied by the UN response and the State Department's attempts at fieldwork, and unable to interest any other organization in undertaking a study, the AAAS finally decided to convene a group of scientists to prepare a plan in which the AAAS itself would participate "within the reasonable limits of its resources." Around the same time Bionetics scientists made a second report on their teratology studies to the National Cancer Institute: the phenoxy acid 2,4,5-T caused birth defects in the offspring of pregnant mice when fed as well as when injected. At a meeting on January 30, 1969, the National Institutes of Health made this information available to regulatory agencies, the National Academy of Sciences, and the chemical industry, among others, but no action followed, and the Bionetics report was passed on to the National Institute of Environmental Health Sciences for further statistical analysis.[47] In March 1969 Pfeiffer and another zoologist, G. H. Orians, under the sponsorship of the Society for Social Responsibility in Science, traveled to Vietnam with the purpose of interesting other scientists in studying long-term herbicide effects while demonstrating that meaningful information could be obtained even on a short, underfunded, understaffed trip in the midst of war.[48] Their foray confirmed Tschirley's observations about the greater damage to coastal mangroves compared with inland forests.[49]

The Mice That Roared

In the summer of 1969 the Bionetics studies drew attention again when the teratology panel of the Commission on Pesticides and Their Relation to Environmental Health asked to see them. This Mrak Commission (so called after its chairman, Dr. Emil Mrak) had been created by the Secretary of Health, Education, and Welfare in response to the rising public concern over pesticides on the one hand and the ire of Republican politicians about the Food and Drug Administration's seizure of 34,000 pounds of pesticide-containing Lake Michigan salmon on the other. Based on the Bionetics findings that the Commission received on September 24, the teratology panel recommended that 2,4,5-T be "immediately restricted to prevent risk of human exposure."[50] At about the same time consumer advocate Ralph Nader was sponsoring a study of the food regulation practices of the Food and Drug Administration (FDA), and one of his researchers happened across a copy of the Bionetics report in FDA files and mentioned it to a friend at Harvard. In early October 1969 the friend in turn mentioned the report to Matthew S. Meselson, a Harvard biology professor.[51]

Told that the Bionetics study was "confidential and classified" when he requested a copy, Meselson got one by an unofficial route. The implications of the findings of birth defects in mice seemed serious to him, so he immediately notified President Richard Nixon's science adviser, Lee DuBridge, who on October 29 announced that "a coordinated series of actions are being taken by the agencies of government to restrict the use of the weed-killing chemical 2,4,5-T." DuBridge promised that by January 1, 1970, the Department of Agriculture would cancel registrations permitting the use of the herbicide on food crops unless the FDA found a safe concentration level in the meantime. DuBridge also promised that the departments of Agriculture, Interior, and Defense would limit their use of 2,4,5-T to unpopulated areas. Finally, DuBridge appointed a panel of scientists "to review all that is known about 2,4,5-T," but he did not publicize this panel until Senator Philip Hart held hearings in April 1970 on the herbicide's environmental and human health effects.[52]

A week after DuBridge's announcement Julius E. Johnson, Dow Chemical's director of research and a member of the Mrak Commission, attempted to persuade the commission that the contaminant dioxin, not the herbicide itself, was responsible for the birth defects. He believed that Dow could alter its production techniques to purify the herbicide of dioxin. (It may be recalled that Dow had adopted a standard of dioxin's presence in 2,4,5-T of 1 ppm in 1965 and that tests of Agent Orange revealed concentrations of 1–2 ppm; the 2,4,5-T used in the Bionetics study, in contrast, had 30 ppm of dioxin, about the concentration found in herbicides used before 1965.) Johnson's theory was not persuasive enough to alter the commission's conclusions, so on November 25 he proposed to the National Cancer Institute that Dow test the theory by feeding pregnant mice "purified" herbicide. On December 5 the Mrak Commission made its report public, including its recommendations to restrict 2,4,5-T use to prevent human exposure. That same month the AAAS appropriated $50,000 toward funding a Herbicide Assessment Commission to go to Vietnam. Meselson was to organize the study.[53]

January 1, 1970, came and went and only the Department of the Interior had acted as DuBridge said it would. Agriculture and Defense were waiting for the results of the Dow tests, while the Defense Department also claimed that spraying herbicide in unpopulated areas had always been its policy anyway.[54] The Dow report, available on January 12, showed that "purified" 2,4,5-T did not cause birth defects in the offspring of pregnant mice. These results stimulated scientists at the FDA and the National Institutes of Health (NIH) to conduct their own tests. Meanwhile, Congress began to take an interest in executive-branch activity (and inactivity) on herbicide regulation. On February 10 Senator Philip Hart (D.-N.Y.) announced that he would conduct hearings on the effects of 2,4,5-T on humans and the environment. On February 24 FDA and NIH scientists presented the results of their teratogenicity tests: even "purified" herbicide caused birth defects. The discrepancies between their findings and Dow's were subsequently partially explained by the government's (and Bionetics') use of higher dosages of 2,4,5-T and a broader definition of "birth defects."[55] In March the Defense Department announced that it was cutting back its herbicide program by 25 percent.

During Senator Hart's two days of public hearings in April the surgeon general cited "new" information that "nearly pure 2,4,5-T was reported to cause birth defects when injected at high doses into experimental pregnant mice, but not in rats," before announcing that the Department of Agriculture would cancel certain registered uses of the herbicide (amounting to about 25 percent of domestic usage).[56] On the same day the Defense Department announced that it had "temporarily suspended the use of 2,4,5-T for military operations pending further evaluation."[57] Dow Chemical and the Hercules Corporation appealed the Agriculture Department's cancellation of 2,4,5-T registration.

Scientists Confirm Defoliation, Not Health Effects

Meanwhile, Meselson had prepared the AAAS's Herbicide Assessment Commission, and in August he and three other scientists toured Vietnam for five weeks. Meselson asked General W. B. Rosson, acting commander of U.S. forces in South Vietnam, for the time and location of defoliation missions and details of the herbicide used,[58] for statistics recently gathered by the army on the number of stillbirths and birth defects in South Vietnam, and for helicopter transport for one or two ground inspections.[59] Rosson denied Meselson the defoliation and health-effect information—it was classified—but provided transport and ground support. Meselson was able to inspect most of South Vietnam from the air, and on the ground he was assisted by South Vietnamese professors and students in collecting samples of plants, fish, hair, and mother's milk, and in interviewing sixty farmers and village officials. The commission got access to army statistics on stillbirths and birth defects when these were published in the United States, but Meselson considered his lack of information on time and location of defoliation "a major disability," because without it he had nothing to correlate with the effects he observed.[60]

Meselson estimated that about half of South Vietnam's mangrove thickets had been sprayed and appeared dead, with "little or no recolonization . . . after three or more years." He also estimated that about a fifth of the hardwood forest had been sprayed, and more than half of that was "very severely damaged. Over large areas, most of the trees appeared dead and bamboo had spread over the ground." Meselson further estimated that about 2,000 square kilometers had been treated to destroy food crops, mainly in the Central Highlands, home to a million Montagnard tribespeople. Meselson also reported that one heavily sprayed province had an unusually high rate of stillbirths and that a large Saigon hospital had an increase in two kinds of birth defects at the time large-scale spraying was carried out. But he cautioned that further study would be required to link these effects to herbicide exposure, since other factors might be responsible.[61]

Sometime during December 1970, the *Washington Post* got a copy of a cable sent jointly by General Creighton Abrams, commander of U.S. forces in Vietnam, and Ambassador Bunker to the White House requesting permission to terminate the crop destruction part of the herbicide program.[62] On December 26, 1970, the opening day of the annual AAAS meeting in Chicago, where Meselson reported the findings outlined above, the Nixon White House announced a phaseout of herbicide operations. The final fixed-wing Agent Orange mission was flown on April 16, 1970, the last helicopter mission on June 6, 1970. And the last herbicides under U.S. control were used in Vietnam on October 31, 1971.[63]

A National Academy of Sciences advisory committee reviewing appeals by Dow and Hercules made its recommendations to the newly created Environmental Protection Agency (EPA) on May 7, 1971: the herbicide was safe provided that its dioxin contamination was reduced to specific low levels.[64] These recommendations were leaked to outside scientists, who, under the sponsorship of the Committee for Environmental Information and Nader's Center for the Study of Responsive Law, criticized them at a news conference in Washington in July. EPA administrator William Ruckelshaus turned to the FDA for further advice and then decided to disregard the advisory panel's conclusions and proceed with public hearings on the chemical companies' appeals. Around the same time the promised review of 2,4,5-T by a panel of the president's Science Advisory Committee was released with its conclusion that defoliation had been useful militarily and that ecological damage was minimal, though the panel had relied on anecdotal evidence in the first instance and entirely neglected to assess the herbicide's detrimental effects on forests in the second. As for the consequences of herbicide exposure for pregnant women, the panel concluded that epidemiological evidence did not support findings of one kind or another, but that theoretical exposure levels could have approached those which produced birth defects in mice and rats. The panel, however, emphasized the unlikelihood of such levels.

Two major herbicide studies were conducted toward the end of the war: one by the National Academy of Sciences (NAS) on ecological and physiological effects, the other by the Army Corps of Engineers on military effectiveness. The Corps of Engineers' study of late 1971 gave a qualified endorsement, recommending that herbicides be included in contingency plans for war in Western Europe, Cuba, Venezuela, Ethiopia, and Korea.[65] For the NAS study 1,500 mandays were spent in Vietnam collecting data. The study committee operated on a budget of $1.4 million and was given access to virtually all documents it requested, though its correlation of herbicide spraying and population distribution was somewhat hampered by the military's slowness in providing data.[66]

The legislation calling for the NAS study required it to be completed by January 31, 1972, but the report was not ready until January 1974. Its conclusions were similar to those of the AAAS study: coastal mangroves experienced the severest ecological consequences, while damage to inland forests was less severe but still significant. As for adverse physiological effects, the NAS found little conclusive evidence. A consulting anthropologist had relayed reports of illness and death among Montagnard children and even adults, which the Montagnards associated with spraying. The committee found these reports "so striking it is difficult to dismiss them," but since the fieldwork was conducted after spraying had ended, researchers had no way to confirm the claims.[67] The NAS concluded that the evidence "does not support the suggestion that herbicide spraying may have engendered birth defects" but was concerned about the dioxin Meselson discovered in fish and shellfish and about the possibility that defoliation had introduced malaria-carrying mosquitoes to new areas.[68] The Defense Department's reading was that "some damage has resulted from the military use of herbicides in Vietnam; however, most of the allegations of massive, permanent ecological and physiological damage are unfounded."[69]

Agent Orange Comes Home

At this point, it should be noted, the only health effects scientists were concerned with were South Vietnamese stillbirths, birth defects, and the possible effects of eating seafood contaminated with dioxin. The thrust of scientists' findings was toward ecological not physiological damage. The effect of herbicide exposure on American servicemen does not appear to have been a source of concern nor, given what was then known about 2,4,5-T and dioxin, was there reason to believe that servicemen were being exposed to unhealthy quantities of herbicides. To our knowledge there is no systematic information on how Vietnam veterans with health problems came to believe these were caused by Agent Orange. There are surely a number of individuals who—independently, by different paths, and for different reasons—came to the conclusion that Agent Orange was at the root of their poor health. But quite clearly a small number of people were able to convince a huge number of Vietnam veterans that Agent Orange had made them sick.

Maude DeVictor is one of these few. A onetime Navy hospital corpsman, DeVictor in 1977 worked for the benefits section of the Chicago regional office of the Veterans Administration (VA). In this capacity she received many phone calls from the wives of terminally ill veterans, inquiring about survivor benefits. One such call was from a Mrs. Charlie Owens. "She was calling right from the hospital because the doctors had just told her he was terminally ill," recalled DeVictor. "All she knew was that he had told her in healthier days that if his death was ever due to cancer, it was because of the chemicals used in Vietnam."[70] This was the first time DeVictor had heard of Agent Orange. After some inquiry she learned more about Operation Ranch Hand and the herbicide program in Vietnam. Then she called up cancer cases on her computer,[71] but was assured by an air force spokesman that there was no proven link between dioxin and cancer. Despite such assurances, DeVictor believed there was a connection and encouraged veterans to make claims for compensation, the first of which was filed in October 1977.

DeVictor also spent some time in a VA hospital over the Christmas holiday, finding a number of Vietnam veterans afflicted with cancers she knew to be rare in people their age. She then began polling widows of veterans who had died of cancer and ex-servicemen who came into her office. She asked about such symptoms as acnelike rashes, frequent headaches, frequent mood changes, loss of sex drive, and whether the wives had miscarriages, stillbirths, or children with birth defects. Veterans who had a majority of these symptoms, she found, also had two other things in common: they were Vietnam veterans and they were in areas supposedly sprayed with herbicides. By early 1978 DeVictor had collected data on over fifty Vietnam veterans.[72]

When DeVictor approached her VA superiors with this information, they told her to stop gathering it. When she got a phone call from Ron DeYoung, a veterans' counselor at nearby Columbia College who was trying to verify some information on an application, DeVictor asked DeYoung if he knew anything about herbicide application in Vietnam. He said he did not. DeVictor spent a half-hour on the phone telling him what she believed and another two hours in his office later that day showing him the reports, government memos, letters, and other documentation she had collected. DeYoung took this information to a college instructor affiliated with WBBM-TV, Chicago's CBS station. Assisted by DeVictor in gathering veterans and experts, anchorman Bill Kurtis then prepared an hour-long documentary, *Agent Orange, the Deadly Fog.*

The program allowed veterans and scientists to talk about a link between the herbicide and birth defects without pointing out that there was no evidence, in either humans or animals, of male-mediated birth defects.[73] Kurtis did say that "officially the Veterans Administration is denying the claims of poisoning by Agent Orange. Their scientists simply feel there isn't any evidence to link defoliation with human problems." His conclusion, however, pointed in another direction: "But after researching this report and listening to the recommendations of the leading dioxin scientists in the country, we feel there is a need for immediate testing of all Vietnam veterans who handled Agent Orange or went into sprayed areas. Not only for the sake of those who have told us of their symptoms, but also for the countless others whose lives and whose children's lives could be blighted by the dioxin poison in Agent Orange."[74] The morning after the program aired on March 23, 1978, DeVictor was besieged in her office by phone calls from other journalists, veterans, and government officials. The VA would not let her take the calls and transferred her to another department, but she persisted with her warnings in every available forum.[75]

In the summer of 1978 an ABC-TV executive called DeVictor. Paul Reutershan had walked into the television studio and announced that he was dying of cancer caused by Agent Orange. Reutershan said that DeVictor could verify his story. She had never heard of him before but told the executive that his claim was probably true, given the information she had gathered. Reutershan was twenty-eight years old and had inoperable cancer of the stomach, colon, and liver. A helicopter crew member in Vietnam, Reutershan became convinced after several weeks of research that his health problems and those of many other veterans were caused by exposure to herbicides. With his sister, Reutershan formed Agent Orange Victims International. He appeared on television and radio programs and gave numerous newspaper interviews. His most publicized statement, on NBC's *Today Show,* was "I died in Vietnam, but I didn't even know it."[76] Several weeks later Reutershan did die, in December 1977, and ABC-TV's *20/20* and the Public Broadcasting Corporation's *For Your Information* featured nationally televised segments on Agent Orange and veterans' health issues.

On his deathbed Reutershan asked Frank McCarthy, a disabled Vietnam veteran, to succeed him as president of Agent Orange Victims International. McCarthy in turn persuaded Victor Yannacone, an attorney who had experience litigating against the manufacturers of chemical substances like DDT, to take over a lawsuit Reutershan had filed before his death. In the fall of 1978 the air force surgeon general promised Congress and the president to investigate the health of Operation Ranch Hand veterans.[77]

Yannacone amended Reutershan's complaint, submitting it as a class action in a U.S. District Court in Manhattan on January 8, 1979. Yannacone initially defined the class as "all those so unfortunate as to have been and now to be situated at risk, not only during this generation but during generations to come."[78] This category potentially included over 2.4 million Vietnam veterans, their wives, and their children.[79] Yannacone named Dow, Monsanto, and three other companies as defendants (more were added later) and sought damages "in the range of $4 to $40 billion." The suit received a great deal of instant publicity. When it was filed, Yannacone recalls, "The phones rang off the hooks all day—150 calls, with the last coming at 3 A.M. from soldiers in Australia."[80]

McCarthy, Yannacone, and some other Long Island lawyers began barnstorming the country, using local veterans' organizations to recruit plaintiffs. Yannacone then convinced their lawyers to sign associate counsel agreements. Equipped with class action forms, Yannacone and local counsel would subsequently march to the nearest federal courthouse to file suit, with the press present to record the event. "While we took every opportunity to cultivate reporters and denounce the callous chemical companies, our opponents for some reason were very unsophisticated about this," recalls David Dean, an associate of Yannacone's. "Only as trial approached did some of them begin sending out press packets explaining their testing results, the tiny dioxin levels, etc., but by then it was too late." "In many news articles Agent Orange was described as a 'killer' or as a 'poison,' without even an 'alleged.'" "From our pollster, we knew that 80 percent of the prospective jurors had heard of Agent Orange and believed that it was harmful."[81]

Yannacone initially resisted taking the case, in part because he knew that it would be very hard to prove that Agent Orange caused the disabilities suffered by his veteran clients. He told them, "We can't win, but I'll at least get you your day in court."[82] Besides dealing with the veterans and the media, Yannacone's major responsibility was to build the scientific case against dioxin.[83] He hoped to do this, in part, by conducting an epidemiological study with the help of his wife, which he believed could be done for under $1 million.[84] His associates refused to give him the money, arguing that although he had greater scientific knowledge than the rest of them, he certainly did not have the kind of credentials that would make him a satisfactory expert witness. They wanted him to assemble such witnesses instead.[85]

The publicity given Vietnam veterans' claims about Agent Orange restimulated scientific inquiry into the link between 2,4,5-T and dioxin. In 1979 some consequential findings were made. An epidemiological study exploring the relationship between miscarriage rates and exposure to 2,4,5-T in Oregon had the greatest immediate impact. The study showed that hospitals reported much higher miscarriage rates in the rural Alsea, Oregon, area where herbicides were used in the forests than in a similar rural area (and an urban area) where herbicides were not used.[86] In light of these findings, EPA administrator Douglas

Costle suspended the use of 2,4,5-T in forestry, rights-of-way, and pastures, which effectively cut back use of the herbicide 75 percent nationwide and 100 percent in Oregon. Costle said, "I am ordering emergency suspension of these uses because I find that they pose an 'imminent hazard' to humans and because I also find that an 'emergency' exists because there is not enough time to complete a suspension hearing before the next spraying season."[87]

Six scientists from Oregon State University severely criticized the Alsea study. They pointed out that differences in miscarriage rates reported by hospitals in different areas could be caused by differences in medical practice across those areas; that reported miscarriages did not "peak" in June for six years but only in one year, which could happen randomly; and that for this reason and because the data on 2,4,5-T spraying were incomplete, there was no significant correlation between time of spraying and time of greatest number of miscarriages.[88] Defenders of the Alsea study blunted this criticism by retorting that the critics had misclassified some rural areas as urban. The net result was that the EPA's emergency suspension remained undisturbed.

That same year, 1979, Swedish investigators reported that fifty-two men with cases of soft tissue sarcoma, a rare cancer, were five to six times more likely than men to which they were compared to have had exposure to phenoxyacetic acids or chlorophenols. The men had worked in forests or saw- and pulpmills where these chemicals were used, and—except for preparations used by two cases and two controls—dioxin was a contaminant in the chemicals.[89] Such apparently significant effects did a great deal to reorient scientific research and to buttress claims that herbicide exposure did have chronic, or long-term, effects on human health. In the same year scientists came to the conclusion that most combustion produces some dioxin, while very high temperatures render it harmless.[90]

In the spring of 1979 Senator Charles Percy asked the General Accounting Office (GAO), Congress's investigative arm, to look into the possibility that ground troops in Vietnam had been exposed to Agent Orange. John Hansen, the GAO employee assigned to make this determination, was able to identify about 5,900 marines who were within half a kilometer of areas sprayed with Agent Orange on the day the spraying took place; some were directly underneath the spray missions.[91] Hansen suggested that the Defense Department did not know that it was spraying its own troops because they were not supposed to move into sprayed areas until four to six weeks after spraying. According to Hansen, the weeks that elapsed between a ground commander's request for defoliation and the Ranch Hands' fulfillment of it may have allowed troops to move into areas destined for spraying and, once there, remain invisible to defoliation crews because of the dense forest canopy. In November the GAO recommended that the government "determine whether a study is needed on the health effects of herbicide orange on ground troops identified in our analysis."[92]

In December 1979 Congress passed legislation directing the VA to identify veterans "exposed to any of the class of chemicals known as dioxins" and determine what health consequences, if any, this exposure had. Congress gave the VA six months to design the study. This legislation was the result of a compromise between the House and Senate committees on veterans' affairs. The Senate committee wanted some other agency to carry out the study because of veterans' complaints about the VA's indifference to their claims of harm from Agent Orange; the House wanted to preserve the VA's responsibility for doing research on veterans' health. The House got its way, but only after agreeing to assign the congressional Office of Technology Assessment (OTA) the function of overseeing the design and execution of the study.[93] That same month President Jimmy Carter created the White House Interagency Work Group on the Possible Long Term Effects of Phenoxyacetic Herbicides and Contaminants.[94]

In January 1980 Congress passed legislation requiring the National Institute for Occupational Safety and Health to study the health of American workers exposed to dioxin, with OTA overseeing the research design and monitoring the conduct of the study. This time, however, the legislation was vetoed by President Carter on the grounds that the OTA had been given an administrative role in violation of constitutional requirements for the separation of legislative and executive functions within government. Carter's veto message also instructed the VA administrator to ignore the OTA's role in overseeing the design and execution of its study of Agent Orange and veterans' health. Senator Alan Cranston, chairman of the Senate Committee on Veterans' Affairs, finessed the constitutional issue and retained OTA's role by reminding the VA administrator that congressional funding of the study depended on his not ignoring Congress's wishes.[95]

Meanwhile, the Interagency Work Group had created a science panel, which concluded that it was impossible to study the effects of Agent Orange on veterans' health. Despite the GAO's identification of marines who had been sprayed or nearly sprayed, the panel felt that it was still too difficult to figure out who had actually been exposed to Agent Orange and to how much they had been exposed. Moreover, there was the problem of separating this herbicide's health effects from those of other herbicides, insecticides, malaria medication, and illicit drugs to which these same veterans might have been exposed. Consequently, it agreed that a study of the effects of serving in Vietnam should be substituted for a study of the effects of exposure to Agent Orange.[96] Congress responded with legislation that allowed the VA to expand the study to include other causes of veterans' poor health.

The Interagency Work Group, however, was not at peace with its recommendation. Members kept trying to figure out a viable way to study the effects of Agent Orange exposure. Jerome Bricker, a science panel member who had helped design the spray apparatus for Operation Ranch Hand, proposed classifying any body of troops known to have been within a specified distance of a Ranch Hand spraying or of emergency dumping, or in a base camp whose perimeter had been sprayed, as having received a "hit" of Agent Orange. Such hits could then be added up, and troops with the most hits could be classified as highly exposed. Work Group members agreed that some classification like this should at least be attempted, and they proposed a study of three groups of veterans: a group classified as exposed to Agent Orange, one classified as unexposed, and a group who did not serve in Vietnam.

In the meantime the VA had contracted with researchers at the University of California, Los Angeles, to design a study. These investigators proposed examining only exposed and unexposed veterans, without studying a non-serving group and thus with no way to isolate the health effects of serving in Vietnam.[97] The OTA rejected the first draft.[98] In August 1981 President Ronald Reagan changed the name of the Work Group to Agent Orange Working Group and made it part of the Cabinet Council on Human Resources, raising its stature and expanding its scope.[99] By September 1982 the OTA had approved the VA study design of Agent Orange's effects on veterans' health. But by this time the science panel of the Working Group had reversed itself again and was pressuring the VA to study only the health effects of service in Vietnam. Eventually, with both House and Senate backing, the responsibility for the study was given to the Centers for Disease Control (CDC).[100]

The CDC acted swiftly where the VA had not. It prepared an outline for the study even before the transfer was completed and submitted a fullblown plan to the Working Group and the OTA by early 1983. The CDC resolved the controversy over whether to do an Agent Orange study or a Vietnam experience study by proposing to do both. In addition, it proposed a more narrowly focused study of the relationship between certain cancers and service in Vietnam and exposure to Agent Orange. These three suggested studies were characterized by the chairman of the OTA review panel as "probably the most expensive, most complex, and largest group of epidemiological studies ever undertaken."[101] The CDC planned to interview 30,000 veterans and to give 10,000 a physical and psychological examination (in five cohorts of 6,000 veterans each, 2,000 of whom would undergo an examination).

Agent Orange on Trial, Round One: Who Knew What When

While the OTA panel was reviewing the CDC's study plan, Judge Paul Pratt was ruling on summary judgment motions made by the defendant Agent Orange producers in the class action suit. Pratt had earlier decided that the best way to try the case was one issue at a time, beginning with the question of whether or not the chemical companies could cloak themselves in the government's sovereign immunity against lawsuits. If Dow and the other defendants could demonstrate that they were acting as government contractors in producing Agent Orange, they would be off the hook. To prove this, Pratt had earlier ruled, the companies would have to show that the government established the specifications for Agent Orange, that the herbicide supplied met those specifications, and that the government knew as much or more than the Agent Orange producers about the human health effects of the herbicide.[102] On May 20, 1983, Pratt ruled that the companies had proved the first two elements. As to the relative knowledge of the companies and the government, this depended on a company-by-company comparison with the government's knowledge. Pratt concluded that "uncontradicted and uncontested evidence . . . reveals that the government and the military possessed rather extensive knowledge tending to show that its use of Agent Orange in Vietnam created significant, though undetermined, risks of harm to our military personnel."[103] Making the comparison, Pratt then found that the Riverdale and Hoffman-Taff companies were ignorant of any hazards and that Thompson Chemicals knew less than the government. Pratt therefore granted summary judgment in their favor, dismissing the veterans' cases against them.

Dow Chemical's relative knowledge was another matter: this was an instance where it did not pay to be an industry leader. First, Dow tried to argue that the only relevant knowledge for purposes of establishing parity was knowledge about hazards that accompanied use of the product. Dow maintained that its knowledge was confined to the hazards of production, where exposures were to much higher levels of dioxin than the 1 ppm found in the finished product. If this was so, the veterans' lawyers wanted to know, then why had Dow cautioned other producers in 1965 about the dangers of repeated exposure to 1 ppm? Dow answered that the problems it was aware of occurred in test animals, not humans. Second, Dow maintained that even if high-level government officials were unaware of Agent Orange's potential health effects, "crucial middle-level government experts . . . had extensive knowledge." The veterans' lawyers countered that even mid-level government experts did not have the specificity of knowledge that Dow did. Finally, Dow insisted that even if it had told the government more, the government would not have changed its defoliation program. Plaintiffs responded that the government would have stopped buying the herbicide and restricted its domestic use had it known what Dow knew. Because of these genuinely disputable factual issues, Pratt rejected Dow's summary judgment motion and that of Uniroyal and T. H. Agriculture and Nutrition, two companies in a similar position with respect to parity of knowledge. Factual issues were for a jury not a judge to resolve.[104]

A little over a month later, on June 30, 1983, air force scientists published the results of their study of the health of Operation Ranch Hand veterans. Annual updates have been made since then, and the air force intends to give extensive medical exams to surviving Ranch Hands through the year 2002.[105] The February 1984 update, for example, tracked mortality and morbidity through December 1983 and found that of the 1,256 men who served in Operation Ranch Hand 54 had died. Proportionally, this is the same death rate as among the 6,171 non–Ranch Hand air force veterans of Vietnam selected for comparison: 265 had died by the end of December 1983. Fewer Ranch Hands died of cancer than did other men studied, while more died of digestive system disorders. Neither of these differences were statistically significant.[106] As the study's reviewers at the National Research Council of the National Academy of Sciences cautioned, 1,256 men is not a big enough group to detect small increases in rare diseases. But if death rates from rare diseases were double those in the comparison group, this increase would show up, which to date it has not.

Meanwhile, in February 1984 the OTA approved the CDC study plan after suggesting some changes. The CDC also had been busy studying birth-defect rates in the children of Vietnam veterans and published the results of this research in 1984.[107] For some years the CDC had kept records of birth defects in children born in the five counties surrounding its Atlanta, Georgia, head-quarters. Of 323,421 babies born between 1968 and 1980 about 13,000, or 4 percent, had major birth defects. When CDC scientists compared the fathers of a subset of these children with the fathers of a sample of children born without defects, they found the same proportion of Vietnam veterans in each group (9 percent), suggesting that service in Vietnam was not related to increased chances of birth defects in offspring.

The CDC also took on the difficult task of making comparisons among Vietnam veterans to see if those who were potentially more exposed to Agent Orange had fathered more children with birth defects. Two indices of exposure were created. Both were based on herbicide use and a veteran's occupation, location, and time of service, but for one index the veteran information was obtained from military records, while for the other it came from interviews of the veterans themselves. Those classified as more exposed on either index did not have a greater risk than controls of fathering babies with birth defects.[108]

The same comparisons were made for each of the 95 birth defects studied, with largely the same results. However, for three defects—cleft lip with or without cleft palate, spina bifida, and "other neoplasms"—fathers classified as having greater exposure on one or another or both of the indices had greater numbers of defects among their offspring than did the controls. These findings quickly led one senator's staff to draft legislation to provide compensation to Vietnam veterans whose children had these birth defects.[109] This potential legis-lation never went any further, however, for the CDC explained to congres-sional staffers and veterans' groups that the method used to estimate opportu-nities for exposure was too uncertain to support an association between Agent Orange and birth defects. Moreover, as the scientists pointed out, there was no evidence that dioxin could have done the things that would have made it cause birth defects: that is, be a mutagen or be present in veterans' sperm in order to enter the mother's womb at the time of conception.[110] There is no plausible biological mechanism for dioxin-caused, male-mediated birth defects. As the CDC scientists concluded, "If there is any increased risk related to exposure to Agent Orange, either the risk must be small, must be limited to select groups of Vietnam veterans, or the increased risk must be limited to specific types of defects."[111]

Agent Orange On Trial, Round Two: No Causation, No Case

While the results of the Ranch Hand and birth-defects studies were coming in, the veterans' case against Agent Orange producers was accelerating toward a $180 million settlement. On October 14, 1983, Judge Pratt, who had been elevated to the New York Court of Appeals, announced that the pressures of his new job would no longer allow him to preside over the case.[112] A week later lawyers for all parties filed into Judge Jack Weinstein's chambers to meet their new trial judge. Weinstein informed them that the case was headed for an early trial—about six months away—and that the pivotal issue was whether Agent Orange caused the injuries claimed, not the relative knowledge of the government and the chemical companies.[113] Subsequently, Weinstein ruled that the government was not immune to certain claims of harm by veterans' wives and children, a decision that brought previously dismissed defendants, like Thompson Chemicals and the government, back into the suit.

At the October meeting Weinstein also told the class action lawyers to select a handful of plaintiffs to represent the injuries of the class, which numbered some 16,000 claimants. Five were eventually approved by the judge but none got their day in court as Yannacone had promised (he had long since been voted off the trial team because of disputes with other members). After living in the courtroom the weekend before the trial, the plaintiffs' lawyers accepted $180 million to settle out of court. (The chemical companies initially set a limit of $25 million, and the veterans had demanded no less than $700 million.)[114] Weinstein decided to allocate liability for the $180 million partially on the dioxin content of the companies' Agent Orange and partially on their share of the Agent Orange market. Under this formula Monsanto had to pay almost half of the settlement (45.5 percent), though it had less than a third (29.5 percent) of the market share. Meanwhile Dow, which had nearly the same market share (28.6 percent) had to pay less than a fifth (19.5 percent).[115]

Weinstein goaded the veterans toward settlement by reminding them of the weakness of their evidence. The Agent Orange manufacturers did not need as much persuasion to settle, for they believed that if the case was decided by a jury, it might grant the veterans a large award, given the serious nature of the plaintiffs' injuries, even if the evidence of causation was inadequate. In a written opinion Weinstein issued in September 1984, after holding hearings on the fairness of the settlement (as required in class actions), he indicated that the chemical companies' evaluation of the causation evidence was probably correct. He noted the generally negative findings of the recently published Ranch Hand and birth-defects studies, and concluded, "This is not sufficient to support a recovery in tort law."[116]

Weinstein's final evaluation of the evidence occurred during the subsequent litigation of the 350 veterans and other claimants who decided to opt out of the class action (and settlement) and pursue their cases individually. On July 24, 1985, during the pretrial phase of the litigation, the companies asked Weinstein for an early, summary judgment in their favor, claiming, among other things, that there was no evidence their product caused the veterans' injuries. This was the same motion the companies had felt it pointless to make in the earlier litigation. Although the law requires a judge to allow only real issues of fact to reach a jury, judges generally follow the rule that doubts about the genuineness of a factual issue must be resolved by allowing the jury to hear the issue. Summary judgment is granted only in cases where there is no dispute about facts or in which plaintiffs produce no evidence, significantly incomplete evidence (when more was available), evidence that contradicts known and incontrovertible facts, or evidence that is internally inconsistent.[117]

To overcome the chemical companies' motion for summary judgment, Weinstein required veterans to produce an expert who would swear as to a particular plaintiff that "in her scientific opinion, there is more than 50 percent probability that [his] particular diseases were due to exposure to Agent Orange in Vietnam."[118] The plaintiffs' counsel, Rob Taylor, wrangled with Weinstein over this standard and many other issues related to overcoming the motion for summary judgment. On January 11, 1986, Taylor asked Weinstein to recuse himself from hearing the other individual cases of those who had opted out of the class action, on the grounds that he was biased and was placing improper pressure on the individuals to return to the class and be bound by the settlement. Weinstein refused, Taylor appealed, and the Second Circuit summarily affirmed Weinstein's position. Taylor went on to produce the plaintiffs' strongest evidence yet of specific causation, the affidavit of the scientist Samuel Epstein, who swore as to 15 individual plaintiffs that their conditions were "much more likely than not to have been caused by exposure to Agent Orange."[119]

On May 8, 1986, Weinstein nevertheless granted summary judgment to the defendant chemical companies, on the grounds that "all reliable studies of the effect of Agent Orange on members of the class so far published provide no support for plaintiffs' claims of causation."[120] "Reliable studies" for Weinstein meant human epidemiological studies, which in his opinion were "the only useful studies having any bearing on causation. All the other data supplied by the parties rests on surmise and inapposite extrapolations from animal studies and industrial accidents." In an earlier decision, on February 11, 1985, Weinstein had dismissed all claims against the government by the veterans, their wives, and their children—in significant part on similar grounds.[121] And since the plaintiffs could not proceed against the government, Weinstein also did not allow the chemical companies to recover their settlement costs from the government. If, as Weinstein finally concluded, the plaintiffs had no case because they could not demonstrate causation, then the government had no obligation to indemnify the manufacturers of Agent Orange against claims of harm or to contribute money toward the settlement.[122]

Studies Indicating Limited Health Effects

In 1988, after the Agent Orange litigation drew to a close, two epidemiological studies of the type Weinstein demanded were published. One, by the husband and wife team of Steven and Jeanne Stellman, found a statistically significant relationship between their exposure classifications and certain types of health conditions in Vietnam veterans.[123] The study was based on 6,810 American legionnaires who responded to a survey in Colorado, Indiana, Maryland, Minnesota, Ohio, and Pennsylvania; 42 percent of respondents were Vietnam veterans. The Stellmans classified them as having low, medium, or high exposure, based on self-reported dates and locations of service in Vietnam and the relationship between these recollections and military records of the dates and locations of Agent Orange missions. The most highly exposed reported significantly more doctor-diagnosed benign fatty tumors, chronic bronchitis, and nervous system disease than those with low exposures. Highly exposed veterans also reported significantly more adult acne, skin rashes with blisters, and sensitivity to light than their low-exposure counterparts.[124]

The CDC epidemiological study was ready in September 1989.[125] Like the Stellmans, CDC scientists classified veterans as having low, medium, or high exposure to Agent Orange, but the CDC relied on service records rather than recollections to establish veterans' time and location of service. This method prevented the possibility of diseased veterans falsely reporting service in a sprayed area. CDC scientists also tried to prevent the converse: veterans who had been in sprayed areas falsely reporting a doctor- or self-diagnosed medical condition. The veterans in this study were all given medical and psychological exams.

What made the CDC study exceptionally powerful was its use of a newly developed method of assessing exposure to dioxin. During the medical exam blood samples were taken from over 600 veterans, including more than 60 percent who were classified as most highly exposed. When the blood lipids were analyzed for their dioxin content, the results showed that no exposure category had higher dioxin concentrations than any other; that is, those classified as highly exposed had no more dioxin in their serum than those classified as having low exposure. In fact, the median dioxin level in the lipids of the most highly exposed was the same as that of a comparison group of non-Vietnam veterans (3.8 ppt). This finding does not mean that those classified as most highly exposed never had higher levels of dioxin in their serum than those classified as less exposed or unexposed to Agent Orange. It does mean that the median exposure level even for those thought to be most highly exposed never could have been very high. Extrapolating from a fairly well documented dioxin half-life of seven years,[126] the median dioxin level of the most highly exposed group could have been 7.6 ppt in 1980, 15.2 ppt in 1973, and 30.4 ppt in 1966, the earliest time at which significant exposures are likely to have occurred. As we will discuss, much higher levels of dioxin exposure are not associated with poor health.

To the CDC the lack of significant differences in the serum dioxin levels of Vietnam veterans meant that there could be no study of Agent Orange's health effects, since exposure classifications could not be used to identify Vietnam veterans whose present health might have something to do with their exposure to Agent Orange. To others the CDC's refusal to do an Agent Orange study seemed like a cover-up. In a July 1989 hearing before the House Subcommittee on Human Resources and Intergovernmental Relations, Chairman Ted Weiss charged that the CDC's serum study was "either politically rigged or monumentally bungled." He cited the Stellmans' research as evidence that a study of Agent Orange's health effects was possible. He also attacked the CDC's exposure classification methodology, though CDC methods of identifying exposed veterans were more rigorous than the Stellmans' and consequently more likely to avoid misclassification.[127] Vernon Houk, director of the CDC's Center for Environmental Health and Injury Control, and Hellen Gelband of the OTA, defended the study's methodology,[128] and no other scientific organization has claimed that the Stellmans did a better job.

The Ranch Hand application crews remain the only identifiable group of Vietnam veterans who have significantly elevated serum dioxin levels. At their most recent physical examination in 1987 the 888 Ranch Hands had a median of 12.4 ppt dioxin, almost three times the median of 4.2 ppt in the other air force personnel who served in Vietnam and were examined for comparison. The median for enlisted Ranch Hand personnel was higher than for officers, as might be expected, since enlisted personnel were more likely to handle Agent Orange spraying equipment directly. The median for the 152 flying enlisted personnel was 17.2 ppt, while that for the 407 ground crew was 23.6 ppt, with the range extending up to 617.7 ppt, the highest dioxin level seen among Ranch Hands.[129]

Air force investigators comparing the overall health of Ranch Hands with that of nonexposed air force veterans of Vietnam found no significant differences. The mortality rate of the two groups was the same. Mortality by specific causes also varied little, with Ranch Hands slightly more likely to have died of circulatory system disease and slightly less likely to have died of malignant neoplasms. Moreover, the distribution of neoplasms by site and morphological pattern was similar for both.[130] Among living Ranch Hands, the only significant health difference was the greater prevalence of sun- and exposure-related skin cancer among them (9.4 percent) than among those studied for comparison (6.9 percent). There were no significant differences in systemic cancers, though the investigators caution that increases in rare cancers cannot be detected in a group as small as the Ranch Hands.[131]

The normal, or background, rates of the rare cancers that some scientists have attributed to herbicide or dioxin exposure are so low that only very large increases will show up in a cohort study. This basic limitation of the contemplated Agent Orange study got lost in all the wrangling over exposure classifications: had there been an identifiable group of highly exposed veterans, the study still would not have been able to answer questions about the incidence of rare cancers, or any rare illness, among them.

To analyze rare cancer rates among Vietnam veterans the CDC performed a case-control epidemiological study, that is, a study in which persons with a particular disease are compared with persons not having the disease. Researchers attempt to find disease-free persons similar to the diseased cases in as many respects as possible. The CDC decided to study men with one of six rare cancers associated in other studies with herbicide or dioxin exposure: non-Hodgkin's lymphoma, Hodgkin's disease, sarcoma, nasal cancer, nasopharyngeal cancer, and liver cancer. Chosen for the study were men diagnosed between late 1984 and late 1988 who were fifteen to thirty-nine years old in 1968 and who lived in an area covered by one of eight cancer registries. These registries encompassed Connecticut, Kansas, Iowa, Miami, Detroit, San Francisco, Seattle, and Atlanta—about 10 percent of the U.S. population. All diagnoses used were confirmed by pathologists' review of tumor tissue samples. Controls were selected from each registry area through random digit dialing and matched to cases.[132]

Regarding five of the six cancers, including soft tissue sarcoma, the Vietnam veterans were not at greater risk than the controls. But veterans were about 50 percent more likely to have non-Hodgkin's lymphoma than men who did not serve in Vietnam. The annual incidence (unrelated to AIDS) of non-Hodgkin's lymphoma among American men aged thirty to fifty-nine is about 10 cases per 100,000, so a rate 50 percent higher would mean that 15 in 100,000 would be expected to develop this cancer.[133] Although investigators conducting the study did not assess the Agent Orange exposure of Vietnam veterans any more directly than in other studies, they were fairly certain that such exposure could not be the cause of the increased incidence of non-Hodgkin's lymphoma for two reasons. First, their analysis of subgroups of veterans revealed that ocean-going navy veterans were more likely than land-based troops to have non-Hodgkin's lymphoma; second, among land-based troops in Vietnam, those in the most heavily sprayed region (III Corp) were less likely to develop this cancer than those elsewhere. In other words, those most likely to have been exposed had less of this cancer than those least likely to have been exposed.[134]

In response to the CDC's study of selected cancers, in March 1990 the Veterans Affairs secretary announced that non-Hodgkin's lymphoma would be treated as a service-related disease; at the same time, the VA refused to link this rare cancer to Agent Orange exposure. This compensate-without-conceding-causality response has been the pattern in Congress as well. In January 1991 legislation was unanimously passed providing permanent disability benefits to Vietnam veterans with non-Hodgkin's lymphoma and soft tissue sarcoma but without attributing these cancers to exposure to Agent Orange.[135] Since the two cancers are rare, the decision to compensate will not have a big fiscal effect.

Meanwhile, interest earned on the $180 million class action settlement with Agent Orange producers has ballooned the award to $240 million, $170 million of which has been set aside to compensate veterans and survivors presumptively injured by the herbicide. The Weinstein court administering the settlement believes that there will be 30,000 eligible veterans (those considered totally disabled under Social Security guidelines) and 18,000 eligible survivor families. These veterans and families will each receive an average of $5,700 but no more than $12,800.[136] Another $52 million has been set aside for grants to social agencies serving veterans. The remainder has gone to attorneys' fees and to Australian and New Zealand veterans who can meet eligibility requirements.[137]

Times Beach, Missouri

When its homes were built in the 1920s and 1930s Times Beach was a summer resort on the Meramec River west of St. Louis, Missouri. By the 1970s its small wood frames housed a working-class community of 800 families and somewhat more than 2,000 people. In 1983 Times Beach was transformed again, this time into a ghost town, its inhabitants evacuated. Currently its homes are being leveled, and for the next ten years its soil will be incinerated. Finally, as we enter the next century, Times Beach will become something else again: a state park.

Paralleling the town's rise and decline is a change in its symbolic meaning. In the 1970s Times Beach, like most small towns, meant something only to the people who called it home. By 1983 Times Beach was a symbol for the nation: it signified the dumping of toxic waste, especially dioxin, and the fear of the by-products of the industrial and technological development in postwar America. It seemed to be an acute example of the devastation that our innocence of hidden dangers could inflict on us. Times Beach had hired a waste-oil hauler to spray its dusty dirt roads, only to find that the hauler had mixed dioxin-laced wastes with the oil and had thereby apparently created a health hazard so threatening that the federal government recommended the town's permanent evacuation.

Today Times Beach means something else again: a sign of overreaction to the dangers of industrial waste. For the dioxin in the roads is not a significant threat to human health; residents could have continued living there without fear of illness, "which," as one commentator put it, "makes this textbook example of the importance of environmental law look like a textbook example of hype, panic, and the use of science for political purposes."[138] Neither textbook example, we believe, captures the truth; viewing Missouri's dioxin history through the prism of Times Beach distorts the lessons of Missouri's dioxin experience.

The most important thing to remember about dioxin in Missouri (or any chemical anywhere) is that its danger depends on the dose received. Dioxin remains deadly in high doses, though the amount that will produce particular health effects varies widely from species to species. The incidents of dioxin exposure in several Missouri horse arenas were serious. Dioxin-contaminated soil killed horses and other animals and severely sickened several children who played in the arenas. These poisonings occurred almost ten years before Times Beach became an issue and resulted from very high dioxin levels and from direct contact with contaminated soils.

In the early 1980s scientists did not have a good idea of what a "safe" dose of dioxin might be, though they should have been able to convey the basic idea that the dose makes a difference. Instead, the mere presence of dioxin became cause for great alarm among politicians and public alike. Today's best evidence indicates that the government overreacted in Times Beach, causing more harm by evacuating the town's residents than the dioxin on its streets ever would. At the same time there was inadequate understanding of the dangers of improper disposal of certain industrial waste products. The history of dioxin's dispersal throughout Missouri clearly demonstrates the need for the regulation of hazardous waste disposal that subsequently developed.

From Agent Orange to Baby Powder

Hoffman-Taff, it may be recalled, was one of the chemical companies that supplied the Department of Defense with Agent Orange during the Vietnam war and one of the defendants in the original class action suit. Judge Pratt released the company from liability early in the litigation because he found it to be ignorant of any of the herbicide's health hazards. One of Hoffman-Taff's Agent Orange production facilities was in Verona, Missouri, a small town on the Spring River, about thirty-five miles southwest of Springfield.[139] The dioxin that eventually wound up on the streets of Times Beach and from there found its way into the headlines of America's newspapers originated with this production plant.

In November 1969 Hoffman-Taff sold and leased part of its Verona production facility to Northeastern Pharmaceutical and Chemical Company (NEPACCO). This company needed the trichlorophenol made by Hoffman-Taff to produce hexachlorophene, a bactericidal chemical used to treat acne and impetigo, to wash babies and clean wounds, and to scrub down patients and medical personnel prior to surgery. Because it is made from trichlorophenol, hexachlorophene is also contaminated by dioxin, though the dioxin in the final product is so small as to be undetectable. NEPACCO's hexachlorophene reportedly contained only 0.1 ppm dioxin.[140] The manufacture and purification of hexachlorophene produces dioxin-contaminated wastewater, filter clay, and "still-bottoms." Of these waste products, the gooey, smelly still-bottom distillations contain the greatest concentrations of dioxin.

In December 1969, a month after NEPACCO moved in, Hoffman-Taff moved out, selling its remaining interest in the Verona location to Syntex Agribusiness, which began manufacturing vitamins, mold inhibitors, and drug intermediaries at the plant. Initially NEPACCO disposed of about 5,000 gallons a day of its dioxin-contaminated wastewater by directing it into Hoffman-Taff's water-treatment system, but leaks into the Spring River were detected by state officials. In 1971 NEPACCO paid the Water and Wastewater Technical School, in Neosho, Missouri, 2.5 cents per gallon to dispose of 225,000 gallons of wastewater. Most of the water was poured into a concrete-lined basin at a sewage treatment plant operated by the school. About 1,000 gallons were kept in a tank on campus, so that students could practice opening and closing the valve on the tank.[141]

The dioxin-contaminated filter clay also had a variety of fates. Initially it was buried in trenches on the plant grounds in Verona; later it was hauled to a dump in Aurora. Some of the clay was also used by local farmers, who thought it helped prevent hoof disease in their cattle. NEPACCO began disposing of the dioxin-saturated still-bottoms by shipping them to Louisiana for incineration. Then, hoping to cut down on costs, NEPACCO searched for other sources of disposal and eventually found Independent Petrochemical and its subcontractor, Russell Bliss, a waste-oil hauler. Bliss hauled away six tanker-truckloads of still-bottoms, a total of 18,500 gallons, between February and October 1971. Most of this waste went to a storage tank on Bliss-owned property in Frontenac, Missouri, where it was mixed with waste oil. Most of the contents of that tank were sold as heating oil, meaning that the dioxin in the still-bottoms probably escaped into the atmosphere, since heating oil is not usually burned at temperatures high enough to destroy dioxin. Some of the still-bottom oil mixture ended up in other places, however, which is why this story does not end in Frontenac.

On May 20, 1971, one of Bliss's drivers had picked up the second of six loads of still-bottoms from NEPACCO when he got a ticket for overloading his truck. Fearing another at the weigh station near St. Louis, he detoured to

Bliss's 200-acre farm in Rosati and dumped 500 gallons of his 3,500-gallon load on a dirt road. Shortly thereafter a dog, seventy chickens, and dozens of wild birds died on the farm. Five days later another of Bliss's drivers sprayed the floor of the Shenandoah Stables indoor horse arena near Moscow Mills with dioxin-contaminated NEPACCO still-bottoms. The stable owners, Judy Piatt and Frank Hampel, paid Bliss $150 for this service, thinking their arena was being sprayed with waste oil, which Bliss had discovered on his own Mid-America Horse Arena kept the dust down for months, unlike water. On June 11 Bliss's men oiled another horse arena, at the Bubbling Springs Ranch, near Fenton. He did this one for free, because the arena was used for charity horse shows by the Shriners. Five days later his men sprayed an arena at Timberline Stables, near New Bloomfield. Also in 1971 Bliss donated the oiling of a dirt driveway at the Community Christian Church, in Manchester, because his mother was a member there.

When the driveway was eventually tested, dioxin was found. But it would be months before soil testing was done anywhere in Missouri, years before dioxin was identified as the poison that was killing animals and sickening people, and still more years before sites like the Community Christian Church were tested for dioxin. In the meantime the Bubbling Springs Ranch and Timberline and Shenandoah Stables were living through a tragic mystery. After Bliss oiled the Bubbling Springs Ranch arena, some people became ill and six horses and many birds died. The ranch manager noted an absence of flies and remembered seeing a swarm of gnats swirl into the arena and then drop to the ground.

Timberline and Shenandoah Stables were plagued by even more severe problems. At Timberline twelve horses eventually died. The three-year-old son of Timberline's managers played in the arena after it was sprayed and developed chloracne, nausea, vomiting, and abdominal pain; his playmate also developed chloracne. The evening after Shenandoah was sprayed, a quarterhorse became ill, and in the next weeks and months many more horses were afflicted. All in all, sixty-two registered quarterhorses died or had to be destroyed. Cats, dogs, and birds also died. Co-owner Hampel spent hours raking up sparrows. Worse, co-owner Piatt's daughters, Andrea, six, and Lori, ten, played in the arena as though it were a sandbox. Hampel, Piatt, and her daughters developed flulike symptoms—diarrhea, headaches, and aching joints and shoulders. Suspicious about Bliss's "oiling," Hampel and Piatt confronted him in mid-July, but he assured them that he had sprayed with regular used crankcase oil. Though Piatt sent her daughters away to her sister on August 1, two days after they returned on August 19 Andrea had to be rushed to the hospital writhing in pain, with her bladder inflamed and bleeding severely. Both girls made many trips to the hospital after that, with symptoms resembling arthritis as well as gastrointestinal problems, nosebleeds, and nausea.

Piatt's veterinarian could not figure out why the animals at Shenandoah Stables were dying, and veterinarians at the University of Missouri at Columbia were clueless as well. Andrea and Lori's doctors at St. Louis Children's Hospital were stumped, too, though they were convinced that Andrea had received a chemical overdose of some kind. They reported Andrea's problems to the Missouri Division of Health, which in turn notified the Centers for Disease Control in Atlanta. The day after Andrea was hospitalized, Hampel stopped the horse shows at Shenandoah Stables and used a backhoe to remove a six-to eight-inch layer of soil from the arena. In late August, three months after the spraying, a CDC physician arrived at Shenandoah, noted the acrid odor emanating from the arena, and took soil as well as human and animal blood samples. Before leaving Missouri, he also visited Bliss's Mid-America Horse Arena, which Bliss told him had been sprayed with oil from the same load used on Piatt's arena, causing Bliss's horses no problems.

In September 1971 Hampel and Piatt sued Bliss. In October Hampel removed a second layer of soil from the Shenandoah arena, this time one foot thick. As before, he dumped this soil in a fill area that was later paved over by a highway. In November he and Piatt attempted to bring horses back to the stables, but these too got sick and died, so they decided to sell Shenandoah. During this time, in the fall of 1971, Hampel and Piatt became their own detectives, wearing wigs and borrowing cars as they tailed Bliss's trucks from their pickups to their dropoffs. Over the next fifteen months Hampel and especially Piatt worked in this way to compile a list of sixteen companies whose waste Bliss had dumped rather than resold and thirty-one sites where they maintained Bliss had dumped or sprayed waste. One of these sites was Times Beach, Missouri.

In January 1972 the Food and Drug Administration banned most uses of hexachlorophene, after talcum powder containing unusually high concentrations of the bactericide killed thirty-six infants in France. (The dioxin in the hexachlorophene is not believed to have had anything to do with these deaths.)[142] This ban caused the collapse of the hexachlorophene market and the closure, also in January, of NEPACCO's Verona plant. Syntex remained and inherited a tank filled with 4,300 gallons of still-bottom waste. Later testing revealed that this tank contained 343,000 ppb of dioxin, which, according to one CDC investigator, was "enough to kill everyone in the United States." At the time of NEPACCO's departure Syntex believed the twenty-foot tall tank had been left empty and did not learn of its contents until two years later, when government investigators came calling.

Oiling Times Beach and Analyzing Shenandoah

Times Beach first hired Bliss in 1972 to oil its twenty-three miles of dirt roads. He charged $2,400 for the service, returning for the next four summers, through 1976, each time spraying about 40,000 gallons of oil. In July 1972

Arthur A. Case, a doctor at the University of Missouri School of Veterinary Medicine, took soil and tissue samples at Timberline Stables. While doing so, the sides of his face were "burned" by what he described as "a gaseous agent." In August 1972 Timberline Stables removed the top foot of soil from its arena to a landfill in Jefferson City, and the CDC reported that it was unable to identify anything in the soil and blood samples from Shenandoah that could have caused death in animals and illness in humans. Late in the year Piatt wrote an eighteen-page report on Shenandoah's problems and on her and Hampel's observations of Bliss's activities. She sent copies to the EPA in Washington, the CDC in Atlanta, and the Missouri Division of Health in Jefferson City.

In March 1973 the Shriners hired Vernon Stout, a road-grading contractor, to remove the top layer of soil from the Bubbling Springs arena. Stout took some of the approximately 850 cubic yards of dioxin-contaminated dirt to a piece of his own property in Meramec Heights where there were two mobile homes. Most of the rest went to help level the back yard of Harold Minker, who lived in Meramec Heights. Some of the soil also was sold to three other property owners who flagged Stout down during the twenty-five to thirty trips it took to complete the removal job. Meanwhile the CDC decided to take another crack at analyzing the Shenandoah soil. Veterinarian Case had taken samples there in August 1971, and in November 1973 he sent these to the CDC at its request. By early 1974 the CDC had confirmed the presence of crystals of trichlorophenol in the soil. When some of this soil was put on the ears of four test rabbits all shortly developed hyperkeratosis, as expected, but within a week two died of liver damage, which was not expected, given the small dose of dirt they had been given. This led CDC chemists to suspect and test for the more highly toxic contaminant dioxin. On July 30, 1974, the CDC completed its gas spectrometry tests. The chemical cause of Shenandoah's animal deaths and human illnesses had been identified. The soil contained over 31,000 ppb of dioxin.

Tracing Dioxin to NEPACCO

On August 2, 1974, the CDC relayed its dioxin discovery to the Missouri Division of Health. Patrick E. Phillips, a veterinarian who led the division's investigation of Shenandoah, in turn notified Piatt, explaining that he had never heard of dioxin himself but that it was also found in Agent Orange, very much in the news at the time. "I'd heard of that," Hampel recalls thinking when Piatt told him about her conversation with Phillips. "I knew they sprayed it in Vietnam." Trying to figure out where the dioxin had originated, the division's chief epidemiologist called Bliss, who said he had no idea where he might have picked up dioxin. CDC doctors Coleman D. Carter and Matthew M. Zack, who arrived in Missouri on August 8, were more successful. Carter went on a hunch that there was a connection to Agent Orange, asking the army for the names of all producers it had contracted with. None had dealt with Bliss. One, Thompson Chemicals, had no need to: Thompson dumped its waste directly from its St. Louis plant into the Mississippi River. But when Carter called Hoffman-Taff, a former supervisor at the Verona plant remembered that NEPACCO had waste hauled by Bliss. In September Piatt and Hampel added two new defendants to their suit against Bliss: NEPACCO and Independent Petrochemical.

Carter and Zack also took additional soil samples for CDC analysis at the Shenandoah, Timberline, and Bubbling Springs horse arenas. That same August the CDC produced the results of its analysis of soil samples collected by Phillips. Soil from Minker's property contained as much as 850 ppb of dioxin; soil from Stout's property as much as 440 ppb. Meanwhile, soil from the Bubbling Springs arena, from which the Minker and Stout soil had been removed, retained only trace amounts of dioxin. The removal had been effective; the same was true of Shenandoah and Timberline. However, at Bliss's Mid-America Horse Arena, where no soil had been removed and where no horse or other animal illnesses had been reported, dioxin was present at up to 150 ppb. The Missouri Division of Health told Minker and the people on Stout's property that dioxin was present, that they should avoid contact with the soil, and that they should not do any gardening.

On March 31, 1975, the CDC recommended some stronger action. In a confidential report released to the EPA and the Missouri departments of Resources and Conservation, the CDC advised the burial of contaminated soil from the Minker and Stout properties in a deep landfill and the destruction of the still-bottoms in the Verona tank. Missouri officials decided not to act on the recommendation regarding the Minker and Stout properties, however, because the CDC also reported that "TCDD (the most toxic of dioxins) is not as stable as originally assumed and present evidence suggests that its half-life in most soils is one year." The officials figured that within a few years the dioxin would have degraded to harmless levels, so they did not even tell the property owners or inhabitants about the CDC recommendation. Instead the Missouri Division of Health focused its attention on the Verona tank. "The only thing on my mind was that tank," Phillips recalled. "I'd lie awake at night worrying that a tornado would take it to who knows where."

The favored way to get rid of the dioxin in the tank was to burn its contents, but Missouri had no incinerator and neighboring states refused to do the job or even to let the waste cross state lines to be burned elsewhere. "We discussed taking it to an incinerator in Minnesota," recalled Earl L. Barkley, a Syntex vice president. "But groups in Iowa said they would call out the National Guard if we went across their state." So Syntex built a concrete dike around the tank, fenced the dike, and set up a committee of twelve academic and industrial experts to advise them on alternative disposal methods. Meanwhile a Missouri assistant attorney general learned of the CDC recommendations regarding the

Minker and Stout properties and urged the Missouri Division of Environmental Quality to act on them. The division defended its inaction by citing the expected rapid degradation of dioxin in the soil and characterized the CDC recommendation as "overly cautious." The regional office of the EPA, in Kansas City, remained similarly inactive, saying they had never received the CDC recommendations.

Federal and state legislators, however, were showing considerable interest. In 1976, during President Gerald Ford's administration, Congress passed the Toxic Substances Control Act, which requires the testing of chemicals that the EPA administrator finds "may present an unreasonable risk of injury to health or the environment."[143] The statute also requires the administrator to prohibit the production, processing, or distribution of chemicals where the administrator "finds that there is a reasonable basis to conclude that the manufacture, processing, distribution in commerce, use or disposal of a chemical substance or mixture, or that any combination of these activities, presents or will present an unreasonable risk of injury to health or the environment."[144]

In 1976 Congress also passed the Resource Conservation and Recovery Act, which amended the Solid Waste Disposal Act to require the EPA administrator to identify different categories of waste, particularly hazardous waste, and to refine a system for tracking these wastes from "cradle to grave," or from their generators to their transporters to their treaters, storers, and disposers.[145] This law was in good part the result of a series of case studies of the hazardous waste problem done by William Sanjour, at that time in the EPA's Hazardous Waste Management Division. His study of Missouri's dioxin troubles was a particular spur to congressional action. Missouri enacted its own hazardous waste management law in 1977, also moved to action in significant part by its dioxin experience.

This was not only a year of legislative activity; 1976 was a year for settling lawsuits. Independent Petrochemical, the company that had contracted with NEPACCO to remove its waste and had subcontracted with Bliss, got hit the hardest. Piatt and Hampel settled their suit against Bliss for $10,000 and against Independent for $100,000. Independent later paid (in 1983) each of Piatt's daughters $1 million, and in 1981 Piatt and Hampel settled for $65,000 from NEPACCO.

Meanwhile, Syntex Agribusiness and its committee of experts finally identified, after five years of searching, a way safely to dispose of the dioxin-contaminated waste in the Verona tank. In 1979 a California waste-management company developed technology that relied on the sun's ultraviolet rays to degrade dioxin. With guidance from the California company, Syntex built a photolysis apparatus and in May 1980 began running the waste through it. By August 1980, within thirteen weeks of around-the-clock operation, all 4,300 gallons of waste in the tank had been treated, and thirteen pounds of dioxin had been more than 99 percent destroyed.

Although state and federal legislators were responsive to the problems surrounding hazardous wastes, the agencies responsible for implementing these laws often took years to do so. Thus, Bliss Oil Salvage continued hauling and spraying waste oil until July 1, 1980, when regulations implementing the Missouri law prohibited Bliss's way of disposal. How and when and to what extent and with what results these various laws were implemented is another story. The same is true for the implementation of the Comprehensive Environmental Response, Compensation, and Liability Act (CERCLA) of 1980, through which Congress instructed the EPA to identify hazardous waste sites and created a multibillion-dollar Superfund to clean up sites where no individual or corporation could be made liable for the pollution (see Chapter 5 for an analysis of the implementation of CERCLA).

How the EPA Tracked Wastes to Disposal Sites

The EPA became attentive to Missouri's dioxin problems when, on October 16, 1979, a former NEPACCO employee told EPA staff about drums of dioxin-contaminated waste that had been buried on the James Denney farm, about seven miles from Verona. NEPACCO paid Denney, a former Hoffman-Taff employee, $150 for the use of his property. Under the EPA's direction Syntex eventually unearthed ninety drums, eleven containing still-bottom waste with dioxin concentrations as high as 2,000 ppm. Under the terms of a settlement agreement, Syntex also paid $100,000 of the EPA's $450,000 response costs. A 1986 court decision interpreting the new Superfund law later allowed the EPA to seek recovery of the balance from NEPACCO. More sweepingly the court held that liability could be imposed for the cleanup of hazardous wastes dumped prior to the passage of the law,[146] a retroactive application that was not spelled out in the CERCLA statute.

After discovering drums of dioxin-contaminated waste at the Denney farm, EPA investigators followed up other leads in southwestern Missouri, finding six more sites where drums were stored or where NEPACCO filter clay had been spread by farmers who believed it would prevent hoof rot in their cattle. Four of these sites were farms in the Verona area. The generally high levels of dioxin found almost a decade after disposal led the investigators to retest the soil from the horse arenas in eastern Missouri. The results confirmed their fear that dioxin had a half-life of much more than one year; it was not degrading at a rate that would render it harmless in short order. In May and June of 1982 EPA scientists took samples from the Minker and Stout properties, from the Shenandoah, Bubbling Springs, and Timberline stables, and from the Bliss farm. Results were announced on August 18: the highest level of dioxin found was 1,750 ppb in some areas at Shenandoah Stables; the highest level on the Minker property was 300 ppb.

The EPA and Missouri officials asked the owners of the stables to close them voluntarily and advised the people living on these properties to avoid exposure to the soil. (By this time the two trailers on the Stout property were unoccupied, but the Minker site had expanded to include six homes.) Regional EPA officials wanted to begin cleaning up the Minker and Stout properties, but Rita Lavelle, an assistant administrator at EPA headquarters in Washington, felt that more data were needed to determine the extent of contamination. In October 1982 the EPA announced that it would be taking six hundred additional samples, which sent the resale value of the homes plummeting while raising residents' fears about their long-term health risks. Yet the EPA was not able to tell them what level of dioxin contamination could be considered safe or what the health consequences of exposure to existing levels might be.

Missouri's Dioxin Troubles as National News

In the fall of 1982 these various Missouri dioxin sites received the sustained attention of the national news media after the Environmental Defense Fund publicized a list of fourteen confirmed and forty-one potential sites (leaked by government sources). The fund also publicized EPA memos showing that the agency was considering relaxing its dioxin cleanup standards in the state. The fears, frustrations, and feelings of betrayal of those who lived near these sites also found their way into the national media via public hearings organized by government officials. Why had they not been told about the presence of dioxin, about what it was, about what it had done to farm animals and to children who had played in contaminated soils? As one neighbor of the Stout site lamented at a hearing: "Our children have been playing in the woods, picking wildflowers, moving rocks. All this for nine years. The Department of Natural Resources knew about it all these years. The EPA knew about it. What have you done to our families?"

The publicity had several effects. Judy Piatt and especially Russell Bliss became household names. Piatt was the featured guest at meetings and talk shows, where she told of her daughters' ailments and her horses' deaths. Bliss was pilloried. Residents of dioxin sites sued him for tens of millions of dollars. Illinois sued him for damages arising from contamination of a parking lot in Sauget where Bliss once had a waste storage tank. Missouri wanted to be compensated for the costs of investigation and cleanup. Missouri also refused to issue Bliss's son a license to haul hazardous waste. Bumper stickers in the St. Louis area read "Ignorance is Bliss." In his defense Bliss said, "They are crucifying me for something I did twelve years ago that wasn't against the law. Back in the '70s nobody knew what dioxin was, nobody knew what PCBs were. You could take whatever you wanted and spread it anywhere you wanted and nobody thought anything about it." Another effect of the publicity was that it generated tips about other potential dioxin sites. Verification required soil tests. The EPA promised that these would be completed in one to two years, but outcry over the wait led them to shorten the period to three to six months. Times Beach residents became so impatient with the delay that they hired a private company to test their soil.

Publicity also attracted the attention of elected officials. The House Energy and Commerce Committee held hearings in mid-November 1982 at which the EPA and the Missouri governor were accused of insensitivity and inaction. In her testimony Rita Lavelle defended the EPA's continued testing of alleged dioxin sites, saying that "this is not an emergency" and that she was "disappointed in the half-truths and half-information leaked by people trying to grab attention."

Lavelle was not alone in questioning the quality of the information reaching the public. In June 1983 doctors from Missouri sponsored a motion to have the American Medical Association (AMA) condemn news reports of dioxin's dangers. Approved by the AMA's 150 delegates, this motion led the group to state publicly that "the news media have made dioxin the focus of a witch hunt by disseminating rumors, heresay and unconfirmed, unscientific reports." The AMA pointed out that there was no evidence of long-term adverse consequences from low-level dioxin exposure. Their diagnosis? Bad journalism: "The lives and well-being of people in areas where dioxin contamination has been found have been unnecessarily and ignorantly damaged by this hysterical malreporting."[147]

A later careful analysis of print and television treatment of the health risks created by dioxin at Times Beach, Missouri, supports the AMA's perceptions. Almost half (46.5 percent) of the statements made by one of nine media sources characterized the risk as significant, requiring corrective action, while barely more than one in ten statements (13 percent) maintained that little or no risk existed. This finding comes from a Media Institute study of the handling of chemical risk information in reporting on EDB in grain products, dioxin at Times Beach, and numerous chemicals spilled by tanker cars in a Livingston, Louisiana, train wreck.[148] The Times Beach portion of the findings is based on a content analysis of coverage from November 1982 through March 1983 in twenty-six evening newscasts by CBS, NBC, and ABC and seventy-six stories found in the final editions of the *New York Times, Washington Post,* and *Los Angeles Times.*[149]

The Media Institute lays most of the blame for the misreporting of dioxin health risks on the mix of sources relied upon. In Times Beach reporting half (49.5 percent) the time (in newscasts) or space (in newspapers) was devoted to statements made by federal, state, or local government officials. And because, according to the institute, "government spokesman asserted that significant risks existed four times more often than they asserted low or no risks, chemical stories took on a tone that assumed the existence of risk—regardless of scientific debate on the subject."[150] More than a fifth (20.7 percent) of coverage was devoted to another source of distortion of health risks: citizens. In print and television coverage of Times Beach and the train wreck, three-fourths of citizen statements speculated on the existence of risk, and more than three-fourths (77 percent) of these statements "alleged that serious risk existed."[151] Moreover, as institute analysts pointed out, these speculative anecdotes probably had an impact greater than the already extensive time and space citizens commanded in coverage of Times Beach. Imagine the effect on viewers of citizen Carol Vickers telling a CBS reporter, "You don't know what it does to you to look at a boy whose nose bleeds and wonder if he's got cancer. You don't know, and it tears you apart."[152]

By contrast with citizen and government sources in Times Beach coverage, public interest groups, independent experts, labor, and industry sources each got less than 2 percent of the available space or time.[153] As analysts noted, "had media outlets given more prominence to scientific experts and informed industry sources, a very different storyline might have emerged," because of their less frequent claims that dioxin posed a significant health risk.[154] "Experts, for example, were only half as likely as government officials (by a 2:1 instead of a 4:1 margin) to assert that risks were significant. Industry sources went even further, with fully half of their statements asserting that risks were small or nonexistent vs. 8.7% that acknowledged significant risk."[155]

Interjecting more scientific data into the reporting on Times Beach without relying on a greater range of experts apparently would have done nothing to alleviate the distorted picture created by overreliance on government and citizen sources. While only 15 percent of coverage was "devoted to input from the scientific community," 83.7 percent of scientific information was used to support claims that a significant health risk existed.[156] It appears that scientists were selected to reinforce the danger message rather than to reflect accurately scientific opinion on dioxin. Why, for example, did the media fail "to put the minute dioxin levels in perspective or stress that dioxin-related illnesses in humans had not been found?"[157] The Media Institute concluded of Times Beach reporting that "(1) the media missed the opportunity to present the legitimate debate that was taking place among government, scientific, and industry experts; and (2) the coverage that resulted most probably overstated actual risk."[158]

Times Beach Sampled, Flooded

When the EPA wanted to begin testing sites where Bliss had disposed of waste between 1970 and 1976, fortunately the CDC had a record of Bliss's disposal from files he had given them before they were destroyed. Times Beach was one of these sites, and Times Beach was the site that the EPA decided to test first, because it was the most highly populated. Testing began on November 30, 1982. In white suits, gloves, and respirators, technicians scooped up their final soil samples a few days later, on December 3. The day after these "astronauts" left, the

Meramec River breached its banks and rose to over twenty feet above flood stage, the worst inundation in Times Beach history. President Reagan subsequently declared the town and other flooded parts of Missouri disaster areas.

At about the same time, the results of the latest round of soil tests at the Minker and Stout properties were being analyzed by the CDC and the EPA. Lavelle interpreted the results to support her original assessment that their dioxin levels did not pose an emergency. Her assistants at the EPA and CDC officials, however, felt that there was an emergency. Debate centered on whether immediate evacuation was necessary and what level of decontamination should be attempted. The EPA had earlier required a New York dump to reduce its dioxin levels to 10 ppt. Lavelle's private science adviser suggested 100 parts per billion (ppb) was a safe level. EPA scientist Renate Kimbrough proposed that 1 ppb would be adequate, but her EPA colleagues challenged even this extreme cleanup goal. Under their worst-case scenarios this level would keep the risk of additional cancers above the 1-in-1-million additional cancers the EPA considered acceptable. Kimbrough therefore proposed to diminish the 1 ppb even further by adding clean dirt to that which had been decontaminated. On December 8 Lavelle reluctantly approved this level of cleanup, on the condition that the standard only apply to the Minker and Stout properties. She also offered to relocate the residents of the Minker site at government expense while the cleanup proceeded.

Permanent Relocation of Residents

After the floodwaters receded, some Times Beach residents returned to make repairs and reinhabit their homes. Meanwhile, results of the various soil tests were coming in. Those commissioned by Times Beach residents established that dioxin was present, which was confirmed by the CDC and the EPA. Potholes and yards along town roads contained dioxin in excess of 100 ppb. The CDC advised the Missouri Department of Health that "residents who have been temporarily relocated (because of the flood) are discouraged from moving back into the area. Residents who have already begun to move back into the area are encouraged to leave." The department publicized this advice along with the EPA test results. Many residents got the news on December 23 at the annual Christmas party in city hall.

The government offered housing to Times Beach residents while it figured out what to do about the flooding and the dioxin. The various agencies that now had jurisdiction over Times Beach were sending mixed signals. The Army Corps of Engineers announced a plan to rebuild Times Beach elsewhere. Lavelle told the Missouri congressional delegation that Times Beach had only a flooding problem, pending further soil tests. White House staffers had been tracking Missouri's dioxin problems since the fall of 1982. Reagan was briefed on Times Beach at least twice during the first week in January, and on January 7, 1983, he announced the creation of a presidential task force on Times Beach, with representatives of the EPA, CDC, Federal Emergency Management Agency (FEMA), and Army Corps of Engineers. Reagan chose Lee M. Thomas, FEMA's deputy administrator, as the task force's chairman.

That same day Thomas talked with the Missouri governor's aides late into the night, and the next day he flew to Missouri, where, accompanied by the governor and other officials, he went to Times Beach. Thomas promised to provide housing and help pay for the removal of flood debris. Back in Washington he found that the primary interagency problems were between the EPA and the CDC; for example, they could not agree on soil testing procedures, which resulted in repeated collection of samples and repeated testing of soils. Thomas finally achieved a coordination of activities by holding daily conference calls with representatives from both agencies.

On February 4, 1983, Rita Lavelle was fired. Her boss, EPA administrator Anne M. Gorsuch, agreed to a buy-out of Times Beach if the CDC and FEMA recommended it. On the weekend of February 20 the task force received the CDC's recommendation that residents leave Times Beach permanently: "In fairly short order," Thomas recalled, "it became obvious that relocation was the only option." Several factors contributed to this conclusion. Among them were the CDC's recommendation, the town's location on a flood plain, the greater cost of a temporary as compared with a permanent relocation of townfolk, the EPA's lack of disposal methods for the large amount of contaminated soil, and the cost of decontaminating it, assuming this could be done. On February 22 the EPA administrator, now Mrs. Anne Burford, announced in Eureka, Missouri, that the federal government would pay $33 million or 90 percent of the estimated $36.7 million cost of buying out the residents and businesses of Times Beach. The state would pay the remainder. Two days after this announcement, Thomas was appointed to Lavelle's former position.

The Buy-Out Fallout

Lavelle remained critical of the buy-out. "Government has really hurt those people," she maintained. "The cost in terms of psychological trauma we cannot measure." There were other critics as well. Among them were Donald G. Barnes and Paul Brown, cochairmen of the EPA's Chlorinated Dioxins Work Group, which Brown claims was left largely unconsulted about the Minker and Stout properties and Times Beach. Because the dioxin contamination of Times Beach was confined to its dirt streets and their shoulders, Barnes believed that paving and "probably widening the streets and building some concrete paths would have done the job." Burford, who was fired along with many of her deputies several weeks after the buy-out, insisted that it "was the best decision we could make given the facts as we had them." Thomas in retrospect seemed to hedge a bit on the "only option." "What I was dealing with over at EPA was a group who didn't know what to do with dioxin. Knowing what I know now, I would have pushed EPA harder to use all of its resources." "The last thing you want to do is disrupt someone's life—and that's what you do with a buy-out."

All in all, 800 Times Beach families, 2,041 residents, were relocated. Some families were forced to move twice as the EPA continued to follow Bliss's dioxin trail through Missouri. The dirt streets of the Quail Run Mobile Manor near Gray Summit turned out to have dioxin concentrations of 1,100 ppb, and the EPA offered to relocate the 28 families living there. Five of them had just been moved to Quail Run from Times Beach. "We've lost something—we've lost that security, that security of a normal, healthy life," said Dan Sharp of his wife and their three children. "Kathy and I used to tell ourselves that this was ours, something nobody could take from us. If we ever left Times Beach, we'd save this house to retire to." Pointing to a problem that could only be resolved by better scientific knowledge, Sharp said, "What bothers me is that there are no answers, no solutions. We have no information about the health effects of dioxin that someone isn't contradicting. And we are caught in the middle." His wife, who was pregnant with the couple's fourth child, shared this concern. "I'd like to know if dioxin is in my body, if it will affect my baby. I'd like to know if the dioxin will be in my milk for my baby. Who will find out for me?"

Studies of potentially exposed people from Times Beach and other Missouri dioxin sites have discovered no serious adverse health consequences that can be linked to dioxin. The first study (done in 1983) of 68 Missourians with the highest potential exposure found no greater amount of disease or disability among them than among a control group of 36 state residents with little or no chance of exposure. "Of note is the fact that no cases of chloracne or soft tissue sarcomas were detected," wrote the CDC's Paul A. Stehr-Green.[159] The design and findings of a second study are particularly salient. Of the nine residential sites sprayed by Bliss, Quail Run had the highest dioxin levels, substantially higher than Times Beach, which is why Quail Run residents were selected for study. Dioxin was detected along the entire length of the road that ran through this trailer park, in concentrations ranging from 39 to 1,100 ppb. More than 1 ppb were found along the road shoulders, in four of eight yards tested and in dust taken from inside twenty-one of thirty-one mobile homes tested. If dioxin had adverse health effects, they should show up among Quail Run tenants.[160]

The researchers classified as exposed anyone who had lived in Quail Run for at least six months between Bliss's spraying in April 1971 and the EPA's cleanup in May 1983. Of the 207 households estimated to have lived in Quail Run during this period, 95 could be identified; in these households 154 of 207 people agreed to participate in the study. The control group was 155 persons who had lived in one of three dioxin-free trailer parks for at least six months.

The exposed and unexposed groups were comparable with respect to age, sex, and race; tobacco and alcohol use; use of pesticides, wood preservatives, and professional herbicidal services; and history of employment that involved contact with chemicals, electrical transformers, capacitators, or the incineration of plastic or wood materials. (Wood preservatives and incineration are possible sources of dioxin exposure; electrical transformers and capacitators may contain PCBs, chemical cousins of dioxin.) The only significant difference between the groups was the lower educational and socioeconomic level of the exposed group. Any explanation of differences in health or subclinical condition, therefore, was effectively narrowed to dioxin exposure, educational attainment, or socioeconomic status.

As the researchers discovered there were only minor differences in health and subclinical condition to explain. Across the two groups the self-reported health histories—verified by medical reports—were remarkably similar. "No cases of chloracne, porphyria cutanea tarda (PCT), lymphoma, sarcoma, or cancer of the liver were reported."[161] "No differences were found between the exposed and unexposed groups in the frequency of reproductive disorders or adverse pregnancy outcomes, such as fetal deaths, spontaneous abortions, and children with congenital malformations." "No difference between groups was observed in the proportion of participants who sought medical care because of numbness or headaches." And "there was no difference in the proportion of each group reporting prolonged infection."[162]

A greater proportion of the exposed group did report numbness or "pins and needles" in the hands and feet (29 percent versus 18 percent of the control group) and persistent severe headaches (26 percent versus 14 percent). These statistically significant differences remained even after the researchers controlled for psychological stress, age, sex, and socioeconomic status. Physical examination also revealed that to a statistically significant degree more members of the exposed group (16) had some kind of dermatitis than did members of the unexposed group (2). Significantly more exposed persons were anergic than unexposed persons, which meant that they did not respond to any of seven antigens placed on their skin, including tetanus, diphtheria, and tuberculin. Of exposed persons 12 percent (compared with 1 percent of unexposed persons) were anergic. Of exposed persons 35 percent (as compared to 12 percent of unexposed) responded to no more than one of the seven antigens, making them relatively anergic.[163] The researchers suggested that these individuals be tracked to see if the apparent impairment of their immune system persisted.

One to one and a half years later 28 of 50 exposed people and 15 of 27 unexposed people agreed to be retested. These study participants did not report any new diseases, nor did any new disabilities show up on physical examinations. Most strikingly, on retesting none of them was anergic; that is, all of them—exposed and unexposed previously anergic persons—responded to at least one of the seven antigens applied to their skin. And only one exposed and one unexposed person were relatively anergic. The researchers suggested several possible reasons for these results but did not believe that recovery from the effects of dioxin exposure could explain them, since previously anergic persons in the unexposed group had also "recovered."[164] In a subsequent study done in 1989 of the health of 41 Missourians with known body levels of dioxin, "no anergy was found, even in those subjects with the highest body levels of TCDD."[165] In fact, on average, the participants' immune system responded to six of the seven antigens. "As in previous studies, no evidence of clinical immunosuppression was found." Moreover, "adipose tissue levels as high as 750 ppt were not associated with adverse clinical health effects." These findings are important because the study participants were classified as having been exposed to dioxin not simply because of their residence near a sprayed road but because dioxin-containing fat had been surgically removed from their bodies. Consequently we know that people with high dioxin levels still have good health and that there was no noticeable difference in health between those with the highest and lowest adipose levels.

Was the Evacuation Unnecessary?

Such studies, along with others undertaken in the decade since the Times Beach buy-out,[166] have moved the CDC to recant its earlier evacuation recommendation and have caused the EPA to reevaluate its risk assessment methodology for dioxin. In May 1991 Vernon Houk, director of the CDC's Center for Environmental Health and in 1982 the official responsible for recommending the relocation of Times Beach residents, said: "Given what we know now about this chemical's toxicity and its effects on human health, it looks as though the evacuation was unnecessary. Times Beach was an overreaction. It was based on the best scientific information we had at the time. It turns out we were in error."[167]

Houk blames the EPA's "exaggerated risk models," which were widely used at the time of his evacuation recommendation, for this "error." "Many scientists now understand that loading an animal with a chemical for a lifetime, then counting tumors and feeding a mathematical extrapolation model does not necessarily predict the chemical's potential for causing cancer in humans." "Scientists now understand that the laboratory rodent is not a small human." Data accumulating from studies of dioxin's health effects (or rather, lack of them) on humans have led Houk to conclude: "If it's a carcinogen, it's a very weak carcinogen and Federal policy needs to reflect that."[168]

Some media reported the new scientific consensus that dioxin was not dangerous at the low doses possibly experienced by residents of Times Beach. Thus the *St. Louis Post-Dispatch*, which in 1983 ran a story under the headline "EPA Calls Dioxin Most Potent Material," in 1991 reported that Houk recanted his own assessment of the risk in the story "Dioxin Scare Called Mistake." The

Post-Dispatch was joined by the *Chicago Tribune,* which printed "On 2nd Thought, Toxic Nightmares May Be Unpleasant Dreams," and the *Los Angeles Times,* whose headline read "Dioxin Joins List of Costly False Alarms," while the *New York Times* reported that "the threat from dioxin had been downgraded from cataclysmic to slight or even nonexistent."[169] But not all the major media tracked these changes. As Accuracy in Media editor Reed Irvine observes in a *Wall Street Journal* opinion piece headlined "The Dioxin Un-Scare—Where's the Press?": "Dr. Houk's turnabout was reported by the *St. Louis Post-Dispatch* under a front-page banner headline, but it got little attention in the East. ABC News reported it; CBS and NBC did not. The newspapers that influence those in Washington who could bring the costly Times Beach boondoggle to a screeching halt buried a small AP story deep on their inside pages."[170]

Asked what he would tell Times Beach residents, Houk replied, "I would tell them when we originally got involved with you, we were very concerned about the substance. But the data we've accumulated in the last 10 or 12 years show we should not be that concerned." "We should have been more upfront with the Times Beach people and told them, 'We're doing our best with the estimates of risk, but we may be wrong.' I think we never added, 'But we may be wrong.'"[171]

Houk's new understanding of dioxin's health effects is in part the result of over $400 million spent by the federal government alone on dioxin research since the Times Beach evacuation. While the fears of former residents about the long-term consequences of dioxin exposure may be eased by Houk's announcement, the leveling of Times Beach and the incineration of its soil is proceeding as planned. Turning the town into a state park is supposed to take ten years and $120 million, $100 million of which is being paid by Syntex Agribusiness. Neither the company nor the government is eager to reverse a decision so painfully taken. "We would even more confuse and alarm the population around Times Beach if the decision was suddenly made that you didn't have to do anything," says Houk. Syntex's Gary Pendergrass agrees: "I don't think there's room for reconsideration."[172]

New Conclusions on Doses and Cancer

In 1991 Marilyn Fingerhut and her colleagues at the National Institute for Occupational Safety and Health published the results of their nearly thirteen-year study of American men exposed to dioxin on the job. They began by identifying U.S. chemical companies that had produced dioxin-contaminated products between 1942 and 1984. At the twelve companies that fit this description, they came up with 5,172 workers, essentially all the occupationally exposed people in the country.[173] Of these 5,172 men, 265 died of cancer, 15 percent more than expected, given cancer-caused death rates among unexposed American males.[174]

A separate analysis of a 3,036-man subgroup with at least twenty years latency since first exposure revealed that most of these additional cancer deaths were concentrated among those who had one year or more of exposure. Those with a twenty-year or greater latency but less than a year of exposure had 48 cancer-caused deaths when 47 were expected. In other words, lesser exposures had no cancerous consequences when these were given plenty of time to manifest themselves. To make sure the high and low exposure classifications correlated with true exposure levels, Fingerhut and her colleagues took blood serum from a sample of 253 workers at two plants. Their analyses of the blood lipids indicate that the higher cancer rate observed among the highly exposed is associated with dioxin doses estimated to be 500 times higher than those received by the general population. By comparison, the group with low exposures and no effects was exposed to dioxin doses estimated to be 90 times higher than those received by the general population.[175]

The Current Dioxin Debate

Dioxin-related developments these days run in a number of directions, all of which serve to obscure the truth that has emerged from the American dioxin experience: this chemical has serious health effects only at extremely high doses. Only those with occupational exposures have anything to worry about, and even for them, elevated cancer rates mean that a few (not most nor all) will develop cancers who otherwise would not have.

One recent and initially promising development was the apparent consensus among researchers that all of dioxin's health effects are mediated by a particular cellular receptor. Moreover, it appears that a number of dioxin molecules are needed to occupy a receptor and a number of receptors are necessary to cause adverse health effects. This discovery implies a threshold for dioxin health effects: below a certain level, little or no harm would occur; thus some body level might be harmless. These findings have led the EPA to begin to reevaluate the way it estimates the risk of dioxin exposure. It did its own analysis of receptor molecular mechanics and concluded that there is no general threshold for dioxin health effects because receptors function differently in different tissues and because, in mice, dioxin gets deposited at different rates in different organs.[176] These are important results for mice and may even give scientists some idea of what effects to look for in humans, but this focus on receptors and thresholds hides from us what we already know: that dioxin causes serious human health effects only at high doses. The is-there-or-isn't-there a threshold debate revolves around low dose effects, and we know that whatever these may be they are not serious.

The same observation could be made about a couple of other dioxin-related developments. One is the hypothesis that dioxin's hormonelike behavior may result in immune system suppression.[177] This hypothesis implies that dioxin's resemblance to a hormone is an unexplored avenue for scientific research, with many questions to be answered before we can know the effects of this newly emphasized characteristic. Scientists actually have known that dioxin acts like a hormone for quite some time, and again, the hormone hypothesis directs our attention away from what we already know: if dioxin's alleged immunosuppression has serious consequences, they will only occur at high doses and they will be none other than those already discovered, whatever the molecular mechanics of their occurrence.

Another recent development is the detection of dioxin in myriad new places. Most incineration, including automobile combustion engines, produces dioxin. It is also found in paper products as the result of chlorine bleaching of wood pulp—a discovery that has resulted in attempts to measure the dioxin content of milk cartons and tampons. These findings are partially the consequence of advances in technology that allow detection of dioxin at levels as low as parts per quadrillion. They are also the result of a seek-and-you-will-find orientation, exemplified by the introduction to a review of research on dioxin as an "environmental hormone": "When a villain starts looking like a friend, it's time to look again."[178] "Looking again" explains the search for new mechanisms for creating effects, for new effects themselves, for dioxin in ever lesser amounts, for dioxin in new places.

It also explains the search for new chemicals that act in "dioxinlike" ways. The EPA has turned what looked like a friend—a single receptor mediating all dioxin's effects, implying a threshold—into an enemy. Instead of trying to identify a threshold for dioxin effects, the EPA is examining the other chemicals that bind to dioxin's receptor and assigning "toxic equivalency factors" to them. In other words, all chemicals whose effects are mediated by the dioxin receptor are being treated like dioxins for purposes of setting safe exposure levels. This classification elevates by definition the amount of dioxin exposure in the population so as to bring it within those high levels that are known to cause serious injury.

Such an expansive approach pursued by the EPA seems to be an indirect way of acknowledging the truth: dioxin itself does serious human damage only at high doses. Yet dioxin and herbicides are regulated as if low doses pose a significant threat. The hundreds of millions of dollars spent on scrubbing air, soil, and water free of only recently perceptible particles of dioxin could be better utilized. When we add the hundreds of millions of dollars spent on researching insignificant dioxin health effects and the hundreds of millions in costs to companies (and consumers) trying to avoid producing the particles that have these insignificant effects, public health has been harmed. We must not forget the hundreds of millions of dollars paid to people who were not injured, nor the hundreds of millions spent regulating inconsequential exposures. Why expend so many resources in the name of public health with so little to show for it?

4

Love Canal: Was There Evidence of Harm?

with Michelle Malkin

"It has all the characteristics of a terrific neighborhood."
—James E. Carr, Planning Director, Love Canal Revitalization Agency, July 26, 1990

Love Canal, New York—the infamous "disease cesspool," "public health time bomb," "toxic ghost town"—has now been resettled. Nearly 1,000 families were evacuated from this Niagara Falls neighborhood following two federal declarations of emergency in the summer of 1978 and the spring of 1980. Huge sums (we could not find a reliable estimate) have been spent on research, remediation, relocation, and litigation.[1] What prompted the costly exodus from Love Canal? And why, over a decade later, has the "revitalization" of the area attracted a waiting list of more than two hundred names to move back in?[2]

Love Canal has long been America's "synonym for chemical tragedy."[3] A search for evidence will open the question of what kind of tragedy it was.

Love and Hooker

The Hooker Chemical Company has been cast as the chemical Goliath in the Love Canal saga.[4] Now a subsidiary of the Occidental Chemical Corporation, Hooker's Niagara Falls plant manufactured pesticides and petrochemicals during World War II. The company initiated feasibility studies in September 1941 to determine the suitability of using an unfinished canal not far from downtown Niagara Falls for disposing of waste chemicals.[5]

The canal's namesake, entrepreneur William Love, had abandoned the site in the early 1900s because of unforeseen economic and technological developments. Love had partially dug a canal to Niagara Falls that was to provide cheap hydroelectric power for a proposed "model city." A recession and the advent of alternating current (which eliminated the need for proximity to the falls) ruined Love's plans. The sixteen-acre tract of land was located in an undeveloped, sparsely populated area. Hooker noted that the canal site was ideal for the disposal of chemical residues, since by design it was built to retain water. Also appealing were the canal's relatively impermeable clay walls.

In 1942 Hooker obtained permission to use the site for disposal and subsequently acquired a 200-foot strip of the property with the canal at its center.[6] Chemicals were placed in large drums, hauled to the site, and buried under several feet of clay material.[7] Tons of chemicals (21,800) were disposed of in this manner until 1953, when Hooker ceased using the canal.

In April of that year the Niagara Falls Board of Education purchased Hooker's land for $1 under threat of eminent domain.[8] Despite numerous written and verbal warnings from Hooker officials, the Board of Education built a school atop the canal. City officials proceeded to develop the neighborhood, and built 100 houses directly next to the canal and another 139 across the street.

Not until the late 1970s would Hooker's name resurface among Love Canal locals. After six years of abnormally heavy rains and flooding, according to residents, the canal overflowed its banks in 1976.[9] Anecdotes about children being burnt, nauseous odors, and bubbling black sludge spread rapidly.[10] After all these years, it was rumored, Hooker's long-buried chemical stew had come back to haunt the Love Canal neighborhood.

In response to residents' complaints, the New York State Environmental Conservation Department began environmental testing on air, surface water, and samples from sump pumps in basements near the Love Canal site. The agency detected low levels of some eighty industrial chemicals, including one known human carcinogen and eleven known or suspected animal carcinogens.[11] Residents' fears were validated—seepage *had* occurred indisputably. Not long afterward, in 1977, Hooker officials began their own study of chemical leachates discovered in residential basements. The company collaborated with Niagara County officials and agreed to pay a portion of remediation costs to collect and treat the leachates.[12] Even though its own storage methods were not directly responsible for the leakage, Hooker sent engineers to trench around the canal. To complete the remediation project, they built a drainage system of tiles and wells to lower the water table.

In August 1978 the Environmental Protection Agency (EPA) conducted environmental monitoring of storm sewers associated with Love Canal. A wide variety of chemicals—mostly chlorinated dibenzodioxins—was detected at six manhole sites. But no individual compound was measured at a toxicity level actionable under the Clean Water Act. That means that although a large quantity of chemicals was present at the canal site, none was found near homes in an amount threatening toxicity. The water was safe to drink. Nor did scientists find that discharge from Love Canal significantly contributed to the contamination of nearby Lake Ontario. The remediation efforts had worked; problem solved, right?

Wrong. Thirty-five years after Hooker yielded ownership of the landfill a federal judge ruled that its parent company, Occidental, produced and stored wastes in a way that would eventually result in toxic leakage. The ruling supported claims that "the Love Canal was simply unfit to be a container for hazardous substances, even by the standards of the day."[13] Yet the "standards of the day" were minimal. As Eric Zuesse has pointed out, "The customary practices then were to pile up such wastes in unlined surface impoundments, insecure lagoons, or pits, usually on the premises of the chemical factory, or else to burn the wastes or dump them into rivers or lakes."[14] A private engineering firm, Conestoga Rovers, concluded that the design of the Love Canal site was well within the standards of the contemporary Resource Conservation and Recovery Act (RCRA).[15] Even William Sanjour, chief of the EPA Hazardous Waste Implementation Branch, commented in 1980 that "Hooker would have had no trouble complying with [today's] regulations. They may have had a little extra paperwork, *but they wouldn't have had to change the way they disposed of the wastes.*"[16]

Hooker is by no means an angel of chemical waste disposal. The company has admitted to improper management of a site in Montague, Michigan, and has paid for cleanup costs and remedial work at two other sites.[17] But in the case of Love Canal, Hooker's actions were both proper and responsible. Not only did the company repeatedly warn the Board of Education of potential hazards, but migration of chemicals was caused by both natural (rains) and unnatural (Board of Education and city development) disruption of the clay cap over the site not of the company's doing.[18] Yet over three decades after deeding the canal to the Niagara school board, Occidental and Hooker would be required to pay $20 million to 1,328 Love Canal residents for health and mental damages incurred by the leakage of chemicals.

The highest award, $400,000, went to Karen Schroeder's daughter, Sheri, who had birth defects and learning disabilities.[19] What was the link between Sheri and Occidental? Hooker chemicals had leaked from the canal—but as the historical record shows, it wasn't Hooker's neglect that caused leakage. In any case, did seepage unequivocally lead to exposure and thus birth defects?

Niagara County, where Love Canal is located, was home to thirty-eight different industrial waste landfills by 1978. County residents were exposed to an unknown amalgam of chemicals, pesticides, and other environmental pollutants in their daily lives. Assuming that a specific migration of chemicals could be traced along ambient, soil, or water routes from Love Canal to residents' homes, even the most current environmental scientific methods could not determine the degree and length of chemical exposure.

As Dr. Renate D. Kimbrough, a scientist with the Centers for Disease Control, points out: "Little is presently known about the effects of multiple chemical exposures, nor is it known whether chemicals measured in soil at Love Canal are actually bioavailable to humans."[20] In plain language, that means that Sheri's birth defects could not be traced to Love Canal. The intracellular concentrations and biological half-lives of Love Canal chemicals are not known with any certainty—nor can the diffusability of the chemicals and their transport to Sheri's cells be established.[21]

Hooker's neglect did not cause seepage from Love Canal; seepage did not necessarily mean exposure. We now turn to several studies conducted on Love Canal residents to test further the claim that exposure to Love Canal meant not only birth defects but also cancer and ultimately death.

Fishing for Damaged Chromosomes

Damage to Chromosomes Found in Love Canal Tests
  —*New York Times,* May 17, 1980

Homeowners at Love Canal Hold 2 Officials until FBI Intervenes
  —*New York Times,* May 20, 1980

Carter Declares Federal Emergency at Love Canal; Evacuations Begin
  —*New York Times,* May 21, 1980

On January 18–19, 1980, the Biogenics Corporation began chromosome testing on Love Canal area residents for the EPA. Because chromosomes are the DNA-containing carriers of hereditary characteristics found in the nucleus of every cell, damage to chromosomes would imply terrible harm. Blood samples were taken from thirty-six residents, analyzed at the firm's Houston laboratories, and compiled by Biogenics' scientific director, Dr. Dante Picciano, for the "Pilot Cytogenetic Study of the Residents of Love Canal, New York." The results were telephoned to the EPA in mid-May, but before peer review of Picciano's work could be initiated, the test results were leaked to the press.

On May 17, 1980, the *New York Times* jumped the gun. "Sources familiar with the study," the newspaper reported, said that eleven of the thirty-six test subjects had "exhibited rare chromosomal aberrations."[22] What exactly did that mean? The article offered only sweeping generalizations from unidentified scientists who linked the alleged chromosome damage to birth defects and cancer.

With the media alarm bells ringing, the EPA was forced to announce the findings in an attempt to control misinterpretive damage by the media. At a noon press conference on May 17, the day the *Times* story broke, EPA officials formally announced that, based on findings of chromosomal abnormalities in eleven residents, it might be necessary to relocate some of the residents who lived in the Love Canal neighborhood.

The announcement backfired. Instead of nurturing a cautious attitude about the preliminary findings, hysteria prevailed. Barbara Quimby, one of eleven residents informed by the EPA that she possessed "chromosomal aberrations," provided great drama for the national media. She was the housewife who blocked two officials from leaving the office of the Love Canal Homeowners' Association (LCHA), rode away with police after the FBI agents "rescued" the "hostages" (only to be dropped off a few blocks away), and ran back to the LCHA office sobbing that she "just wanted to be a housewife" again.[23]

The study that gave rise to this episode in the Love Canal melodrama was subsequently reviewed by peer panels and individual scientists. Unresolvable differences surfaced over Picciano's identification of chromosomal damage, lack of suitable controls, and interpretation of uncertain epidemiological evidence. Follow-up studies have cast great doubt upon the study's design.

CYTOGENETICS 101: A BRIEF INTRODUCTION

Before we examine the Picciano findings, it is necessary to have some information on cytogenetics, the study of heredity through the methods of cytology and genetics (the studies of cells and of genes). Genes are hereditary units that carry instructions essential for the life of a cell. Cytogenetic analysis is concerned with the structure, number, function, and movement of chromosomes, which carry genes.[24]

Every human has twenty-three pairs of chromosomes. Through karyotyping, cytogeneticists are able to classify all of the forty-six chromosomes and to identify breaks and other anomalies. Photographs of white blood cells (lymphocytes) are taken, fixed, stained, and arranged on a slide for microscopic examination. Damage appears in the karyotype in the form of aberrations or "sister chromatid exchange" (SCE), which refers to the individual strands of the chromosome structure. A chromosomal aberration involves the loss, duplication, or rearrangement of genetic material. Breaks can occur spontaneously in nature or as a result of mutagens—physical or chemical agents that change the structure of genes. Anything from viral infections to natural radiation to marijuana may be mutagenic. The entire chromosome may break, causing phenomena such as "marker chromosomes." Or chromatids may break; in SCE, for example, chromatid "legs" are broken and exchanged.[25]

These low-level mutations occur spontaneously in all living things. The alterations are difficult to pin down to any specific cause: random environmental effects, copying errors, or radiation, for example, may cause the breakage of strands or fragmentation of genetic material. Finally, white blood cells (which Picciano used to examine the chromosomes of the Love Canal subjects) are not likely themselves to give rise to cancer, and they cannot contribute to birth defects because they are neither sperm nor egg cells (only these sex cells can contribute directly to birth defects).[26] If chromosomal aberrations were so severe as to be detectable, the aberration would tend to lead to cell death and therefore would not be passed on to succeeding generations.[27]

WHAT PICCIANO DID AND DIDN'T DO

As stated earlier, Picciano drew blood samples from thirty-six test subjects. The samples were immediately transferred to a culture laboratory at the University of Buffalo, then treated, incubated, fixed, and stained. The cells were "pooled," which means that they were analyzed as a group. Data were not recorded on a per subject basis, so the average number of abnormal cells per individual in the group could not be analyzed. That was one small deficiency in Picciano's work. Even scientists sympathetic to Picciano, like EPA toxicologist Sidney Green, pointed out that the pooling of data weakened the study.[28]

Even more glaring—and serious—mistakes were made very early on by Picciano and his assistant, cancer researcher Beverly Paigen: in the selection of the test subjects and the lack of contemporary controls. How were the residents selected and with whom were they compared? Those questions bear heavily on the credibility of the findings.

Picciano arrived at his findings by conducting a pilot study, a low-cost endeavor used to determine whether studies on a larger scale should be conducted. To maximize the possibility of finding results, Picciano and Paigen sought out Love Canal residents who lived in wet areas, those in homes with air readings showing detectable levels of toxic chemicals, and women who had borne children with birth defects since moving to the area.[29] While it has been maintained that such a procedure is a "standard scientific" one, some scientists, like Robert S. Gordon of the National Institutes of Health, see in it a serious methodological flaw: "The Biogenics report selected the Love Canal residents in a way that would increase the chances of finding defects . . . residents who were chosen were *already known to have health problems* . . . that might be accompanied by chromosomal damage that may or *may not* have been caused by the toxic wastes at Love Canal."[30]

Further, Picciano failed to gather a contemporary control group with which to compare the Love Canal residents. Instead he used forty-four nonexposed "historical controls," or subjects that had been used in a previous study. Had the study been conducted in a "standard scientific" manner, cells of a suspect population would have been compared with those of a contemporary control population, the cells of both groups being cultured at the same time.[31] Controls would have been matched closely to exposed subjects in age, sex, medical history, smoking history, and geographical area (factors that can affect the number of aberrations). Finally, an attempt would have been made to ensure that only cells dividing for the first time in a culture would be scored; this is absolutely necessary to carry out a quantitative study of chromosome damage.[32] Picciano failed to take any of these measures.

He claimed that the EPA failed to provide him with the authority and the funds to obtain contemporary controls, "though he made every attempt to do so."[33] He has also said that since there was not enough time to select such controls for the four-month study, he was "forced" to take control data from other published studies on the background rate of chromosome abnormalities in normal, healthy populations.[34] Paigen, who selected the thirty-six residents to be tested, claimed she had also selected simultaneous controls but that the EPA rebuffed her.[35] For whatever reason, the study was conducted without such controls.

With these flaws in mind we can look at the main findings that so galvanized the public and the press. From the historical "control" group 8,800 cells were examined and compared with 702 cells from the thirty-six Love Canal residents. The following breakages and aberrations were scored.[36]

	Love Canal resident (%)	Controls (%)
Chromatid breaks	0.83	1.1
Chromosome breaks	0.55	0.35
Marker chromosomes	0.10	0.06
Supernumerary acentric fragments	0.20	0.00
Total abnormal cells	1.10	1.40

From these results Picciano observed an increase in the distribution of Love Canal residents with chromosome breaks (47.2 percent) as compared to the controls (40.9 percent). Though the frequencies of chromosome breaks and marker chromosomes were not significantly higher for Love Canal residents than for the controls, Picciano saw these results as indications that Love Canal residents might have an increased number of cells with aberrations.[37] Nowhere did he emphasize, however, that the overall number of abnormal cells in Love Canal residents was significantly lower than that of the controls (78 versus 123). Furthermore, all of these changes are within the normal variability that results from repetitive sampling of the same group.[38] (That is, several tests of a single group of cells might demonstrate the same type of quantitative changes that Picciano deemed significantly abnormal.)

What most alarmed Picciano—and thus the residents—was the existence of "supernumerary acentric chromosomes" (extra chromosomes lacking a specialized part called a centromere) detected in eight of the thirty-six Love Canal residents. In his "own experience with the examination of over 6,000 individuals," Picciano claimed, this condition should normally occur in only one out of a hundred individuals.[39]

But assuming this condition did exist in eight of thirty-six residents, it is still doubtful that cause for alarm was justified. Only 7 out of 7,000 pooled cells that were examined exhibited the condition, whereas only 3 would have been expected; the difference is not statistically significant. Moreover, it is estimated that each of us has approximately 5–6 million chromosomally aberrant cells in our bodies.[40]

Why did the EPA commission the chromosome study? EPA officials readily admit that it was less for science than for litigation—a fishing expedition to be used as a basis for the lawsuit against the Hooker Chemical Company. The suit, filed in cooperation with the Department of Justice in December 1979, sought $117.5 million in cleanup costs and $7 million as reimbursement for relocation costs during the first evacuations of August 1978. On May 22, 1980, a *New York Times* editorial assailed the study as the "scientific equivalent of a prosecutor's brief rather than a fully objective study"—after presenting it to the public as the latter.[41] From its genesis, then, the chromosome study was a hybrid of litigious intent and half-hearted science. The Justice Department and the EPA merely wanted proof that a health risk existed in the Love Canal area. Picciano provided it in his official summary to EPA: "It appears that the chemical exposure at Love Canal may be responsible for much of the apparent increase in the observed cytogenetic aberrations and that the residents are at an increased risk of neoplastic disease, of having spontaneous abortions, and of having children with birth defects."[42]

What became all too apparent, however, was the scientific evidence against this claim—and the doubts cast by outspoken officials from the New York State Department of Health (NYSDOH): "The study should be regarded as indeterminate."[43] "The findings did not necessarily mean that the chromosome abnormalities were directly related to current toxic exposures in the Love Canal area . . . hair dyes, occupational hazards, or smoking could be factors in the results."[44] Members of two scientific panels that reviewed Picciano's work made stronger negative statements:

[The study] provides inadequate basis for any scientific or medical inferences from the data (even of a tentative or preliminary nature) concerning exposures because of residence.[45]

[We do] not know whether the degree of chromosomal injury claimed, even if confirmed, is in itself a reason for alarmed predictions concerning cancers or congenital defects . . . Such a poorly designed investigation as this one should not have been launched in the first place.[46]

And cytogeneticist Margery Shaw, who agreed with some of Picciano's findings about possible chromosomal damage, was unwilling to make a conclusive assessment: "The results are neither positive nor negative because of the absence of contemporary controls."[47]

Scientists generally agreed that Picciano's study had come up empty. Despite the study's mild caution for "prudent" interpretation, it had produced an aura of causal relationship between toxic exposures and genetic damage on the one hand and genetic damage and disease and birth defects on the other. From tentative language such as "it appears" and "may be responsible" the study had moved to more assertive phrases like "are at an increased risk." It was never stressed—not in Picciano's report and not in the media—that on the average, a population with damaged chromosomes may have more cancer and more birth defects than otherwise expected, but the individuals in the population whose chromosomes are damaged are not necessarily those who will suffer these ill effects.[48] Picciano demonstrated "a total lack of understanding of what the chromosomes mean in terms of disease in the individual in which they occur and in future generations."[49]

As a tool for demonstrating health effects, chromosome studies are notoriously difficult to interpret.[50] Although aberrations can be an indicator of exposure in a population, they cannot be used to predict what will happen to a given individual.[51] Some scientists even doubt that cytogenetic findings give an informative measure of exposure: "The use of aberration studies [has] not been useful in monitoring for chemicals exposures," with the exception of SCEs.[52]

HOW TO DO IT RIGHT

In a follow-up cytogenetic study conducted by the Centers for Disease Control, blood samples were drawn between December 1981 and February 1982. Clark Heath and colleagues concluded that chromosome alteration frequencies were the same in Love Canal residents as in residents elsewhere in Niagara Falls.[53] In contrast to the Picciano pilot study, Heath's carefully designed study included geographic controls to identify varying levels of exposure of organic chemical levels as measured in basement air samples; screening of study participants by questionnaire; and screening for confounding factors such as smoking and recreational contact with the waste site.[54] In addition, control groups were matched with regard to socioeconomic conditions, sex, marital status, and age.[55]

The experimental group of Love Canal residents was divided into two subgroups: the first consisted of twenty-nine residents who lived in seven of twelve homes directly adjoining the canal in 1978 where air, water, and soil showed elevated levels of chemicals; the second consisted of seventeen residents who participated in Picciano's 1980 test.[56] In neither group did frequencies of chromosome aberrations differ significantly from control levels.

The control group consisted of forty-four matched subjects from a neighborhood north of the Love Canal where no dump site was located.[57] Controls were randomly selected and carefully matched. All subjects were screened by questionnaire for confounding factors, such as attending an elementary school that adjoined the site. A history of current cigarette smoking was significantly associated with increased SCE; none of the other factors—alone or in combination—were associated with significant increase in chromosome damage. Love Canal residents may have been especially interested in the findings on supernumerary acentric fragments. That condition was observed in 20 percent of the residents and in an even higher percentage of the controls (32 percent).

Heath and his colleagues concluded their report with a lengthy elaboration of the limitations of cytogenetic assays. They stated frankly that "although the presence of an increase in chromosome alterations could indicate acute exposure to chemical agents that cause chromosome damage, the absence of an increase does not establish the absence of such exposure."[58] Yet they observed, "Had evidence of increased chromosome damage been found, it would still have been impossible to know whether those findings might predict later clinical illness in individuals."[59]

Women and Children First

"I fear that my kids might be dying."
　—Patricia Sandonato, quoted in *New York Times*, May 18, 1980

"We want out. We want out."
　—chant adopted by Love Canal children

On August 2, 1978, New York State Health Commissioner Robert Whalen declared Love Canal "a great and imminent peril to the health of the general public."[60] Soon after the announcement the "general public" balked at one of Whalen's recommendations. He had singled out pregnant women and children under the age of two living nearest the canal and suggested that they relocate temporarily as soon as possible. Residents questioned why their health was any less threatened by the canal site than that of their children and childbearing counterparts.[61] Others were concerned about the devaluation of property that leaving immediately might cause. Their anger forced a widening of the scope of Whalen's recommendation from an anticipated 25-odd family members to nearly 240 entire families. Five days after the NYSDOH declaration, President Jimmy Carter proclaimed "a national emergency" and released Federal Disaster Assistance Administration funds to aid in the bailout. New York Governor Carey sanctioned permanent relocation for the affected families and promised to buy their homes.

The major impetus for evacuation was a preliminary report by NYSDOH which found that "residents living in the first ring of homes at the southern end of the Canal had a much higher than normal incidence of miscarriages (1.49 times the expected rate),"[62] and that among families living in the first two rings, there was a higher than normal frequency of birth defects. At the same time, however, there was no discernible relationship between the larger than usual effects and exposure to chemicals coming from Love Canal. NYSDOH's chief epidemiologist, Dr. Nicholas Vianna, who began health studies at Love Canal in June 1978, filed this preliminary report.[63]

Dissatisfaction with Vianna's methodology (and results) prompted Love Canal residents to enlist the aid of Dr. Beverly Paigen (who had worked with Picciano and since then befriended several of the Love Canal residents) to conduct their own study in the fall of 1978. Based on the recollections of older residents and aerial photographs, Paigen and members of the LCHA hypothesized that historically wet areas of the canal neighborhood, "swales," were carrying contaminants to houses through underground routes. Paigen contacted the NYSDOH as early as September regarding the survey, but the LCHA president, Lois Gibbs, leaked preliminary findings of increased miscarriages and birth defects among Love Canal neighbors to the press in November, prior to full analysis and peer review of the data and conclusions. Public furor after Gibbs's leak prompted a reevaluation by the NYSDOH of their data.

On February 8, 1979, New York State Commissioner of Health David Axelrod announced higher than expected rates of miscarriages, birth defects, and low-birth-weight infants in swale areas. The declaration was based on both the reevaluated Vianna data and the Paigen survey results. He authorized the evacuation of pregnant women and children under two years of age from the swale areas; a total of 100 families relocated.[64]

Both the Vianna and Paigen studies contained several self-recognized flaws. Inadequate controls and small sample sizes prevented either investigator from making clear-cut statements regarding the relationship between pregnancy outcomes and chemical exposure. Yet the two reports have become embedded in Love Canal lore as "incontrovertible," "distressing," and "shocking" proof of health damage caused by toxic chemical exposure.[65]

A "SWALE" IDEA

The first fundamental phase in conducting an epidemiological study is the documentation of the nature and extent of chemical exposure.[66] Such documentation is necessary in order to hypothesize about health effects. A specific mode of exposure must be identified to generate possible epidemiological implications. Both the Vianna and Paigen teams set out on geologic missions to identify possible routes of chemical migration. The LCHA-Paigen team gathered anecdotal evidence from older residents about the existence of dried-up underground streambeds that cut through the canal. These natural drainage ways apparently ran for long distances through the area, possibly carrying chemical leachates with them.[67] LCHA members also examined aerial photographs of their neighborhood and constructed a map of historic water areas. They placed circles on the maps to represent a host of self-reported illnesses including nervous breakdowns and urinary disease in addition to miscarriages, crib deaths, and birth defects. It was clear both to the LCHA and to Paigen that the geographical clustering of illness around the swales was directly related to the canal chemicals.[68]

While the LCHA and Paigen focused on one, and only one, hypothesis—that swales carried contaminants and contaminants caused illness—Vianna and his colleagues advanced several hypotheses. By considering more than one possibility of exposure, Vianna ensured a more accurate assessment of the probability that clusters might be due to chance.[69] He enlisted the aid of the Cornell University School of Civil and Environmental Engineering and also mapped out a route of swales. Based on additional considerations of housing development patterns and topography of the study area, they generated four possible explanations for the distribution of spontaneous abortions, birth defects, and low birth weights in specific areas of the study area (97th and 99th streets, adjacent to the canal; historically wet areas; historically dry areas):

1. An excess of some or all spontaneous abortions, birth defects, or low birth weights (referred to as "indicators") might be demonstrated only among pregnant females from 97th and 99th streets—where the likelihood of chemical contamination is greatest.

2. The entire study area (excluding the two streets adjacent to the canal) might have an excess of some or all indicators. If a gradient effect were detected by examination of each indicator by single street as distance from the canal increased, some general mechanism of contamination might be present.

3. An excess of certain indicators might be demonstrated in water areas, where chemicals might have migrated from the canal.

4. Regardless of geographic distribution, occurrence of indicators might be related to temporal factors and the nature and concentration of chemicals to which pregnant females were exposed.[70]

The results of the Vianna study demonstrated no unequivocal support for any one hypothesis. The report concluded. "We have not yet been able to correlate the geographic distribution of adverse pregnancy outcomes with chemical evidence of exposure. At present there is no direct evidence of a cause-effect relationship with chemicals from the canal."[71]

THE VIANNA REPORT

The New York State Department of Health surveyed the Love Canal neighborhood door-to-door over a seven-month period for three specific "indicators," or "endpoints": spontaneous abortions, congenital malformations, and low-birth-weight infants. Spontaneous abortions represented miscarriages before the twenty-eighth week of gestation and excluded induced abortions and stillbirths. Only those verified by a physician or by hospital records, or those occurring with evidence of a pregnancy and subsequent history compatible with the adverse outcome, were included in the statistical analysis. Congenital malformations—birth defects—were medically confirmed, and low birth weights (less than 2500 grams [g]) were culled from birth certificates and vital statistics records. Baseline data for low birth weights were taken from an average of all white live births in New York State from 1950 to 1977, excluding

New York City. Only indicators among women who had resided in the study area as of June 1978 and had lived there during their entire pregnancy were included.[72]

In addition to these bias-reducing measures, field investigators were "blinded"—they were without knowledge of the hypotheses being tested. Interview respondents underwent a validity check built into the questionnaires, which asked them the same questions repeatedly to verify answers, and the questionnaires themselves were reviewed by an independent statistical unit.[73]

Despite these procedures, the study was rejected by the journal *Science* and received negative peer reviews. Skepticism over two of the control groups utilized in the spontaneous abortion analysis fueled the dispute. Vianna had selected three different control groups, the first a historical set taken from a study by Dorothy Warburton and F. Clarke Fraser in 1963.[74] The Warburton and Fraser data were appealing because of the large sample size—over 6,000 pregnancies were observed, and spontaneous abortion frequency was tabulated by birth order and maternal age. Because of possible demographic and geographic differences, a second set—a contemporary control—was selected from Colvin Avenue, a neighborhood north of the canal and matched in terms of single-family houses near an industrial area. A confounding factor—higher education among the Colvin group as compared to those in the entire study area—was controlled in statistical analysis. Finally, a table relating birth order and maternal age was constructed based on the spontaneous abortion experience among women in the dry ("nonwater") areas; this internal comparison group was used to test the hypothesis of excess endpoints in water sections.[75]

How does the report hold up under scrutiny? One reason offered for *Science*'s rejection of the study was that Vianna "did not identify which of the two control groups was more appropriate for comparison with the exposed population."[76] Vianna had already explained in the report, however, that "considering the limitation of each comparison group, the results obtained from all three should be considered in formulating conclusions."[77]

Beverly Paigen criticized the use of the Warburton and Fraser report. While the Vianna team included only hospital and medically verified cases of spontaneous abortions, she observed, the Warburton and Fraser study allowed self-reported cases in their analysis. Comparison between the two, Paigen charged, minimized the frequency of spontaneous abortions.[78] As noted, however, Vianna's data also included those spontaneous abortions "occurring with evidence of a pregnancy with subsequent history compatible with the adverse outcomes." In other words, it is likely that Vianna did include legitimate self-reported cases.

The Colvin contemporary control group fell under criticism because its residents might have been "contaminated." While group members did not live on or near a dump site, it was not clear whether they had been exposed to chemicals migrating from Love Canal or other nearby disposal sites. This flaw, too, was addressed by Vianna. The possibility of contamination was not only recognized but emphasized in the report.[79]

Field interview study techniques were criticized as "limited by a lack of details about the number of eligible individuals, of those approached, and of the acceptance rate."[80] Yet the information on miscarriages included in the study, and on those excluded because of false claims or unverified information, is readily available in the report.[81]

In his evaluation of the Vianna report, Clark Heath warned that "no clear pattern of increased risk has yet been seen in relation to closeness to the canal or its potential drainage channels," and that "interpretation [of Vianna's data] is limited by the relatively small number of births in the exposed cohort and the absence of any means for directly measuring the cumulative exposure of individual persons to Canal chemicals."[82] Once again, this was but an echo of observations made by Vianna in the report: the small numbers involved in statistical analysis were "an obvious limitation."[83]

Author Adeline Levine (whose own study we discuss below) charged that "the document offered little beyond the information on excess miscarriages apparent by the summer of 1978, except for a startling disclosure that miscarriage rates among Love Canal's pregnant women had soared to 50 percent during the mid-60s."[84] The disclosure is not so startling, however, given the recognition by Vianna of recall bias. Indeed, as noted by Joe Grisham, "as the study was conducted against a background of intense local concern, it would be hardly surprising that residents living close to the Love Canal recalled more miscarriages ten or more years previously (the period in which the reported excess was most marked) than did those living further away."[85]

The New York State governor's panel noted that "the investigators thought there might be some increase in miscarriages and infants with LBW [low birth weight], but the data cannot be taken as more than suggestive."[86] From the outset Vianna acknowledged this: "We present evidence that spontaneous abortion and low birth weight might be good initial indicators of human toxicity to multiple chemical agents but frequency of spontaneous abortions is exceedingly difficult to measure. Various congenital malformations are difficult to diagnose and birthweight alone is not a sufficiently sensitive nor specific measure of toxicity."[87] In sum, the Vianna report suffered less from misinterpretation at the hands of its creators than from misreading by its peers. This was not *Science*'s finest hour. If the reviewers for *Science* did not do their job, what on earth were the residents of Love Canal, let alone the general citizenry only peripherally involved yet worried by scare talk, to think of these conflicting assessments?

THE LCHA-PAIGEN REPORT

When Lois Gibbs learned that nineteen families who had requested temporary relocation for health reasons were denied assistance by the New York State

Department of Health, she launched her own informal survey: "Feeling rather desperate, because she wanted out for herself, for her family, and for all the residents who felt as she did, she worked throughout the night at her kitchen table."[88] The members of the Love Canal Homeowners' Association, who frequently chatted with friends and neighbors about problems and illnesses, kept telephone-call records of reports by residents of various illnesses. "When people said they had a health problem," Gibbs related, "we wrote it down."[89]

From these records Gibbs constructed a series of maps that seemed to show clusters of illness around swale beds. Originally Gibbs said that she herself "drew a swale that old-time residents had told me about." She later revealed that it was the "old-time residents" that mapped them out: "Actually, my neighbors drew the line for the swales."[90] Gibbs contacted Paigen for help through her brother-in-law, a biologist at the State University of New York at Buffalo.

Paigen believed the "eager volunteers in the LCHA" were fully capable of conducting a legitimate study, and she helped the women devise a more "systematic telephone questionnaire" and procedure that would be scientifically acceptable.[91] She instructed the LCHA members to "have people tell exactly what illnesses they had and whether they had been to the doctor for diagnosis."[92] Through the survey LCHA members obtained informal information about employment, length of residence, and family history. Paigen left the LCHA members to verify reports by making repeated phone calls to survey participants. After data analysis, Paigen and the LCHA concluded that toxic chemicals were migrating through swale beds, resulting in a clustering of various illnesses. Unlike the investigators in Vianna's team, the LCHA interviewers were far from "blinded" to the hypothesis at hand. No validity checks were made; nor was there an independent statistical review.

Paigen recognized several glaring flaws in the study: First, layperson-to-layperson conveyance of information results in bias and overreporting of disease by information collectors and reporters; some people may not understand the true nature of their illnesses. Second, both the people reporting and the people collecting the data had a vested interest in the outcome because they wished to be relocated. And, third, "no resources were available to verify reports of disease with physician records."[93]

Unlike in the Vianna study, control groups were not actively selected. Instead, "as data analysis began, we realized that an internal control existed."[94] The aim at the outset of any epidemiological study is to describe a relationship between exposure and outcome that cannot be explained by extraneous differences between at least two groups.[95] If there is nothing to compare your observations to, they are meaningless.

Paigen garnered praise for her concern from a group of scientists (the Rall Panel) convened to review both the Vianna and LCHA studies: "Dr. Paigen is to be commended for her humanitarian interests and her willingness to devote her expertise to this problem."[96] In *Hazardous Waste in America* Samuel Epstein quoted out of context one scientist's assessment of Vianna's study as acceptable by current "state of the art" standards and fashioned it into an appraisal of Paigen's.[97] Epstein quoted the following passage: "By current 'state of the art' standards in epidemiological field studies, the methodology [of the Paigen study] is acceptable and the quality control exceeds usual standards." Dr. George Lumb noted that "although the numbers are small, there do seem to be some important correlations."[98] Alas, for this view, there are almost always correlations. To be meaningful, causes have to be tied to effects, both by theory and by observation. Without expectations generated by theories against which to test findings, the existence of correlations shows only that if you do enough calculations there are bound to be correlations. Neither was applicable here.

On the other side, the Rall panel cautioned that the quality of Paigen's data was weak: "[It] may not be complete and may contain potential biases."[99] The Thomas Panel, appointed by the governor of New York, was not as sparing: "Her data cannot be taken as scientific evidence for her conclusions. The study is based on largely anecdotal information provided by questionnaires submitted to a narrowly selected group of residents. [It is] literally impossible to interpret . . . [and] has the impact of polemic."[100]

The Topic of Cancer

Imagine that you are a resident of 96th Street, three blocks away from the Love Canal. It is June 1980, a month since the Picciano chromosome study was released and the second federal emergency in your neighborhood declared. You have learned that a neighbor three doors down has bladder cancer; three more houses down, another neighbor has throat cancer. Six women on your block have had mastectomies because of breast cancer and you are one of them. This was the predicament of Phyllis Whitenight. Like her, you might complain to the press that "you don't have to be a doctor to know there's something wrong here."[101] And you might agree with local *Niagara Gazette* reporter Michael Brown that "it seems too coincidental that the increase in cancer rates has so closely paralleled the increase in rates of chemical waste production."[102] You might be angry, frightened, or even panicky. And you might seek an outside expert opinion, like that of Dr. Paigen, to confirm your worst alarms: "I know the fear in the pit of your stomach when you think about getting cancer. There is a definite risk of cancer here."[103]

What would your reaction be, one year later, to an epidemiological study that found "no evidence for higher cancer rates associated with residence near the Love Canal toxic waste burial site in comparison with the entire state outside of New York City"?[104] Before you react, let us take a closer look at cancer, epidemiology, and the study itself.

BACKGROUND

"No word in the English language is more chilling than cancer." So wrote Congressman David Obey in his foreword to Samuel Epstein's *Politics of Cancer*.[105] Indeed, while the thought of cancer chills many a human heart, it has also chilled minds. Few of us think clearly enough to pay heed to the American Cancer Society's own observation that "the overall incidence of cancer decreased slightly since the 1950s." (We present up-to-date findings on cancer incidence in Chapter 8.) Headlines told a different story: "Children's Sleepwear Treated with Cancer-Causing TRIS," "Cancer Hazard in Plastic Wrap," "Cancer Hazard in Plastic Soft-Drink Bottles."[106] Journalist Larry Agran wrote in 1977 that "what we are witnessing is the unmistakable emergence of a national cancer epidemic. An epidemic of frightful proportions. A cancer pox."[107] Ralph Nader decried America's entrance into the "carcinogenic century" and bemoaned "corporate cancer."[108] Cancer was everywhere, exposure unavoidable, death-by-disease inevitable unless corporate pollution was abated.

The topic was not merely on the minds of Love Canal residents in the late 1970s; it was an issue of national concern. When Gibbs appeared on the *Phil Donahue Show* in the fall of 1979, her fellow guest was Dr. Samuel Epstein, who in a book published that year had spread the message that "cancer is now a killing and disabling disease of epidemic proportions" and "the plague of the twentieth century."[109] He assailed the "carnage of chemical cancer" and advised readers to stay away from "innumerable sites where hazardous wastes have been improperly dumped," for example, Love Canal.[110] Larry Agran echoed the theme that man had created most human cancer.[111]

THE JANERICH STUDY

The study "Cancer Incidence in the Love Canal Area," published in 1981 by Dwight Janerich and colleagues, utilized data from the New York Cancer Registry.[112] The study was conducted as part of the NYSDOH's two-year research project at Love Canal. The investigation included environmental, animal, and human studies and Vianna's work on reproductive outcomes. Based on the known effects of certain chemicals identified at the landfill (including benzene and trichlorethylene), the investigators selected rates of liver cancer, lymphoma, and leukemia for special attention. In addition to these three "most likely" cancers, other categories were examined, for example, digestive, respiratory, bladder, and breast cancers. Janerich used census tract data to define the population base at Love Canal. According to the 1970 census, the tract contained 4,897 people, including 700 individuals who lived in houses on and adjacent to the dump site. Cancer incidence in the Love Canal tract was adjusted for age and sex, and these were compared to county and state rates (excluding New York City). In comparing observed versus expected rates, no evidence of an increase in lymphoma or leukemia was observed. In the period from 1955 to 1965, 2 cases of liver cancer in women were observed, though 0.3 were expected.[113] Janerich noted that the women's residences were not in close proximity to the dump.

Among the other types of cancer studied, only respiratory cancer showed elevated rates. Between 1966 and 1977, 25 cases in men were observed and 15 expected; 9 cases in women were observed and 4.6 expected.[114] Only 4 of the 34 cases were located in homes on or adjacent to the canal. Although Love Canal rates of respiratory cancer did not differ significantly from those of the tracts in the city of Niagara Falls, the investigators recommended that lung cancer rates be monitored in the area in the future, because the city in general had a high lung cancer rate, and the magnitude of the increased frequency was within the range of those shown to be associated with factors like smoking.[115]

The report emphasized that questions of latency had not yet been addressed; that given the large number of new cases reported each year, the degree of completeness and accuracy of the New York Registry was not known precisely; that confounding factors such as smoking and socioeconomic status were not assessed; and that effects on people who had moved from the area were not studied. Why, then, did researchers choose census tract and registry data? Standardization of reporting minimized information bias and, more important, the data did not depend on patient recall.

The Janerich study has been variously classified as "descriptive," "demographic," and "ecologic." In contrast to cohort and case-control studies that gather data on specific individuals, the descriptive study collects data on populations to identify the distribution or patterns of disease. No control group is matched to the experimental group and exposure is not considered. According to Clark Heath, the use of an entire census tract to define the "exposed" Love Canal group diluted the potential degree of exposure in the study population and diminished the power of the study to detect differences in disease frequency between exposed and nonexposed residents.[116] Hence, the population at risk was insufficiently defined.[117] Criticism from Epstein and his group of researchers focused on the small number of cancer cases studied; less weighty was their complaint that "the study failed to consider a wide range of other adverse effects besides cancer, such as miscarriages and birth defects."[118]

A sharp criticism of the Janerich study pertained to its "unjustified categorical negative inferences."[119] But recall for a moment Phyllis Whitenight's predicament. She told the *Detroit Free Press* that she was having "more sleepless nights than she could count."[120] Perhaps the limitations of the Janerich study caused such anguish. It is more likely, however, that informal studies did greater damage. After instructing six Love Canal residents to "list the illnesses they knew of on their block," the reporter observed that there seemed to be "many cases of cancer among the women."[121]

The Canal Womens' Tales

The women of Love Canal captured a special place in the media—and it was usually located on the front page. Karen Schroeder, a primary source of information for Michael Brown's *Laying Waste,* recounted her troubles for many newspapers and magazines. Her in-ground pool had popped up, ruining the family's backyard. As already mentioned, her daughter, Sheri, suffered tragic handicaps: a cleft palate, deformed ears, a hole in the heart, impaired learning ability, and a double row of teeth.[122] Both the *New York Times* and the *Washington Post* featured the plight of Phyllis Whitenight. Her daughter had throat infections and an ugly rash; six of her pet birds died in the Whitenight's basement; and she herself had endured both a mastectomy and a miscarriage.[123] Whitenight and her husband appeared separately on two network news programs in May 1980 after learning that they had chromosomal damage.[124] Luella Kenny, cited in *Insight* and the *Village Voice,* suffered worse—the death from kidney failure of her seven-year-old son, Jon.

Infections, rashes, cancer, birth defects, loss of a child. These occurrences happen in all neighborhoods—often inexplicably. Yet the press reported those from Love Canal residents with special vigilance. In a ten-day period in May 1980 there were thirty-one separate articles in the *New York Times,* including eight front-page stories and three editorials; network news programs devoted 31.5 minutes to the case.[125] The mountain of articles generated by the women's tales had several common characteristics: lead paragraphs painted a desolate picture of the neighborhood—"eerie" and "ghostlike" peppered the prose of *Time, Newsweek,* the *Atlantic,* and the Congressional Quarterly's publication, *Environment and Health.* The metaphorical pitch of a ticking time bomb appeared in the title of a report to the governor of New York ("Love Canal: Public Health Time Bomb"), and the *Journal of the American Medical Association* warned that Love Canal was "one time bomb that will keep on ticking."[126] Photos of mothers and infants, or of angry young fathers, accompanied the newspaper features, their signs reading, "Give Me Liberty, I've Got Death."

Love Canalers were depicted as victims, trapped and suffering the consequences of corporate callousness. A 1978 article in *Newsweek* lamented: "The only certainty is that the people of Love Canal are simply the latest victims of a burgeoning menace to technologically sophisticated man—toxic chemicals. In a grim update of the Faustian legend, the products and by-products of industrial efforts to improve consumers' standards of living are threatening those same people with disease and death."[127]

Not until *Reason* magazine published Eric Zuesse's investigative piece, "Love Canal: The Truth Seeps Out," did any probing of the validity of Love Canal residents' claims appear in the press. Why not? Martin Linsky observed that "the *New York Times* and the *Washington Post* did run some pieces on the scientific aspects of [Love Canal], but the main stories were written by reporters who could not be expected to have understood in scientific detail what was involved in demonstrating a connection between a particular health defect in a certain percentage of the population and the presence of toxic chemicals in the neighborhood."[128]

Another possibility is that journalists were perfectly capable of such scientific understanding but were not expected to educate themselves or their audiences about such matters, given deadlines and competition for dramatic stories.

A basic scientific foundation was missing from Love Canal coverage in the lay press. When results of informal surveys were discussed, no mention of their limitations appeared. Nor were background incidence rates on diseases supplied so that readers could put findings in proper perspective. A typical example is the lead paragraph of Nader and Brownstein's article in the May 1980 *Progressive:* "In the neighborhood where Lois Gibbs lives, only one of the fifteen pregnancies begun last year ended in the birth of a healthy baby, according to a survey of the homeowner's association. Four ended in miscarriage. Two babies were stillborn. Nine others, including a pair of twins, were born deformed."[129]

Were the birth defects and miscarriages medically validated? Did the mothers have a history of medical problems in their families? How about alcoholism? How broadly defined is "deformed"? And what percentage of pregnancies end in stillbirths, miscarriages, or birth defects in the total population? None of these questions were answered, which confirms Zuesse's proposition that they were never asked.[130]

The media's failure to research the facts resulted in the public's miseducation about Love Canal. Nowhere was this more apparent than in the original leaked story on chromosome damage. Irwin Molotsky of the *New York Times* wrote that "most scientists in this field [cytogenetics] believe that such chromosome changes are frequently linked to cancer and should be taken seriously as a harbinger of the disease and that in adults they could lead to genetic damage in offspring."[131] He asserted further that "scientists believe that such chromosome damage could lead to severe birth defects, and some residents of the area have charged in the past that children born there have suffered such defects." Molotsky quoted no cytogeneticist making such claims. He did, however, "contact by telephone" Love Canal resident Barbara Quimby. The news of possible chromosome damage from toxic materials buried in the canal was no surprise to her, since her "first daughter has three congenital birth defects."[132]

A year later, June 20, 1981, *New York Times* editors took a sly turn in their piece "A Second Look at Love Canal." Questionable studies inflamed public fears, they wrote. The scientist who said he found chromosome damage among Love Canal residents rang an alarm that generated "scary headlines," they asserted. But of the *Times*'s own responsibility, nothing was said.

Some disagree. Author Levine praised newspaper reportage: "Newspaper reporters were on the scene day after day, week after week. I found that their accounts were usually highly accurate, and, since I was close to the events, I could check the occasional discrepancies. I realized that the newspaper articles provided detailed documentation of key events by skilled observers, producing an informal current history for my use."[133] She did not go into the reporting about cause and effect.[134]

In our view the media placed far too much weight on self-interested personal reports and pilot and preliminary findings. Science journals fell into a similar trap. *Science News, New Scientist,* and *Science Digest* followed the "Disaster on 99th Street" with great interest, but with little probing. Gina Bari Kolata's piece in the June 3, 1980, issue of *Science,* "Love Canal: False Alarm Caused by Botched Study"—analogous to Zuesse's *Reason* article in the lay press—provided the first critical assessment of the conduct of scientists in the area.

Adding to Love Canal's paper trail are a number of case studies, sociological works, and personal accounts. Most prominent among the literature is *Niagara Gazette* reporter Michael Brown's book, *Laying Waste: The Poisoning of America,* which garnered a Pulitzer Prize nomination. Brown examined toxic waste disposal sites from Iowa to New Jersey to California, but Love Canal was his centerpiece. A native of Niagara Falls, Brown painted the before and after picture of his neighborhood with great poignancy but little documentation. While a *Newsweek* reviewer lauded Brown for producing "a book which should be required reading for every aspiring investigative reporter,"[135] columnist Richard Cohen saw otherwise:

> [The book] is dismal. It is poorly written and horribly edited . . . The word "seems" appears over and over again: "there seemed to be many cases of cancer among the women . . . It seemed that some officials were trying to conceal their findings . . . Laurie seemed to be losing some of her hair." Deaths that Brown hears of get mentioned in just that fashion: "I was told that one elderly woman, seriously depressed upon hearing the state's plans, had died of heart failure." Testimony is mentioned, but not where it was given. People suddenly appear and then drop from sight and jargon percolates through the text like the chemical waste through the soil of Love Canal. "The dioxin potency of this isomer was nearly beyond imagination."[136]

Case studies of Love Canal, including Lewis Regenstein's *America the Poisoned,* Samuel Epstein's *Hazardous Waste in America,* and Russell Mokhiber's *Corporate Crime and Violence,* are disarmingly indefinite. Informal surveys and recollections of residents become damning evidence of health damage. Isolated cases become representative of the suffering of all residents.

Adeline Levine, a Buffalo sociologist, produced a study that won much praise: *Love Canal: Science, Politics, and People.* She "reconstructs virtually the entire public record of the crisis"[137] in a generally "balanced and accurate" way. But like Brown's work, Levine's was "flawed by a periodic absence of important relevant information."[138] The people part of Levine's analysis is insightful, in our view, but the politics is incomplete and the science shallow. How, we ask, can a person be on the side of the residents if that means accepting as true fears that are unfounded?

In October 1982 Martha Fowlkes and Patricia Miller filed a report on sixty-three in-depth interviews with Love Canal residents for the Federal Emergency Management Agency entitled "Love Canal: The Social Construction of Disaster." The study documents and analyzes views of residents heretofore unseen and unheard in either the media or literature:

> I don't know why these people left. Like these people down the block, they claim their kids were born without fingers, without teeth. I think some of them inherited it from their grandparents, it wasn't from Love Canal. You know, the people are crying, you know, several things, like little children having different difficulties. If it was the chemicals, why wouldn't the whole area be affected. We're all drinking the same water.

> [Neighbors] started complaining all of a sudden, "I can't breathe, I can't do this or that." I said, "Aw. Come on. You've been there 20 years and never had a problem." They're what you call complainers. You know, some people just thrive on being sick for some reason. It makes you want to throw up.

> Everybody knew what was going on and when you got right down to it, nobody knew what was going on. Everybody had their opinion, but nobody had any hard fact; but I think that's the whole summary of the Canal.[139]

It also serves as a good summary of most of the literature about Love Canal.

One last Love Canal opus merits attention—Lois Gibbs's *Love Canal: My Story*. Gibbs was a six-year resident of Love Canal in 1978 when news leaked from the pen of Michael Brown that chemicals were oozing from her neighbors' basements. Brown's work for the *Niagara Gazette* is a testament to the influence of the press on the lives of ordinary citizens. How did Gibbs, the founder and president of the Love Canal Homeowners' Association, become aware of the Love Canal situation? Not by direct experience but, as she testified before Congress, "from reading a series of articles that were being printed in a newspaper in the area." When asked if she had difficulty believing that there was a hazard, Gibbs hedged: "Well, I believed there was a hazard immediately after reading the articles."[140]

Before a congressional committee, she was "shy, but plucky" and, with the help of her brother-in-law, a local professor, she became adept at capturing media attention. At the conclusion of her testimony in 1979 the Interstate and Commerce Committee recommended that "she should get an honorary degree in biology for the tremendous amount of information that she not only accumulated, but mastered." The Goldman Environmental Foundation awarded her $60,000 and praised her for "speaking out for the environment at great personal sacrifice."[141] After the second federal evacuation of Love Canal in 1980, Gibbs moved her family to Arlington, Virginia, where she established the Citizens' Clearinghouse for Hazardous Wastes.

Government

Two days before the first-ever declaration of a man-made federal disaster by President Carter in August 1978, an entourage of local, state, and federal officials toured Love Canal. They included staff members from the Niagara Falls mayor's office, the New York governor's office, the state's two senate offices, the EPA, and the Federal Disaster Assistance Administration (FDAA). A vanguard of residents led FDAA chief William Wilcox and his troops to a neighbor's swampy backyard, in the midst of a "week-long media blitz." As officials stooped to examine the fields, reporter Donald McNeil observed that there was "some hesitation on the part of the [group] to step into the two-inch deep mud."[142]

The episode is a fitting analogy for all levels of government involvement in Love Canal. It was a Niagara Falls employee who wrote to a local congressman, John LaFalce, in 1977 about concern over "dangerous chemicals."[143] After visiting homeowners LaFalce launched a personal crusade that ranged from writing the Niagara city manager to dining with President Carter. And it was Gibbs, LCHA president, who galvanized Congress, the EPA, and the Carter administration from 1978 to 1980, when the situation reached a crescendo. Michael Brown added fuel to the LCHA fire with his stream of articles for the *Niagara Gazette*.

LOCAL

Congressman John LaFalce was perhaps the most enthusiastic of governmental actors. He had "little scientific expertise," but when he examined "basements and backyards," it was obvious that "something was seriously wrong."[144] Reports by residents who claimed they had witnessed dumping in the canal by army trucks prompted LaFalce to take action, which included calling for a Defense Department investigation. His efforts led to the creation of the New York State Assembly Task Force on Toxic Substances. The Defense Department denied charges of dumping. LaFalce organized a review panel of members of the EPA and the Department of Health, Education, and Welfare. Headed by David Rall, the panel evaluated data submitted by both Beverly Paigen and Nicholas Vianna.[145] The report found there was cause for concern, given the available data on pregnancy outcomes. LaFalce became a local hero and trusted member of the Love Canal community.

State Senator Thomas Bartosiewicz also contributed. A "personal investigation" into health studies in the area prompted him to write Governor Carey that there was a "cover-up."[146] He then wrote to all New York state senators.

Bartosiewicz's campaign led to the formation of a scientific review panel appointed by Governor Carey.

THE CARTER ADMINISTRATION, OR, WHEN CAN A HOUSE BE A HOME?

One month before the presidential elections of 1980 President Carter traveled to Love Canal and signed an unprecedented agreement between federal and state governments to buy out hundreds of homes of remaining residents.[147] Together they paid $17.2 million for 463 homes—which the state has been left to maintain and sell.

The environmental monitoring report written in May 1982 was issued by the EPA in July in three hefty volumes. Its purpose was to determine the extent of contamination directly attributable to migration of substances from Love Canal to nearby residential areas.[148] Researchers collected and analyzed approximately 6,000 samples of soil, air, sump water, and groundwater from the emergency declaration area, the outer area beyond the canal that extended past the two closest rings of homes evacuated during the first emergency declaration. The sampling sites included ten subregions of the area, a region directly adjacent to Love Canal, and a control region that included selected sites throughout the Niagara Falls area.[149] EPA's Nicholas Riordan noted that the control sites were a "sufficient distance from the former canal so as to be free from potential contamination related directly to Love Canal," and not known "to be located near any other known hazardous waste landfill areas." Upon analysis, the EPA concluded that "the data revealed no evidence of environmental contamination in the residential portions of the area encompassed by the emergency declaration order that was directly attributable to the migration of substances from Love Canal."[150] "No evidence" is exactly right.

Dispute arose, however, over the adequacy of the control sampling, the quality of the study design, and the incompleteness of the EPA's data set. Concern about the study's scientific integrity was hotly debated at an August 1982 hearing before a subcommittee of the House Committee on Energy and Commerce. The Environmental Defense Fund lodged complaints over the lack of statistical power of the data; according to the defense fund, the reason no significant difference was found between the declaration and control areas was that the number of control sampling sites was inadequate.[151] There also was internal dissent within the EPA over the study's conduct.[152]

After consultation with statisticians from the National Bureau of Standards, the Department of Health and Human Services (HHS) confirmed that the emergency declaration area was "as habitable as the control areas with which it was compared," provided that Love Canal be "constantly safeguarded against future leakage" and that "cleanup [be] required for existing contamination of local storm sewers and their drainage tracts."[153] The "provideds" are sensible, but the findings are stark and, we believe, justified not only by the HHS study but also by the evidence. There was no significant contamination where people were living; hence there was no known harm to health from this cause.

The Office of Technology Assessment (OTA) joined the habitability hubbub in June 1983 with its own technical memorandum criticizing the HHS report. Uncertainties over possible synergistic human health effects of multiple toxic chemicals present at low concentrations, levels of toxic chemicals detected, and possible levels of chemicals not detected prompted the OTA to place doubt on the HHS decision.[154] What the OTA said was right, but was it relevant? It was tantamount to saying that if its study could not rule out a negative, there was still a possibility that unknown synergistic effects at undetectable low levels caused harm. The trouble is that this assessment always applies to every case.

In response to the report, the EPA established the Love Canal Technical Review Committee to oversee the final remediation and habitability of the area.[155] Environmental sampling for Love Canal indicator chemicals was conducted to test for migration or movement of chemicals known to have been disposed of in the canal: 887 samples were collected; 781 were successfully analyzed.[156] In four of the subregions of the area, indicator chemical levels were "not in any consistent way significantly higher than levels in soil from Niagara Falls comparison sites." Two other regions showed "quite low" levels, and although another region showed "relatively high" levels, the area "may be used for . . . [purposes other than normal residential use] i.e. commercial, industrial, without remediation."[157]

After the bureaucratic papermill on habitability ceased, residents clamored to move back. Shrugging away thousands of pages and hours spent on making the habitability decision, an interested buyer commented, "It's probably one of the safest places to live in Niagara Falls by now. [Anyway], there are problems no matter where you live in the world."[158]

If the first declaration of emergency was a child of the ineptitude of scientific studies, the second was a product of governmental errors. Congressman John LaFalce commented that the second evacuation "may have been too extensive." Judgments were made on soft data or no data at all, and the entire decision "got out of hand."[159]

Why were so many so seriously misled to believe in probable damage from proximity to Love Canal, whereas a judgment of "highly improbable, though not quite impossible" was in far better accord with the evidence? The conflicting claims, the ensuing derogation of expertise, the hurly-burly make this misapprehension understandable. It is easy to see what was missed: the importance of controls, the invalidity of self-reported sicknesses, the knowledge of the prevalence of certain misfortunes, the importance of checking on statistics, the weakness of relying on single studies, the stage of peer review. We might add unanimity to the list of missing items; after all, if it is science, it is supposed to be objective, and doesn't that mean everyone reaches the same right conclusion?[160]

Yet, apart from some governmental actors, to whom we will return, it is hard to fault any of the participants. They gave their all. Residents called in the authorities, complained to their representatives, listened to different accounts of what had happened to them, brought in rival experts, and thereby sought to inform themselves. Indeed, with the aid of Beverly Paigen, they went further than most citizens have gone by organizing their own study. They consulted older residents, drew maps of suspected transmission belts, interviewed residents, and otherwise tried to produce a reliable scientific study. True, they did this work not to determine if they might be wrong but to prove they were right. But this motive alone would not necessarily differentiate their biases from those of professional scientists. In short, the active residents of Love Canal acted like citizens in a democracy.

What went wrong? What, in a book about and for citizens in a democracy, can be learned from Love Canal? What might the residents have done to come closer to the truth, to worry less about extremely unlikely harms, to concentrate more on mitigating whatever hazards there might be, and to avoid the vast disruption and consequent harm of being forced to move?

Their first error was to rely too much and too soon on scientific experts to inform them, which is to say that they first needed to know something about the subject themselves before making use of experts. The science of cytology provides an apt illustration. Had interested residents of Love Canal read up on cytology, once the pilot study results were made public, they would have quickly discovered its requirements for evidence and the prevalence of broken chromosomal fragments in the general population. Similarly, by reading about epidemiology they would have recognized the onerous requirements for proof with small examples. Given their heightened fears of numerous horrors, they should have been calmed by evidence that the doses needed to create these illnesses would have to be inordinately large.

Always the controls bear investigation. Are the residents really different from other people similarly situated, so they have good reason for concern, or are they pretty much alike? It is in the residents' interest—their mental as well as physical health—to have as closely matched controls as possible. Those who urge good controls are not trying to put them down or trivialize their traumas but instead are offering the best help they can give. As we shall argue in Chapter 13, "Citizenship in Science," lay citizens serve themselves best by following good procedures.

Armed with this much understanding, residents of Love Canal might have decided not to move at all. Or they might have sought a more discriminating approach. There were only a limited number of pathways through which suspect chemicals could have entered their homes. These could have been tested and if found to have more toxicity than was common elsewhere, the pathways could have been closed down or, if possible, remediated. The mass exodus was uncalled for.

Some governmental officials knew better. They did not have to maximize the dangers. Of course, that is asking a lot under the circumstances—resisting well-publicized public outrage. Perhaps, like EPA Director William Ruckelshaus on DDT, they relaxed their standards, hoping to stay in office to fight another day. Had he insisted on preponderant knowledge about the effects of DDT, he might not have been around to lend his credibility to an EPA under attack during the Reagan administration. Should President Carter or Governor Carey or some other public official have risked their careers to determine and tell the truth about Love Canal? As things stand, no illness, not even a cold, can properly be attributed to living next to Love Canal.

5

Superfund's Abandoned Hazardous Waste Sites

with David Schleicher

Cleaning up abandoned hazardous waste sites has come to be one of the most visible and costly undertakings of U.S. environmental policy. At the center of this effort is the federal Superfund program, which has to date targeted more than 1,200 sites nationwide for cleanup. The Environmental Protection Agency (EPA) has so far spent $10 billion on the cleanup effort, and private industry has by most estimates spent at least this amount. These sums are dwarfed by potential future costs: most projections see thousands more sites being added to the cleanup list, with costs reaching into the hundreds of billions of dollars over the next several decades.[1]

The financial burden of the cleanup campaign is widely shared. In setting up the remediation program, Congress attempted to "make the polluter pay" by assigning cleanup cost liability to parties linked to polluted sites, and by financing the backstop "public superfund"—for use when responsible parties cannot be found or are unable to pay—primarily with a special tax applied to the chemical industry. But the expenses imposed on the chemical industry by the superfund tax are in large measure passed on to consumers in the form of higher prices on a wide array of products.[2] And the EPA's efforts to pinpoint liability at the site level have produced a list of more than 1,700 "potentially responsible parties," including not only numerous businesses but also town, city, and county governments.[3] Legal wrangling has drawn banks, savings and loans, and insurance companies into the liability web as well.[4]

Representing as it does a significant commitment of national resources, the Superfund program has attracted a great deal of critical scrutiny. Charges of various sorts of inefficiencies are common. Recent General Accounting Office (GAO) studies have cited Superfund as one of several federal government programs most vulnerable to "fraud, waste, and abuse."[5] Still more galling to observers is the amount of public and private money that is diverted to "transaction costs" associated with cleanup actions—principally lawyers' fees. The

Office of Technology Assessment (OTA) recently concluded that transaction costs comprise nearly one-half of total Superfund expenditures.[6]

Attention has rightly been devoted to program inefficiencies and to administrative reforms that might pare down costs. But the more fundamental question in considering whether Superfund monies are well spent is what is being gained by cleaning up abandoned chemical waste sites. It is generally assumed that cleanups guarantee health benefits, making efficiency the only remaining consideration. But are there indeed health benefits and, if so, are they substantial enough to justify the huge expenditures? Answering these questions requires consideration of the evidence of public health threats stemming from these sites. We consider first the evidence as it appeared to legislators who established the cleanup program in 1980; we then look to epidemiological studies conducted in the years since, and finally to how the EPA gathers and interprets evidence of hazard at particular sites.

The Comprehensive Environmental Response, Compensation, and Liability Act of 1980

In one of its final acts the Ninety-sixth Congress in 1980 overwhelmingly passed the Comprehensive Environmental Response, Compensation, and Liability Act (CERCLA) to deal with hazardous waste storage and disposal facilities abandoned by their operators and apparently contaminating their environs. These waste sites, President Jimmy Carter had warned, posed "some of the most significant environmental and public health problems facing our Nation."[7] A committee report accompanying the bill to the Senate floor suggested that it addressed what might be the biggest public health challenge of the coming decade.[8]

This same depth of concern was shared by the American public of 1980. A series of headlines and media images through the late 1970s brought about a popular conviction that abandoned hazardous waste sites were a deadly menace. At center stage was Love Canal, which first captured national attention in the autumn of 1978, having been declared a "Public Health Time Bomb" and granted state of emergency status by New York State officials. Alarming images of lawns and basement walls saturated with brightly colored liquid waste and interviews with worried residents dominated the nightly network news. A *Time* magazine cover story and network documentaries soon followed. A second wave of media attention and public alarm swelled in the spring of 1980, with the completion of studies linking the noxious Love Canal wastes to chromosome damage and increased rates of miscarriage and birth defects among local residents. Now cast as a modern-day tragedy, Love Canal was at or near the top of nightly newscasts for a month.[9]

Meanwhile, other reports fueled popular fears that Love Canal was emblematic of a far-reaching danger. National media play was given to visually dramatic sites such as Kentucky's "valley of the drums," where hazardous liquid wastes were leaking from tens of thousands of corroded barrels.[10] *Time* pronounced "The Poisoning of America," while regional media hunted for and publicized local "time bombs." By 1980 polls showed 80 percent of the public wanting swift federal action to identify and clean up potentially dangerous abandoned waste sites.[11]

Hazardous Waste, 1980

In 1980 the EPA considered hazardous any waste that "may cause or significantly contribute to serious illness or death, or that poses a substantial threat to human health or the environment when improperly managed."[12] Wastes were designated hazardous if they tested positive for one or more of four characteristics: ignitability, corrosivity, reactivity (wastes that "tend to react spontaneously, to react vigorously with air or water, to be unstable to shock or heat, to generate toxic gases, or to explode"), or toxicity (wastes that "when improperly managed, may release toxicants in sufficient quantities to pose a substantial hazard to human health or the environment").[13] Wastes designated hazardous by the EPA in 1980 numbered in the thousands, with classes such as phenols, organic chlorine compounds, and heavy metals prominent. These wastes were generated—and had been for decades—principally by private industry in the manufacturing of fuel, automobiles, plastics, paper, clothing, paints, pesticides, medicines, and countless other products. Hospitals, research laboratories, and the military also contributed to the flow. In 1980 the United States generated 57 million metric tons of hazardous waste, in varied physical forms including liquids in containers, bulk liquids, semiliquid sludges, and bulk solids.[14]

The end of the road for most of this waste was land-based disposal, a solution long preferred by waste producers as convenient and relatively inexpensive (an inclination ironically strengthened by federal regulations that, starting in the 1950s, sharply limited the discharge of toxic chemicals into the air or surface water).[15] Common forms of land disposal were open dumps, containers buried in landfill, above-ground bulk containers, and surface impoundments (pits, ponds, and lagoons for liquid and sludge wastes). Reactive, corrosive, and ignitable wastes were usually segregated in containers, ranging from 1-gallon cans through 55-gallon drums to 1,000-gallon bulk tanks.[16] Disposal sites generally operated for a limited time, from a few years to a few decades,[17] after which they were often left with little active management or deserted outright. It was to hazards posed by such "abandoned" (or "unmanaged" or "inactive") sites that the Superfund program would respond.

Evidence of Acute Risks from Abandoned Hazardous Waste Sites

In fashioning the Superfund program, Congress saw and sought to redress two distinct classes of threats: those requiring emergency measures and those requiring long-range remedial efforts. For the first of these, an "emergency removal" program authorized the EPA to respond rapidly to near-term hazards—sites that might catch fire or explode or that in some way might result in acute human exposure to toxic substances. EPA ameliorative actions were to be sharply focused and limited in implementation time (not to exceed six months) and cost (not to exceed $1 million).[18]

The belief that unmanaged sites could pose acute threats to human health derived in significant part from a pair of highly visible incidents, one occurring shortly before Congress began consideration of the Superfund bill and one during the late stages of deliberation. The first of these involved a Chester, Pennsylvania, disposal site that caught fire in 1978. The site housed over 30,000 barrels of industrial waste, much of which was toxic, ignitable, or explosive. Most of the barrels were strewn above ground, and many were corroded and leaking, allowing the intermingling of chemicals intended to remain segregated. The fire, determined later to have started spontaneously, emitted a toxic cloud of smoke that forced the closing of a nearby highway. Forty-five firefighters required medical attention, most for skin and lung irritation from the chemical fumes.[19]

A second dramatic accident occurred in April 1980, at an Elizabeth, New Jersey, disposal site that included over 40,000 drums of hazardous wastes, including live viruses, nerve gas, and nitroglycerine in corroded, leaking drums.[20] This time the result was a powerful explosion: "a flash lit up Elizabeth's sky. There was a thunderous boom and chemical drums shot 200 feet into the air, exploding like skyrockets. Soot and fumes rained over a 15-square-mile area, stinging eyes and throats."[21] Favorable winds kept the cloud from drifting toward New York City. Again, firefighters required medical care for toxic effects.

The Chester and Elizabeth episodes demonstrated vividly that unmanaged hazardous waste sites could pose near-term, acute threats to public health and welfare. At the same time it was clear that disposal sites commonly contained ignitable, explosive, or reactive wastes that required segregation, and that the barrels intended to segregate such wastes were sometimes crushed during site operations or, more often, corroded with time.[22] It was also known that abandoned sites often lacked adequate fencing to bar entry by nearby residents, including children, who might gain access to toxic runoffs and soils.[23] The CERCLA "emergency removal" program was thus a focused response to a readily discernible hazard.

Evidence of Chronic Risks from Abandoned Hazardous Waste Sites

The second threat targeted by the Superfund law, and the one toward which the vast majority of the program's resources would be directed, was that of chronic health effects resulting from long-term exposure of populations near unmanaged waste sites. It was this perceived threat, raising the specter of miscarriages, mutations, birth defects, and cancers that most frightened the public and most concerned officials and lawmakers. Thus while Superfund's removal program redressed acute hazards, the legislation's centerpiece was a "remedial action" program to undertake the long-term and costly cleanups necessary to eradicate chronic exposure threats. It was this mission that led the sponsors of CERCLA to cast it as the most important public health measure to emerge from Congress in years.[24]

The grand estimation of threat underlying the remedial action program derived from the coupling of two distinct convictions: that a great many abandoned sites had subjected or could in the future subject people to sustained low doses of an array of industrial chemicals, and that chronic low-dose exposures to industrial and particularly synthetic chemicals could cause cancer and other grave illnesses. The first judgment was grounded in a patchy but compelling body of evidence; the second was essentially a prior conviction that was not rigorously examined during congressional consideration of the Superfund law.

Three types of evidence combined to convince Congress in 1980 that at a great number of inactive waste sites around the country chemicals could migrate into the environment. The first type was a cluster of high-profile cases, like Love Canal and Kentucky's "valley of the drums," that vividly demonstrated that wastes could escape from landfill or corroded containers and leach into underlying soil. The lawns and basements of Love Canal offered colorful testimony to the mobility of such leachates. Analytic tests on the Love Canal ooze identified two hundred different industrial chemicals.[25]

A second set of cases suggested that pollutants from waste sites could make their way into groundwater, an environmental medium of particular concern since about 40 percent of Americans draw their drinking water from underground aquifers. Near Denver EPA sampling found a thirty-square-mile area of groundwater to have been infiltrated by synthetic chemicals, the legacy of years of disposing of pesticide wastes into unlined ponds.[26] In the twin towns of Toone and Teague, Tennessee, water supplies had been found contaminated in 1978. Leachates from a disposal site housing 350,000 drums of pesticide wastes—shut down and inactive since 1972—had seeped into the aquifer from which local drinking water was drawn.[27]

Were these cases typical or atypical? In passing judgment on this important question Congress drew on a third type of evidence: broad surveys designed to determine the overall contours of the inactive hazardous waste site problem in the United States. For instance, the EPA compiled a list of 250 inactive sites around the country that were deemed "dangerous" by virtue of such criteria as volume of waste, manner of containment, and age.[28] EPA field analysts detected waste chemicals in groundwater at 130 of these facilities.

This survey was complemented by reports that sought to gauge the total number of inactive waste sites in the country, especially those that might cause groundwater contamination. An extensive survey directed by the EPA and executed in large part by the states located 11,000 industrial sites with a total of 25,000 surface impoundments of liquid wastes. More than half of these pools contained wastes classified as hazardous by the Resource Conservation and Recovery Act. A third of these impoundments were found to be unlined, atop permeable soils and over "usable" aquifers. Groundwater monitoring around these sites was virtually nonexistent.[29] In another report the EPA maintained that 90 percent of the hazardous waste disposed of had been handled in an "environmentally unsound" fashion, with environmental and public health considerations typically taking a back seat to economy and convenience in choice of site location and design. High-profile cases, the EPA concluded, were likely "but a tip of the hazardous waste iceberg."[30]

A couple of considerations should have tempered the conclusions drawn from the body of evidence presented to Congress. For one, the EPA, in citing "contamination" of groundwater near Denver or at many surveyed sites, used the term in its most encompassing sense, denoting only infiltration of a medium to a scientifically detectable level—at a time when advances in detection techniques were allowing the recognition of ever-smaller trace quantities of chemicals.[31] Evidence of widespread "contamination" in its more onerous sense—the exceeding of some critical health threshold—was lacking. A second tempering consideration should have been that the estimates of site numbers and conditions were produced principally by the EPA and the states, both of whom had an interest in the passage of a far-reaching cleanup measure.[32] But even with a skeptical disregard for some of the grander estimates, one could in 1980 reasonably conclude that at many inactive sites around the country chemicals were slowly migrating into the environment.

Congressional belief that public health could be jeopardized by low-dose exposure to commercial chemicals actually preceded the CERCLA deliberations by many years. As early as 1958, in the Delaney Amendment to the Pure Food and Drug Act, Congress had judged that high-dose effects in animal studies were indicative of low-dose dangers to humans and thus banned outright any food additive found carcinogenic in laboratory animals. In 1972, in the stringent section 307 of the Clean Water Act, Congress mandated that public health be fully protected against "hazardous" pollutants—designated as such principally on the basis of animal testing suggesting carcinogenic and other toxic effects—without regard to cost (whereas "conventional" pollutants were to be regulated in a manner that balanced costs and benefits).[33] Exacting toxic substances mandates within the Resource Conservation and Recovery Act of 1976 and the Clean Air Act Amendments of 1977 further demonstrated congressional conviction that a wide array of industrial chemicals could even at low dosage endanger human health.

With this conviction as backdrop, CERCLA scientific deliberations focused

on whether inactive waste sites had caused or could in the future cause environmental exposures to such chemicals. There was, however, a smattering of fresh
evidence cited on behalf of the notion that such exposures could be harmful.
One manner of affirmation came from new reports addressing the general
problem of contamination of the environment by toxic substance. Most prominent here was a 1980 surgeon general's report concluding that "while at this
time it is impossible to determine the precise dimensions of the toxic chemical
problem, it is clear that it is a major and growing public health problem. We
believe that toxic chemicals are adding to the disease burden of the United
States in a significant, although as yet ill-defined, way. In addition, we believe
that this problem will become more manifest in the years ahead."[34] The
National Institute of Environmental Health Sciences offered an equally grave
assessment of the overall toxic chemical threat. Its director suggested in testimony before a Senate committee that "the entire population of the United
States" could be at risk from toxic synthetic chemicals in the environment.[35]

At the same time, however, Congress was apprised of the weakness of evidence linking environmental contamination and adverse effects on human
health. A 1980 Congressional Research Service report, for instance, stressed
that "the low-level, long-term effects of individual chemicals, let alone combinations of them, remain obscure."[36] The principal tools then employed in
research, epidemiological studies and animal testing, had "significant limitations, particularly at low levels of exposure."[37] The study concluded of twenty
chemicals common to waste sites that "chronic health effects [from exposure
to these substances] are unknown, or at best, only suspected,"[38] and cautioned
against sweeping proclamations of a synthetic chemical peril.

Meanwhile, evidence of actual health effects from chronic exposure to inactive waste was scant. The two studies that had purported adverse health effects
at Love Canal were almost immediately discredited by a high-level scientific
panel convened by New York governor Hugh Carey. The panel concluded of
one study that it "could not be taken seriously," while the other was deemed
"botched and misinterpreted."[39] An EPA survey claimed twenty-seven sites to
be "associated" with adverse effects such as cancer and spontaneous abortions,[40] but these associations were largely anecdotal, as little had yet been
done in the way of actual epidemiological studies of abandoned waste sites. A
Congressional Research Service review of six such studies found that serious
chronic health effects were in each case reported as "none" or "unknown."[41]

Epidemiological Evidence of the Public Health Threat

During the late 1970s and early 1980s, as official and public concern about
inactive hazardous waste sites grew, epidemiologists began to study waste site
communities in search of adverse health effects. A comprehensive review and
analysis of these studies was conducted in 1986 under the direction of the

Universities Associated for Research and Education in Pathology. A collaborative effort of forty academic scientists, *Health Aspects of the Disposal of Waste Chemicals: A Report of the Executive Scientific Panel* is still cited as a definitive work. Participants were charged with reviewing and critically assessing "scientific evidence of association between demonstrated exposure to chemicals from disposal sites and occurrence of human diseases and disorders."[42] The panel's preliminary report was subjected to peer review, and a subsequent draft was made available for public comment before being published.

The twenty-nine studies turned up by the panel, and the twenty-one sites investigated, composed a nicely representative sample.[43] Chemicals of concern varied—PCBs, pesticides, metals, many solvents—as did exposure pathways, which included ingestion of food and contaminated drinking water as well as inhalation. Health endpoints encompassed cancer incidence, reproductive outcomes, neurological changes, chromosomal abnormalities, and general physical ailments. Study designs included cohort (investigation of health consequences in a group known to have been exposed), case-comparison (investigation of exposure histories of "clusters" of persons with an identified health condition), and cross-sectional (attempts to link exposure patterns and health endpoint patterns across a broad population). Nearly three-quarters of the studies were conducted ten or more years after the waste facility had commenced operations, with nearly one-half after the facility had been operating for twenty-five or more years. Most of the investigations were executed or at least orchestrated by government agencies.[44]

The panel's intensive review of this corpus of studies produced the conclusion that "with one exception, evidence for causal association with occurrence of disease is weak."[45] The one case of a methodologically sound finding was an instance of acute exposure to arsenic. Thirteen people in Perham, Minnesota, ingested well water drawn from an aquifer contaminated by an arsenic-based pesticide. Several were hospitalized with gastrointestinal disorders, and one lost the use of his legs for six months.[46] The epidemiological report of this incident was deemed "exemplary" by the panel, which noted that "a dose-response relationship was evident by the severity of the effects and was further substantiated when the presence of arsenic was detected in the hair of the individuals with the most severe exposure."[47] It should be stressed that this incident and its investigation occurred in the mid-1970s—prior to the initiation of the CERCLA emergency removal program designed to detect and redress acute exposure hazards. The panel discovered no cases of acute health effects after 1980. In regard to chronic health effects, the panel concluded that "to date, none of the investigations has provided sufficient evidence to support the hypothesis that a causal link exists between exposure to chemicals at a disposal site and latent or delayed adverse health effects in the general populace."[48]

Concurrent with this study a separate attempt at a comprehensive, critical review of existing site studies was undertaken by Gary Marsh and Richard Caplan, of the University of Pittsburgh, and published in 1987 as part of Julian Andelman and Dwight Underhill's *Health Effects from Hazardous Waste Sites.* Limiting themselves to published studies, Marsh and Caplan assayed investigations of fifteen different waste sites. They concluded, much as the panel did, that "the exposure-health outcome linkages that were examined are, for the most part, weak or inconclusive."[49]

Most of the community health studies reviewed by the panel and by Marsh and Caplan reported no adverse health effects. A minority reported associations between exposure and health but were dismissed as unconvincing by reviewers because they failed to meet several standards of sound epidemiology:

1. *Accurate specification of timing, duration, and level of exposure among the population studied.*[50]   Precise and reliable characterizations of exposure facilitate the drawing of nuanced links between degree of exposure and degree of effect. The more closely such a pattern is drawn, the more convincingly implicated is the chemical agent at issue. In waste site contexts, however, exposure levels and patterns are usually difficult to gauge with precision or reliability.[51] In many instances site studies rely on blunt estimation techniques to characterize exposure patterns, such as dividing populations into a series of concentric circles radiating from the site. Also commonly utilized are exposure history questionnaires, regarded by most epidemiologists as a tool of last resort because respondents' recollections may be inaccurate or may be distorted by fear or—where lawsuits are pending—by self-interest.[52] In the panel's collection of studies, actual body level measurements, the most precise and reliable indicators of chemical exposure, were obtained at only seven of twenty-one sites, with approximation techniques and questionnaires sufficing for the remainder.[53]

2. *Careful and specific measurements of health effects, using standardized methods.*[54]   Satisfaction of this criterion both builds confidence that the putative effects are genuine and facilitates the inference of precise exposure-effect relationships. By contrast, waste site studies purporting exposure-illness linkages showcased patient reports of ambiguous, subjective discomforts such as fatigue, headache, nausea, and shortness of breath. Because community fears and anger can be expected to lead to "hyperreporting" of vague ailments, the panel considered studies that could not be objectively confirmed to be unconvincing.[55]

3. *Careful identification of and control for potential confounding factors.*[56]   The panel observed that the problem of confounding—"the *mixture* of the effect of the exposure variable with the effects of other variables"—may lead to the misreading of data from site studies.[57] In particular, studies that fail adequately to consider and control for confounding variables may attribute to chemical exposure public health patterns that stem from other causes.

Common confounding variables include age, sex, marital status, cultural and personal habits, drug intake, and preexisting medical conditions. Confounding can theoretically be controlled for by careful identification of a comparison population that matches the studied community in all important respects (thereby isolating the waste site as a causal factor); in practice, however, this proves difficult owing to the large number of variables that must be controlled for.

Because existing studies either report negative findings or report positive findings without meeting the above standards, peer reviewers have concluded that epidemiological evidence supporting conjectured public health risks from abandoned hazardous waste sites is very weak. At the same time, we are kept from jumping to any sweeping *negative* conclusions, in part because so few studies have been done relative to the total number of sites, and in part because those studies completed to date have fallen well short of two additional standards of sound epidemiology:

1. *Studying populations large enough to register the suspected effects.*[58] The populations investigated in the twenty-nine studies surveyed by the Executive Scientific Panel were generally too small to have manifested the low level effects that were hypothesized. In the panel's sample 40 percent of the studies addressed populations of less than 100; 75 percent looked at populations of less than 1,000; the largest population studied was just over 3,500. The panel concluded that "the power of statistical tests used to detect differences in serious health effects, such as increased mortality, cancer, or reproductive abnormalities, was extremely limited because of the size of the populations," and noted that small populations were cited by several studies as a possible explanation for negative findings.[59] Michael McDowell, in his *Identification of Man-Made Environmental Threats to Health* (1987), argues more broadly that while exposure to environmental pollutants can be expected to increase the incidence of any disease by only a small percentage, if at all, "epidemiology can rarely pick up an increase of 25% or less, [in part] because of the very large samples which would be required." He goes on to question the scientific value of investigations addressed to small populations, doubting whether "studies are worth undertaking if the sample size calculations indicate that insufficient cases are exposed for any but a very large excess risk to be identified."[60]

2. *Undertaking studies with the most promise of furthering knowledge.* In broad epidemiological efforts generally, the panel points out, "the appropriateness of proposed investigations is largely controlled through the external peer review system by approval or disapproval of funding; thus marginal investigations are restrained and restricted," and knowledge gradually advanced.[61] By contrast, selection of waste site communities for study often proceeds by the squeaky wheel principle. For example, many of the investigations surveyed had

been initiated in response to residents' complaints of foul odors. Since odors typically affect only very small populations in close proximity to the source, these studies were in the panel's view more "a response to an over-anxious public . . . than an exercise of responsible scientific judgment."

Citing such failings, the panel cautioned that existing studies "cannot be interpreted to indicate that no risks exist but, rather, that no serious health effects have been identified to date by the study methods employed."[62] Wary that overreaching generalizations might be drawn from its critical survey, the panel urged that "the absence of demonstrable effects on human health should not be taken as proof that no such effects exist."[63]

Given the difficulty that site studies have in meeting all or even most of the standards enumerated, some critics have grown pessimistic about epidemiology's ability to shed more than a dim light on the question of waste site risks. One observer sees a "malaise" setting in, exacerbated by the willingness of some researchers to commit epidemiological "malpractice" to produce positive findings.[64] Others are more hopeful and urge more resources, more studies, better site selection and study design, and innovative methodologies.[65] Pessimism or optimism aside, tangible progress has not been forthcoming. A search through nearly a thousand titles of epidemiological studies published in refereed journals from 1988 to 1991 turned up virtually nothing regarding hazardous waste sites. Most recently National Research Council investigators bemoaned the general inadequacy of epidemiological study of and knowledge about abandoned hazardous waste sites.[66]

Despite such failings and frustrations, a pair of important inferences may be drawn from the work of epidemiologists during the first decade of Superfund. First, the more alarmist images and metaphors that held sway as the Superfund program was initiated were exaggerated. The Executive Scientific Panel report, while cautioning against drawing sweeping negative conclusions from a collection of flawed studies conducted under trying conditions, allows that "even in circumstances unfavorable to the detection of an effect . . . negative studies can help to determine the probable upper limits of risk."[67] The panel did not venture an estimate as to just what "probable upper limit" ought be inferred from the negative studies that it reviewed. But it seems reasonable to surmise that this limit is comfortably below the level of risk connoted by the ticking time bomb metaphor that haunted policymakers and the public in 1980. If there were in fact decades-old time bombs tick-ticking away across the land, we would expect by now some "explosions" detectable to epidemiologists.

Second, and more fundamental, community health studies have thus far failed to substantiate the premise that launched and sustains the cleanup campaign: inactive hazardous waste sites endanger those who live near them. That we have nevertheless spent tens of billions of dollars to clean up these sites and are contemplating spending hundreds of billions more raises questions about how risks to public health are gauged and ameliorated at Superfund sites and directs attention to the EPA's implementation of the Comprehensive Environmental Response, Compensation, and Liability Act.

Designating Superfund Sites

Charged with identifying and cleaning up the most dangerous of the nation's inactive hazardous wastes sites, the EPA developed and set in motion a three-stage ranking procedure:

1. Site identification. Suspect sites are reported to the EPA's National Response Center, usually by local officials or sometimes by former operators or concerned individuals, and entered into the EPA national site inventory system.

2. Preliminary assessment. EPA personnel examine local and state records and interview site owners to determine whether an inspection is merited.

3. Site inspection. Limited site sampling is conducted and further information gathered. The site is scored according to the EPA Hazard Ranking System to determine whether inclusion on the National Priority List is merited.[68]

Fewer than one in ten sites entered into the EPA's inventory system ultimately qualify for the priority list.[69] It is the application of the Hazard Ranking System (HRS) that does the final winnowing and determines the size of the National Priority List (NPL). The ranking system was developed for the EPA by a contractor shortly after the passage of the Superfund law. Designed to provide a rapid, low-cost way of gauging relative risks, the assessment in most cases requires only one visit to the site by EPA field personnel.[70] For a given site three "hazard modes" are assessed: migration (of waste chemicals through air, surface water, and groundwater), fire and explosion, and direct contact. Scores are assigned to each mode, based on such factors as waste characteristics (volume, toxicity, reactivity, ignitability, solubility), nature and condition of containment structures, degree of extant release, distance to underlying aquifers, and proximity and size of local population. Each of these factors is scored and weighted; ultimate scores for the three hazard modes are expressed as a percentage of the worst possible score. Sites scoring high on fire and explosion or direct contact modes are entered into the emergency removal program. Sites scoring 28.5 percent or higher on the migration mode are proposed for the National Priority List. Nearly all proposed sites will, after a period of public comment, be designated an official NPL or Superfund site and entered into the remedial action program.

The 1,200 sites on the NPL merit more concern than the 20,000 inactive waste sites assessed and denied Superfund designation, but being on the NPL in itself tells us nothing about the magnitude of the harm posed by a site. This is so because the all-important cutoff point, the 28.5 HRS score that determines the fate of individual sites and the size of the NPL, has little public health meaning. The cutoff level is instead an artifact of an initial congressional mandate that the NPL include at least 400 sites. In a recent evaluation of the ranking system, the EPA reflected that at Superfund's outset the agency "chose the 28.50 cut-off score . . . because it yielded an initial NPL of at least 400 sites as required, not because of any determination that the cut-off represented a threshold of unacceptable risks presented by sites."[71] This cutoff level has been maintained because it has proven "an effective management tool,"[72] but the EPA stresses that the ranking system can provide only "a measure of relative rather than absolute risk."[73]

Assessing Superfund Site Risks

It is during the subsequent "baseline public health evaluation," conducted as part of the Remedial Investigation, which typically takes several years to complete, that the EPA renders its full judgment as to the nature and severity of the threat posed by a given site. These risk assessments are performed by EPA regional office personnel, with considerable assistance from contractors. Coherence and consistency is given to this far-flung process through adherence to two key guidance documents: the *Superfund Exposure Assessment Manual* (1988) and the *Superfund Public Health Evaluation Manual* (1986).[74] Although the hundreds of NPL sites vary in many ways and are investigated and pronounced upon by disparate personnel, insights into the general character of evidence underlying evaluations may be gained by attending to these guidance manuals.

Superfund site exposure analysts face the imposing task of determining who has been and who will in the future be exposed to how much of what chemicals through which environmental media. The EPA's exposure assessment guide breaks this task down into five principal steps:

1. Selection of target chemicals. If sampling at the site identifies more than a dozen waste chemicals, as is often the case, a set of "indicator chemicals" is selected for exposure assessment. These are chemicals that "pose the greatest potential health risk at a site," as determined by volume present, mobility (for example, as airborne gases or as leachates through soil), persistence (to last in the environment in essential form, as opposed to biodegrading), and toxicity.[75]

2. Contaminant release analysis. Point-specific release rates for each indicator chemical to each environmental medium are identified and then the sum rate from all release points of each chemical to each medium is calculated.

3. Contaminant transport and fate analysis. Working from release rates, investigators estimate current and future migration patterns and concentration levels for indicator chemicals, taking into account area sampling results as well as atmo-

spheric, meteorological, and hydrogeologic conditions, and properties such as solubility, reactivity, and biodegradability.

4. Exposed populations analysis. Here judgments must be made about what groups of people will be affected by migrating chemicals, through which media, when, and at what points of dispersal.

5. Integrated exposure analysis. This step provides point-of-exposure concentration levels for each indicator chemical through each final medium (for example, ambient air, drinking water, shower water, fish).

In the course of this multistage process, site analysts do a good deal of testing and generate a considerable amount of data. But as exposure assessment is a "young and developing field,"[76] and vastly complicated at Superfund sites, the final judgments are in important respects conjectural. The EPA notes three broad categories of uncertainty that mark the typical site exposure assessment:

1. *Input variable uncertainty.*   This category refers to the site-specific values that must be plugged into models used to predict contaminant migration patterns. Generating data of sufficient scope and quality to allow a confident determination of such values is difficult, especially when groundwater is involved. As a rule, "the analyst will not be able to determine the value of [model-supporting] parameters with absolute certainty."[77] Assumptions and rough estimates will thus play a critical role in assessments of chemical migration at Superfund sites.

2. *Modeling uncertainty.*   Whether models employed in depicting extant and predicting future contamination migration are in fact representative of a given site is "a large source of potential uncertainty."[78] Analysts are encouraged to rely on monitoring data rather than simplifying models whenever possible,[79] but modeling is nevertheless an integral part of an exposure assessment, especially insofar as long-term migration scenarios are required. And even the most nuanced of models may obscure or misrepresent important site features. The guidance manual notes, for instance, that "degradation is not accounted for in some models. If the chemical is extremely refractory . . . this limitation will not materially affect the answer. If the contaminant degrades quickly, however, this limitation will cause the model results to be in substantial error."[80] The error in this instance would be a significant overestimation of the amount of the chemical that populations would encounter at ultimate exposure points.

3. *Exposure scenario uncertainty.*   Prominent among these imprecisions is the "inability to define exposed populations with confidence."[81] The uncertainty can be especially deep and its handling especially consequential when groundwater is at issue and the time horizon extended. Will an aquifer that is not now used for drinking water perhaps be so utilized in the future? Questions of this sort cannot be answered with confidence but must be addressed nonetheless.

Faced with points of ignorance or uncertainty, site analysts most often proceed by using "conservative," or projected risk-enlarging, assumptions and estimates. The contemporary common wisdom on such matters—if there is to be error, better that it be on the side of "prudence"—is embraced here. The EPA notes that conservatism in the face of exposure uncertainties is "traditional."[82] But when such a posture instructs decisions throughout a complex and multistaged analysis wherein "uncertainty may be a factor at each step,"[83] there is potential for final assessments that reflect less an amassing of evidence than a compounding of cautious assumptions. The guidance manual recognizes as much and warns that "use of reasonably conservative assumptions at each step may produce cumulative assessment results that are overly conservative and thus unreasonable."[84] Site analysts are encouraged to generate data rather than rely on conservative "default values" (assumptions provided by the EPA for use at common points of uncertainty) and to try to balance exposure-inflating assumptions and estimates with exposure-deflating ones. Final assessment documents are to include an "uncertainty analysis."

Often, however, such admonitions either go unheeded or are attended to in a perfunctory manner that has little effect on the final assessment. Thus a former Superfund project manager expressed frustration with cleanup actions driven by "fanciful scenarios that assume underground soil in industrial areas becomes evenly distributed topsoil in residential gardens, or that salty groundwater near [the ocean] becomes the sole source of drinking water for a large population."[85] More broadly, independent commentaries and government-conducted evaluations alike often count far-fetched exposure scenarios among the cleanup program's flaws.[86]

Generating more site-level data and relying less on default values might result in more modest estimates of potential exposure, but pervasive use of default values can be expected to remain the norm as a consequence of several forces. One is the conservatism common to mid- and low-level public health bureaucrats, reflecting the structure of rewards and penalties within which they operate. As an observer suggests, "No bureaucrat has ever been fired for going with the most healthy conservative estimate of risk."[87] A second force is the political pressure under which assessments are conducted. Generating data takes time. Faced with congressionally mandated deadlines and local calls for less study and faster cleanups, analysts understandably make use of EPA-sanctioned assumptions and estimates whenever possible. That using default values is in the short run less costly than generating additional data further strengthens the tendency.

Once exposure point concentrations have been estimated for all indicator chemicals, assessment of public health risks is guided by the *Superfund Public Health Evaluation Manual*. Analysts first compare postulated concentrations with standards associated with other environmental regulatory regimes. These may include air quality standards, maximum contaminant levels (MCLs) developed pursuant to the Safe Drinking Water Act, state-developed quality criteria for particular water bodies according to the federal Clean Water Act, and standards promulgated under state environmental laws.[88] The public health risk is characterized by expressing the exposure-point concentration as a percentage of the identified standard (for example, "vinyl chloride in groundwater is 150 percent of the Safe Drinking Water Act MCL").

At most sites, however, standards are not available for the full set of chemicals and exposure pathways of concern, and for such combinations risk analyses are used. The first step is to move from exposure-point concentrations to estimates of actual bodily intake of suspect chemicals.[89] At this point analysts must speculate on how frequently people will actually be within the range of contaminated air and how much of it they will breathe in, how much contaminated water they will drink, how much contaminated fish they will eat, and so on. To bridge uncertainties, the EPA provides an array of default values, again acknowledged to be conservative. For inhalation of contaminated air, for instance, "it is assumed that exposure occurs 24 hours per day for the entire period that contamination is present."[90] The manual suggests that such assumptions "can be modified based on site-specific information to the contrary."[91] But again, because relying on default values is easier, faster, and cheaper than generating local data, chemical-intake analyses can be counted on to add another dollop of "prudent" overstatement to already conservative exposure assessments.

After making intake assessments, analysts assign toxicity values to the chemicals at issue.[92] In judging the low-dose toxicity of a wide array of waste chemicals,[93] analysts proceed by consulting EPA Health Effect Assessments (HEAs), a document series supporting the *Public Health Evaluation Manual*. These documents codify the EPA's judgments about subchronic, chronic noncarcinogenic, and chronic carcinogenic risks posed by chemicals commonly found at Superfund sites. Because hypothesized carcinogenic risks drive most Superfund remedial designs,[94] it is instructive to consider the bases of the HEA carcinogenicity profiles.

EPA Health Effects Assessments rate the potential harm of chemicals common to Superfund sites by assigning each a "carcinogenic potency factor," denoting lifetime cancer risk per milligram per kilogram body weight per day. Although the evidence supporting these potency profiles varies in the particulars from chemical to chemical, two important generalizations may nevertheless be made. First, the EPA carcinogenicity profiles are derived principally and in some cases exclusively from tests performed on laboratory animals. That the profiles lean heavily on animal test results is suggested by the EPA's own "weight-of-evidence for potential carcinogens" classificatory system, which designates four categories:

Group A—Human carcinogen. Sufficient evidence from epidemiological studies to support a causal association between exposure and cancer in humans.

Group B1—Probable human carcinogen. Limited evidence of carcinogenicity in humans from epidemiological studies.

Group B2—Probable human carcinogen. Sufficient evidence of carcinogenicity in animals; inadequate evidence of carcinogenicity in humans.

Group C—Possible human carcinogen. Limited evidence of carcinogenicity in animals.[95]

The EPA offers no figures on the distribution of chemicals among these categories, but the numbers can be gleaned from the capsule HEAs provided within the guidance manual. Fewer than 10 percent of the waste chemicals designated potential carcinogens fall into the Group A—Human Carcinogen category rooted in human studies. A small number are classified as Group B1. The great majority of Superfund carcinogens—nearly 85 percent—are designated Group B2 or Group C, indicating an evidentiary base composed solely of animal test results.[96]

We should not be surprised that waste site risk assessments tend to be driven by inferences from animal tests, given the lack of human studies supporting the notion that numerous industrial chemicals are toxic at the very low exposure levels associated with waste sites. A comprehensive survey of studies of workplace exposure to chemicals common to Superfund sites, conducted by the Executive Scientific Panel referred to earlier in connection with site community studies found little in the way of observed toxic effects on workers from low-dose exposures.[97] Also reviewed as part of this exercise were accidental releases to the environment of PCBs, trace metals, organic pesticides, and other substances common to waste sites. The episodes surveyed had resulted in contaminated air, drinking water or food. The reviewers found that, "in general, as for occupational studies, reported effects were dose-dependent . . . Only in relatively high-dose situations were clear clinical illnesses apparent."[98]

The second generalization we can make about these profiles is that they share a particular manner of inferring human cancer risks from animal tests.[99] In moving from observed high-dose effects in animals to predicted low-dose effects in humans, the EPA employs the "no threshold, linear multi-stage" model, widely regarded as among the most conservative of statistical extrapolation methods. Because it presumes that "any exposure, no matter how small" will cause harm, such an approach "tends to overestimate calculated risks."[100] In relating doses administered to small mice and rats to hypothesized doses ingested by large humans, the EPA employs surface area ratios, which yield greater human risk estimates than do the body weight ratios used by the Food and Drug Administration (FDA). The FDA also employs an alternative high-dose/low-dose extrapolation model. Between these two differences in inferential method, the FDA's human carcinogenic risk estimates will be one-tenth those of the EPA when both use the same animal test data.[101] The EPA further elevates its waste site carcinogenic potency factors by basing them on 95 percent confidence limits.[102] All told, few would question the EPA's observation that its Superfund carcinogen profiles express "an upper bound estimate of cancer risks."[103]

Once carcinogenic potency factors have been assigned to each indicator chemical, analysts complete the baseline public health evaluation by gauging total carcinogenic risks to specified subpopulations in the area of the waste site. Risks to a given subpopulation from each indicator chemical are figured for inhalation and ingestion and these are added: risk from chemical X through inhalation is added to the risk from chemical X through ingestion, and then the sum risks from chemicals X, Y, and Z are added together to yield lifetime cancer risk projections for each exposed population.

Cleaning Up

Following baseline public health evaluation, site analysts examine remedial alternatives and select a final cleanup plan. The rigorousness of remedial actions varies somewhat from site to site, depending, for instance, on local interpretation of ambiguous federal directives and on whether the cleanup is funded by public monies or by settled-upon contributions from responsible parties.[104] In general, cleanups are strikingly thorough, taking years or decades and tens of millions of dollars to complete. Three principal forces push remedial actions in this direction.

1. Stringent public health standards. The initial Superfund law mandated that remedial actions "prevent or minimize the release of hazardous substances so that they do not migrate to cause substantial danger to present or future public health."[105] The EPA's National Contingency Plan, which governs the cleanup procedure, declares that "the national goal of the remedy selection process is to select remedies that are protective of human health."[106] More concretely, two public health standards—detailed in the plan—have guided the EPA's choice of cleanup procedures.

First, all "applicable or relevant and appropriate requirements" from other federal and state environmental regulatory programs are to be satisfied.[107] The EPA was pushed into adopting this approach initially by the courts, when in 1984 a settlement of a case brought against the EPA by the Environmental Defense Fund and the state of New Jersey elicited a commitment that Superfund remedial actions would meet at least these standards.[108] Congress reinforced the "applicable or relevant requirements" standard by writing it into the 1986 Superfund reauthorization law.[109] Meeting such standards in actual site situations can be difficult and costly, since contaminated environmental media have to be restored to purity levels that were in most cases established with the relatively simpler and cheaper task of pollution prevention in view.[110]

Second, total residual (post-cleanup) carcinogenic risk to any exposed individual is not to exceed 10^{-6}.[111] This risk implies that an individual would have a 1 in 1 million chance of contracting cancer because of the site. Since the "background" cancer rate for Americans is 1 in 4 (meaning that 1 in 4 Americans will contract cancer in their lifetime from all causes combined), a 10^{-6} risk stemming from the site means that the exposed individual's prospects for getting cancer are 1 in 4 plus 1 in 1 million, versus 1 in 4 for a person not exposed (or 250,001/1,000,000 versus 250,000/1,000,000). A site that upped an individual's total cancer risk to, say, 250,010 out of 1,000,000 is by this standard unacceptably dangerous.

This 1-in-1-million risk limit, reflecting popular thinking as to what constitutes a "vanishingly small" probability, commonly instructs EPA regulatory standards for particular chemicals.[112] But since hypothesized cancer risks at Superfund sites often emanate from many chemicals and multiple exposure media simultaneously, a tolerated residual cancer risk of 1 in 1 million will have to be apportioned—assuming, as the EPA does, additivity of risks across chemicals and media—among a potentially large number of chemical-medium combinations. Thus the risk from single combinations will often have to be reduced to levels well *below* 1 in 1 million. (At a hypothetical site where a neighborhood is exposed to trichloroethylene and toluene in groundwater, and lead and chromium in air, the projected harm from these four chemical-medium combinations should not *total* more than 1 in 1 million.)

2. *Cautious exposure and intake estimating techniques and upper-bound toxicity values.* When a linear, no-threshold model is used for toxicity estimates rooted in animal tests, getting down to a 1-in-1-million risk level for a given chemical will commonly require that exposures be limited to about 1/400,000th the dose at which lab experiments were conducted.[113] Worst case projections of future exposure patterns also weigh heavily. A recent Office of Technology Assessment (OTA) study found that highly conjectural long-term exposure scenarios commonly steer EPA choices about what degree of cleanup to require and suggested that preoccupation with such "speculative" risks constituted a misappropriation of scarce program resources.[114] Similarly, a recent analysis of the Resource Conservation and Recovery Act (RCRA) "corrective action program," which cleans up active sites in much the same manner as the Superfund program approaches inactive sites, suggested that enormous savings could be achieved if more reasonable scenarios, rooted in current land and resource use rather than long-term, worst-case projections, were employed.[115]

3. *Preference for use of permanent treatment technologies.* There are three principal approaches to cleaning up hazardous wastes evaporating into air, running off into surface water, or leaching into deep soil and groundwater: *removal and transport* to another site, *containment* on-site, or *treatment* on-site to reduce toxicity. The last of these is the most time consuming and costly (pump and treat remedies for contaminated groundwater—which repeatedly extract water, treat it to reduce contaminant levels, and reinject it into the ground until purity targets are reached—may take decades or longer).[116] The

National Contingency Plan mandates that treatment is to be the remedy of choice "wherever practicable." "Minimiz[ing] untreated waste" is cited as a main goal of the remedial program, right after "protect[ing] human health."[117]

As with the mandate to "satisfy relevant regulatory standards," the EPA embraced the "treatment whenever possible" approach with a strong push from Congress. The 1986 Superfund reauthorization bill required the EPA to provide public notice and justification in cases where treatment was eschewed.[118] Congress took this action in the face of reports suggesting that both on-site containment and removal and transport (typically to RCRA-governed landfills) were in the long run unsatisfactory, since containment structures might leak with time, and sites that received transported wastes might someday have problems of their own.[119] Alarmed by such scenarios, legislators steered the cleanup program in the direction of costly but presumably permanent treatment.[120]

Site Profiles

A review of several site histories brings into sharper focus the character of risk assessment and remediation in the Superfund program. These cases are reflective of the range of sites—in terms of geographic location, facility type, and assessed severity of danger—encompassed by the program.

OLD SPRINGFIELD LANDFILL, VERMONT

Springfield is a town of 10,000 in southeastern Vermont.[121] Precision tool and die manufacturing has been its primary industry since the 1800s. In 1947 the town began operating a landfill that received both municipal and industrial wastes, the latter including degreasing solvents from local machine shops. Operations ceased in 1968. Two years later the twenty-seven-acre facility was covered over and turned into a trailer park, where sixty people in thirty-eight trailers eventually resided. The park community received its water from the municipal system, drawn from surface waters elsewhere in Springfield, while some homes adjacent to the site relied on private wells for drinking and household water. In 1972 a complaint by a private well user that his water smelled and tasted bad drew the attention of local officials. The EPA began testing at the site in 1976. In 1982, shortly after the passage of the Superfund law, Old Springfield Landfill was placed on the National Priority List. Eight years later, the EPA announced a final cleanup for the site that would cost about $20 million and take ten to twenty years to complete.

Cleanup plans for Old Springfield Landfill were based on an assessment of public health risks completed in 1988. At the same time, analyses were conducted by consultants hired by the parties that had early on been identified by the EPA as liable for cleanup costs (including the town itself, since it had operated the landfill). Because there was no indication that the site had caused health problems among local residents, analysts focused on the possibility of future effects.

Sampling of soil, surface water, groundwater, and air at the landfill identified seventy-five waste chemicals. In keeping with agency risk assessment guidelines, EPA field analysts selected from this group twenty-four indicator chemicals (based on volume at the site, mobility potential, and putative toxicity), including PCBs and volatile organic compounds such as vinyl chloride, trichloroethylene (TCE), and toluene. Three principal pathways concerned the EPA: exposure of skin to or ingestion of contaminated soils, inhalation of air at the site and around its periphery, and drinking of site groundwater. The agency concluded that the exposed soil was in many places contaminated enough to be dangerous to young children who might ingest it. The town and the other parties responsible for the cleanup costs did not contest this conclusion and agreed that some measures ought be taken, such as covering the contaminated areas with a thick layer of clean soil.

For the two other exposure pathways, air and groundwater, gauging the threat was more problematic. Contamination of the air occurs when liquids seeping to the surface give off chemical gases, principally benzene, TCE, and chloroform. Because EPA analysts were by admission lacking adequate data to gauge confidently the rate at which such gases would form, their exact chemical composition, their migration and dissipation in ambient air, and the frequency and duration of human contact with them, they had to lean heavily on models and default values in assessing the inhalation dangers. They developed two inhalation scenarios. In the average exposure scenario, excess cancer risks to residents atop and immediately adjacent to the site—excess, that is, beyond the national background rate of 1 in 4 persons contracting cancer—were projected to be 9×10^{-5}. This level implies that a population of 100,000 exposed persons might see 25,009 cancers rather than the 25,000 that we would expect based on background rates. This risk projection itself has important elements of conservatism built into it, including figuring individual chemical toxicities by using animal test "carcinogenic potency factors." The EPA also figured a reasonable maximum exposure scenario that assumed the worst at all points of uncertainty. This tack produced an inhalation risk level of 5×10^{-3}, or an excess risk of 1/200 over the 1 in 4 background level. It was this calculation that the EPA employed in setting cleanup requirements for the site.

Inhaling the air at the Old Springfield Landfill appears dangerous, and then moderately so, only if worst case assumptions are substituted where evidence is lacking. The threat posed by groundwater at this Superfund site is similarly conjectural. Sampling has determined to the satisfaction of all interested parties that the groundwater in the shallow sand and gravel and in the deeper bedrock underlying the landfill is unfit for drinking. Vinyl chloride concentrations exceed Safe Drinking Water Act MCLs—standards devised with tap water in mind—by more than 200 times (420 parts per billion versus 2 parts per billion);

PCBs exceed MCLs by 140 times, and benzene levels exceed MCLs by 110 times.[122] But at present no one draws water from this source; municipal water is available to all residents, and the few private wells in the vicinity were shut down in the 1970s after EPA testing. But because the EPA allows that the contaminated aquifer may be drawn on for drinking water at some future point, Safe Drinking Water Act MCLs become "applicable or relevant and appropriate regulations" that must be satisfied by the chosen cleanup plan. Hence the EPA determined that the groundwater under the Old Springfield Landfill, though it currently poses a threat to no one, should be cleaned up to potable standards.

The town of Springfield and the machine shops liable for site cleanup costs objected strenuously to the EPA's assessment of inhalation risks and its speculative characterization of groundwater ingestion risks. The agency's calculations of the public health threat were, they insisted, inflated by "erroneous assumptions . . . and highly unrealistic exposure scenarios."[123] The ambient air inhalation scenario, for instance, assumed that residents would be breathing in the contaminated air twenty-four hours a day for seventy years. Town officials felt the threat from groundwater contamination to be "nonexistent." The bedrock aquifer beneath the landfill passed under an adjacent parcel of land and emptied into the Black River. Because this land was undevelopable (owing to its terrain and geology), there was no chance that the aquifer would in the future be needed for water supplies. Institutional controls such as deed restrictions could bar future residents of the trailer park from drawing up groundwater from directly beneath the site.

In response, the EPA defended its assumptions for the inhalation scenario and groundwater migration as "conservative, yet acceptable." Supplying cautious assumptions where data were lacking was "standard risk assessment practice that has been used by EPA at many other sites" and provided a "plausible upper bound" of threat.[124] Regarding the potential tapping of the tainted groundwater, the EPA contended that at least some of the land under which the aquifer passed en route to the Black River was developable and maintained that institutional controls such as deed restrictions were "less reliable and less preferable to active measures."[125]

The remedy selected by the EPA calls for placing a clay "cap" over an eight-acre area to prevent human contact with the contaminated soil and to block infiltration of rainwater that would otherwise leach through the soil and further contaminate underlying groundwater; a gas collection and treatment system to draw up and completely detoxify whatever gases build up beneath the cap; and a groundwater extraction system with shuttling of the tainted water to Springfield's treatment works for cleaning. It is expected to take from twenty to thirty years to clean the groundwater to the point at which it will satisfy the Safe Drinking Water Act. This remedy carries a price tag of $16 million. The town, which has a limited income base and an annual operating budget of only $4.5 million, must pay for the cleanup's operating and maintenance costs, which are expected to be $130,000 a year for up to thirty years.

REICH FARM, NEW JERSEY

In the summer of 1971 Mr. and Mrs. Samuel Reich agreed to rent a corner of their farm in Dover Township, New Jersey, to an individual who said the space would be used for temporary storage of used fifty-five-gallon drums.[126] Several months later the Reichs checked up on the parcel of land and discovered several thousand waste drums, most still full of liquids and many leaking. Most of the drums were marked Union Carbide Corporation and had identifying labels such as "lab waste solvent" and "solvent waste." It appeared that liquid wastes had been dumped in open trenches as well. Together with Dover Township the Reichs filed suit against the individual and Union Carbide. The company denied culpability, claiming that it had contracted with the individual to transport and dispose of waste in approved facilities, but agreed to remove the drums from the site. This removal was completed in 1972. Some buried drums were later discovered; these and more than a thousand cubic yards of soil were removed by Union Carbide in 1974.

By the time of Carbide's final removal action, nearby residents were complaining of foul smelling and tasting well water. After sampling detected petrochemicals such as toluene and phenols at higher than background levels, the Dover Township Board of Health closed all private wells around the periphery of Reich Farm. Concerned about soil and groundwater contamination, the EPA in 1982 designated a three-acre portion of Reich Farm a Superfund National Priority List site.

In 1986 and 1987 EPA contractors conducted field tests at the farm. Soil sampling revealed that down to ten feet volatile and semivolatile organic compounds could be detected but did not exceed state guidelines for soil quality (known as New Jersey soil action levels). Samples from deeper soils, from ten to thirty feet below the surface indicated "hot spots" where soil action levels were exceeded for substances including ethylbenzene and chlorobenzene. Groundwater sampling found many chemicals at levels that exceeded New Jersey maximum contaminant levels for drinking water. Many of these substances—nickel, lead, cadmium, chromium—were detected at greater-than-maximum levels at wells scattered about the area, including upgradient from Reich Farm, and were not thought to be associated with the site. Other contaminants, particularly volatile organic compounds including TCE, 1,1,1-trichloroethane (TCA), and tetrachloroethylene (PCE), were detected in patterned, elevated levels that pointed to Reich Farm as the source.

In gauging potential public exposure to these chemicals, the EPA considered air, soil, surface water, and groundwater as potential media for both current use and future use scenarios. In evaluating the public health threat, the EPA "used the maximum concentration of each indicator chemical detected" in field sampling,[127] instead of the average or mean concentration of the chemical.

Despite this hazard-maximizing method of representing site contaminant levels, EPA analysts found that Reich Farm poses no current threat to anyone. Because the top ten feet of soil are not contaminated to levels in excess of New Jersey's guidelines, direct contact with surface soil is not considered a threat, nor is groundwater ingestion. For the same reason, air above the site and surface runoff from rainwater were deemed harmless. The town's municipal well field, one mile downgradient from Reich Farm, has been monitored by EPA analysts, who have concluded that "there is no indication that these wells have been affected by contaminants from the site."[128] A monitoring well halfway between the farm and the town's well field has given no indication of elevated levels. The EPA concluded that "the analyses performed . . . gave no evidence that Reich Farm is currently impacting private drinking wells."[129] Groundwater directly below the site is not currently used for drinking water.

Only "potential future ingestion of groundwater on-site" poses a projected public health risk. Organics such as TCE, TCA, and PCE in the groundwater presently exceed Safe Drinking Water Act or New Jersey maximum contaminant levels—at least at their highest sampled concentrations, which the EPA's risk assessors took as representative of the entire underlying plume. Contaminants currently in deep soil could eventually leach into and further taint the groundwater; the EPA maintained, for instance, that "ethylbenzene and chlorobenzene would pose a significant health risk if they attain their maximum concentrations in the groundwater"—as projected by a worst-case leaching model—"and if this water was then used for drinking purposes."[130]

Union Carbide, which will bear the costs of the Reich Farm cleanup, objected vigorously to the EPA's assessment of "future use" site risks. The agency's projection of migration of contaminants from deep soil to groundwater, for instance, the company rejected as "not realistic" because of ultraconservative assumptions and models. Union Carbide pointed to the EPA's use of maximum detected contaminant concentrations as representative of the entire site when sampled concentrations varied widely and were often zero, and to the omission from the calculations of processes such as biodegradation and dispersal that could be expected to lessen concentrations over time. The company also questioned the appropriateness of drinking water safety standards as benchmarks for assessing risks from groundwater, since "effective administrative controls are in place" to prevent wells from being dug on-site.[131]

The EPA defended its risk analysis methodologies at Reich Farm as "conservative but not unreasonable" and common to other Superfund sites. The selected remedy for the site entails excavating and cleaning about 1,500 cubic yards of soil, to prevent leaching of deep subsurface chemicals into underlying groundwater, and extracting and treating groundwater beneath the site until it meets all drinking water standards. The cleanup will cost approximately $6 million and take ten years.

PALMERTON ZINC, PENNSYLVANIA

Palmerton, Pennsylvania, a town of 5,400, is located twenty-five miles north-west of Allentown.[132] The community grew up around a zinc smelting opera-tion established in 1898. The smelting enterprise came to be one of the largest in the nation and was particularly active during the world wars. By the 1970s and 1980s, however, employment at the facility was dropping markedly; by 1990 operations had virtually ceased.

Wastes from the smelting process, called slag, have since 1913 been dumped into an above-ground pile. The slag pile now covers 200 acres and weighs approximately 33 million tons. Since mid-century municipal wastes have also been deposited in the pile. Running along one long edge of this mountain of smelting waste is Aquashicola Creek. Rising up behind the opposite side of the mound is Blue Mountain. Emissions of lead, zinc, cadmium, and sulfur diox-ides from the smelting operation have gradually defoliated more than 2,000 acres of Blue Mountain. This mass defoliation caught the attention of the EPA, which in the 1970s tagged the facility's operators with several violations of Clean Air Act emission standards. Concerned about the gigantic heavy metal-laden slag pile, mass defoliation on Blue Mountain, and tests suggesting ele-vated levels of cadmium in Aquashicola Creek, the EPA in 1983 listed Pal-merton Zinc as a Superfund National Priority site.

The Palmerton community greeted the designation with hostility. Fearing great expense to the zinc company, as well as to the town itself, and skeptical about any public health threat, Palmerton residents, businesspeople, and offi-cials wrote to the EPA protesting its "intrusion."[133] Local doctors pointed out that life expectancy in Palmerton was actually higher than the national average. A state senator made a futile effort to get the EPA to reverse its listing decision. The site remained on the NPL, however, and the EPA set out to assess the public health threat posed by the operation. By 1988 EPA analysts had com-pleted the major parts of their risk assessment and had tentatively proposed a cleanup plan that could cost upwards of $200 million.

Testing at the slag pile revealed that the many places where surface water runoff made its way down the mound and into Aquashicola Creek contained highly elevated levels of heavy metals such as zinc and cadmium. Sampling found cadmium at an average of 10 times the natural background levels and zinc at 24 times. When testing wells were dug deep into the slag mound analysts found that underlying groundwater also contained elevated levels of zinc and cadmium. Much of the groundwater underpassing the slag discharged into Aquashicola Creek. The soils from the defoliated area of Blue Mountain, having been blanketed by a century of smelting emissions, were even more contaminated: sampling found levels up to 2,600 times regional background levels for cadmium, 2,000 times background for lead, and 400 times background for zinc. Rainwater passed over this soil and, like the runoff above and the groundwater below the mammoth slag pile, wound up in Aquashicola Creek.

No one lived on the smelting-facility side of Blue Mountain or, obviously, atop the slag heap. The threat to public health, by the EPA's reckoning, came from Aquashicola Creek. Testing of creek waters found in the worst areas zinc concentrations 40–80 times natural levels and cadmium at 10 times background concentrations. Studies suggested that these levels had taken their toll on fish in the creek. The U.S. Fish and Wildlife Service found high levels of metals in fish in the stream, while EPA studies found higher than average mortality rates among the fish population, with the highest in segments with the greatest levels of zinc contamination.

One source of potential exposure to waste metals from the smelting facility, then, was through consumption of fish caught in Aquashicola Creek. A public health study performed by the Agency for Toxic Substances and Disease Registry in 1987 concluded that it would be unsafe to eat fish from the creek more than once a week.[134] As a second exposure pathway the EPA foresaw people drinking water from wells drawing on groundwater plumes charged by Aquashicola Creek. At that time, the few such wells that existed had been shut down, and municipal water, which comfortably satisfied all safety standards, was available to all residents in the area. Still, the EPA risk assessment cast potential groundwater ingestion as a significant health threat.

The zinc company, principally responsible for paying for cleanup costs, suggested that an advisory issued to fishermen would effectively (and inexpensively) mitigate the threat of excessive consumption of local fish. The EPA rejected this approach, arguing that simply issuing a health advisory "would not resolve the problem" and "would only sidestep the main concern which is the elimination of future fish contamination due to metals."[135] Such an approach would also do nothing to lessen the potential for future exposure to contaminated groundwater. The EPA preferred to stem the flow of contaminants into Aquashicola Creek.

To do so it would be necessary, first, to revegetate the defoliated areas of Blue Mountain so that rainwater would no longer stream down the mountain face, laden with contaminated soil, into the creek. Revegetation would also serve the Superfund aim of redressing environmental damage at waste sites. The EPA here opted for a novel approach that involves spreading across the contaminated soil a mixture of sewage and potash on which new plant life can take hold. The revegetation program is now well under way, has cost very little, and is widely regarded as a success.

Far more troublesome is the task of preventing runoff of metal-laden surface water atop the 200-acre slag pile and discharge of groundwater underneath it into the adjacent creek. The EPA's currently proposed remedy calls for collecting and treating surface runoff, recontouring the pile to soften steep faces, covering those parts of the pile where municipal wastes have been deposited with a clay and soil cap, and covering the whole mound with an ash and sludge layer that would facilitate revegetation. This remedy would prevent direct contact with the slag pile and block rainwater from infiltrating the pile and leaching into underlying groundwater. It would also satisfy Pennsylvania regulations governing the closure of facilities that receive municipal wastes. The project will cost from $200 to $250 million. Much of the expense will go toward extinguishing perpetual fires that burn within the crust of the slag mound. Because of the tremendous cost of its preferred remedy, first put forth in 1988, the EPA has allowed the parties responsible for the cleanup—who were dismayed by the proposed price tag—an opportunity to submit an alternative proposal.

SELMA PRESSURE TREATING COMPANY, CALIFORNIA

Selma, California, is a small agricultural town fifteen miles south of Fresno.[136] Just outside town is the Selma Pressure Treating Company Superfund site, which includes a four-acre wood preserving and treatment complex and fourteen surrounding acres of vineyards. From 1942 to 1981 chemical wastes from the preserving and treating processes were discharged to an unlined pit, to dry wells, to ditches, and to open ground. In 1981 an EPA study of the site, under the authority of the RCRA program, concluded that groundwater was likely being contaminated by the operation and ordered the treatment company to change its disposal practices. Instead, the company filed for bankruptcy. Another treatment enterprise bought the property from the bank in 1982 and adopted disposal methods that satisfied RCRA guidelines. The following year the EPA placed the site on the Superfund National Priority List.

In 1986 and early 1987 the EPA conducted field tests at the Selma Pressure Treating Company site (named for its pre-1982 operator). Sampling found higher than natural background levels of chromium, arsenic, copper, phenols, and dioxin in soils at seven different areas within the eighteen-acre property. Only chromium and arsenic in four of these areas were detected at levels near or exceeding EPA soil quality guidelines. Testing of the groundwater beneath the site found that only chromium significantly exceeded background levels; other contaminants in the underlying soil were relatively immobile and had not leached into the aquifer. Chromium levels, however, were quite high: as compared with natural backgrounds of 2–7 micrograms per liter (μg/liter), or parts per billion, of groundwater, site sampling found chromium at 8,710 μg/liter in groundwater directly under the wood-treatment facility and at 326 μg/liter at testing wells up to 1,000 feet downgradient. A chromium-infiltrated plume extended downgradient from the complex for approximately 1,200 feet, stretching southwest, away from the town of Selma.

In characterizing the public health risks posed by the site, analysts developed current use and speculative future use exposure scenarios. For each, an average case and a plausible maximum case were considered. For the average case, mean detected contaminant levels were used, along with the "most likely (though conservative) exposure conditions." For the plausible maximum calculation, "the highest measured concentrations [were] used, together with high estimates of the range of potential exposure parameters relating to frequency and duration of exposure and quantity of contaminated media contact." The record of decision for the site notes that the risks the EPA attributes to the site "generally . . . are associated with the plausible maximum scenario, rather than the average case."[137]

In EPA's "current use, plausible maximum" scenario putative risks are limited to those associated with surface soil contamination. There is no public health threat from groundwater at present, since the chromium-infiltrated plume "does not currently affect any municipal, private, irrigation, or industrial wells in the vicinity."[138] Selma is supplied water by a private company whose wells, half a mile to the north and upgradient of the facility, have been tested and show no elevation of chromium levels. Surface soils, meanwhile, are said to pose a present threat both to workers at the facility and to area residents who may frequent the vineyard lands. The "plausible maximum" scenario postulates a total excess carcinogenic risk to site-visiting residents of 3×10^{-4}, or 3 additional cancers for every 10,000 exposed persons. This figure adds together dermal exposure, ingestion, and dust inhalation risks for both arsenic and chromium in soil. The corresponding risk level for the facility workers—again under plausible maximum assumptions—is calculated to be 4×10^{-3}, or an excess 4 cancers for every 1,000 people exposed.

Chromium contamination of groundwater threatens public health only in a future use scenario wherein new residences spring up atop the tainted plume and draw drinking water from it through private wells. These hypothetical residents would, under plausible maximum assumptions, have a daily intake of chromium 49 times the reference dose. Reference doses are daily maximum intake guidelines employed by federal health agencies for noncarcinogenic toxics (the EPA considers chromium to be a carcinogen when inhaled, a noncarcinogenic toxic when ingested).

Cleanup at the site is thus driven by the plausible maximum soil exposure scenario and by a future use scenario in which the contaminated plume of groundwater becomes a source of drinking water for new residents. In the most contaminated areas 16,000 cubic yards of soil will be dug up, treated with a fixative, reburied, and covered with a multilayered clay, polyethylene, sand, and soil cap. The cap will prevent contact with the contaminated soil as well as minimize the amount of rainwater that reaches the soil and leaches through to the groundwater. The fixative will further impede leaching from either rainwater that penetrates the cap or fluctuations in the groundwater table that might bring shallow groundwater up to the level of the deeper contaminated soil. The cleanup also entails the extraction and treatment of approximately 3 billion gallons of groundwater until chromium levels are brought down to the Safe Drinking Water Act chromium standard. The cleanup is expected to take five to ten years and cost upwards of $11 million.

MOTCO, TEXAS

La Marque, Texas, is a town of 30,000 located in a strip along the Gulf of Mexico that is dominated by petroleum refineries and chemical processing plants.[139] The MOTCO site, an eleven-acre abandoned industrial waste disposal and recycling facility, sits two miles southeast of town. The facility handled local chemical and petroleum wastes from 1959 to 1974. The site was not actively managed, with wastes deposited in several large unlined pits. In 1974 MOTCO Corporation purchased the site and closed it down after a failed attempt to recycle tars from the waste pits. In 1976, under orders from the Texas Water Quality Board to clean up the site, MOTCO declared bankruptcy.

The MOTCO site hosts approximately 15 million gallons of liquid organic chemicals and contaminated water and 18,000 cubic yards of tars and sludge in seven unlined pits and lagoons and a series of above-ground tanks. Most of the surrounding usable land is employed for industry and agriculture. There are, however, two residential areas within a half mile of the site, with 3,000 people within a one-mile radius and 12,000 within a three-mile radius. At the site's southeastern edge is a coastal marsh that drains to Galveston and Jones bays, then on to the gulf. The site sits a mere five feet above sea level and is within the 100-year flood plain. Rainwater coupled with overfilling has led to recurrent overflows of the liquid waste pits, which have caught the attention of local and ultimately federal officials. La Marque officials had declared the site a health hazard by 1968. In 1980 Coast Guard officials acting under the authority of the Clean Water Act fenced in the facility and extended dykes around it. In 1981, having found shallow groundwater under the site to be contaminated with heavy metals, the EPA proposed the site be placed on the Superfund National Priority List.

Of immediate concern to EPA officials was the prospect of recurrent overflowing of the pits, which would send liquid wastes cascading over the grounds and potentially oozing beyond the site's borders. The EPA mitigated this threat through a series of actions under the Superfund emergency removal program. Some of the liquid waste and contaminated water in the open pits was removed along with waste in several unreliable above-ground tanks.

The EPA's site risk assessment, which would inform the long-term cleanup plan, focused on twenty indicator chemicals selected from the seventy-one waste chemicals identified in the pools, sludges, and soils. Of these chemicals, ten were considered to be carcinogens and ten noncarcinogenic toxics. Among those classed as carcinogens were arsenic, benzene, vinyl chloride, and PCBs.

Testing of soils just beyond the site borders found elevated levels of several of the indicator chemicals, suggesting that some migration had occurred by way of regular surface runoff and the occasional overflowing of pools during past storms. Groundwater beneath the site was found to be contaminated with vinyl chloride that at some locations exceeded its Safe Drinking Water Act maximum contaminant level by 340 times, as well as benzene at 7 times its MCL, and 1,1-dichloroethylene at 13 times its MCL.

In gauging the threat to public health, EPA analysts considered first a current use scenario, which indicated that with the completion of the emergency removal actions the site posed little threat. They found air inhalation to be of no concern. Since there were no private wells in the vicinity, and water supplies for the nearest communities came from a municipal system that drew from surface waters far from the site, the chemical-infiltrated groundwater beneath the site also posed no threat. Of greatest current concern was the contaminated soil outside the site's borders and fencing, which could conceivably be ingested by children from the nearest neighborhoods (which, again, were about a half-mile away).

Also contemplated was a future residential scenario, a worst-case vision in which the areas adjacent to the site, now either vacant or used for industrial or commercial purposes, became new neighborhoods. It was further postulated that these areas, rather than receiving water from the municipal system would draw their water from wells dug at the site's edge. If in the future a hypothetical resident living at the edge of the site drew drinking water from a well at the site's edge and drank two liters a day for seventy years, he or she would incur an excess cancer risk ranging from 3 in 100 to 1 in 10, depending on the exact location of the well. This risk level—assessed from EPA carcinogenic potency factors and presuming additivity of risks across chemicals—reflects high levels of vinyl chloride and arsenic in groundwater beneath the site and adjacent terrain.

Parties responsible for the cleanup objected to this speculative element in the EPA assessment. Even if the areas adjacent to the MOTCO site were to some day become neighborhoods, they argued, groundwaters "would unlikely . . . ever be used as drinking water sources due to low yields, high concentrations of naturally occurring dissolved solids and chlorides, and readily available alternative sources." Institutional controls such as deed restrictions could provide insurance against this unlikely contingency.

Nevertheless, the EPA maintained that the groundwater beneath the site and adjacent terrain had to be considered "potable" (apart from the contamination) and that institutional controls alone would leave future residents vulnerable to "significant levels of risk" from groundwater ingestion. The agency also recalled the Superfund law's expressed preference for treatment and for reducing the volume of hazardous materials at sites, which worked against relying on institutional controls alone in redressing the groundwater ingestion threat. Consequently, the cleanup will remove and treat groundwater in an effort to bring it up to drinking standards. In addition, soils from the site and immediately adjacent areas will be excavated, consolidated, and covered over with clay to prevent further leaching of contaminants to the groundwater below. The projected cost of the cleanup effort is $40 million.

Evidence and Action

In evaluating the Superfund program a decade and a half after its inception, one must be careful to distinguish between its "emergency removal" and "remedial action" elements. The Superfund removal program was a closely focused response to a readily discernible public health threat of limited scope. Aged, inactive hazardous waste sites in some identifiable instances had caught fire, or were too easily accessible to children, or at least in one case had caused acute contamination of well water. Hazards of this sort could be identified with confidence and eradicated at limited expense by cleaning up spilt ignitables, erecting fences, or closing down contaminated wells. By most accounts the Superfund removal program has been successful in this effort. Even observers critical of the higher-profile remedial campaign sometimes offer a tip of the hat to the removal effort: a scientist who pronounced the overall cleanup regime "nauseating" allows that "to its credit, Superfund has allowed EPA to act expeditiously in emergency removals,"[140] while another close observer, now assistant secretary of energy in the Clinton administration, concludes that "EPA has made great strides in eradicating the worst threats that were posed by hazardous waste sites when the program began . . . Its removal program has probably eliminated most of the *immediate* health risks posed . . . To date little credit has been given to EPA for the risk reduction these removal actions achieve."[141] We agree.

Current evidence of chronic health risks from inactive hazardous waste sites, toward which Superfund's centerpiece remedial program is directed, can be summarized as follows:

1. There are thousands of inactive waste sites around the country where waste chemicals have to some degree migrated into the surrounding environment. At many of these sites, chemicals have made their way into underlying aquifers.

2. There is no peer-supported epidemiological evidence of inactive waste sites having caused chronic illnesses such as cancer in surrounding communities.

3. There is no occupational-study evidence of serious illness stemming from chemical exposure levels as low as those associated with waste sites.

4. Administered at dosages thousands of times higher than would be encountered by populations in the vicinity of waste sites, some common waste chemicals have caused cancer in mice and rats.

At each of the five sites described in our case history section trace amounts of waste chemicals have migrated from containment structures into the environment, including groundwater. At none of these sites has public health been found to have been harmed. Though most of the sites were several decades old by the time EPA analysts made risk assessments, there were none showing evidence of health effects attributable to exposure to waste chemicals. According to EPA current use/average exposure scenarios, which extend current land and resource use into the future and base exposure projections on mean concentrations, public health will not in the future be threatened by these sites. These "average" or "most likely" risk projections are benign, despite their reliance on highly conservative animal test inference techniques.

These sites begin to appear moderately threatening only in maximum plausible exposure/residential use scenarios, which take the highest concentrations of chemicals to be representative of entire areas of soil or plumes of groundwater and project that new neighborhoods could pop up on or adjacent to the site and draw their water from underlying aquifers rather than accessible municipal systems. In each instance this worst-case conception of risk guided EPA decisions about the degree of cleanup necessary.

In the administration of Superfund the contemporary inclination to "err on the side of safety" at points of uncertainty meets with a risk situation in which gaps in knowledge are legion; the result is hazard scenarios that multiply unlikelihoods and cleanup plans that sometimes cross the border from the prudential into the absurd. If we had limitless resources, this would not be cause for concern. But we do not, of course, and the billions of dollars we spend ensuring against remote possibilities at abandoned waste sites must be seen as billions that might otherwise be invested, publicly and privately, in health, wilderness protection, education, and other things that we value.

The Superfund program's most recent reauthorization, in late 1990, provoked little debate; observers pointed to congressional and interest group exhaustion in the wake of the Clean Air Act reauthorization fight earlier in the session.[142] Deliberations over the program's next reauthorization promise to be far more rancorous. As cost projections for the cleanup campaign have soared, and as Superfund's liability web has grown to ensnare not just chemical waste producers but local governments, insurers, and banks, calls for reassessment and reform have taken on greater urgency. Questions about how to distribute costs will be pressed on legislators as different groups try to limit their share of the burden. But Congress can best serve the public by carefully reconsidering whether the costs of zealously cleaning up inactive chemical waste sites are justified by benefits to public health.

6

No Runs, No Hits, All Errors:
The Asbestos and Alar Scares

I. Is Asbestos in Schoolrooms Hazardous to Students' Health?

with Darren Schulte

From 1940 to 1978 asbestos-containing materials were used extensively in U.S. schools for their superior thermal and acoustical insulation as well as fire-retardant properties. Since the turn of the century scientists have known of the dangers attributable to high occupational exposures to asbestos, but the possibility of health hazards from the low-level exposures found in schools was not believed worthy of consideration. In a document publicly circulated in 1980 the Environmental Protection Agency (EPA) calculated that "approximately 100 to 7,000 premature deaths" would occur as a result of exposure to asbestos-containing materials over the next thirty years.[1] That year Congress passed the Asbestos School Hazard Detection and Control Act to assist schools in their asbestos-mitigation programs, advise school administrators of possible substitutes, and provide loans.[2]

Background

Asbestos is not a specific substance but rather a term applied to six naturally occurring fibrous silicate minerals that exhibit similar physical properties. Rock forms containing asbestos minerals are widely dispersed on five continents; in the United States alone there are twenty-four states that have high concentrations of these minerals.[3] Natural exposure to asbestos results when these rocks are broken down by weathering.

Since the time asbestos was first incorporated into pottery, as early as 25 B.C., 1 million tons of minerals have been extracted from the earth for use in thousands of applications.[4] Properties such as incombustibility and resistance to moisture, microorganisms, chemical attack, and decay, as well as its ability to insulate against noise, heat, electricity, and wear make asbestos extremely valuable. The flexible fibers of asbestos can be spun or felted into fabrics: Romans 2,000 years ago were known to have woven asbestos into tablecloths that could be thrown into the fire for cleaning.[5] Modern textile and building material manufacturers frequently incorporate asbestos fibers in their products for strength, insulation, and fire retardation.

Asbestos minerals can be divided into two groups: serpentine and amphibole. Of the six principal asbestos minerals only one, chrysotile, may be characterized as serpentine. The other five are considered amphibole minerals: amosite, crocidolite, anthophylite, tremolite, and actinolite. Even though the two groups of minerals have some common properties, there are definite differences in chemical composition, physical characteristics, and properties.

When magnified under an electron microscope chrysotile (white asbestos) resembles hollow scrolls that are curly and pliable. These fine, silky fibers normally occur in bundles like balls of yarn. Even though chrysotile fibers appear to be delicate, they have a tensile strength equivalent to steel wire and are known to withstand temperatures of over 5,000° F. Chrysotile accounts for 90 percent of the world's production and about 95 percent of the asbestos used in the United States, of which 93 percent is mined in Quebec Province, near the towns of Thetford Mines and Asbestos.[6]

In contrast to serpentine chrysotile, the amphibole fibers are smooth, solid rods that are relatively inflexible. Amosite (brown asbestos) fibers, for example, are significantly thicker than chrysotile. Amosite is used more extensively than chrysotile for insulation where greater chemical resistance is desired. Only about 2 percent of the asbestos used in the United States is amosite, imported exclusively from the Transvaal and Cape provinces of South Africa.

Structurally similar to amosite, crocidolite (blue asbestos) is extensively used in asbestos-cement pipes for water transportation because of its chemical stability. Of the total asbestos used in the United States about 3 percent is crocidolite, imported mainly from South Africa and Australia.

Hazards Linked to Asbestos

Pliny the Younger (61–114 B.C.) commented on the sick condition of slaves who worked with asbestos,[7] but not until the beginning of the twentieth century were the specific health hazards resulting from occupational exposure first identified. The major medical problems associated with long-term inhalation of asbestos fibers are pleural thickening and calcification, pulmonary fibrosis (asbestosis), lung cancer, and mesothelioma.

Pleural thickening and calcification results from the formation of plaque in the lining of the chest cavity, which generally does not impair breathing. But if continual high exposure to asbestos persists (usually for decades), the increased plaque stiffens the walls of the airspaces within the lung and begins to affect respiratory functions; then pulmonary fibrosis sets in. In its severe forms, asbestosis can cause death, because either the lungs are unable to provide enough oxygen to the body or the heart is unable to pump blood through the badly scarred lungs, causing cardiac arrest.[8]

ASBESTOSIS

In 1924 Dr. W. E. Cooke published "Fibrosis of the Lungs Due to Inhalation of Asbestos Dust" in the *British Medical Journal,* which reported the first clear case of death due to pulmonary fibrosis.[9] The term "asbestosis" was coined by Cooke in a more detailed study undertaken in 1927.[10] Between 1929 and 1930, Dr. E. R. A. Merewether, medical inspector of factories for the British Home Office, examined 363 asbestos textile workers and found that 26 percent definitely had asbestosis.[11] Merewether also demonstrated that the incidence of asbestosis increased in direct proportion to the length of exposure, reaching 81 percent among a group of workers who had been employed for twenty years or more. As a result of this report Parliament passed legislation in 1931 requiring factories to institute methods to lower the amounts of asbestos dust in the asbestos textile industry.[12]

LUNG CANCER

In 1935, eleven years after the first report of asbestosis, the first account was published of bronchogenic carcinoma (lung cancer) possibly related to asbestos exposure.[13] The case involved a fifty-seven-year-old man who had worked for twenty-one years in an extremely dusty asbestos mill as a weaver. For five years before his death he had been aware of some shortness of breath and complained of a pain on the right side of his stomach. His autopsy revealed lung cancer in his right lung and extensive asbestosis, which explained his stomach pain. Because of the long latency period of lung cancer (fifteen to thirty-five years), however, there was no concrete evidence for a positive correlation with exposure to asbestos.

In 1955 Dr. Richard Doll completed the first epidemiological investigation demonstrating that occupational exposure to asbestos can definitely cause lung cancer.[14] He reported on coroners' autopsies performed since 1935 on 113 men who had been exposed to asbestos for at least twenty years in a British asbestos textile factory. Doll concluded that "the average risk among men employed 20 or more years had been 10 times that experienced by the general population."[15] In the United States Dr. I. J. Selikoff and others have further documented the relationship between lung cancer and occupational exposure to asbestos.

MESOTHELIOMA

In 1960 Dr. J. C. Wagner and his colleagues added mesothelioma to the list of asbestos-related diseases.[16] Mesothelioma is a malignant cancer originating

in the lining of the chest or the abdominal cavity. When in the chest, mesothelioma tends to cover the surface of the lung like an armor plate composed of
dense, often leathery, tissue. In the abdomen, this tumor covers the intestines
and tends to bind them together, causing intestinal obstruction. Mesothelioma
has a long latency period (ten to forty years), but once it strikes it is rapidly
fatal with no known cure. A total of 4,539 malignant mesotheliomas have been
reported worldwide (across twenty-two countries) between 1959 and 1976.
Lung cancer causes over 100,000 deaths in the United States annually; mesothelioma accounts for only 0.001 percent of U.S. deaths.[17]

CIGARETTE SMOKING AND OCCUPATIONAL ASBESTOS EXPOSURE

In 1968 Selikoff and colleagues released a study based on a group of 370
asbestos insulation workers who demonstrated an excess risk of lung cancer
owing to the combined effect of cigarette smoking and exposure to amosite
asbestos.[18] Cigarette smoking is prevalent among asbestos workers, as it is
among workers in other blue-collar occupations.[19] Selikoff and E. C. Hammond later revealed that "asbestos workers who smoked cigarettes had roughly
90 times the risk [of lung cancer] for similar men who neither smoked nor
worked with asbestos."[20]

HOWELL SCHOOL DISTRICT DISCOVERY

In the fall of 1976 considerable public attention was drawn to six New Jersey
schools in the Howell district that had loose, flaking asbestos-containing material (ACM) falling from ceilings. There were claims of children developing
respiratory problems (though the most publicized case turned out to be caused
by mononucleosis), and parents threatened the school district with a strike and
general boycott of classes.[21] As a consequence, the New Jersey Department of
Education requested that school districts report the presence and condition
of ACM in all school buildings within the state. A few months later, under
overwhelming pressure from parents and faculty, the Howell school district
decided to close temporarily the six schools and completely remove all ACM
at a cost of around $200,000.[22] Later we will examine whether the decision
to remove all ACM from schools was necessary based on the actual risk that
these sprayed-on materials posed to schoolchildren as determined by contemporary scientific knowledge.

ACM Characteristics and Uses

Asbestos-containing construction materials such as cement products, plaster,
fireproof textiles, vinyl floor tiles, and thermal and acoustical insulation were
widely used in schools from 1940 through 1978.[23] Sprayed-on material was
used mainly for thermal insulation, decorative purposes, and noise control in
auditoriums, libraries, and classrooms. Hard ACM, such as cement products

and vinyl floor tiles, generally does not create exposure problems; it is mostly sprayed-on insulation that releases fibers into the air when damage from normal wear and tear, vandalism, fire, or water makes the material friable (easily crumbled).[24]

Schools built or renovated from 1940 to 1973 were required to have asbestos insulation as a fire safety measure,[25] even though the EPA had listed asbestos as a "hazardous air pollutant" under its 1970 Clean Air Act, as a result of quantitative evidence establishing asbestos-related carcinogenic effects obtained from numerous studies of workers.[26] In 1973 the EPA banned the spraying of materials containing more than 1 percent asbestos to prevent its introduction into the air.[27] Measurement from 1969 to 1970 of chrysotile asbestos fiber concentrations within a half-mile of construction sites that used sprayed-on asbestos materials had revealed levels 100 times greater than the ambient air.[28]

Nevertheless, the EPA allowed schools to keep using sprayed-on asbestos materials for decorative purposes until 1978, when it implemented a complete ban.[29] No one then thought to see if the air outside schoolrooms contained any more or less asbestos than the air inside.

Public Anxiety and Legislative Action

After the finding of ACM in the New Jersey schools, the Massachusetts Public Interest Research Group began to conduct nationwide surveys to assess the extent of the problem. The Centers for Disease Control alerted state health departments in 1977 that asbestos in schools might be hazardous.[30] In August 1978 Secretary of Health, Education, and Welfare Joseph A. Califano, Jr., wrote the governors of all fifty states to notify them of the problem that sprayed-on asbestos materials posed in schools.[31]

In 1978 a Draft Summary released by the National Cancer Institute (NCI) and the National Institute of Environmental Health Sciences (NIEHS) stated that past occupational exposure to asbestos "is expected to result in over two million premature cancer deaths in the next three decades," or "roughly 17 percent of the total cancer deaths in the next three decades."[32] On one hand, this estimate, if true, revealed a significant unacknowledged threat that demanded massive emergency action. On the other hand, if as many as 17 percent of all cancers could be controlled by asbestos removal alone, that was the best health news for a long time.

This figure came from a paper titled "Estimates of the Fraction of Cancer Incidence in the United States Attributable to Occupational Factors," which was formally released shortly after Secretary Califano alerted the governors. One month after its appearance, Philip Abelson, editor of *Science,* informed readers that the paper was widely condemned by scientists, including one of the world's foremost epidemiologists, Richard Doll of Oxford University, who called it "scientific nonsense."[33] Epidemiologists commonly estimate that all known occupational carcinogens, asbestos included, cause approximately 1 to 5 percent of the total cancer burden, which makes the assertion of occupational asbestos exposure causing 17 percent of all cancer deaths highly unlikely.[34] (Later, in 1981, Doll and his colleague Richard Peto prepared an extensive analysis which concluded that occupational carcinogens of all kinds constituted only 4 percent of all cancer deaths.)[35]

Meanwhile, an EPA telephone survey taken in October 1978 indicated that about 15 percent of the nation's schools contained some degree of friable ACM. In response, the Environmental Defense Fund petitioned the EPA to initiate rule-making procedures on the issue of ACM in schools.[36] The EPA decided in March of 1979 that a voluntary state and local "technical assistance program" was a better method of dealing with the problem and formally denied the petition.[37] This program was designed to help school districts inspect friable ACM and, if necessary, suggest ways to correct the problem using either encapsulation, removal, or closure as abatement techniques.[38] No federal monies were provided to actually mitigate asbestos problems. Considering technical assistance inadequate, the Environmental Defense Fund sued the EPA in May 1979 to begin rule-making procedures. Under intense public and political pressure, the EPA decided to grant the EDF's original 1978 petition.[39]

By the summer of 1979 the EPA's Office of Pesticides and Toxic Substances had provided a set of guidelines for all school districts to help cope with asbestos.[40] Through an EPA Voluntary Asbestos Survey Report circulated to all districts in 1979, it was estimated that 2,992,347 students and 269,311 faculty and staff were "at risk" from asbestos exposure.[41] "At risk" meant that the school had friable ACM capable of leaking asbestos. The potential health hazard posed by asbestos, as determined by the EPA, persuaded the House Education and Labor Committee's Subcommittee on Primary, Secondary, and Vocational Education to conduct hearings on legislative action, which resulted in the Asbestos School Hazard Detection and Control Act of 1980.

Corrective actions cost anywhere from $10,000 per school for small jobs, to $20,000 per school for larger cleanups, to as much as $275,000 for one high school, which in certain cases represented more than a school's annual capital expenditure,[42] but parents and faculty, not to mention many environmental organizations, considered the ultimate safety of children the highest priority. In certain instances parents threatened school districts with lawsuits.[43]

To avert possible bankruptcy, some schools entered the legal arena to try to recover the costs of detecting, containing, or removing asbestos. Under the authority of the 1980 asbestos act, the attorney general launched an investigation in 1981 to determine what legal actions could be taken to force manufacturers of sprayed-on asbestos materials to compensate schools for their abatement expenditures.[44] With the belief that any trace of asbestos was harmful to children, more and more schools opted to remove ACM, deluging the courts with product liability suits that were not certain to be successful. There was panic, there was action, but what was the truth behind these responses?

Scientific Knowledge

Accurate assessment of the potential health risks faced by children in schools with friable ACM depends on the crucial measurement of airborne asbestos concentrations in terms of their ability to become respirable. To understand the scientific evidence concerning the risk of asbestos exposure, one must first become acquainted with the respiratory system and its clearance mechanism with respect to asbestos fibers.

RESPIRATORY DEFENSE MECHANISM

Since all air contains an abundance of suspended particles, humans have developed fairly efficient self-cleansing processes that can remove about 98–99 percent of the inhaled dust from the respiratory tract to guard against possible disease.[45] If an individual smokes, the ability of the clearance mechanisms to remove potentially harmful particles can be significantly reduced,[46] which probably explains the apparent synergism between cigarette smoking and occupational exposure to asbestos.

Because there are several tiers of defense within the human respiratory system, the size of an asbestos fiber determines whether it enters and remains in the lung or is removed. When inhaled, most large asbestos fibers that have a diameter of 10 micrometers (μm) or larger (1 μm equals 1/25 inch) become trapped in the mucus and hair of the nasal passages. Fibers between 5 and 10 μm in diameter can enter into the upper respiratory tract but are cleared away by the continuous movement of mucous fluid toward the throat, resulting either in immediate expectoration or swallowing of the fibers. Smaller fibers, particularly those less than 3 μm in diameter, are able to penetrate much deeper until they come upon the tiny air sacs located deep in the lung. However, if these smaller fibers are less than 5 μm in length, they can be completely engulfed or "swallowed" by scavenger cells (microphages), which are then cleared out of the respiratory tract by processes that are not completely understood.[47] If the fibers are longer than 5 μm (but less than 100 μm) and have a diameter of 1.5 to 2 μm or less, they are considered to be the most significant in inducing carcinogenic effects in humans because they are able to dodge all respiratory defenses and permanently reside within the lung.[48]

Even though an asbestos fiber may exhibit the physical dimensions necessary to bypass the self-cleansing processes, the various asbestos minerals possess different aerodynamic properties depending on their physical structure, which can affect the time spent airborne as well as inhibit the extent of penetration into the lung. In contrast to crocidolite, amosite, and other amphibole fibers, which are stiff needlelike rods, chrysotile fibers are curly and flexible. This curliness promotes clumping into bundles, which allows the fibers to be airborne for a shorter amount of time (relative to amphiboles) and to have a greater resistance to becoming airborne again through agitation. In addition, bundles of chrysotile can easily become intercepted at airway bifurcations, making penetration difficult, whereas individual amphibole fibers can align their major axis in the direction of air flow and then penetrate the peripheral lung more easily.[49] Numerous examinations of chrysotile-exposed workers have shown "an appreciable burden of amphibole fibers, which were only used for brief periods in the workplace."[50] Besides being physically stopped at forks in airway passages, bundles of chrysotile fibers are more likely than amphiboles to be cleared out of the lung by coming into direct contact with the mucous lining of the respiratory tract.

Through animal inhalation and injection studies, it has been shown that fibers below 5 μm appear to be much less toxic than longer fibers because of the presence of scavenger cells within the lung.[51] Very short fibers of 1-2 μm may not even be carcinogenic at all.[52] It should not be assumed that every airborne fiber is bound to become permanently lodged in the lungs. When measuring the concentration of asbestos fibers in the air, scientists look only at those greater than 5 μm in length, and regulatory agencies have adopted this guideline when measuring asbestos levels in buildings.[53]

ACM MEASUREMENTS IN SCHOOLS

In addition, most regulatory bodies count fibers that have a length-to-width (or aspect) ratio of 3:1 or greater, realizing that the aerodynamic diameter is essential for respirability. But there are many analysts, including those at the Occupational Safety and Health Administration (OSHA), who argue that the aspect ratio should be a minimum of 5:1 to be properly classified as asbestos fiber, for fibers of that size or larger have a definite tendency to align themselves with the direction of the airstream.[54] At an aspect ratio of 3:1 it is possible for a tiny mineral sliver to be mistaken for an asbestos fiber. No epidemiological, clinical, or laboratory studies have demonstrated that a mineral sliver can be harmful,[55] so using this lower aspect ratio could lead to an overestimation of airborne fiber concentrations in schools.

Most of the earlier studies (pre–1980) that measured fiber concentrations in school buildings concentrated on the potentially more severe exposures and therefore were not based on representative samples.[56] Taking two estimates of average airborne fiber concentrations can help to demonstrate the extent to which the recent results deviate from the earlier. In both cases the number of fibers per milliliter (f/ml) of air was measured (roughly the amount of air in a thimble), and only fibers longer than 5 μm and having an aspect ratio of 3:1 or more were counted. In 1979, William Nicholsen and colleagues reported that the chrysotile concentrations found in twenty-seven school samples, supposedly indicative of U.S. schools generally, averaged 0.0072 f/ml in comparison with the average reported by B. T. Mossman and colleagues in 1990, which for seventy-one school samples was 0.00024 f/ml.[57] The 1990 figure is exactly three times less than the 1970 one. Recent studies suggest that a very high asbestos concentration in schools is about 0.001 f/ml, which is indicative of heavily damaged ACM.[58]

RISK ANALYSIS

The epidemiological data relating the amount of asbestos exposure to the excess risk of asbestosis, lung cancer, and mesothelioma have come mainly from studies of workers subjected to asbestos over several decades. To assess health hazards from low-level, nonoccupational exposure scientists are forced to extrapolate from the hazards determined from heavy occupational exposures that occurred in the past. The "estimation of potential risks by extrapolation rather than by direct observation necessarily results in considerable uncertainty in the risk estimates, especially when extrapolation is performed over several orders of magnitude," which is true for school building asbestos exposures.[59] The studies that detail the specific hazards from occupational exposures to chrysotile asbestos are most relevant, because friable ACM found in schools contains 95 percent chrysotile.[60]

Most of the analyses involving workers are in the form of cohort studies. A cohort, a group with certain defined characteristics of exposure, is selected and followed to determine the number of members who die within a specified time, which is usually the average latency period of the particular cancer. Typically in occupational cohort studies, the rate of occurrence of death or disease is compared with the rate in some other appropriate external population closely resembling the cohort in age, gender, and race. The results of cohort mortality studies are usually expressed as a standardized mortality ratio (SMR), defined as the observed deaths in the cohort divided by the expected deaths in the external population. When the number of expected deaths due to a particular disease, such as mesothelioma, is so low that it is only barely detectable among 1 million people, the SMR for relative risk cannot be used. An SMR greater then 1 indicates that excess deaths have occurred in the exposed cohort.

The danger of contracting asbestosis at the low level exposures normally found in schools with ACM is nil. As the amount of asbestos exposure decreases to these levels, the "development of incapacitating fibrosis slows down" to the extent "that no person with otherwise healthy lungs would develop significant disability before reaching an age when he was likely to die of other causes."[61]

As already mentioned, the combination of cigarette smoking and occupational exposure to asbestos drastically increases the risk of developing cancer. In 1973, Hammond and Selikoff found that among the 73 deaths in a cohort of 2,066 asbestos insulation workers who did not smoke cigarettes, studied from January 1, 1967, to December 31, 1971, only 2 deaths resulted from lung cancer (versus 5.98 expected).[62] Of these 2, 1 had admitted smoking a pipe or cigars. In sharp contrast, among a cohort of 9,590 workers in the same study who had a history of cigarette smoking, there were 134 lung cancer mortalities (versus 25.09 expected), which was appropriately described as exhibiting an excess risk. In addition, of the 429 deaths observed by Selikoff and others in a later study, only 44 were nonsmokers. Based on the dramatic results, the authors concluded that "lung cancer is uncommon among asbestos insulation workers who have no history of cigarette smoking."[63]

Nonoccupational studies deal primarily with two types of cohorts: those who live in neighborhoods surrounding asbestos factories, mills, or mines and those who live within the household of an asbestos worker who presumably has carried asbestos dust back home on his clothing.[64] These studies usually involve exposure to a single type of fiber, such as chrysotile from Quebec mines or crocidolite from South African mines.

The incidence of mesothelioma among residents of Quebec chrysotile mining regions can provide an accurate assessment of the hazard encountered from nonoccupational exposure. The amount of chrysotile present in the ambient air of the town of Thetford Mines due to normal mining operations has been estimated to be about 0.36 f/ml.[65] To put this into perspective, 0.36 f/ml is 5.5 times less than the 1976 OSHA regulatory standard of 2 f/ml for occupational asbestos exposure and 360 times higher than the fiber concentration of 0.001 f/ml recorded in schools with badly damaged ACM. A 1980 study of all known mesothelioma deaths that occurred in the entire province of Quebec from 1960 to 1978 included only two residents who lived within two miles of the chrysotile mines and mills.[66]

To obtain a deeper understanding of the low risks posed to neighborhoods around the Quebec mines, Dr. R. Pampalon reported the mortality figures for 1966–1977 of women who lived in Thetford Mines (population 20,000) and Asbestos (population 10,000).[67] Since 1900 these mines have produced 4 million tons of chrysotile. An overwhelming proportion of the miners and millers are men, so it can be assumed that the women had not had occupational exposure to asbestos. It was reported that fewer women died than expected (based on death figures for the general population): 1,225 observed compared with 1,356 expected. Furthermore, fewer deaths were attributable to cancers than expected (292 observed versus 321 expected). The most interesting finding was that there were no excess deaths due to lung cancer among these women (23 observed versus 23 expected). This is especially remarkable since the women were exposed to significant amounts of chrysotile in the ambient air, and some of them could have smoked or lived with a miner or miller, thus increasing their risk of lung cancer.

These findings make clear that chrysotile poses no risk to residents living around mines who inhale fiber concentrations that are hundreds of times higher than any reported in schools with friable ACM, composed almost entirely of chrysotile. In direct contrast to the low amount of fatal mesotheliomas in neighborhoods surrounding Quebec chrysotile mines, there is a significant increase in mortality due to mesothelioma among residents who live near crocidolite mines in South Africa.[68]

ASBESTOS IN THE ATMOSPHERE

Asbestos—mainly chrysotile—is present in the atmosphere to such an extent through natural and industrial sources that it is constantly inhaled by most people living in the United States. Rocks containing asbestos are prevalent in half of the lower forty-eight states. At New Idria, California, it is estimated that about 1 million tons of chrysotile escape into the air and the water every year from asbestos containing rock.[69] Industrial sources, such as factories manufacturing asbestos products, contribute to the amount of asbestos in the ambient air. In 1976 the Department of Health, Education, and Welfare published some estimates of asbestos fiber concentrations in the air of major U.S. urban areas.[70] Manhattan and Philadelphia residents were breathing concentrations that were the same or worse than the amount found in schools with damaged ACM: Manhattan—0.001 f/ml, Philadelphia—0.002 f/ml, Washington, D.C.—0.0007 f/ml, San Francisco—0.0008 f/ml, and Los Angeles—0.0009 f/ml exposure. A jogger in downtown Philadelphia may inhale as many as 1.3 million asbestos fibers while exercising.[71]

ACM REMOVAL

The minute health risk in schools containing ACM can paradoxically increase if a school district decides to remove completely any trace of asbestos, for a removal job that requires extensive scraping or tearing will necessarily disperse asbestos into the air. The congressionally chartered National Institute of Building Sciences noted in 1984 that "whether the removal process involves wet or dry disruption of the in-place asbestos, data shows that a substantial quantity becomes resuspended and recirculated throughout the building."[72]

The quality of the building contractor will ultimately dictate whether these airborne fibers are removed during an asbestos abatement project. Certain devices, such as negative-pressure air filtration systems, in conjunction with extensive cleaning and lockdown of asbestos debris, can help clear the air.[73] But the effectiveness of these systems is dependent on the worker's attention to detail and the experience of the contractor. The increased demand for abatement projects in schools has led to a phenomenal rise in underqualified contractors who want to take advantage of the potentially lucrative profits available. In 1986 the EPA estimated that as much as 75 percent of all asbestos abatement work was being conducted improperly, which means that the amount of airborne fibers actually increased in those buildings.[74] Removal of asbestos material has been known to cause fiber concentrations to soar above 100 f/ml. Since the health risks of asbestos exposure in schools are so low in the first place, removal of ACM may not always be advisable.[75]

How Much Concern Is Too Much?

In June 12, 1990, William K. Reilly, administrator of the EPA, gave a talk at the American Enterprise Institute titled "Asbestos, Sound Science, and Public Perceptions: Why We Need a New Approach to Risk." He recalled a conversation upon his appointment with Senator Daniel P. Moynihan, who said, "Above all—do not allow your agency to become transported by middle-class enthusiasm!" What Moynihan meant, Reilly told his audience, was, "Respect sound science; don't be swayed by the passions of the moment."[76] As a conservationist, he continued, he had experience with "the law of unintended consequences."[77]

After meeting with various school officials including a delegation from the U.S. Catholic Conference and absorbing severe criticism of the EPA in the media, Reilly continued, it became clear to him that "a considerable gap has opened up between what EPA had been attempting to say about asbestos, and what the public has been *hearing*." He claimed that the agency had been trying to tell the public that it was desirable to manage asbestos in place by covering it up rather than ripping it out, but school districts were instead beginning wholesale programs of tearing asbestos out, thereby adding many more fibers to schoolroom air. Though perhaps others were also to blame, Reilly thought the EPA deserved its share for not stating clearly enough what the danger was and how it should be dealt with. As a result, he was commissioning a major management review of EPA communications.[78]

EPA regulations and recommendations certainly had not been clear. For example, earlier in his speech Reilly had noted that "most recently, the unusually compelling medical evidence on asbestos led to my decision last year to phase out virtually all *remaining* uses of asbestos in consumer products, in order to prevent the introduction of additional asbestos into the environment."[79] But why? If small amounts of asbestos could be safely left on pipes and walls and ceilings in schools, why had there been a total ban? Our reading of prior EPA regulations is different from Reilly's in that we think school districts could reasonably have concluded that they had to rid their schools of asbestos.

No doubt Reilly was concerned about the type of headlines his agency was getting; one stated, "Risk of One Type of Asbestos Discounted, Health Experts Say Billions May Be Wasted by Removing Banned Insulation Material." What is this about different types of asbestos? Could it be that the type of fibers present in most school buildings are safe at nonoccupational levels? EPA officials were quoted as replying that there were " 'some studies that give a hint that some of these fiber types are less likely to cause cancer,' but that given scientific uncertainties 'we treat them as of equal concern.' "[80] In that hint of a distinction among fibers hangs a tale.

Blue asbestos (crocidolite) comes from South Africa and Australia; brown asbestos (amosite) from South Africa. There is a much higher incidence of cancer of the linings of the lung and abdomen among miners of blue and brown than among workers handling only white asbestos.[81] These two, brown and blue, are classified as naturally occurring amphibole minerals; in the United States chrysotile is also a naturally occurring mineral. Once we realize that asbestos is naturally found everywhere, the task of ridding ourselves of low-level exposures becomes quixotic.

Epidemiologists Doll and Peto concluded that low-level exposures to asbestos increase the risk of dying of cancer by "approximately one death a year" in the whole of the British Isles.[82] (For comparison, playing high school football increases the risk of death by ten deaths per million players.) Hans Weil and Janet Hughes of Tulane University give a conservative estimate of deaths from low-level asbestos exposure of 0.25 deaths per million people.[83] A World Health Organization study done in 1986, "Asbestos and Other Natural Mineral Fibers," disposes of the matter by saying, "In the general population, the risks of mesothelioma and lung cancer attributable to asbestos cannot be quantified reliably and are probably undetectably low."[84] Similarly, the "Report of the Royal Commission on Matters of Health and Safety Arising from the Use of Asbestos in Ontario" (1984) states that "even a building whose air has a fiber level up to 1 times greater than that found in typical outdoor air would create a risk of fatality that was less than one-fiftieth the risk of having a fatal automobile accident while driving to and from the building." Furthermore, "asbestos in building air will almost never pose a health hazard to building occupants."[85]

None of this kept popular magazines from ringing the alarm in stories such as *Good Housekeeping*'s "I Saved My Family from Asbestos Contamination."[86] By contrast, the *New England Journal of Medicine* in 1989 published the balanced conclusion of Brooke Mossman and Bernard Gee: "It remains uncertain whether any type of asbestos acting alone can cause lung cancer in non-smokers." The authors determined that "in the absence of epidemiological data or estimations of risk that indicate that the health risks of environmental exposure to asbestos are large enough to justify high expenditure of public funds, one must question the unprecedented expense on the order of $100 billion to $150 billion that could result from asbestos abatement."[87]

Evidently concerned that this article and a wealth of supporting studies had not reduced the growing costs of asbestos abatement in schools (then running at $4 to $5 billion), a team of five scientists wrote a review of the literature in the January 19, 1990, issue of *Science*.[88] They specifically came out against mass removal of asbestos from buildings. Their review article resulted in an editorial in *Science* condemning existing practice and thus led indirectly to Reilly's speech.[89]

After citing various studies, the five scientists noted with some asperity that *"fiber concentrations* from recent studies *in buildings* [in the United States, France, and Britain] *are comparable to levels in outdoor air,* a point surely relevant to assessing the health risk of asbestos in buildings."[90] Why remove a mineral from buildings that is present in equal concentration in the outside air? In reply, Robert C. McNalley, chief of abatement programs in the EPA's Office of Toxic Substances, wrote to an asbestos abatement trade journal, *Asbestos Issues,* that the EPA was concerned about periodic peak exposures; besides, no level of asbestos could be said to be safe since scientists did not agree on this matter. Yet he added: "Almost every day, we are exposed to some prevailing level of asbestos fibers in buildings or experience some ambient level in the outside air. And, based on available data, very few among us, given existing controls, will ever contract an asbestos-related disease at these low prevailing levels."[91]

The scientists had ended their paper by saying, "Even acknowledging that brief, intense exposures to asbestos might occur in custodians and service workers in buildings with severely damaged ACM, worker education and building maintenance will prove far more effective in risk prevention for these workers."[92] Their conclusion was this:

> The available data and comparative risk assessments indicate that chrysotile asbestos, the type of fiber found predominantly in U.S. schools and buildings is not a health risk in the nonoccupational environment. Clearly, the asbestos panic in the U.S. must be curtailed, especially because unwarranted and poorly controlled asbestos abatement results in unnecessary risks to young removal workers who may develop asbestos-related cancers in later decades. The extensive removal of asbestos has occurred less frequently in Europe.[93]

Does one go with preponderant evidence or does one ask for proof positive that no damage can ever occur from white asbestos? With the second choice—essentially the view that there is no threshold—virtually nothing could be said to be safe.[94]

It is likely that the banning of asbestos for multiple uses, given its extraordinary qualities, will have adverse effects on health. We can also expect increased costs for various products (when other substances are substituted), which leads to a lower standard of living, which in turn reduces health.[95] There are those, like Malcolm Ross of the U.S. Geological Survey, who blame the Challenger spaceshuttle disaster directly on the federal paranoia over asbestos. Asbestos was a prime ingredient of a special putty used to protect the O-rings from the burning rocket gases; when the special Fuller-O'Brian putty was removed from the market because of its asbestos content and replaced by an inferior putty, the hot gases passing through this putty burned out the O-rings. Others point to the consequences of removal of asbestos from brake linings and innumerable other products that may well create more accidents and greater fire hazards and whose replacement fibers may well be more carcinogenic than asbestos at low levels.[96]

Risk Comparisons

In our opinion there is no hazard in classrooms with asbestos fibers unless there is a lot of flaking. The reason there appears to be any hazard at all is that analysts have adopted low-dose linearity, that is, the idea that damage is proportional to the dose and that even the tiniest dose must be rated above zero and assumed to have some effect. Consider average exposures, as calculated by Ralph D'Agostino, Jr., and Richard Wilson.

Insulation workers: average 50–500 f/ml; maximum 2,000 f/ml

Textile workers: average 30 f/ml; maximum 300 f/ml

Asbestos miners and mill workers: average 10 f/ml

Asbestos cement workers: average 6 f/ml up to 60 f/ml

Office workers: average 0.003 f/ml; maximum 0.05 f/ml

School children: average 0.0005 f/ml; maximum 0.01 f/ml[97]

Assuming low-dose linearity, even the infinitesimal amounts of asbestos fibers to which the average person on the street is exposed would have to be rated somewhere above zero, though at the lowest possible level. How low is that? Too low to fathom, especially if we consider D'Agostino and Wilson's risk comparison.

For example, the *annual* risks of death for driving a car (200×10^{-6} per year) are 15 to 100 times higher than those from the asbestos exposure, and thousands of times higher if calculated over a lifetime of driving ($15,000 \times 10^{-6}$). Some people reject such comparisons because automobile driving is voluntary, whereas asbestos exposure in schools is not. However, a pedestrian's risk of being killed by a car is also much higher ($2,000 \times 10^{-6}$ per lifetime). The risk is also higher for a person who drinks chlorinated tap water in a typical U.S. city (200×10^{-6}, calculated by standard EPA methodology), although the numbers here are much more variable. The risks from the asbestos exposure are a factor of 10,000 below that of child-hood death among blacks and minority groups (5×10^{-2})—which suggests where society can best spend money or direct its concern. From such calculations, many experts conclude that there is, in most cases, no significant risk from asbestos in buildings if it is in good condition.

Selikoff . . . accepts this conclusion for teachers and children, but suggests that school maintenance workers might suffer much higher risks. In a study of New York school custodians, Selikoff . . . found many radiological abnormalities and pleural plaques, and suggested that these may indicate asbestos exposure. He admitted that there were many problems in his study, and endeavored to correct for them. Most of the custodians had previous history of asbestos exposure, and most smoked cigarettes, which also scars lungs. Moreover "there was no relationship evident between mixing, fixing, and removing asbestos and the prevalence of asbestos abnormality," which makes his interpretation puzzling . . . Custodians of course must limit their exposure to asbestos, just as they must pay attention to other potential hazards, such as electrocution.[98]

If the behavior of Americans over trace amounts of chrysotile asbestos in buildings were compared to that of a prehistoric tribe warding off evil by sacrificing children, would we come out better?

The Asbestos Epidemic That Never Was

When in 1978 Health, Education, and Welfare Secretary Califano claimed that 17 percent of all cancers in the future would be caused by asbestos, he relied on the unpublished paper "Estimates of the Fraction of Cancer Incidence" (sponsored by the NCI, NIEHS, and National Institute for Occupational Safety and Health). The paper used data provided by Selikoff indicating that there would be anywhere from 5,000 to 58,000 deaths caused by asbestos involving gastrointestinal and lung cancers and mesothelioma. Thus Califano began what became known as the great asbestos epidemic.

Why were the government estimates so far off and so high? The author of the unpublished paper, Dr. Marvin Schneiderman, who was then at the National Cancer Institute, has said that it was a mistake to apply Selikoff's data, which concerned workers heavily exposed, to the much larger population, where exposures were far smaller. "We made the inappropriate estimate that short-term exposures were just as nasty, as carcinogenic and deadly as long-time exposures," Schneiderman stated. "Now it looks as if you have to have fairly continuous exposure" to bring about the worst effects. The worst mistake "was to use Selikoff's estimates without questioning them." While agreeing that the 1978 estimates may have spawned undue concern, Schneiderman still felt that there was sufficient risk, especially among smokers, to justify Califano's publicity program.[99]

By 1986 Congress had passed the Asbestos Hazardous Emergency Response Act, which mandated inspection of material containing asbestos in all schools and required that, if such material were found, schools notify parents and develop management plans, thus initiating a hugely expensive process without sufficient cause.[100] Table 6.1 exaggerates the health consequences of asbestos in school buildings in order to compare it with other known causes of cancer. As Melvin A. Bernarde, associate director of Temple University's asbestos abatement center wrote to the *New York Times* in 1989: "The fear that a few asbestos fibers are deadly or cancer-producing is unwarranted and wholly

Table 6.1   Life's risks

Risk	Expected deaths per 100,000 people by age 65
Smoking (all causes)	21,900
Smoking (cancer only)	8,800
Motor vehicle accident	1,600
Frequent airline travel	730
Coal mining accident	441
Indoor radon	400
Motor vehicle–pedestrian collision	290
Environmental tobacco smoke, living with a smoker	200
Diagnostic X rays	75
Cycling accident	75
Consuming Miami or New Orleans drinking water	7
Lightning	3
Hurricane	3
Asbestos in school buildings	1

Source: Harvard University symposium, Aug. 1989, rpt. and ed. Gary Slutsker, "Paratoxicology," *Forbes,* Jan. 8, 1990, pp. 302–303. Reprinted by permission of *Forbes* magazine. © Forbes, Inc., 1990.

unsupported by current evidence . . . This erroneous message is needlessly exacting a frightful toll, both mental and economic."[101]

A similar price was paid during the Alar scare, as apple growers lost valuable produce and parents were wrongly led to believe that their children were in grave danger.

II. Does Alar on Apples Cause Cancer in Children?

with Maria Merritt

What wondrous life is this I lead!
Ripe apples drop about my head.

    —Andrew Marvell, "Thoughts in a Garden"

The apple scare of 1989 began with an episode of the CBS television program *60 Minutes,* which aired on February 26, reaching an audience of some 40 or 50 million viewers. The image of a red apple overlaid with a large skull and crossbones dominated the background, while host Ed Bradley said, "The most potent cancer-causing chemical in our food supply is a substance sprayed on apples to keep them on the trees longer and make them look better. The EPA's acting administrator, Dr. Jack Moore, acknowledged that [the Environmental Protection Agency] has known about the cancer risk for 16 years."[1]

Presumably Bradley meant to imply that the EPA had knowingly stood idle while a great harm, of a kind the agency had a duty to prevent, was being visited on the public. In fact, the agency had taken several measures to reduce harm from the chemical well before 1989 and had even tried to take it off the market in 1985. An important distinction should be kept in mind as one reads this section: the question of whether regulatory practice violates a legal standard is different from the question of whether harm is being done. It could be that a given legal standard is too stringent, and failure to uphold it causes no harm; or, in the opposite case, a given legal standard could be not stringent enough, so that compliance with it may not ensure protection. Standards are not necessarily truths.

The chemical in question was daminozide, trade name Alar, a growth regulator developed and manufactured by Uniroyal Chemical Company, Inc., and first registered for use on apples in 1968.[2] In a brief on-camera appearance, EPA administrator Moore confirmed Bradley's charge that the pesticide statutes of the day made it difficult to remove a chemical from the market if it had not been a suspected carcinogen when first registered, even if it had in the meantime come under suspicion. (Although daminozide is a *growth regulator,* not a pesticide, it falls under the legal definition of a pesticide in the relevant piece of legislation, the Federal Insecticide, Fungicide, and Rodenticide Act.) Congressman Jerry Sikorski also put in an appearance, saying that if viewers were wondering whether the benefits of using Alar might make it worth the risks, they should go to the cancer ward of a children's hospital to see the "bald, wasting away kids" (who in fact had no connection to Alar).

Next up was Janet Hathaway, attorney for the Natural Resources Defense Council (NRDC). "What we're talking about," she said, "is a cancer-causing agent used on food that the EPA knows is going to cause cancer for thousands of children over their lifetime." Bradley added that Hathaway's group, NRDC, had just completed "the most careful study yet on the effect of daminozide." He was referring to the NRDC report *Intolerable Risk: Pesticides in Our Children's Food* (which was not released to the scientific community, the press, or the general public until the day after the broadcast).[3] Hathaway said the study had found that from eight pesticides taken together, children faced a cancer risk of about 250 times the level the EPA considers safe. She explained, "What that means is that over a lifetime one child out of every 4,000 or so, of our preschoolers, will develop cancer just from these eight pesticides." What significance do these figures have, one may ask? How does the alleged cancer risk from pesticides compare with total cancer risk?

Later in the show Moore expressed the EPA's view that Alar should eventually be taken off the market but was not an imminent hazard and so did not require an immediate ban. Hathaway chided him: "If EPA doesn't think that the most potent cancer causing chemical in our food supply is grounds enough to declare it an imminent hazard and remove it from food, well, I don't know what kind of risk it takes then to declare an imminent chemical hazard."

Rounding off the segment was Ed Groth, whose group, Consumers Union, publishes the magazine *Consumer Reports*. He implied that apple processors who claimed they were not accepting Alar-treated apples from growers had been less than honest. Consumers Union, he said, had analyzed thirty-two samples of apple juice in 1989 and found only nine with no detectable daminozide. (These findings were published in *Consumer Reports* in May 1989.) "It's supermarket roulette," he warned. "You don't know. The consumer can't tell by looking at the bottle whether it's got daminozide in it or not. And unfortunately the consumer can't tell by depending on the manufacturer's assurance that they're using daminozide-free apples because we've shown that isn't valid in all cases." Since Alar is absorbed into the systems of apple trees to which it is applied, it may exist in trees in residual form for several years after application has ceased. Therefore, the presence of trace amounts in the apple products examined by *Consumer Reports* does not mean that any pledges dating from the mid-1980s were broken.

Allegations of deadly peril and dishonesty make a potent combination. Not only were children being harmed but apparently industry, agriculture, and government were coconspirators in this poison plot. Senator Joseph Lieberman of Connecticut spoke for many Americans when, in a Senate hearing, he described his and his wife's reaction to the *60 Minutes* broadcast: "My wife and I . . . were shocked to hear the story about the pesticide Alar and its effect on apples, particularly apple products such as apple juice and apple sauce . . . we went to our food cabinets and, frankly, after the program, threw out all the applesauce and applejuice we could find. We were concerned in doing that not so much about its effect on our own health, but on its potential effect on our baby daughter, Hana, who, coincidentally . . . is 1 year old today."[4]

The next morning the NRDC held news conferences in twelve cities, where it released its report *Intolerable Risk*. On March 7 at a Washington news conference, actress Meryl Streep announced the formation of an activist group called Mothers and Others for Pesticide Limits. With her were several board members of Mothers and Others, including nationally respected pediatrician T. Berry Brazelton and the president of the National Parent and Teacher Association, Manya Unger.[5] The story was covered by many of the leading popular newspapers, magazines, television news shows, and talk shows. It might have seemed to consumers that they were suddenly confronted on all sides by warnings about the dangers of synthetic chemicals in food generally and of Alar in apples specifically.

If consumers did feel bombarded by such warnings, it was no accident; they were in fact the targets of a skillfully organized public relations campaign. In October 1988 the NRDC had enlisted the services of David Fenton and his company, Fenton Communications, to help bring *Intolerable Risk* into the public eye. Fenton wrote in a memo:

> Our goal was to create so many repetitions of NRDC's message that average American consumers (not just the policy elite in Washington) could not avoid hearing it—from many different media outlets within a short period of time. The idea was for the "story" to achieve a life of its own, and continue for weeks and months to affect policy and consumer habits . . . Media coverage included two segments on CBS 60 Minutes, the cover of Time and Newsweek (two stories in each magazine), the Phil Donahue show, multiple appearances on Today, Good Morning America and CBS This Morning, several stories on each of the network evening newscasts, MacNeil/Lehrer, multiple stories in the N.Y. Times, Washington Post, L.A. Times and newspapers around the country, three cover stories in USA Today [one headline read "Fear: Are We Poisoning Our Children?"], People, four women's magazines with a combined circulation of 17 million (Redbook ["Shocking Report: The Foods That Are Poisoning Your Children"], Family Circle ["Forbidden Fruit"], Women's Day ["Fruits and Vegetables That Can Poison Your Kids"] and New Woman), and thousands of repeat stories in local media around the nation and the world.[6]

The announcement by Meryl Streep of the formation of Mothers and Others for Pesticide Limits was arranged to follow one week after the release of *Intolerable Risk*. Fenton explains: "The separation of these two events was important in ensuring that the media would have two stories, not one, about this project. Thereby, more repetition of NRDC's message was guaranteed."[7]

The bottom fell out of the market as apples and apple products across the country were pulled from stores and banned from school lunchrooms.[8] Doctors' offices and poison-control centers were besieged with telephone calls from concerned parents, including a caller who wanted to know whether it was safer to pour apple juice down the drain or take it to a toxic waste dump.[9] When apples and apple products were admitted back into the New York and Los Angeles school systems, it was only on the assurance that no trace of Alar could be found in them.[10] In an effort to calm rampant fears, a joint statement was issued on March 16 by the EPA, the U.S. Department of Agriculture (USDA), and the Food and Drug Administration (FDA), assuring the public that "the Government believes it is safe for Americans to eat apples."[11]

There are no precise figures on the numbers of apple growers forced out of business, but according to the USDA, growers in Washington State alone— where about 6 percent of the U.S. apple crop is produced—suffered at least $125 million in losses during the six months after the scare began.[12] The apple market in general had rebounded by the fall of 1990. Some growers banded together to bring a lawsuit *(Auvil v. CBS)* for product disparagement against the NRDC, Fenton Communications, and CBS News. Eleven families, who claimed to represent all growers in the state of Washington, accused the defendants of publishing "false, misleading, and scientifically unreliable statements about red apples."[13] The USDA later bought $15 million worth of apples in an effort to cut the surplus.[14] Reeling under the blows of consumer reaction, the apple industry announced on May 15 that growers would voluntarily stop using Alar on future crops starting that summer, even though spokespeople for the apple industry trade association maintained along with the EPA that Alar did not pose an imminent health hazard.[15] Meanwhile, further punishment was dealt to the apple market by the article in *Consumer Reports,* which repeated the news—recall the words of Groth on *60 Minutes*—that Alar was detectable in the products of processors who had in the mid-1980s pledged to boycott Alar-treated apples.[16]

On June 2 the manufacturer of daminozide, Uniroyal, announced that it would halt domestic sales of Alar (which was usually applied to apple trees in late June or early July). Uniroyal said that Alar was safe and overseas sales would continue. James A. Wylie, a Uniroyal vice president, said that although there was "no scientifically valid reason to stop using it," the company was pulling Alar from the domestic market to eliminate consumer fears. Uniroyal's director of communications added that the domestic sales halt was also intended to help apple growers who presently had "a severe marketing problem."[17]

In May Senator John Warner of Virginia had introduced a bill to ban the distribution, sale, and use of daminozide,[18] which was withdrawn after the Uniroyal decision to stop domestic sales. A Senate subcommittee report says that the EPA and Uniroyal began discussions about voluntary withdrawal of Alar from the market in "direct response to the introduction of legislation to ban Alar."[19]

In July 1989 the EPA announced it would propose revisions in legislation and regulation governing pesticides, making it easier to cancel the use of a chemical in light of new data suggesting that it might be carcinogenic.[20] In addition, the revised rulings would narrow the definition of economic benefits that could justify allowing such chemicals to remain on the market. Prominent in the agency's rationale for the proposal was concern about future panics of the Alar kind. "The Alar case illustrates that allowing continued use of pesticide products for years after significant health risks are identified," the draft proposal said, "results in loss of public confidence in Government's ability to protect their health."[21] That public confidence declines may well be true. But is it true that Alar in apples is harmful to children's health?

What is Alar? What is the evidence for claims that its residue in apples and apple products is carcinogenic to humans? How does the release of the NRDC's alarming report in February 1989 relate to the long and tortuous regulatory history of Alar? The wave of panic about Alar fed on public fears that government pesticide regulators, charged with protecting consumers' health, could not be trusted. Were these fears justified?

Two allegations did most to fuel the apple scare of 1989: that Alar presented a high cancer risk to humans, particularly to young children, and that the EPA had dangerously, perhaps negligently, underestimated the true risk. Regarding the first allegation, within three years most independent scientists were able to agree that Alar as used on apples is *not* dangerous to humans. For instance, in late 1991 a United Nations panel of experts from most of the industrialized countries reviewed the most recent data and concluded that as used in growing apples Alar presents no danger to people, including children, who eat apples and apple products.[22]

However, since the data that now seem to have settled the matter were not available at the time the commotion over Alar took place, the opinions on which opponents of Alar acted should not immediately be dismissed as misguided. To assess the allegations properly, we must consider in some depth the scientific background as it existed at the time, laying out the relevant studies and examining the interpretations that were placed on them. Along the way, we will also have to delve into Alar's regulatory history.

What Is Alar?

"Alar" refers to the chemical daminozide, the common name for a substance with two synonymous technical names: butanedioic acid mono (2,2-dimethyl-hydrazide) and succinic acid 2,2-dimethylhydrazide. The latter is sometimes abbreviated to SADH. (In dealing with the scientific and regulatory background of Alar, we will always refer to it as daminozide.) At room temperature daminozide is a white crystalline solid, odorless or nearly odorless.[23]

One of the breakdown by-products of daminozide is 1,1-dimethylhydrazine, also known as unsymmetrical dimethylhydrazine, or UDMH. It is estimated that humans metabolize about 1 percent of ingested daminozide to UDMH. In addition, heat accelerates the breakdown of daminozide, so that heating used in the production of apple juice and applesauce converts about 5 percent of daminozide residues in apples into UDMH, which remains in the processed food product.[24] At room temperature UDMH is a liquid with a density of 0.79 grams per milliliter (g/ml) and a boiling point of 63–64°C. UDMH itself has various technological applications, most notably as a component in rocket fuel.[25]

Daminozide is used as a plant growth regulator for food and ornamental crops. It causes the cells to grow more densely.[26] On fruit like apples, this density affects fruit set and maturity, firmness and color, fruit drop, and thus the quality of salable fruit at harvest and during storage.[27] Its greatest economic benefits for apple growers and consumers are reduction in fruit drop and increased storage life.[28] Avoiding preharvest fruit drop is important for growers, because some types of apples drop soon after they ripen and individual apples ripen at different rates, so without a way to keep apples on trees longer, growers need to make a series of harvests rather than just one. A series of harvests makes for higher labor costs and wasted apples.[29]

When used by apple growers, daminozide is dissolved in water and applied as a spray from ten days to three weeks after the trees have reached full bloom. Thus it is incorrect to say of daminozide, as Ed Bradley did in the *60 Minutes* segment, that it is sprayed directly onto apples. The daminozide-water solution is absorbed through the foliage. The label recommends that trees be sprayed only once in a growing season.[30]

Daminozide was first registered by Uniroyal in 1963 for use on chrysanthemums and first registered for food use on apple trees in 1968. By 1989 daminozide had food-use registrations not only for apples but also for other orchard crops—cherries, nectarines, peaches, pears—and peanut and tomato plants.[31]

The Claims in Question

The claims that apparently had the greatest influence on consumers during the 1989 apple scare were those quoted or paraphrased from the 1989 NRDC report, *Intolerable Risk*. Accordingly, the claims we have chosen to assess are quotations from that report. First note that although the growth regulator Alar is a "pesticide" by legal definition only, it is made guilty by association in the subtitle of the NRDC's document: *Pesticides in Our Children's Food*.

Claim 1. "These staples of children's diets [fruits and vegetables] routinely, and lawfully, contain dangerous amounts of pesticides, which pose an increased risk of cancer, neurobehavioral damage, and other health problems."[32]

This amounts to saying that one of two things is the case: either the regulatory authorities are acting in bad faith when they set legal standards, or they are acting in good faith but are ignorant or neglectful of some crucial element(s) that, if taken into account, would lower the legally permissible limit considerably.

The NRDC report charges the EPA with failure to recognize at least two factors that together would contribute to a greater risk estimate: actual dietary exposures higher than those used in EPA risk assessment and, more important, greater susceptibility of children to lifetime cancer risk from pesticide exposure than EPA risk models allow for.[33] How much do these technical differences contribute to the difference between the EPA's moderate caution and the NRDC's uncompromising fear? We need some scientific background to consider the question thoughtfully and will provide that later.

Claim 2. "Preschoolers are being exposed to hazardous levels of pesticides in fruits and vegetables. Between 5,500 and 6,200 (a risk range of 2.5×10^{-4} to 2.8×10^{-4}) of the current population of American preschoolers may eventually get cancer solely as a result of their exposure before six years of age to eight pesticides or metabolites commonly found in fruits and vegetables. The potent carcinogen, unsymmetrical dimethylhydrazine (UDMH),[34] a breakdown product of the pesticide daminozide, is the greatest source of the cancer risk identified by NRDC. The average preschooler's UDMH exposure during the first six years of life alone is estimated [by NRDC] to result in a cancer risk of approximately one case for every 4,200 preschoolers exposed. This is 24 times greater than the cancer risk considered acceptable by EPA following a full lifetime of exposure."[35]

These fearful-sounding risk projections report the figures the NRDC calculated using different dietary exposure estimates and a different method of risk modeling from those used by the EPA. The language of the report here, and in other claims as well (see Claim 5 below), implies that the figures represent actual harm—that they can be used to make fairly reliable predictions about what will happen.

Claims 3 and 4. The following statements appear in Table 1-1 of the NRDC report: Pesticides and Metabolites Evaluated in NRDC Study and Their Potential Health Effects: "Daminozide: 'Probable human carcinogen' [citing EPA classification]; also contains and breaks down to UDMH (see below); causes multiple tumors at multiple organ sites (lung, liver, kidney, reproductive and vascular systems) in animals."

"UDMH: 'Probable human carcinogen'; causes multiple tumors at multiple sites (lung, liver, pancreas, nasal tissue, vascular system) in animals; mutagen."

The sources the NRDC cites for these statements include EPA Registration Standards, Pesticide Fact Sheets, and Special Review Position Documents. In other words, the NRDC does not in this instance question the EPA's judgment but rather relies on it for support. To assess these claims we need to assess the relevant EPA methods and reasoning.

Claim 5. "Out of the current preschool population, as many as 5,300 children may develop cancer at some time in their lives as a result of only their preschool exposure to UDMH."[36]

Again, what sense should we make of this number? And what does it mean to say "may develop cancer"? Does this mean we can be quite sure they will, or that there is some indeterminate possibility that they will? Does that possibility include the probability they will not?

Claim 6. "Apple products, including fresh apples, apple sauce and apple juice, were the greatest contributors to the risk, contributing over 74% of the risk."[37]

We will not examine this claim. There is no dispute that the predominant food use of daminozide was on apple trees. We have included this quotation only to suggest how the chemical the NRDC cited as posing by far the greatest cancer risk, within the report's group of suspected carcinogens, could have come to be so easily linked in the media and in the public mind with such a potent symbol of health, well-being, and natural goodness as the apple. The imaginative association was, as it were, ripe for the picking.

Animal Studies and the Regulatory History of Daminozide

Animal tests for cancer-causing potency have been performed for both daminozide and its metabolic breakdown product UDMH. For each study, where possible, we have calculated the experimental dose in milligrams per kilogram of body weight of the test animal per day (mg/kg/day).

As early as 1961 Mary F. Argus and Cornelia Hoch-Ligeti reported that at a dose which, we have calculated, started at about 3.5 mg/kg/day and (as the test animals grew to adulthood) ended up being around 0.7 mg/kg/day, UDMH did not induce tumors in rats. Margaret G. Kelly and colleagues reported in 1969 that UDMH given in a weekly dose of about 34 mg/kg had "no demonstrable carcinogenic activity in mice." However, F. J. C. Roe led a study in 1967 in which mice given UDMH at a dose of 33 mg/kg/day showed an increased incidence of lung tumors as compared with controls.[38]

In a paper published by Bela Toth in 1973 mice given UDMH at a dose of about 23 mg/kg/day (our calculation based on his reported dose of 0.01 percent solution of UDMH in drinking water) developed high incidences of blood vessel, kidney, and lung tumors as compared with controls. Toth in 1977 reported that mice given in drinking water a 2 percent solution of daminozide (according to our calculations, a dose of around 5,700 mg/kg/day) had high incidences of blood vessel, kidney, and lung tumors as compared with controls.[39] (Weights are not reported in the Toth experiments; our calculation of dose in mg/kg/day is based on an assumption that the mice weighed the same as in other experiments—30 g.) In another paper of 1977 Toth observed that hamsters given UDMH developed bowel and blood vessel tumors, whereas only 1 of 200 untreated controls developed a tumor. The UDMH solution the test hamsters were given was ten times more concentrated than the solution given to mice in Toth's 1973 experiment.[40]

In 1978 the National Cancer Institute tested daminozide in both rats and mice. Rats of both sexes were given daminozide at doses of 500 mg/kg/day, while mice of both sexes received 1,500 mg/kg/day. Other rats and mice in the study were given half these amounts. Only female rats (higher incidence of uterine tumors than controls) and male mice (higher incidence of liver tumors than controls, in a dose-related trend) presented evidence for the carcinogenicity of daminozide.[41]

In 1980, going mainly on the evidence published on daminozide in Toth's 1977 study and the NCI's 1978 study, the EPA planned to begin a Special Review process for daminozide.[42] This is a procedure EPA may follow to consider banning a previously registered chemical for which new suspected harms, not known when the chemical was initially registered, have come to light through a "valid laboratory study."[43] If such findings apply to a chemical that is already widely used, the government bears the burden of proving that the chemical presents an unacceptable risk. A Special Review can be called for when the EPA has found that a chemical meets one or more of the criteria given in title 4, part 162, of the *Code of Federal Regulations,* governing the EPA's implementation of the Federal Insecticide, Fungicide, and Rodenticide Act (FIFRA).

Rather than proceed immediately with a Special Review in 1980–1981, the EPA conducted a comprehensive evaluation of all available findings on daminozide, particularly studies suggesting that it had cancer-causing effects in lab animals. In reaching a preliminary conclusion that daminozide residues on food crops did present an unreasonable risk to human health, the EPA placed considerable weight on the belief that UDMH, daminozide's breakdown by-product, is carcinogenic to lab animals.[44]

On August 25, 1983, the agency issued a Special Data Call-In Notice to the registrants of daminozide, requiring them to perform further studies. The two registrants were Uniroyal Chemical and Aceto Chemical Company, Inc. Aceto's registration was suspended in 1984 because it did not comply with the data call-in. Uniroyal voluntarily canceled use of daminozide on plums and prunes, brussel sprouts, peppers, and cantaloupes, instead of doing the required residue studies for these crops.[45]

Apparently the early Special Review plans were suspended after meetings between the EPA and Uniroyal. These meetings were closed to the public, a fact that prompted the NRDC to introduce litigation on the matter of private meetings about regulation between the EPA and pesticide manufacturers. The settlement required the EPA to hold open meetings on regulation of pesticides.[46]

On July 18, 1984, the EPA formally opened a Special Review for daminozide, citing the results from the three Toth studies and the NCI's 1978 study.[47] The summary of their Position Document says that "EPA has determined that daminozide and its hydrolysis product unsymmetrical 1,1-dimethylhydrazine (UDMH) are oncogenic in laboratory animals. Daminozide and UDMH have been found in both raw agricultural commodities and processed food: therefore, a risk may be present to human health."[48]

Typically, the next step in a Special Review process involves accepting and evaluating further information and comments, performing a risk and benefits assessment, and then coming up with a proposal for regulatory action. The assessments along with the proposal are written up in a draft Position Document, which is then submitted for comment to the registrants, the secretary of agriculture, the public, and the FIFRA Scientific Advisory Panel (SAP).[49] The SAP was established by Congress in order to subject to scientific review any EPA proposed actions that may be harmful to business. Its members are selected by the White House from candidates nominated by the National Science Foundation and the National Institutes of Health.[50]

In the draft Position Document for daminozide, the EPA proposed cancellation of all food uses. The SAP meeting to consider the proposed regulation and its rationale was held September 26–27, 1985. According to several accounts the testimony before the panel was a dramatic confrontation between EPA and Uniroyal witnesses. The gist of the EPA's strategy was a sort of commonsense view of the studies in aggregate: "one cannot ignore the fact that there were about seven studies here on the parent [daminozide] and the metabolite [UDMH], and that there appears to be a common thread that runs through most of these studies, in terms of a certain tumor being seen."[51] By contrast, Uniroyal's strategy was to attack point by point and "to discredit each study individually, so as to leave no credible support for the proposition that either daminozide or its metabolite UDMH were carcinogens."[52] This strategy carried the day with the panel. The resulting decision against the proposed ban on daminozide was an embarrassment to the EPA, because it appeared to discredit the agency's ability to pass credible judgments on the scientific evidence.[53]

On the question of the EPA's *qualitative* assessment of the oncogenic potential of daminozide, the SAP said: "The data available are inadequate to perform a qualitative risk assessment. The Toth Alar studies do give rise to concern over the potential oncogenicity of daminozide. The Panel feels that the other studies are equivocal." On the issue of *quantitative* risk assessment—whether the animal studies are adequate as a basis for extrapolation of numerical estimates of human risk—the SAP responded: "None of the present studies are considered suitable for quantitative risk assessment." In general, the panel expressed "reservations over the technical soundness of the Agency's quantitative risk assessments as regards daminozide and UDMH in both raw and processed food."[54]

The panel's objections centered on various deficiencies in scientific procedure. The NCI study was criticized for the low number of its controls and the unclear or marginal statistical significance of its oncogenicity results. As for the three Toth studies, even if the panel conceded that they showed a statistically significant relationship between tumor development in test animals and treatment with daminozide or UDMH, their suitability for regulatory decision making was undermined by several factors, two of particular importance: (1) only a single dose level was used in each study (without more than one data point charting the relation between dose and response, no dose-response curve can be plotted, so there is little or no basis for extrapolating to responses for exposures at lower doses); and (2) the dose given to the animals was so high that the possibility of tumor induction by sheer toxicity could not be ruled out. Consequently, "after careful consideration of the recommendations of the Scientific Advisory Panel . . . EPA determined that the data base was inadequate to support evaluation of the carcinogenicity of daminozide and UDMH."[55] The EPA did, nonetheless, impose conditions on the continued food uses of daminozide. Most notably, Uniroyal was required to conduct its own animal cancer studies and to test residues in retail food products. The EPA also required that application rates and allowable levels of daminozide in foods be reduced.[56]

The EPA announced in January 1986 that it would not ban Alar and stirred up a certain amount of negative publicity at the time, particularly on morning television news programs. Interestingly, this did not have as marked an effect on consumers' concerns as it did on apple processors who *feared* consumers' concerns. Major processors such as Mott's, Veryfine, and Red Cheek told growers and suppliers that they would refuse to buy any Alar-treated apples from the 1986 crop. Heinz, Beech Nut, and Gerber rejected Alar-treated apples for processing into baby food. A comment by John Brock, an executive in the company that owns Mott's, is representative: "[The boycott] has nothing to do with the safety of Alar or the position that the E.P.A. has taken. The reason is simple. There has been so much negative publicity that continued use of apples treated with Alar could hurt consumer confidence in the Mott brand name."[57]

Lawrie Mott of the NRDC—who later edited the NRDC's 1989 report, *Intolerable Risk,* and who has, we assume, no family relation to the Mott's brand of apple products—commented: "It's a scandal that the E.P.A. allows a suspected carcinogen to remain in our food supply. The processors' stand is laudable—it is an interesting demonstration of how the market works, the processors acting before the E.P.A."[58] One may ask, however, whether "the market" could have worked as it did without the flurry of negative publicity, perhaps promoted by groups like the NRDC, on popular news programs. It looks as though the processors, anticipating consumer reaction, were more sensitive to this publicity than the consumers themselves turned out to be in 1986. The allegations made by Alar's enemies did not change substantially between 1986 and 1989. Could it have been their revamped publicity campaign alone that made the difference?

Complying with the earlier Data Call-In Notice, in August 1988 Uniroyal submitted the results of studies that tested daminozide's cancer-causing potential in rats and mice. In the rat study, where both sexes were fed daminozide for two years at doses ranging from 5 to 500 mg/kg/day, there was no evidence that daminozide caused cancer. In mice, given daminozide for two years at doses ranging from 50 to 1,500 mg/kg/day (these are rough estimates), there was a statistically significant dose-related trend in increased numbers of liver blood vessel tumors, but there was no statistically significant increase by comparison of treated animals with control groups. At the highest dose among male mice, the combined incidence of benign and malignant liver blood vessel tumors was ten out of forty-eight mice, as compared with six out of fifty control mice; for female mice, the highest-dose incidence was seven out of fifty mice as compared with four out of forty-six controls.[59]

The EPA gave the following reasons for its conclusion relating the high-dose increase in tumors to treatment with daminozide: (1) the high-dose tumor incidence is nearly twice that of controls tested at the same time; (2) the high-dose tumor incidence is outside the range of historical controls for this kind of tumor in this strain of mice (compared with control animals of the same species in other experiments, the high-dose mice in this experiment had significantly more tumors); (3) the incidence of this kind of tumor in the control group for this experiment is *also* outside the historical range (we cannot see how this could count as a convincing reason for attributing the observed effect in the high-dose animals to treatment with daminozide); and (4) this is the same type of tumor observed in the early studies that prompted the Special Review to begin with.[60]

Under the EPA's requirements, Uniroyal was also conducting oncogenicity studies for UDMH on rats and mice.[61] In the mouse study, mice were given UDMH in drinking water for two years at concentrations of 1, 5, 10, and 20 parts per million (ppm). One-year interim results showed no significant increase in tumor formation for treated animals as compared with untreated controls, but there were some signs suggestive of liver toxicity.[62] (The final results too showed no UDMH-related oncogenic effects in male mice. An increase in lung tumors among females at the highest dose was regarded as only equivocal evidence for the oncogenicity of UDMH.)[63]

Based on the 1987 results of preliminary UDMH toxicity studies by Uniroyal, the EPA required the company to run an additional mouse study at higher doses,[64] because in the agency's view the doses of 10 and 20 ppm were not reaching the maximum tolerated dose (MTD). A brief explanation here may be useful. It has been standard regulatory practice for the highest dose of a test to reach the MTD of a substance for a given animal; that is, a dose high enough to elicit some signs of toxicity but low enough so that the animals do not die earlier than they would if untreated. If no tumors appear even at this dose, regulators assume that rules protecting the public against a chemical's toxicity in general will also protect it against cancer risk from that chemical.[65] But what if increased numbers of tumors do appear in the animals at the MTD? Do tumors or other ill effects at the MTD result from carcinogenic activity of the test substance or are they symptoms of poisoning caused by sheer overload?

Uniroyal and the EPA had conflicting opinions about the MTD for UDMH in mice.[66] Uniroyal presented a case for its being at 20 ppm, but the EPA disagreed and required a second, high-dose, UDMH mouse study with doses at 40 and 80 ppm. Uniroyal contended that the MTD was exceeded at both those dose levels, citing "excessive mortality and clinical pathology results in treated animals that are indicative of severe hepatotoxicity and toxicity to the hemo-poietic system."[67] The one-year interim results showed an increased incidence of blood vessel tumors in the liver and of lung tumors in male and female mice. The EPA's quantitative cancer risk assessment, and its regulatory action of 1989, were based on the malignant blood vessel tumor incidence in the group receiving 80 ppm.[68]

An expert panel appointed by the British government, upon reviewing the interim results in December 1989, concluded that the small amounts of dami-nozide and UDMH in food presented "no risk to health." The chairman of the panel was Colin Berry, a pathologist at London Hospital Medical College and president of the European Society of Pathologists. He endorsed Uniroyal's judgment that the mice were given too much UDMH in this study for any sensible extrapolation to human exposure. "We don't always make the assumption that the animal data are transferable to man," Berry explained, "particularly in the absence of pharmacokinetics that make it clear that the compound is handled in the same way at massive doses as it is at low doses."[69]

On the basis of the high-dose UDMH mouse study, the EPA proceeded toward cancellation of daminozide through the Special Review, in the mean-time extending the permissible use of daminozide for another eighteen months, since human exposure over this time would not constitute sizable risk. The agency published its decision in the *Federal Register* on February 1, 1989, over two weeks before the *60 Minutes* episode that broke the "story" on the EPA's allegedly scandalous casual attitude.

Risk Assessment

Risk assessment as an adjunct to regulatory policy making involves estimating probabilities for various events of concern to regulators. Imagine that you are considering the probability of some event in the absence of relevant informa-tion. The most you can say to begin with is that the probability of the event's occurrence lies somewhere between zero (it can never happen) and unity (it will certainly happen). Thus risk assessment is not so much a way of making reliable predictions as it is a discipline designed to cope with lack of informa-tion. In keeping with this emphasis, it can be argued that the task of risk assess-ment has two key components: (1) "to use whatever information is available to obtain a number between zero and one for a risk estimate, with as much precision as possible,"[70] and (2) to give an estimate of the imprecision. For any particular risk assessment, then, the presentation of results would ideally include the probability estimate, an account of the types of information avail-able and the way the information was used to arrive at the probability estimate, and a reminder of the imprecision involved.

The principal question the EPA addresses is this: does exposure to the sub-stance have the potential to increase the incidence of human cancer? Evidence taken into account in making this judgment includes human epidemiology, toxic effects, short-term tests for genetic toxicity, and, primarily, long-term animal studies at or near the maximum tolerated dose. Under the 1986 guidelines, lack of positive results in the short-term tests "does not provide a basis for discounting positive results in long-term animal studies."[71] Moreover, in considering the results of animal studies, a "positive carcinogenic response in one species/strain/sex is not generally negated by negative results in other species/strain/sex."[72] The EPA, which follows these conservative practices so as to maximize the probability of finding cancers, concluded by classifying both daminozide and UDMH as "probable human carcinogens."[73]

The next step after hazard identification is dose-response assessment, which involves extrapolating from high-dose animal studies to low-dose human exposure, as well as considering the validity of the extrapolation. EPA guidelines acknowledge the major uncertainties in extrapolating both from animals to humans and from high to low doses.[74] However, the uncertainties are not commonly mentioned in public announcements of EPA regulatory decisions, nor are they commonly presented alongside the numerical risk estimates the agency cites in support of its decisions. Three types of decisions come up in dose-response assessment: selection of data, choice of mathematical extrapolation model, and choice of equivalent exposure units among species.

Selection of data.   When there are no human studies, data from a species responding most like humans are preferred. If several animal studies are available, involving different species, strains, sexes, and dose levels, the data from studies showing greatest sensitivity to the substance should weigh the most heavily as evidence. (Such data will be used as long as they are "biologically and statistically acceptable.") The rationale for the bias toward the most sensitive test animal is the possibility that human sensitivity is as high as that of the most sensitive animal group studied.[75]

The agency selected the data on animals receiving the highest dose from the interim results of the 1989 high-dose UDMH mouse study. The type of tumor selected was hemangioma/hemangiosarcoma (blood vessel tumor, both benign and malignant tumors taken into account) of the liver. These selections were made for the following reasons: first, hemangiosarcomas are uncommon malignancies and occur at a low background rate in the strain of mice tested; second, hemangiosarcomas are the same type of tumors as those observed in the earlier UDMH and daminozide studies; and third, since similar malignancies were found in mice given doses of 40 ppm and killed for observation after one year, a dose-response relationship would be likely for this tumor type at the completion of the two-year study.[76]

Was it appropriate to use these data and this reasoning as the basis for numerical risk assessment and for regulatory action?[77] If making a reasonable decision means conforming to legal and regulatory mandate, then the EPA's decision was reasonable. If making a reasonable decision means deciding in accord with the preponderant scientific evidence, the reasonableness of the decision is doubtful.

Choice of mathematical extrapolation model. EPA guidelines recognize that different mathematical models may fit the observed data equally well and yet result in large differences in hazard projections for low doses. The model recommended by the guidelines, when there is not adequate information about the mechanism of carcinogenesis to count against it, is the "linearized multistage procedure." This model (mentioned in the section on asbestos) is based on the "one-hit" theory, that "a single exposure to a carcinogen at a cellular target can initiate an irreversible series of events that eventually lead to a tumor."[78] To adopt the one-hit theory for the purposes of risk assessment is to assume, as the EPA generally does,[79] that a zero response for human carcinogenesis occurs only at a zero dose, and that the probability of cancer increases linearly as a function of dose at all points above zero.

The 1986 EPA guidelines acknowledge that an estimate derived from the linear model "does not necessarily give a realistic prediction of the risk. The true value of the risk is unknown, and may be as low as zero."[80] This acknowledgment shows that the EPA builds policy into its risk assessment procedures, choosing the model that, if inaccurate in its depiction of actual danger, will always err on the side of predicting greater danger, sometimes much greater.

Choice of equivalent exposure units among species. The EPA handles interspecies extrapolation by expressing doses in units of milligrams per square meter of body surface area per day. This measurement is based on an assumption that "different sized animals are not equally sensitive to equal concentrations."[81] In particular, we may add, the assumption is that larger animals (humans) are more sensitive. This is considered a conservative assumption, because other factors (such as differences in metabolism and in susceptibility among species) that vary independently of body size may affect comparative sensitivity. Thus, because EPA scales base extrapolations on relative surface area, whereas the FDA's are based on relative weight (mg/kg body weight), the EPA's estimates of human risk based on animal data are higher than the FDA's.

All of the factors worked out in the stage of dose-response assessment are used in a formula to calculate the "cancer potency estimation," or Q^{*1}. This is the estimated potency per unit exposure to the substance, expressed in units of $(mg/kg/day)^{-1}$. In 1989 the EPA did not calculate a Q^{*1} for daminozide, opting instead to calculate one for UDMH (the calculation being based, as we saw, on data from the Uniroyal high-dose mouse studies) and to peg regulatory action to that. EPA investigators believed that responses to daminozide are actually caused by the UDMH already present in daminozide and the additional UDMH generated through metabolic conversion in test animals. Moreover, they had relatively good metabolism and exposure data for UDMH. This is why the EPA regulated for UDMH and not daminozide per se. The Q^{*1} for UDMH was estimated using the Crump Global 86 linearized multistage model at 0.88 $(mg/kg/day)^{-1}$.[82]

Here is a very important point at which the EPA and the NRDC differed in 1989. In deriving the risk assessments published in its report *Intolerable Risk,* the NRDC used the EPA's earlier Q^{*1}, or cancer potency estimation for UDMH, calculated in 1984.[83] This earlier EPA estimation was calculated from the results of Toth's 1973 study (in which mice were given UDMH at a daily dose of about 23 mg/kg), and came out to be 8.7 $(mg/kg/day)^{-1}$—higher than the 1989 EPA Q^{*1} by a factor of almost exactly ten.[84]

Recall that in 1985 the Scientific Advisory Panel, and the EPA following the panel's recommendation, had rejected the Toth studies as suitable bases for quantitative risk assessment, the two main reasons being that the animals were tested at only one dose and that the MTD might have been exceeded. By comparison, the high-dose Uniroyal mouse study undisputably met the first of the two standards not met by Toth's 1973 study, while the EPA and Uniroyal differed over the second point. If we admit the possibility at all of using animal studies to extrapolate to human risk, and if we set those two criteria on the admissibility of animal data for such purposes, then the data from the 1973 Toth study fail to meet both, while data from the Uniroyal high-dose mouse study may fail to meet one. Imagining that it is reasonable to pose the choice at all, it would seem to make most sense to take the lesser of two evils and give credence to the 1989 Q^{*1} over the 1984 Q^{*1}.

Another substantial point at which the EPA and the NRDC differed was in the choice of extrapolation models. Whereas the EPA's linear model is time-independent, the NRDC uses a time-dependent linear model. The NRDC model yields a higher estimate of risk for children. The EPA's method of risk estimation gives special consideration to children only to the extent it takes children to be a more susceptible subpopulation in terms of dietary exposure. Children consume more food, thus more pesticide residue, relative to body weight.[85] The scientific issue as to whether, in general, the best model is time-independent or time-dependent is at present unresolved. The National Academy of Sciences is considering whether pesticide residues on food pose a special risk to children in this respect.

Finally, there is the summing-up step known as "risk characterization." For any suspected carcinogen a cancer potency factor, calculated at the earlier stage of dose-response assessment, is multiplied by the estimated human exposure. The product is the risk estimate itself—the probability that somebody who would not otherwise have gotten cancer will get cancer from lifelong exposure to the substance. The EPA's estimate of average individual lifetime risk was 4.5×10^{-5}, or 0.000045. This figure means that, on the conservative assumptions we have described, an individual exposed to UDMH (via exposure to daminozide) over a lifetime (seventy years) could be expected to increase his or her risk of getting cancer by odds of 45 in 1 million.[86] The figure, moreover, represents a 95 percent upper-bound risk, which means there is a 95 percent certainty that the risk would be no greater than 4.5×10^{-5}.[87]

The NRDC estimated an individual's increased cancer risk *solely from childhood exposure* to UDMH/daminozide at 1 in 4,200, or about 24 in 1 million.[88] Since the EPA's usual benchmark for relative safety is an increased risk of 1 in 1 million, we can see where the NRDC gets the expression "240 times higher than EPA considers safe."

It is worth pointing out that the NRDC's estimate of added lifetime risk from *childhood exposure* and the EPA's estimate of added lifetime risk from *lifetime exposure* differ by less than a factor of 6. In other words, something other than a difference in the numbers must explain the dramatic difference between the two organizations' attitudes toward Alar.

Someone could object that if the added risk from childhood exposure alone is as great as the NRDC estimate suggests, then the added risk from a whole lifetime of exposure must be horrendous. But if the point of the NRDC estimate of risk is to say how much worse exposure in childhood is than at other times of life, we would expect rather that the risk tapers off so that most of the lifetime risk comes from childhood exposure. Indeed, the authors of *Intolerable Risk* say exposure in the first six years of life accounts for "more than 50% of a person's lifetime cancer risk from exposure to carcinogenic pesticides used on fruit."[89] It would make sense, then, to think of the EPA's estimate of lifetime risk from total lifetime exposure, on the one hand, and the NRDC's estimate of lifetime risk from childhood exposure alone, on the other hand, as roughly similar estimates of lifetime risk.[90]

The final difference between the NRDC and the EPA was over how many apples and apple products children typically eat. (The NRDC used the same residue figures the EPA had derived from Uniroyal market basket surveys of 1986 and 1987.)[91] The consumption survey the NRDC used was more recent than the EPA's but surveyed only 489 children.[92] Both samples have advantages and disadvantages. It is likely that the EPA's data, on 2,000 children ages one to six, constitute a better statistical sample since the net is cast wider, but the data the NRDC used might reflect more accurately the supposed increase in fruit consumption among children over the years separating the two surveys.[93]

The Claims Revisited

In the case of Alar, let us return to Claim 1, which states that apples and apple products routinely, and lawfully, contain dangerous amounts of daminozide and UDMH, which pose an increased risk of cancer. Whether we accept this as true or reasonable against the scientific background will depend primarily on how we understand the key term "dangerous" and secondarily on how we understand the expression "increased risk."

By EPA standards, people with different views of what danger is and what things are most dangerous may answer the question in different ways. For example, a person's annual risk of death from a motor vehicle accident has been estimated at 2.4×10^{-4}, which means that *every year* a person has a 2.4 in 10,000 risk of death from an auto accident.[94] Consider too that this is an *actual* risk not merely a theoretical one. The NRDC gives roughly the same figure, 2.4×10^{-4}, as the theoretical estimate (extrapolated from animal data) of the increased risk of cancer—spread out over a whole lifetime—that a person might face because of childhood exposure to Alar.[95] Is eating Alar-treated apple products during childhood, then, as dangerous as driving a car for a year?

Some people may feel more indignant about risks they do not intentionally or voluntarily take, and it may be argued that risks imposed by the use of agricultural chemicals are especially objectionable for this reason. One could concede this point and still insist on the importance of remembering that such value-charged perceptions do not increase *actual risks*. If I am angrier about a minuscule risk I did not choose to take than I am about a sizable risk I did choose to take, nonetheless the greater risk still poses much more danger to me than the smaller one. The EPA mandate is to protect people from actual dangers, not to make them feel happy.

What of Claim 1's implication that the EPA was reprehensibly failing to consider crucial factors that have a particularly damaging effect on children? In our judgment the scientific background suggests that it is wrong to charge the agency with negligence or callousness.

Claim 2 is that UDMH is a potent carcinogen, especially to children, resulting in a greatly enhanced probability of cancer in exposed children. Our main criticism of this claim is that it does not present the degrees and kinds of uncertainty surrounding the quantitative risk assessment. For a numerical risk assessment to be meaningful, it should always be accompanied by a brief explanation of associated uncertainties. If we had been advancing this claim, we would have followed or preceded it immediately with an explanation of the assumptions that had gone into the calculation, a numerical representation of the remaining uncertainty, if possible, and a qualifying reminder of the arguments that can be made against extrapolating at all from high-dose animal studies to low-dose human exposures.

In fact, the authors of the NRDC report precede the claim with a statement that the estimate is "based on scientifically conservative risk assessment procedures," without further illumination. A reader of the report may just as well surmise that the term "conservative" indicates an *underestimation* of the actual risk rather than what is quite probably an extreme *overestimation*.

Claims 3 and 4 say that daminozide and UDMH are "probable human carcinogens." To accept this one would have to put a lot of weight on the assumption that effects on the most sensitive animal tested at high doses are relevant to human health from exposure at low doses.

Claim 5 holds that of the current preschool population, as many as 5,300 children may develop cancer at some time in their lives as a result of only their preschool exposure to UDMH. This claim is the most misleading. To begin with, the language implies that it is a pretty sure thing these extra cancer cases will occur. When the potency factor is mentioned, it is explained as representing "how many cancer cases the chemical will cause for a given dose."[96]

The number 5,300 is not given in an informative context. How large is the current preschool population? Suppose it is 22,000,000. Our fears may increase if we think, as Claim 5 encourages us to do, of the 5,300 possible cases in absolute and individual terms, but the fearfulness of this figure may diminish somewhat if we think of it in comparative terms. An important factor in comparative thinking would be the number of cancer cases otherwise expected. If this number is 1 in 4, then 5.5 million people in the current preschool population can be expected to get cancer at some point in their lives anyway.

We overlook much of the significance of the Alar scare if we dismiss it merely as a misperception of the technicalities surrounding the use of a single, isolated agricultural chemical. However inappropriate from the technical point of view, belief in the danger of Alar may be part of a more general belief: that it would be best to use as small a quantity of agricultural chemicals as possible compatible with high standards of food production. The reasons for this belief need not be limited to worries about consumer health risks. Other reasons include the desire to reduce occupational health hazards, the desire to minimize air and water pollution in chemical manufacture and use, and the awareness that since pests are increasingly resistant to chemicals, it may be inefficient to depend too heavily on chemical pesticides. If so, proponents of the general belief should argue for it straightforwardly in terms of occupational exposure, pollution prevention, and agricultural efficiency, rather than resort to scaring parents and other consumers with dire warnings about dietary exposure—as well as harming apple producers.

Some practitioners of "integrated pest management"—a method aiming to minimize the application of agricultural chemicals—have pointed out that not using Alar actually increases a grower's need for other chemicals. Without Alar's prevention of fruit drop, for instance, the apples have to be protected chemically against pest infestation on the ground. Thus an Alar ban may undermine efforts to decrease the total use of chemicals.[97]

The NRDC has developed several point-by-point rebuttals to criticisms of *Intolerable Risk*.[98] To the general objection that it is not obviously appropriate to base quantitative risk assessments for carcinogenicity on any animal data, the NRDC responds that "U.S. federal regulatory agencies, the National Academy of Sciences, the International Agency for Research on Cancer (IARC), and the National Toxicology Program (NTP) have all concluded that laboratory evidence is a valid basis for identifying carcinogens and estimating the magnitude of human risks."[99] This is, we think, a good enough answer. Its weakness is that it cites authorities without entertaining the possibility that these authorities could be mistaken. The NRDC needed to show that use of animal tests was based on good theory and appropriate evidence, not just that it was official policy.

In answer to the more specific objection that the particular cancer potency estimate used was invalid, the NRDC observes that the Scientific Advisory Panel's ruling of invalidity on the early studies (the basis for the NRDC's cancer potency estimate) was "not consistent with other analyses by EPA concerning these studies."[100] This is not a good answer. The mere fact that there was disagreement between one administrative body and another cannot in itself do anything to vindicate one or the other party.

A more promising answer to the objection is the NRDC's observation that the later EPA cancer potency estimate, lower than the earlier one by a factor of ten, could be uncertain because it was based on interim data. This means that upon completion of the studies the cancer potency estimate could well have been revised upward by one to two orders of magnitude, as the EPA itself anticipated. Fair enough. By itself, however, that does not justify an alarmist reaction, given that the interim results, in the EPA's judgment, did not demonstrate that Alar was an imminent hazard. (In fact, the agency's final analysis of the completed studies reduces UDMH's carcinogenic potency factor by about half, to 0.46 from 0.88.)[101]

The most disappointing rebuttal is against the point that the "NRDC's study inappropriately portrayed estimates derived from risk assessment models as actual cancer cases that will eventually develop."[102] In response, the NRDC excuses its presentation by saying that the information thus presented was "the best available government information." Now even if this was so, it does not constitute an excuse for presenting the best available information in a misleading way. No better are the excuses that follow: a collection of conjectures as to why things might be even worse than the risk figures imply. These have mainly to do with limitations on the number of pesticides the report could consider and on other sources of residue consumption besides fruits and vegetables. Again, even if whatever makes things possibly worse was not considered in the estimate, that does not justify giving a misleading view of what *was* considered. Recall that the United Nations panel concluded that Alar does not cause cancer in mice, that its breakdown product UDMH causes no increase in tumors in mice below doses of 3.9 mg/kg/day, and that, therefore, any trace amounts of either in food are not dangerous to humans.

If we had been presented with the scientific evidence about daminozide and UDMH that already existed at the time of the controversy, with a full interpretation of the risk assessments that were derived from that evidence, we would not have worried about eating Alar-treated apple products or letting children eat them. In the words of Gary Flamm, a former FDA staffer, "One has to conclude the publicity [Alar] got really had nothing to do with the science."[103] We agree with Daniel Koshland's editorial for *Science,* charging that "a clearly dubious report about possible carcinogenicity by a special interest group was hyped by a news organization without the most simple checks on its reliability or documentation."[104]

Regulation without Evidence of Harm

If we accept Ralph Keeney's estimate that a decline in living standards of around $7.5 million is responsible for one early death,[1] the billions spent on removing asbestos from schoolrooms for no health benefits created a lot of destruction outside as well as inside the buildings. It took a long time for a few simple facts—for instance, that white chrysotile asbestos does no harm in small exposures—to catch up with the public. The entire episode is an indictment of the "if it could conceivably be harmful, spare no expense in removing it" approach. Billions for schools, we say, but not one cent for nonsense. Imagine the consternation of those imposed upon for millions of dollars in removal costs to discover that if they had measured asbestos fibers outside their schoolroom, the concentration would have registered pretty much the same as inside. Where were the engineers and the material scientists in those neighborhoods? Congratulations to Malcolm Ross of the U.S. Geologic Survey, to Brooke T. Mossman and her scientific colleagues, and reporter Michael Bennett, from whose series in the *Detroit News* we first learned these facts, for speaking out with evidence. But where were the other knowledgeable people? Silence in the face of fallacy, even well-intentioned fallacy, harms us all.

Words fail on Alar. The most charitable interpretation is that an environmental group, observing promises but not action from the EPA, and frustrated by the agency's issuing statements on how bad Alar was while refusing to ban the chemical, decided to take matters into its own hands by writing a report and orchestrating its release to the media in so forceful a manner as to compel governmental action. The syndrome its report played out is by now distressingly familiar: a few suggestive tests involving tiny quantities raised way above the actual amount by extreme assumptions about children's eating habits, expanded further by statistical manipulation, extrapolated against huge populations to create row-upon-row of child cancer victims.

The denigration of government that gratuitously accompanies these scares is worrisome. If citizens took these accusations literally, they would view government as a conspiracy to harm them and their children. These charges, we know, are false. But they have their effect.

7

How Does Science Matter?

I. Is Arsenic in Drinking Water Harmful to Our Health?

with David Schleicher

In the popular imagery arsenic is an exotic and deadly poison that the killer discreetly slips into his unwitting victim's drink. In fact, arsenic is all around us, a widespread element in our natural environment. We ingest tiny amounts of arsenic every day in the food we eat and the water we drink. Such trace quantities, ingested chronically, are by most accounts benign; some studies have even suggested that complete absence of this element from the diet could be detrimental to health.

At the same time, there has been concern among public health agencies that constant intake of somewhat elevated levels of arsenic may result in toxic effects including cancer. The Occupational Safety and Health Administration (OSHA) thus limits on-the-job exposures to arsenic, such as those at smelting facilities, which produce airborne arsenical dusts thought to cause lung cancer. The Environmental Protection Agency (EPA) has sought to contain environmental exposures, principally by promulgating an arsenic drinking water standard that governs waste site cleanups and other activities that impact groundwater quality.

The case of arsenic in drinking water nicely illustrates the dictum that the dose makes the poison. Future risk detectives take note.

Arsenic in the Environment

Arsenic (As) is a naturally occurring element distributed widely in the earth's crust. Pure arsenic is a gray-colored metal; but this form is rare in the environment. Rather, As is commonly found in combination with elements such as oxygen, sulfur, and chlorine (collectively referred to as inorganic arsenic), or in combination with both carbon and hydrogen (organic arsenic).[1] Arsenic trioxide (AsO_3), the most common inorganic environmental arsenic, is formed when As-containing igneous rocks, such as copper, iron, and cobalt, weather and erode. The dustlike AsO_3 may be retained in soil or dissolve in water, where it can migrate widely. Interaction of AsO_3 with aqueous organisms can yield a variety of organic arsenical compounds.[2]

That small amounts of inorganic arsenic are broadly distributed in the environment is evidenced by a host of field surveys. Arsenic is found in U.S. soils at an average level of 5 parts per million (ppm). There is arsenic in sea water at an average of 3 parts per billion (ppb); in lakes and streams (average 65 ppb); in groundwater (average 10 ppb); and in plants and animals (from about 300 ppb for land plants and animals to 30 ppm in brown algae and shrimp).[3]

Humans add to this natural background of inorganic arsenic principally through the smelting of copper, cobalt, lead, zinc, and gold. When ores are "roasted" at smelting facilities, AsO_3 is created as a by-product. Some of this dust escapes to the outside air; most is trapped in dust collectors. Some facilities discard the dust as waste, while others process it to produce a more purified form of AsO_3, for which there is an industrial market.

Purified AsO_3 has been used principally in herbicide production and wood preserving. Four arsenic-based herbicides are commonly used in cotton growing in the deep South, Texas, and southern California. Arsenic-based pesticides were widely used in the first half of the twentieth century, but after World War II they came to be replaced by newly available organic chemicals. The wood-preserving industry employs two types of copper arsenate. By-product arsenic from smelting operations can also be processed to create a purified arsenic metal. Use of this metal in the semiconductor industry makes for a small but growing market.[4]

The main source of arsenic intake for most people in the United States is food: the average American ingests between 25 and 50 micrograms (µg) a day. Ingestion by way of drinking water is more variable. On average the levels of arsenic in U.S. drinking water are quite low: about 2 µg per liter (l) of water. But the EPA has specified three locales where elevated levels of As in drinking water, tens or hundreds of times the national average, may be found:

1. Areas where there are natural mineral formations that have high arsenic content (mostly in the western United States).

2. Areas surrounding chemical waste dumps that contain arsenical wastes from smelting, herbicide-producing, or wood-preserving facilities.

3. Areas with long histories of intensive pesticide and herbicide use.[5]

Whether elevated levels of arsenic in drinking water are a cause for concern is a question that has received considerable attention from public health researchers and officials over the past three decades. That some evidence is positive and some negative, of course, need not mean that they are of equal worth.

Health Effects from Elevated Levels of Inorganic Arsenic in Drinking Water

Arsenic[6] was first isolated as an element, reportedly, by Albertus Magnus in 1250. It was discovered soon thereafter that arsenic dust was a potent poison: a small amount mixed into food or drink would do someone in swiftly and reliably.[7] We now know that a mere tenth of a gram or so of AsO_3 (100,000 µg) is adequate for this purpose.[8] We know as well that doses shy of this can cause acute effects short of death, such as temporary paralysis or loss of feeling in limbs.

That this form of arsenic could do harm to human health at considerably lower doses, if consumed chronically, first came to be believed by medical researchers in Britain at the turn of the twentieth century. In 1903 the Royal Commission of London established a drinking water standard for arsenic of 108 µg/l. The commission was responding to two outbreaks of "arsenical" skin disease characterized by large scaly patches on the hands, feet, and torso. The first was in Reichenstein, Silesia, where town water supplies were found to contain arsenic at levels up to 12,000 µg/l. When this water was replaced by water containing about 15 µg/l, the epidemic expired. A second outbreak of skin poisoning occurred among regular patrons of beer halls in a region of England. Beer supplied to these pubs was found to be particularly high in arsenic content, containing 270 µg/l; when beer containing 216 µg/l was brought in, the symptoms receded. Figuring that those beer drinkers who exhibited symptoms of arsenic poisoning had been imbibing two liters of brew a day, the Royal Commission judged that about 450 µg/day was the maximum safe dose for ingested arsenic; from this they derived a moderately conservative health standard of 108 µg/l for drinking water. In setting this standard the commission also had in view a report by Sir Jonathan Hutchinson in 1888 of unusual skin cancers occurring among persons who regularly took arsenic-containing medications.[9]

CONTEMPORARY STUDIES REPORTING POSITIVE FINDINGS

Tseng et al., 1968. W. P. Tseng and his colleagues conducted an extensive study of residents of thirty-seven villages on the rural southwest coast of Taiwan where elevated levels of arsenic had been detected in well water.[10] The arsenic in the groundwater underlying these villages was thought to stem from natural mineral formations. A good-sized population had been ingesting unusually large doses of arsenic throughout their lives, for deep wells had been dug for village water during the early 1900s.

The 40,421 villagers examined were categorized by age and by the level of arsenic in their particular water source: 0–300 µg/l; 300–600 µg/l; and 600+ µg/l (recall that the U.S. average for drinking water is 2 µg/l and the lethal dose, about 100,000 µg). Since there was little migration in this rural region, subjects' ages were taken as indicative of their length of exposure to the arsenic-containing drinking water.

The principal health effect of interest to the researchers was a form of skin cancer that had been associated with arsenic since the 1888 reportings of Hutchinson. Readily distinguishable from ultraviolet-ray-induced skin cancer, arsenical skin cancer was characterized by multiple lesions on parts of the body, such as the trunk and the feet, that were not often exposed to the sun.

A control population of 7,500 persons from nearby villages that did not have high levels of arsenic in their drinking water was also examined. Tseng and colleagues maintained that these nearby villages constituted an effective control population since they mirrored the exposed villagers in all important respects: age distribution, socioeconomic status, occupational profile, and diet.

There were two principal observations. First, in the exposed population of just over 40,000 persons, 428 cases of arsenical skin cancer were discovered, while in the control population not a single skin cancer was detected. Second, within the exposed population, skin cancer was clustered among those who had been drinking water containing more arsenic (300–600 µg/l and 600+ µg/l) and among older people who had been ingesting the water for a longer time. Skin cancer incidence was highest among the elderly who had been drinking from the most contaminated wells. Furthermore, skin cancer was found to be higher among male villagers—who, when working in the fields and on the docks, drank large amounts of water—than among female villagers. This confluence of findings led Tseng to support the notion that ingestion of unusually high amounts of arsenic in drinking water could cause skin cancer.

Cebrian et al., 1983.   In 1983 Mariano Cebrian and his colleagues examined residents of two rural towns in northern Mexico.[11] Those living in one town had long been exposed to elevated levels of arsenic in their drinking water, believed to reflect natural mineral substrata and possibly local use of arsenic-based pesticides prior to World War II. Drinking water drawn from wells contained on average 411 µg/l of arsenic. A second nearby town was selected to serve as a control: the two populations were comparable in age distribution, socioeconomic status, occupational profile, and diet, but in the "control" town arsenic levels in drinking water averaged only 5 µg/l.

In each town the Cebrian team of researchers visited every third home and examined each member of the household. In the "exposed" town this amounted to 296 persons, constituting 30 percent of the population; 312 persons, 29 percent of the population, were examined in the control town. The team looked in particular for symptoms of arsenical skin poisoning and skin cancer. These included hypo- and hyperpigmentation (light and dark patches on the skin); palmoplantar keratosis (small, scaly, elevated nodules or "corns" on palms and on the soles of feet); papular keratosis (flat, reddish, scaly patches on other parts of body); and ulcerative lesions. Although the researchers did not perform biopsies, citing resistance on the part of subjects, they considered papular keratosis and ulcerative lesions to be indicative of skin cancer. In justifying this categorization, Cebrian and his cohorts noted that autopsies performed on deceased villagers had found these two types of lesions to be cancerous.

The following numbers indicate persons among the two populations afflicted with each particular malady (note that a resident may exhibit more than one).[12]

	Control (312)	Exposed (296)
Hypopigmentation	7 (2.2%)	52 (17.6%)
Hyperpigmentation	6 (1.9)	36 (12.2)
Palmoplantar keratosis	1 (.3)	33 (11.2)
Papular keratosis	0	15 (5.1)
Ulcerative lesions	0	4 (1.4)

In addition to finding strikingly higher skin lesions among the population with greater arsenic exposure, the Cebrian team observed that prevalence of these health effects was greater among older residents of the village, who had been drinking village water longer. They also found that the most serious afflictions—the papular keratoses and ulcerative lesions—were manifest mostly among older residents who, by their recollection, had earlier suffered the lesser symptoms of skin discoloration or corns.

Fierz, 1965. V. U. Fierz and his associates conducted a retrospective study of Germans who had received long-term treatment with an arsenic-based medicine known as "Fowler's solution" between 1939 and 1959.[13] A total of 1,450 patients were identified as having been administered orally a 1:1 dilution of Fowler's solution containing 3.8 grams per liter (g/l) of arsenic for ailments including psoriasis, chronic eczema, and severe acne.

Invitations for a free medical examination were mailed to these 1,450 patients; 262 of them responded and were examined by the Fierz team. From the patients' medical records the researchers were able to determine the cumulative dose of arsenic that each subject had ingested during the course of treatment. Of the 262 patients examined, 106 (40 percent) were afflicted with hyperkeratosis (dark reddish patches on the skin). Of the 21 subjects diagnosed as having skin cancer, 13 had multiple carcinomas and 16 exhibited palmoplantar keratosis, or "arsenic warts" on the palms and soles. Incidence rates of both hyperkeratosis and skin cancer increased with the cumulative dose ingested, as did the size of the hyperkeratosis. When dose was controlled for, skin cancers were not correlated with the original ailments for which Fowler's solution was prescribed. No control group was studied by the Fierz team. Again, it seemed that relatively modest amounts of ingested arsenic had serious effects. Yet other studies proved negative.

CONTEMPORARY STUDIES REPORTING NEGATIVE FINDINGS

Morton et al., 1976. In 1976 William Morton and his associates reported on their extensive study of the relation between ingestion of arsenic-contaminated water and skin cancer in Lane County, the second most populous county in Oregon, encompassing both Eugene and Springfield.[14] An arsenic-rich geologic formation underlies the south-central region of the county. During the mid-1960s sporadic cases of acute arsenic poisoning from intensely contaminated wells had been reported. The Morton research team hypothesized that a retrospective study of skin cancer rates in Lane County would turn up higher rates in the arsenic-rich south-central region than in other parts of the county.

Skin cancer incidence in Lane County from 1958 through 1971 was determined by a search through tissue pathology files for the county. That a single group of pathologists served all the hospitals and physicians in the county made the researchers' task somewhat simpler. They pinpointed place of residence for each case of skin cancer and then tallied the cases for each census tract in the county.

Morton and his associates calculated mean arsenic levels in drinking water supplies for each census tract based on samples taken by local researchers from 1968 to 1974. The average contamination for the "high arsenic" south-central region turned out to be only 33 µg/l. Though some isolated private wells had As levels as high as 2,150 µg/l, the majority of the 186 samples taken from this region registered less than 50 µg/l. Morton surmised that reports of acute arsenic poisoning from well water in the region back in the early and mid-1960s had led many people to close down wells and hook up to systems that tapped local surface waters. In other parts of Lane County, arsenic levels ranged from 3 to 11 µg/l.

Comparing pathological data with water sampling data, Morton concluded that skin cancer rates in Lane County from 1958 to 1971 "showed little or no correlation with . . . observed arsenic levels." When controlled for age of population, the rates in south-central Lane County were no higher than those found elsewhere in the county. Contrary to their expectations, the researchers professed that "our data showed no evidence of water arsenic influence on skin cancer incidence."[15]

Harrington et al., 1978. J. Malcolm Harrington and his colleagues in 1978 reported on an investigation into the small community of Ester Dome, just outside Fairbanks, Alaska, where elevated levels of arsenic in well water had been detected by university geographic researchers.[16] The arsenic was linked to local mineral substrata and, in the minds of some, to a nearby gold-mining operation. The average contamination level in Ester Dome wells was 224 µg/l.

Harrington and his associates examined the members of thirty households in Ester Dome and thirty households from a nearby town of similar socioeconomic status, where arsenic levels in drinking water were low. A total of 119 persons constituted the "exposed" population; 113 persons made up the control group. Giving each person a questionnaire about possible symptoms and a physical examination, the researchers found no cases of arsenical skin poisoning or skin cancer in either the exposed population or the control group.

Southwick et al., 1983. In Millard County, Utah, J. W. Southwick and associates examined 145 persons from the small towns of Hinckley and Deseret who had drunk water containing elevated levels of arsenic.[17] A sampling of Hinckley's water supply found a mean arsenic concentration of 180 µg/l, while Deseret's water averaged 210 µg/l. Researchers selected a matched control group of 105 persons from the neighboring town of Delta who had been consuming water containing only 20 µg/l.

The Southwick team found no statistically significant difference in health effects of concern among the two populations and reported that "the participants with higher arsenic exposure did not show evidence of health problems any more than did participants with lower arsenic exposure."[18]

Valentine et al., 1992. Jane L. Valentine and her colleagues in 1992 reported on a study by health questionnaire of 181 persons spread across three small communities in Nevada and one in California where levels of arsenic in drinking water, ranging from 51 µg/l to 393 µg/l, had been detected.[19] A socioeconomically comparable control group of 93 persons was questioned as well. The researchers reported that despite substantial differences in chronic arsenic intake, "Comparison of the exposed groups to the control group found no difference in symptom response."[20]

ANIMAL TESTING

Laboratory animal tests conducted with arsenic have proven negative. Arsenic has been administered to rats and mice in food and in drinking water, through injection and through skin painting; arsenic has also been administered to rabbits, chickens, pigs, and dogs. Results have been consistently inconclusive or negative. To date, arsenic has not been shown to cause cancer in laboratory animals.[21]

Having taken a first look at the evidence on arsenic, we next consider how it appeared to the regulatory authorities.

EPA Conclusions and Actions: Qualitative and Quantitative Risk Assessment

There are three key dimensions to the EPA's current posture toward arsenic in drinking water:

1. The EPA has deemed ingested arsenic a Class A or "known" (as opposed to "probable" or "suspected") human carcinogen. Arsenic is one of the very few chemicals that the agency places into this category.

2. The EPA has assigned ingested arsenic a carcinogenic potency factor of 1.5×10^{-3}. This number purportedly represents the cancer risk associated with intake of 1 microgram per kilogram of body weight per day (μg/kg/day) of an element over the course of a lifetime. A 1.5×10^{-3} potency factor for ingested arsenic expresses the EPA's estimate that a person ingesting 1 μg/kg/day of this substance—about 70 μg/day for the average-sized American male—would be subject to a 1.5 in 1,000 chance of contracting skin cancer.

3. Under the guidelines of the Safe Drinking Water Act, a maximum contaminant limit of 50 μg/l has been announced for arsenic in drinking water.

QUALITATIVE RISK ASSESSMENT: IS INGESTED ARSENIC
A HUMAN CARCINOGEN?

The classification of ingested arsenic as a known rather than a merely probable or suspected carcinogen reflects the kind of studies done: unlike the situation with the majority of chemicals that the EPA regulates, evidence of arsenic's carcinogenicity comes from human population studies rather than high-dose animal testing. For this reason both the International Agency for Research on Cancer and the World Health Organization consider arsenic a known human carcinogen. On what evidence is this judgment based?

The centerpiece of contemporary official assessments of ingested arsenic's carcinogenic potential is the Tseng team's study of villagers in southwest Taiwan. The Cebrian study from northern Mexico and the studies of Germans taking arsenic-based medications are usually cited as supporting the conclusions inferred from the Tseng investigation, which seems to present a compelling correlation between arsenic ingestion and skin cancer.

The Tseng study has been widely cited by academic researchers and governmental agencies and is acclaimed as an unusually powerful epidemiological investigation.[22] A number of its features have led most observers to consider it a reliable indicator of the skin-cancer-causing potential of ingested arsenic: the unusually large size of the exposed population examined (over 40,000); the juxtaposition of the exposed population against a large and well-matched control population; the strikingly higher incidence rate of skin cancer in the exposed population as compared to the rate in the control population (which makes less likely the possibility of confounding variables or random variation accounting for the difference); and the appearance of a strong dose-response relation *within* the exposed population (subgroups that had been drinking more severely contaminated water for longer exhibited higher rates of skin cancer).[23]

Two flaws that might detract from the Tseng study's reliability are sometimes cited. First, the researchers who examined the Taiwanese villagers were not "blinded"; that is, they were not ignorant as to which group of subjects had been exposed to the suspect agent and which was the control or unexposed group (the same was true of the Cebrian study).[24] Thus the researchers could have brought expectations and biases into their examinations and diagnoses that made them more likely to see symptoms among the exposed group. Second, the water supplies of the exposed and control populations were tested only for arsenic content. There is the possibility that some other contaminant could have been at work in the drinking water of those villages that exhibited the elevated rates of skin cancer.[25]

These weaknesses have been taken to be inconsequential by the EPA and other official health organizations given the extent of the difference in cancer incidence between the exposed and control populations. The EPA points out that although the Taiwan researchers were not blinded, the symptoms they were searching for were uncommon and easily detected, and the standard international protocols were followed in identifying and registering these symptoms. The EPA also considers extremely unlikely the possibility that some other unidentified element in the exposed population's drinking water was responsible for the full difference between the villagers' skin cancer incidence. It is conceivable that some other chemical contributed to the effects, though to date no substance other than arsenic has been associated with the sort of skin problems at issue. In any case, the presence of another element would bear on the quantitative, dose-response inferences that might be drawn from the Taiwan study, not on the basic qualitative classification of arsenic as a human carcinogen. Finally, the EPA notes the implausibility of presuming that undetected substances were also at work in north-central Mexico and among Fowler's solution users.[26]

What of the negative findings in U.S. epidemiological studies and in laboratory tests on animals? Do these cast doubt on the implication of human carcinogeity that seems so apparent in Tseng's, Cebrian's, and other positive studies? At least one analyst has inferred from these negative investigations that ingested arsenic should not be classified a human carcinogen.[27]

To take up the matter of negative animal studies first: most analysts are unwilling to allow negative findings from laboratory animal tests to supersede respectable human epidemiological findings (we may note the irony that many of these same analysts are willing to allow *positive* animal test results to supersede *negative* epidemiological investigations). In the particular case of ingested arsenic and skin cancer, the EPA in dismissing the negative results of animal testing has argued that humans appear to be more sensitive than animals to this substance. For instance, it takes a considerably higher dose of arsenic to kill an animal than it does to kill a human: reported lethal doses for laboratory animals have been 10–300 mg/kg body weight, while doses of 0.6–2.0 mg/kg have been known to prove fatal to humans.[28]

As regards the negative epidemiological findings from U.S. studies, the EPA and other analysts, including the authors of these studies themselves, have not viewed these findings as conflicting with or calling into the question the reports of Tseng or Cebrian for two reasons. First, most U.S. studies have examined small populations (which might not necessarily manifest effects that would appear in a more sizable study group). Second, and more important, the exposures of the American study populations have not been as great as those of the Taiwan, Mexico, and Fowler's solution populations.[29]

It is possible that the difference in findings between the positive foreign results and the negative U.S. studies had to do with differences in level of exposure, a possibility that raises the issue of quantitative risk assessment.

QUANTITATIVE RISK ASSESSMENT: HOW POTENT A CARCINOGEN IS ARSENIC?

In performing a quantitative carcinogenic risk assessment for ingested arsenic, the EPA relied on the data reported by Tseng and his team. Extrapolating from their data on drinking water containing hundreds of micrograms of arsenic per liter, the EPA first calculated the risk of skin cancer to be expected from ingesting water with 1 µg/l. Employing a linear no-threshold or "one-hit" model of carcinogenesis (see the discussion on asbestos in Chapter 6), the EPA came up with a projected risk of 5×10^{-5} associated with drinking two liters of water a day with 1 µg/l of arsenic in it over a seventy-year lifetime.[30] Along this same dose-response line the EPA located a carcinogenic potency factor— the risk attendant from chronic intake of 1 µg/kg/day, or about 70 µg/day of a given substance—of 1.5×10^{-3} (1.5 in 1,000).

The EPA itself acknowledged that it had derived an "intentionally conservative"[31] quantitative risk assessment for a number of reasons. For instance, the previously noted lack of blinding by the Tseng team, while not of great enough consequence to cast doubt on the qualitative classification of ingested arsenic as a human carcinogen, might detract from attempts to draw precise quantitative inferences from the study. Of greater concern is the possibility that the villagers who exhibited elevated incidence of skin cancer may have ingested significant quantities of arsenic from sources other than their water supply, in particular, from food grown in their arsenic-rich region.[32] If in fact the villagers were getting inordinate amounts of arsenic from their food as well as from their water supply, then their response pattern would stem from higher-than-reported arsenic dosages, and dose-response calculations based on the reported intakes alone would overestimate the potency of this substance.

Another consideration is that the villagers studied were known to have diets low in protein and certain vitamins. To the extent that better nutrition might provide some resistance against the effects of arsenic ingestion, a dose-response relation derived from the Taiwanese study would overestimate risks to the comparatively well-nourished American populace.[33]

The EPA risk assessors acknowledged these concerns but did not incorporate them into their risk projections. The most serious measure of conservatism in the EPA's assessment, however, is the use of a linear no-threshold dose-response model despite evidence that there is a threshold dose below which arsenic does not cause skin cancer or any other ailment. Gerhard Stohrer has argued recently that arsenic should be considered a member of an increasingly well understood and recognized class of "indirect carcinogens" for which linear no-threshold dose-response extrapolations are inappropriate.[34]

According to Stohrer, toxicokinetic tests for inorganic arsenic cast doubt on the EPA's default presumption of a genotoxic carcinogenesis mechanism. During the 1950s it was discovered that some chemical carcinogens acted through damage to genes, as was previously known to be the case with ionizing radiation. The experimental dose-response for gene damage from radiation and apparently for certain chemicals was different from any other biological effect previously detected; under certain conditions, response was proportional to dose down to the lowest limits of measurement. It was surmised that a single "hit" from this class of chemicals could cause gene damage and ultimately cancer. This one-hit theory of genotoxic carcinogenesis inspired the absolutist restrictions of the Delaney Clause, which banned food additives found to cause cancer in lab animals.[35]

But a second class of carcinogens has since become known to toxicologists: indirect carcinogens, such as alcohol consumed at chronic high doses, which work not through one hit but through cumulative stress to genes. Stohrer reports that "while the hallmark of the genotoxic carcinogen is its ability to cause gene damage and mutations, the non-genotoxic indirect carcinogens and tumor promoters have a multitude of stress-related effects that ultimately induce or otherwise affect the expression of genes."[36] Biochemists have shown that a minimum effective dose of such substances is required before these stress-induced genetic effects appear; in other words, there is a dose-response threshold.

Toxicokinetic testing on arsenic has yielded results that seem to set this substance apart from the bulk of those studied by toxicologists: arsenic fails to induce point mutations in genes (that is, damage to a single nucleotide of a DNA molecule). According to Stohrer, results of laboratory studies of arsenic are consistent with the profile of an indirect carcinogen.[37]

Epidemiological studies reporting correlations between arsenic ingestion and adverse health effects, such as Tseng's and Cebrian's, suggest a cutoff point of about 400 µg of total ingested arsenic per day below which there are no findings of either skin cancer or other skin disorders. In the Tseng study, for instance, the lowest dose for which arsenical effects of any kind were reported was about 100 µg/l. Given that water consumption in tropical climes averages 4 l a day (in comparison to 2 l a day in the United States), this figure translates to a daily intake of 400 µg As (plus whatever As was consumed in food). All the recorded cases of arsenical skin poisoning and skin cancer were associated with this amount or, in most instances, considerably higher doses; and among the 7,500-person control population in the Tseng study, those exposed to 19 µg/l or about 75 µg/day of arsenic, not a single case of arsenic poisoning or skin cancer was reported. This same threshold of about 400 µg/day was exhibited as well in lower profile studies from Chile and India.[38]

The many negative U.S. studies would seem to provide further confirmation of a threshold. In none of these studies were the levels of arsenic ingestion as great as in the Taiwanese and Mexican reports. Among the U.S. communities examined, Ester Dome, Alaska, and Deseret, Utah, had the highest average arsenic levels in drinking water, both a shade over 200 µg/l; estimating 2 l/day of water consumption, this figure translates to intake of about 400–440 µg/day among people in these communities. Average arsenic levels in south-central Lane County, Oregon, and in the several communities studied in Nevada were somewhat lower. No statistically significant findings of arsenical skin poisoning or skin cancer were reported in any of these studies. In a special forum held in 1987 to work through some of the peculiarities of the evidence linking arsenic ingestion to skin cancer, EPA risk assessors acknowledged that a case could be made for arsenic being an indirect carcinogen. They observed, for instance, that "since arsenicals do not appear to induce point mutations, one rationale for assuming low-dose linearity and using the generalized multistage model might not apply."[39] And the Agency for Toxic Substances and Disease Registry later echoed these analysts in allowing that numerous negative U.S. epidemiological studies of populations somewhat less exposed to arsenic than those in positive foreign studies "could suggest that arsenic-induced cancers have a threshold dose."[40]

But this evidence proved insufficient to move the EPA from its default presumption that "cancer-causing agents can increase risk even at very low exposures."[41] Of the toxicokinetic findings that seem to cast doubt on the classification of arsenic as a "one hit" carcinogen, those at the 1987 forum judged that "on balance . . . there is a paucity of information on the mechanism of carcinogenic action or the pharmacokinetics of arsenic that leads to confidence that any particular extrapolation approach is more appropriate than another . . . In the absence of fully persuasive evidence for any of the possible mechanisms, a generalized multi-stage model that is linear at low doses was used to place an upper bound on the expected human cancer dose-response."[42] Thus the EPA places the burden of proof squarely on the shoulders of any who would suggest that a given substance does *not* exhibit a linear dose-response function all the way down to zero dose. In the case of arsenic, the EPA finds the evidence "intriguing" but not sufficiently overwhelming to warrant abandoning this default assumption.[43]

SETTING STANDARDS

Currently on EPA books is a maximum contaminant level (MCL) for arsenic of 50 µg/l. Under the Safe Drinking Water Act, public water supplies are not to exceed this threshold, and private wells in excess of this ceiling can be shut down. Under the Superfund regime, potable groundwater near chemical waste sites that surpasses this MCL has to be cleaned up. In promulgating a standard of 50 µg/l, the EPA carried forward the judgment of its predecessor, the Federal Water Quality Administration, which first set this ceiling in 1942.[44]

The EPA appears to be of two minds in regard to the threat from ingested arsenic. Typically, drinking water MCLs, like many other EPA environmental quality targets, are set to approximate a 1-in-1-million residual cancer risk. But by the agency's quantitative carcinogenic risk assessment for ingested arsenic, as we have seen, the risk associated with 1 µg/kg/day arsenic, or about 70 µg/day for the average sized person, is 1.5×10^{-3}. Since a 50 µg/l drinking water standard allows for the ingestion of 100 µg/day of arsenic (figuring 2 l a day for water consumption), the EPA's current MCL, according to its own risk assessment, tolerates a residual cancer rate considerably higher than 1-in-1-million.

A number of factors account for this relatively tolerant EPA standard. First, the skin cancer associated with arsenic ingestion is not nearly as serious as most other forms of cancer. Arsenical skin cancer is easily detected, rarely fatal, and in many cases reversible with proper treatment.[45] Second, as the EPA and the Agency for Toxic Substances acknowledge in their 1989 public health assessment of environmental arsenic, elevated cancer risks from arsenic in drinking water have not been detected in the United States despite numerous chronicled cases of communities exposed to levels somewhat in excess of the current MCL.[46]

The EPA is also aware of recent tests on laboratory animals which suggest that low levels of arsenic intake may be beneficial to health.[47] Lab animals fed diets completely lacking in arsenic lost weight, did not give birth as often as normal, and produced underweight offspring. Whether these findings have relevance for humans is not yet known. But EPA risk assessors acknowledge that an argument might be made that "taking action to reduce the level of arsenic below some critical level . . . may reduce any potential cancer risk only at the expense of other decrements to human health."[48]

Postscript

The evidence of public health risks from environmental arsenic is atypical. Tests on laboratory animals, the foundation of most modern carcinogenic risk assessments, have consistently proven negative. But because arsenic is widely and somewhat unevenly distributed in the environment and because it has been employed in medicinal solutions, there have been unusual opportunities to study its effects among actual exposed human populations, some of them quite large. Evidence from these studies shows that we should be concerned about highly elevated levels of arsenic, above the 250–300 μg/l range, in drinking water supplies. This same body of evidence, supplemented by progress in pharmokinetic research, shows that we need not be alarmed about less elevated arsenic levels.

Environmental arsenic cannot be considered a significant public health threat in the United States, because water arsenic levels in this country rarely match the levels associated with illnesses in foreign studies. Even communities with severe enough arsenic problems to attract curious epidemiologists have had arsenic intakes significantly below those in the Taiwanese, Mexican, and German studies, and they have exhibited no signs of ill health.

Groundwater near inactive chemical waste sites is sometimes said to exceed the EPA maximum contaminant standard for arsenic (which is about five times less than amounts known to cause harm), though this threat might be discounted, since highly conservative methods are used to gauge levels of contamination at these sites and groundwater there is rarely tapped for drinking or any other use. Monitoring water supplies, particularly private wells, near arsenic-containing waste sites and taking sensible action when water that is actually being used significantly exceeds the 50 μg/l standard should suffice to contain this limited public health nuisance.

Currently the EPA is giving internal consideration to reducing its MCL for arsenic to bring the standard more in line with the quantitative risk assessment that it contrived for this substance. A recent analyst reviewed the evidence on ingested arsenic and concluded that the present standard of 50 μg/l was "reasonable and conservative."[49] One may hope that the agency maintains the "reasonable" part of this equation by declining to lower its current arsenic standard.

The arsenic case shows that we should use caution in generalizing from one prominent study. The episode involving nitrite in food further illustrates how circumspection can prevent us from jumping too quickly to frightening conclusions.

II. Whom Can You Trust? The Nitrite Controversy

with Michael T. Sweeney

Nitrite, mainly sodium nitrite, is used in about 7 percent of the nation's food supply.[1] It is used most often in curing meat and poultry products, because it prevents spoilage and food poisoning, helps maintain the pinkish color of certain meats, and imparts a distinct flavor. In 1978 the Food and Drug Administration (FDA) and the U.S. Department of Agriculture (USDA) released the results of a study that found sodium nitrite carcinogenic in laboratory animals. The agencies soon made public their intention to implement a gradual ban on the use of sodium nitrite. In light of serious doubts about the validity of this study, the agencies reversed their position two years later and canceled plans for a ban. Sodium nitrite remains in use today.

Background

For hundreds of years saltpeter (potassium nitrate) has been used to preserve meat. In the nineteenth century it was discovered that the bacterial reduction of nitrate (NO_3) formed nitrite (NO_2), and around 1900 nitrites were shown to be responsible for the pinkish color of cured meats. In the 1920s it was first recognized that nitrites inhibit the growth of certain bacteria such as putrefactive anaerobes, which cause the spoilage of meat, and *Clostridium botulinum,* which causes the potentially fatal disease of botulism.[2]

In 1925 the USDA formally authorized food processors under federal inspection to use sodium nitrite to cure meat products such as bacon, ham, sausages, and frankfurters. This authorization limited nitrite levels to 200 ppm in the meat after processing.[3] In the 1930s and 1940s increased use of mechanical refrigeration enabled meat processors to use milder cures; this trend toward less nitrite use was aided by the development of both oxygen-impermeable and vacuum packaging. In the early 1950s definitive studies verified sodium nitrite's ability to inhibit *C. botulinum.* At the same time, however, some concern began to surface about the possible toxicity of certain nitrite-related compounds.

The N-nitroso compounds, such as N-nitrosamines, are formed when nitrites or nitrates combine with nitrosatiable substrates such as amines. Speculating that one such nitrosamine, N-nitrosodimethylamine (NDMA) might be the cause of cirrhosis and ill effects in industrial workers, J. M. Barnes and P. N. Magee in 1954 tested the substance on several animal species; in 1956 they demonstrated that NDMA caused cancer in rats.[4]

The Delaney Clause's stipulation on food additives does not apply to substances that received government sanction prior to the adoption of the Food Additives Amendment. Since nitrites had been awarded approval for meat products in 1925, their use is exempt from the Delaney Clause.[5]

Although nitrites are best known as additives to pork and beef products, it was their use in fish that initiated the first serious inquiry into their safety. In the early 1960s outbreaks of botulism caused by smoked fish resulted in the widespread addition of nitrite to fish products. Several studies linked Norwegian outbreaks of liver toxicity in mink and sheep to the presence of NDMA in their feed, which contained herring meal treated with nitrite. Other studies in the late 1960s suggested that nitrosamines could be formed within the human stomach.[6] In response to concerns raised by these studies, the FDA awarded a contract in 1971 to the Massachusetts Institute of Technology (MIT) to determine whether continuous exposure to a dietary nitrite would induce tumors in rats and hamsters.[7]

The First MIT Study: 1971–1974

Conducted by R. C. Shank and P. M. Newberne, of MIT's Department of Nutrition and Food Science, and performed on 1,566 Sprague-Dawley rats and 327 Syrian golden hamsters, the study involved the administration of diets containing either N-nitrosomorpholine (NNM) or a combination of sodium nitrite (NaNO$_2$) and morpholine.[8] At one time morpholine (an amine) was used as an anticorrosive agent in boiler water; hence it represented an unintentional additive to many boiled food products. Ultimately Shank and Newberne concluded that dietary sodium nitrite and morpholine induced liver and lung tumors identical to those induced by NNM.

The study involved twelve test groups of rats fed an agar-gel diet consisting of various concentrations of nitrite, morpholine, or NNM, and one group that received food containing neither NNM nor the nitrite and morpholine combination. Nine groups received diets containing levels of sodium nitrite at 0, 5, 50, or 1,000 ppm and levels of morpholine also at 0, 5, 50, or 1,000 ppm. The tenth group was the control, and the eleventh and twelfth received diets containing NNM at levels of 5 and 50 ppm. In addition, twelve groups of hamsters were fed diets matching those of the twelve rat groups. The rats were fed for 125 weeks or until they died. The hamsters were fed for 110 weeks.

Among the rats receiving the highest levels of both nitrite and morpholine, Shank and Newberne reported correspondingly high levels of cancerous tumors, which included liver-cell carcinomas, liver angiosarcomas, and metastases from liver to lung. According to the researchers, these tumors were morphologically similar to those induced by the diet containing a high level (50 ppm) of NNM.[9] Although Shank and Newberne noted that the Syrian golden hamsters had been a poor choice for the experiment because relatively few of them developed tumors of any kind, they concluded that there is a clear dose-response relationship between nitrite and morpholine concentration on the one hand and cancerous tumors on the other.

The study's most controversial claim concerned the group of rats that was fed a diet containing no morpholine but a high level of sodium nitrite. According to the researchers, 27 percent of this group developed lymphomas (lymphatic tumors), the highest rate of any of the twelve rat groups. In the rat group that was fed a diet containing neither sodium nitrite nor morpholine, only 6 percent developed lymphomas.[10] Thus the MIT scientists were the first to suggest that sodium nitrite alone might be carcinogenic.[11]

The Second MIT Study: 1974–1978

Newberne conducted a new study involving approximately 2,000 rats divided into eighteen groups. Groups 1 to 5 were fed an agar gel containing 0, 250, 500, 1,000, or 2,000 ppm of sodium nitrite. Groups 6 and 7 were fed agar gel but received the sodium nitrite (1,000 or 2,000 ppm) in their drinking water. Group 8 was a "positive control" that received agar gel containing urethane (2,000 ppm) instead of sodium nitrite. Groups 9 to 11 ate Ralston Purina Chow containing 0, 1,000, or 2,000 ppm of added sodium nitrite. Group 12, another positive control, ate the Ralston Purina containing 2,000 ppm urethane. Groups 13 and 14 were fed the agar gel diet in dry form with 0 or 1,000 ppm sodium nitrite. The respective diets of groups 1 to 14 were given to the mothers of these rats starting five days before they gave birth. The rats in groups 15 and 16 were the mothers of rats in groups 1 and 4. Groups 15 and 16 received 0 or 1,000 ppm sodium nitrite in agar gel. Finally, groups 17 and 18 received 0 or 1,000 ppm sodium nitrite in agar gel but not until after they were weaned. Certain rats were killed at predetermined intervals of six, twelve, eighteen, and twenty-four months. The rest were allowed to live until they died or until their group had diminished to 20 percent of its original number.[12]

According to Newberne's research, lymphomas increased in all groups fed nitrite. The incidence in groups that were not fed nitrites was 5.4 percent; in the combined nitrite-treated groups it was 10.2 percent. Newberne argued that the "pattern of tumors suggests that the carcinogenic effect of nitrites was through a mechanism other than the formation of nitrosamines," since nitrosamines "characteristically cause cancers to develop at a number of sites." Because he observed no tumors other than lymphomas and because feed samples were not found to contain any preformed nitrosamines, he concluded that nitrite alone increases the incidence of lymphoma.[13] In May 1978 Newberne conducted a two-hour briefing of senior FDA and USDA officials on his findings, convincing Donald Kennedy, FDA commissioner, that nitrite causes cancer.[14] According to Kennedy, the Newberne study "looked like a sound and well-conducted piece of research."[15]

Initial Reaction to the MIT Study: 1978

After the briefing by Newberne, Commissioner Kennedy initiated a confidential review by a task force composed of FDA personnel of his selection. According to FDA officials, Kennedy chose this approach over the usual review procedures for two reasons: he believed that the claim that nitrite causes cancer, if true, would present the FDA with one of the most difficult regulatory issues ever; he was also concerned that MIT's findings, if disclosed prematurely, might create public panic. The FDA had previously turned to a task force to review food additives such as saccharin and Red Dye No. 40.[16] Known as the Project

Advisory Group (PAG), the task force operated in unprecedented secrecy during the summer of 1978, prompting complaints within the FDA about Kennedy's exclusionary policy and also about the wisdom of having a task force lacking experience in carcinogenicity studies.[17]

In a memorandum of July 11, 1978, the PAG criticized the second MIT study on several grounds: its questionable statistical analyses, its failure to distinguish male and female rats, and its failure to provide photomicrographs of certain important lesions. Nevertheless, on July 27, 1978, the PAG recommended that the Newberne report "be accepted as received."[18] Thus the FDA prepared to release the study and a plan to phase out nitrites.

Four Bureau of Foods scientists assisted the FDA in preparing for questions from the media and industry. These scientists soon began to identify problems with the Newberne study, including (1) failure to use appropriate control groups, (2) potential "litter effect" caused by too many test animals being selected from each litter, (3) lack of a predictable dose response, (4) possibility that nitrosamines caused the reported tumors, and (5) failure to use appropriate statistical methods.[19]

All of these points threatened the validity of the Newberne study. The first two methodological problems related to the way rats were assigned to test groups. Rather than distribute them randomly, Newberne sorted the rats sequentially as they arrived at the lab over the course of six weeks. Any difference between a control group and a test group threatens statistical significance. Since membership in a certain test group was correlated with date of shipment, control groups often differed significantly from comparable test groups in terms of their shipment group.[20] The "litter effect" is a phenomenon in which the results of a study appear to be statistically significant but are actually not because the animals in the test group are too closely related. In the case of the Newberne study, members of the same litter were more likely to be members of the same shipment group.

As a result of these flaws, Howard Roberts, the acting director of the Bureau of Foods, requested that Kennedy form a working group to review Newberne's data.[21] On August 8, 1978, the Interagency Working Group was established, including experts in toxicology, pathology, chemistry, risk assessment, statistics, and residue evaluation from the FDA, USDA, National Cancer Institute, and the National Institute of Environmental Health Sciences.[22]

Nevertheless, on August 11, 1978, the FDA and USDA released the results of the Newberne study. In a joint statement the agencies said they would assess "several options" as they "weigh[ed] the risks associated with nitrite added to food against the health risk from not adding it." Consumer groups were critical of the noncommittal nature of the statement. Referring to the Delaney Clause, Ralph Nader said, "[Assistant USDA Secretary] Carol Foreman and Donald Kennedy should go back to the law and read it because they are violating it."[23]

On August 17, 1978, Foreman and Kennedy issued another joint statement, saying, "When a regulatory decision has been selected from the options being considered, a full description of our plans to carry it out will be presented for public and congressional criticism."[24] On the same day, the *Washington Post* obtained a forty-nine-page "action plan" developed by the two agencies that called for a "phased removal of nitrite over 'several years' to give the food industry time to develop other ways to preserve cured meats." The plan was submitted to the Justice Department for review. Foreman was quoted as saying, "If Justice approves, that's the way we intend to go. I don't really anticipate any other trouble."[25] The man who ignited the nitrite controversy was a proponent of this phaseout strategy. In a letter sent to Donald Kennedy on August 25, 1978, Paul Newberne wrote that his study "leaves no responsible alternative except to substantially reduce nitrite as a food additive or, if feasible, to eventually eliminate it entirely."[26]

In an interview with *U.S. News and World Report,* Kennedy reiterated that a phaseout plan was being prepared. He stated that "we have been developing a strategy for phasing out the use of nitrite over a period of time, dependent, of course, on scientific confirmation of the MIT study."[27] Acknowledging that the study would have to be reviewed, Kennedy asserted, "With any study of this caliber, the initial presumption has to be that the results are correct. The work was done by a scientist of excellent reputation, Dr. Paul Newberne, at an institution of high quality. The design of the experiment is viewed as good even by people who don't like the results."[28] In an adjacent interview, the president of the American Meat Institute, Richard Lyng, argued that "scientists who are now reviewing the study are finding so many unanswered questions in the test, so many flaws, that it's clear to me the study is invalid and should not be given much consideration."[29] Other critics of the Newberne study argued that the levels of nitrite fed to the rats were the human equivalent of eating 586 pounds of cured meat daily.[30]

The Controversy Heats Up: 1979

The Justice Department never approved Donald Kennedy's phaseout plan. Rather, it decided that if sodium nitrite was a carcinogen, it must be banned immediately, not phased out.[31] In 1979 there was considerable pressure for such a ban: certain consumer groups were actively involved in efforts to ban nitrites even before the release of the second MIT study. Citing previous studies linking nitrosamines to cancer in laboratory animals, these groups hoped to ban nitrites under the Delaney Clause. Their strategy was to argue that since nitrite had been approved in 1925 only as a color additive, its current use as a preservative would not have "prior sanction" status.[32] One such group, the Community Nutrition Institute, had been petitioning the USDA to ban nitrite since 1977.[33] In early 1979, citing the Newberne study, the institute asked the public to engage in a write-in campaign to pressure Congress to ban sodium nitrite.[34]

Prompted by the fears generated by the Newberne study, there was a proliferation of nitrite-free meat products in 1979. On September 20, 1979, new USDA regulations made it legal for uncured meat products to be sold under the same common name as their cured counterparts. Ellen Haas of the Community Nutrition Institute described the new regulations as a victory for meat processors who "have been hassled by USDA" when they tried to get approval to market nitrite-free meats.[35] At the same time, the USDA became concerned about the potential dangers of nitrite-free meats. Thus, in the fall of 1979 it issued a stern warning to consumers stating that nitrite-free products must be treated like fresh meat since they keep for only four to seven days instead of the two to three weeks their cured counterparts last. Carriers of brown-bag lunches were warned that sandwiches containing nitrite-free cold meats should be frozen overnight to prevent spoilage over the course of the next day.[36]

The Rehabilitation of Nitrites: 1980

By the end of 1979 Kennedy had resigned from the FDA and Jere Goyan had replaced him. Kennedy later stated that he had accumulated enough doubts about the Newberne study by late 1979 that he "would have been reluctant to use it as the only finding invalidating nitrite use."[37] Early in 1980 others were raising serious questions. At the request of seven members of the House of Representatives, the comptroller general of the United States issued a report on January 31, 1980, that criticized the FDA for not monitoring Newberne's laboratory until the testing phase was nearly over. He also cited the failure of the FDA's initial review to include a verification process for pathology diagnoses. Nor was there reexamination of either the animal tissue slides or the notes from the physical examinations of the animals, he noted.[38]

On August 15, 1980, two years after its establishment, the Interagency Working Group on Nitrite Research (IAWG) released its review of the Newberne study. Ultimately the IAWG came to the conclusion that "insufficient evidence exists to support the conclusion that sodium nitrite per se fed to rats causes cancer, based upon the MIT study."[39] The working group identified several confounding factors in the experiment. First, there was the possibility of preformed nitrosamines in the rats' diets. In late 1979 and early 1980 MIT scientists replicated the diets that had been used, had them tested, and found in the nitrite-treated Ralston Purina Chow diet relatively high levels (5–71 ppb) of dimethylnitrosamine.[40] A second confounding factor was the use of urethane as a positive control. The IAWG asserted that the high levels of urethane in the diets of some of the rats might have vaporized and been transferred to other parts of the lab.[41]

The IAWG also criticized Newberne for not taking into account positive correlation between litter mates in his statistical analyses.[42] Because most previous feeding studies involving sodium nitrite emphasized its interaction with amines, the working group concluded that it was really impossible to compare Newberne's study with other experiments. In those previous studies, the rats fed nitrite but not amines were used only as controls.[43]

The pathology review led to the IAWG's most indicting criticism of the Newberne study. The working group and Universities Associated for Research and Education in Pathology, Inc. (UAREP), examined the approximately 50,000 tissue slides from the study, a review described by the *Los Angeles Times* as "a detailed study of another scientist's work that may be unprecedented in the annals of science."[44]

UAREP pathologists studied the lesions described by Newberne as "immunoblastic cell proliferation" and dismissed them as "extramedullary hematopoiesis" and "plasmacytosis" (accumulations of blood).[45] UAREP reported an incidence of lymphoma of 1.5 percent in the combined control groups and 1.15 percent in the combined experimental groups,[46] in contrast to the percentages reported by Newberne, 5.4 and 10.2 percent, respectively.[47] The incidence of lymphoma reported by UAREP is similar to the incidence usually found in the breed of rats used in the experiment.[48] Most of the tumors that Newberne had described as lymphomas were classified by UAREP as "histiocytic sarcomas," malignant tumors of a rare type. Because there were similar levels of these conditions in both the control and treatment groups, UAREP could find no statistically significant relationship between these tumors and nitrite treatment. UAREP reported relatively high levels of angiosarcomas, liver neoplasms, pancreas tumors, pituitary tumors, ear duct tumors, and mammary neoplasms in both treatment and control rats. However, they could find no statistically significant relationship between any of these conditions and ingestion of sodium nitrite.[49]

On August 19, 1980, Foreman and Goyan issued a joint statement, saying, "There is no basis for the FDA or USDA to initiate any action to remove nitrite from foods at this time."[50] The FDA also announced that it would have the National Academy of Sciences evaluate available data on nitrites and issue recommendations about their future use.[51] A government scientist described the statement as an example "for the first time of the FDA eating crow."[52] On the day of the announcement, Goyan even went so far as to reveal, "I had a hot dog for lunch."[53] Opponents of nitrites were predictably upset with the substance's apparent vindication. The Consumer Nutrition Institute remained steadfast in its opposition, with Ellen Haas arguing, "The release today did not give a clean bill of health to nitrites . . . there is still the problem of nitrosamine formation."[54]

Defusing the Controversy: 1981–1982

In September 1980 the USDA and FDA entered into a contract with the National Academy of Sciences (NAS) to examine the health effects of dietary nitrate and nitrite and to evaluate possible alternatives to nitrite.[55] The work done by the NAS is considered by some to have been the final defusing factor in the nitrite controversy.[56] For two years the NAS Committee on Nitrite and Alternative Curing Agents in Food studied vast amounts of data on nitrites, nitrates, and nitrosamines. In late 1981 the committee released its first report,

The Health Effects of Nitrate, Nitrite, and N-Nitroso Compounds. In mid-1982 it issued a second report concerning possible alternatives to nitrites.[57]

In the first report the committee acknowledged the benefits of sodium nitrite, citing estimates of the number of deaths from botulism that might occur as a result of a nitrite ban, some exceeding eighty per year. The committee also concluded that nitrite is not by itself carcinogenic in animals.[58] Yet the committee found the hypothesis that nitrosamines were closely linked with stomach cancer to be "plausible."[59] Cured meats, particularly fried bacon, were identified as a source of nitrosamines.[60] At the same time, the committee admitted that cured meats were far from being the average American's chief source of nitrosamines.

The committee estimated average daily levels of exposure to nitrosamines from the following sources: cigarette smoking (17 µg/person), automobile interiors (0.20–0.50 µg/person), beer (0.34–0.97 µg/person), cosmetics (0.41 µg/person), and cured meats (0.17 µg/person).[61] These estimates are significantly lower than levels of occupational exposure: in the leather tanning industry, daily inhalation of nitrosamines might be as high as 460 µg/person; workers in the rubber industry could be exposed to 430 µg. The study also suggested that the formation of nitrosamines in vivo (in one's body) might be unlikely in individuals who have diets with adequate levels of ascorbic acid (vitamin C), which inhibits nitrosation.[62]

In its second report, *Alternatives to Current Use of Nitrite in Foods,* the NAS committee analyzed possible additives to cured products that would inhibit the formation of nitrosamines. Research suggested that sodium ascorbate and a-tocopherol if added to bacon could inhibit nitrosation during frying.[63] The committee also looked at alternatives to sodium nitrite itself. It found that irradiation, sodium hypophosphite, and certain lactic-acid-producing organisms might be suitable alternatives to nitrite for inhibiting botulism and preventing spoilage.[64] But the committee admitted that further research would be necessary to determine whether these processes were as effective in preservation as sodium nitrite in all cured meat products.[65] The committee also conceded that the ability of alternative processes to impart the same flavors and colors as sodium nitrite was questionable.[66] By the time the second NAS report was issued, however, interest in the subject of nitrites had rapidly declined. As a result, there was little reaction to the recommendations of the NAS committee.[67]

Media Coverage

Media attention to the nitrite issue clustered around the two major events of the controversy: the release of Newberne's results and the FDA's later reversal of its decision to phase out nitrite use. In general, the media initially responded to the Newberne study with acceptance: the claim that sodium nitrite was carcinogenic appeared on the front pages of most major American newspapers on August 12, 1978. The *Washington Post,* for example, reported that "sodium nitrite, a chemical additive used in most processed meats, causes cancer, according to a study conducted for the Food and Drug Administration."[68] This article recorded the reactions of two of nitrite's opponents, Ralph Nader and Environmental Defense Fund lawyer Anita Johnson, but it did not include quotes from the chemical's defenders. The article referred only to supporters of nitrites insofar as it mentioned that meat industry officials had declined to comment on the study. A commentary in the *Washington Post* derided the planned phaseout of nitrites as "disappointing," because "we have test results showing that nitrites cause cancer by themselves, without combining with other chemicals."[69] Popular news magazines treated Newberne's study with a comparable level of respect. An article in *U.S. News and World Report* described the Newberne study as "the first solid evidence that nitrite by itself can cause cancer."[70]

The FDA's reversal of its position on nitrites also received a good deal of media attention. The story appeared on the front pages of most major newspapers on August 20, 1980. Although the newspapers acknowledged that the Newberne study had been largely discredited by the Interagency Working Group on Nitrite Research, most were not prepared to declare that the nitrite controversy was over. An article in the *Los Angeles Times,* for example, contended that "other scientists believe that there is sufficient evidence against nitrites without the MIT study."[71]

Other segments of the media depicted the FDA announcement in a similar fashion. *Newsweek* reported that the meat industry had "hailed" the government decision; at the same time the article asserted that "what worries scientists most is that even if nitrites are not a direct cause of cancer, they can be transformed in the stomach into substances called nitrosamines, which are known to be potent carcinogens."[72] Others in the media spoke of the FDA's reversal in almost apocalyptic terms. An article in the *Nation* argued that the "rehabilitation of nitrites may represent the dawn of a new era in which science abdicates its primary responsibility to protect the health of the public in favor of deregulation. Now some scientists and their allies in industry are even challenging the idea of public health."[73]

The Limits of Science

One of the most important principles of scientific advancement, if not the most important, is replicability. A corollary of replicability is keeping good notes so that other investigators can reproduce what was done. This principle and its corollary are violated by acceptance of a single study as a basis for action, however prestigious the investigator or the institution.

Is replicability a principle only for pure science? Should it also be required of regulatory science? Proponents of the safety-first doctrine think not and keep the focus on averting harm. But there would be harm also in losing nitrites as a preservative without a better alternative in view.

The pressure to act is transparent: various groups believe that a given chemical is, as Ibsen put it, an enemy of the people and that those who fail to regulate it are likewise enemies. Fortunately, when asked to implement the ban on nitrates, FDA officials noticed anomalies and reconsidered. This turn toward science speaks well of the competence and integrity of agency personnel who were able to discern errors and to counter the recommendations of a researcher from a renowned university, a special team of luminaries, appointed by the director, and the director himself. The arsenic case also reveals the defect in relying on a single study. Some people drew overreaching conclusions from the Taiwanese research and presumed a threat in the United States at Superfund sites and other areas where soil and groundwater arsenic exceeded average levels. But it became clear, eventually, that the U.S. cases were different from the celebrated Taiwanese case: the exposures were lower and the people better nourished. A chemical that caused harm in one circumstance proved benign in others.

We deliberately chose the two subjects of this chapter to highlight different scientific methods—epidemiology (for arsenic in water) and animal cancer tests (for the nitrite preservative). Each method has advantages and limitations. Of neither the arsenic case nor the nitrite case could it be said the knowledge was complete. Why, then, given that everyone would agree with this characterization, is there so much disagreement? Disagreement over science is normal; the competition of ideas is what keeps it going. What is abnormal, in our view, is the demand for total knowledge of tiny causes and minuscule effects. Science, though not all-knowing, is far from ignorant. We know more than enough to make reasoned judgments. But science cannot tell us whether to be concerned about remote and improbable dangers.

8

Do Rodent Studies Predict Cancer in Human Beings?

with Leo Levenson

This chapter explains what the experts know—and do not know—about whether a particular chemical exposure might cause cancer in people. We discuss the fierce debate in the United States over how to design regulations regarding chemicals linked to cancer in animal tests and in epidemiological studies. And we offer our own conclusions on how our governmental agencies should evaluate available evidence to decide which chemical exposures to regulate.

Essential Elements of Toxicology

The fundamental observation of toxicology is that the poisonous effects of chemicals depend on the dose.[1] Exposing people or animals to large enough amounts of synthetic or natural chemicals or minerals can cause toxic effects such as liver damage, cancer, or even death. As exposure is reduced, toxic effects decline and eventually disappear. Reactions may vary greatly among individuals within a species as well as among different species, owing to variability in metabolic rates, size, defense mechanisms, processes by which cancer is caused, and many other factors.

The earliest toxicologists were killers and healers (and sometimes engaged in both professions at once). Murderers and hunters studied how much of a chemical was needed to kill people and animals. Healers tried to find therapeutic doses of potent chemicals that could do more good than harm to their patients. Both groups experimented, administering potent natural substances to pets, prisoners, patients, or prey, until they could establish a pattern of effects associated with different doses.[2]

In recent years toxicologists have been asked to do something very different from killing or healing; they have been asked to tell us whether the synthetic chemicals we encounter at work or in the home are safe and in particular whether these substances can cause cancer. With increasing frequency since the 1940s Congress has passed a series of laws requiring federal agencies to set maximum residue or exposure levels for industrial chemicals in food, water, air, and soil, to protect the public and the environment from adverse effects.

Evidence of harm to humans may come from reports that workers handling a particular chemical are suffering rashes, dizziness, or other symptoms. In response, new guidelines are created requiring extra ventilation or other measures to bring chemical exposures down far below the levels that cause complaints. If chemicals have not been in widespread use for a sufficient length of time to gather reliable human evidence on health hazards, regulators turn to animal tests to help them set safety standards. Experimenters expose groups of animals to various chemical doses to see whether they suffer more harmful effects than animals not exposed. The highest chemical dose that consistently fails to cause harm in experimental animals is known as the no-observed-effect-level (NOEL) or, more precisely, the no-observed-adverse-effect-level (NOAEL).

Standards for "safe," "permissible," or "acceptable" levels of chemical exposure have commonly been determined by applying "safety factors." In case people are more sensitive than animals to a given chemical, common practice in the United States has been to set maximum human exposure limits at 1/100 the NOAEL in animals (unless some additional biochemical or toxicological evidence suggests that the safety factor should be adjusted up or down).

The hundredfold safety factor between animal NOAELs and human exposure standards has been explained as involving two divisions by a factor of ten: one division to allow for the possibility that the animal species may be less sensitive to the chemical than humans, and another to protect unusually sensitive people.[3] This safety factor is attributable to the happenstance that we feel comfortable working with powers of ten. However, in the absence of further evidence on the effects of the chemicals in humans, this calculation is arbitrary.

The Delaney Clause

Since the passage of the Delaney Clause of the Food, Drug, and Cosmetic Act by Congress in 1958, one group of chemicals has been treated differently by regulators—those that appear to cause cancer in people or in animals. The Delaney Clause banned the addition to food of chemicals "found to induce cancer when ingested by man or animal." In effect, it required the Food and Drug Administration (FDA) to extend the safety factor for carcinogens from 100 times below doses that caused no apparent harm to infinity. No matter how small the exposure to a potentially carcinogenic chemical, regulators would have to regard it as unacceptable.

In 1959 the cranberry scare provided the first public example of the Delaney Clause in practice. As described in Chapter 1, the Department of Health, Education, and Welfare recommended a temporary halt to cranberry sales when small residues of a weed-killing chemical were found on a few batches of the fruit. If the usual safety factor had been applied to evaluate the chemical residues, there would have been no problem, for likely human exposures would have been more than 1 million times less than the lowest amount seen to be toxic to animals in long-term tests.[4]

Why treat chemicals that cause cancer in animals differently from those that cause other toxic effects? The philosophy behind the Delaney Clause grew out of a specific idea of how chemicals might cause cancer. Known as the one-hit or no-threshold hypothesis, it holds that cancers develop from a single "transformed" cell, which divides repeatedly to form a group of cells with an abnormal growth pattern. Any chemical that causes genetic damage in cells might happen to cause a mutation in growth-regulating genes that would make the cell "precancerous." The greater the chemical exposure, the more mutations and the greater the risk that an unlucky mutation would instigate the cancer development process. Even a single molecule of such a chemical might make an unlucky hit on a gene, which would lead to cancer. This fear that tiny amounts of chemicals might raise cancer risks by increasing the cellular mutation rate underlay arguments that the federal government should set as a public health goal the elimination of such chemical exposures.[5]

In the years after the Delaney Clause was passed, critics raised two types of arguments against the "zero-tolerance" rule for animal carcinogens. First, many animal carcinogens appear to cause cancer through indirect mechanisms (not by directly causing mutations) that are active only at high exposures to the chemicals. For example, high doses of the herbicide involved in the 1959 cranberry scare were suspected of causing thyroid tumors in rats by blocking iodine uptake, provoking excess production of the rat's own thyroid stimulating hormones. If that is how the herbicide caused cancer in rats, then the much lower, intermittent exposures to herbicides on the cranberries experienced by consumers would not have been sufficient to affect iodine uptake in humans, excite their hormones, or do anything to increase their cancer risk.[6]

Second, even if we suspect that an animal carcinogen is directly mutagenic at low doses, our cells are well-equipped with DNA-repair enzymes and other defenses that protect us against low-level rates of mutation. The danger posed by very small amounts of these chemicals is likely to be negligible compared with the vast amounts of other natural causes of mutation, such as sunlight. During the cranberry episode critics of the scary publicity about the herbicide noted that concentrated natural extracts of common vegetables could also promote thyroid cancers when high doses were administered to animals over a prolonged period. Why should we be worried, they asked, about tiny amounts of synthetic chemical residues that are no more potent than extracts of vegetables?[7]

The original Delaney Clause pertained only to chemicals deliberately added to food, including synthetic flavors, food colorings, and preservatives. It seemed reasonable to believe that if one particular chemical appeared to cause cancer in animals, suitable replacements could be found. Applying the clause to food additives was not unduly costly; but it served as a precedent that would be incorporated into much more extensive and expensive laws governing water, air, and soil.

Quantitative Carcinogenic Risk Assessment (QCRA)

In 1962 Congress modified the Delaney Clause to make it explicitly applicable to veterinary drugs and feed additives given to food-producing animals. The new amendments permitted continued use of potentially carcinogenic veterinary drugs only if the FDA determined that no residues could be found "by methods of examination [it] prescribed or approved."[8]

The FDA ran into practical difficulties implementing the new law, especially in setting the detection limit required for residue testing. Any time an animal receives some tiny amount of medicine, the substance is likely to remain in its tissues after slaughter. If the agency allowed drug companies to use tests that could detect only high levels of residues, then lower levels could remain in meat and end up being ingested by consumers. The more funds available for improving residue testing, the lower the agency could get the detection limit, until all feed additives that caused animal cancers at some high dose would have to be banned.

Lacking guidance from Congress on how to establish detection limit requirements or residue testing methods, FDA administrators constructed their own rationale. In 1973 they proposed a statistical model to quantify the cancer risk posed by tiny chemical residues in order to set a "de minimus" standard, a standard below which harm to humans would be negligible.[9] As modified in 1979, the proposal required tests to be designed to discover residues that could result in up to a 1-in-1-million extra lifetime cancer risk to consumers.[10] If such residues could be found, the use of the feed additive that led to the residues would be banned.

How could the FDA quantify a 1-in-1-million risk level? There was no way to observe such a small increase in the cancer rate in animal tests. Even using hundreds of rodents, the most precise quantitative estimates of cancer risk the laboratory scientists could provide established only that some large dose of a chemical could give 1 out of 10 animals cancer; any smaller increase in the animal cancer rate would usually go unnoticed given the natural variation in tumor development among the animals.[11]

To extrapolate from high-dose animal tests giving 10 percent or more of the animals tumors to a hypothetical 1-in-1-million cancer risk for humans, the FDA recommended performing two basic calculations: dividing the animal test dose by 100,000 or so and adjusting it for the difference in weight between humans and rats. This step (with the addition of a few twists) was the essence of the linear extrapolation model that the FDA proposed to use. The drawback was that there was no scientific basis for assuming that a dose 1/100,000 the test dose posed 1/100,000 the cancer risk. As the FDA itself noted after reviewing a range of proposed statistical procedures for conducting QCRA:

> The same scientific and technical limitations are common to all [methods]. Specifically, because the mechanism of chemical carcinogenesis is not sufficiently understood, none of the procedures has a fully adequate biological rationale. All require extrapolation of risk-dose relations from responses in the observable range to that segment of the dose-response curve where the responses are not observable.[12]

"Fully adequate" could cover a number of meanings, including no biological rationale whatsoever.

Trying to find some justification for choosing the 1-in-1-million cancer risk model as the equivalent of a 0 risk, the FDA asserted that any stricter risk standard "would not significantly increase human protection from cancer," while a more relaxed standard "might significantly increase human risk." Fair enough—if true. But how can we possibly know, given that these risk calculations emerge from models with "no fully adequate biological rationale"? This is like trying to get a quantitative prediction of the odds that intelligent aliens will land on earth this year: 0? 1 in 10,000? 1 in 100,000? 1-in-1-million? If there are no other space-traveling beings in the universe, then the odds are 0, but how can we be sure? Without experience, we have no way of estimating the odds.

Questions like "What is the dose of this chemical that would give a person an extra 1-in-1-million chance of getting cancer?" or "What are the odds that aliens will land this year?" have been called trans-scientific. Although they may be asked in the language of science, they cannot be answered scientifically, since pertinent evidence cannot be gathered.[13]

The FDA document assumes that providing a more conservative (that is, stricter) maximum risk of getting cancer will necessarily be more protective of public health. Is this necessarily so? If low-level residues are below a carcinogenic threshold, then the more conservative standard would make no difference in public health, but it could force expensive actions by farmers or chemical companies. These actions could drive up the cost of food, resulting in less adequate diets for many people, especially those with low incomes. Thus, if food prices rise, or if workers lose their jobs when a chemical plant closes down, mental and physical health might actually be harmed by the conservative residue standard.[14]

The FDA was trying to use QCRA to lower the economic cost of compliance with the Delaney Clause by avoiding banning chemicals completely. The only way to comply without imposing a ban was to claim that the cancer risk numbers that emerged from their extrapolation models had some bearing on reality. But FDA scientists had the integrity to insist that the fundamental uncertainties inherent in the extrapolations had to be acknowledged. Hence we get a regulation that says there is no way to measure cancer risk at low doses and then proceeds to prescribe a method to do just that.

The Environmental Protection Agency and QCRA

Legislation passed during the 1970s and 1980s directed the Environmental Protection Agency (EPA) to draft regulations to protect the public from dangerous levels of chemicals in air, water, and soil. The legislation left it largely up to the EPA to decide what level of chemicals would be considered dangerous.

EPA administrators made an early decision to maintain the Delaney Clause philosophy of treating possible carcinogens as a category separate from all other toxic chemicals but to join the FDA in applying quantitative carcinogenic risk assessment methodologies. They ended up using a "linearized multistage" QCRA extrapolation model that added a few bells and whistles to the FDA model but included many of the same unverifiable assumptions.[15] The EPA stated that "since only rarely do we know for sure that an agent is indeed a human carcinogen, the first step involves an evaluation of all the relevant biomedical data to determine the weight of evidence that an agent might be a human carcinogen . . . in terms of rough estimates for current exposures as well as estimated exposures for various regulatory options."[16] The agency also borrowed the 1-in-1-million hypothetical cancer risk standard pioneered by the FDA. Although this number was never established formally in legislation or regulations, it became the EPA's preferred benchmark for setting cleanup levels for chemicals in air, water, and soil.[17]

Comparing the 1-in-1-Million Standard to the Safety Factor

According to a 1984 study, the cleanup standards for potentially carcinogenic chemicals set by the EPA, using the 1-in-1-million risk level, are designed to regulate human exposures as small as 40,000–4 million times less than the maximum tolerated dose (MTD) used in animal tests.[18] The MTD is the highest dose that does not cause serious toxic effects in short-term tests (generally ninety days) and thus is not associated with a significant increase in early death among animals in long-term tests from causes other than cancer. The MTD is equivalent to a short-term NOAEL.

In contrast, when dealing with noncarcinogenic chemicals, the FDA and EPA generally allow exposures up to about 1,000 times less than the short-term NOAEL, or 100 times less than a NOAEL derived from long-term tests. Thus, according to QCRA, high-dose animal carcinogens are regulated in a fashion that is on average about 40–4,000 times stricter than the regulation of noncarcinogenic chemicals. In fact, much of the billions of dollars spent on toxic waste site cleanups may be attributed to the required removal of tiny residues of carcinogens that would be regarded as completely unimportant if a 1,000-fold safety factor were used, rather than the stricter limits created by QCRA and the 1-in-1-million hypothetical risk standard.[19]

Before evaluating the defects of animal cancer tests, we can address that perennially helpful question—compared with what?—by first considering the weaknesses of studies on human beings. Evaluating epidemiology will help answer the question of whether, despite their flaws, rodent cancer studies are the best available alternative for predicting whether chemicals are likely to cause human cancers at different levels of exposure.

The Limits of Epidemiology

Epidemiology is a science dealing with the exposure of populations to various suspect agents and the occurrence of diseases that may or may not be traced to them. The great virtue of large human epidemiological studies is that they are about what we want to know—the incidence and causes of disease—so that once a damaging agent has been identified fewer problems of interpretation remain. Nevertheless, epidemiology has limitations that have led to a search for alternatives.

Given the moral objection to experimenting on human beings, various exposures have to be determined after the fact rather than when administered by trained personnel. Determining exactly what the subjects have been exposed to is not easy. Retrospective data are subject to recall bias, in which people sometimes remember more frequent exposures to the things they think caused their diseases. It is thus difficult to assemble data on large numbers of people, whether it be the extent of their illnesses or disabilities or the amounts and duration and kinds of exposures.

Even when it is possible to have some confidence about damage and exposure, it may not be possible to learn enough about confounding variables—other causes that may be responsible for the phenomena in question. That is why it is exceedingly important to control for such variables as age, sex, smoking, and variations in exposure to the alleged source of distress. Large numbers of subjects and carefully matched control groups yield the best results.

With so many people and so many possible causal factors, it is difficult to discern effects that are small proportionately but may nevertheless, when applied to the general population, affect a large number of people.[20] Insofar as cancer is the object of concern, moreover, its long latency period, which may last decades, requires that studies of people suffering from cancer be able to answer questions about their occupation and food and medicinal exposures, as well as their personal habits, for the last twenty or thirty years. Needless to say, people have faulty memories and may well talk about what they should have done rather than what they did, or leave out habits they surmise the investigator may not approve of. In addition to the considerable differences in susceptibility to exposures among human beings, people's habits differ greatly. It makes a big difference, Dale Hattis and David Kennedy tell us, whether people breathe through the mouth or nose, how quickly and deeply they breathe, and how thick their skin is.[21]

There are also insuperable obstacles in the way of testing the thousands of chemicals developed each year. The offending substance may not have had a long enough opportunity to produce adverse effects. This limitation is most serious for occupational exposures. It is less serious for ephemeral exposures to trace elements in the general population for which we already have evidence from several years of occupational exposure to comparatively massive doses of the same chemical, which can still tell us a good deal.

The choice we face is stark: should we rely on epidemiological studies, which are more difficult to do, take longer, but are more conclusive than animal tests, or should we rely on animal tests with their numerous (to be discussed) flaws? If we rely on epidemiology, there is a small chance that very dangerous substances will escape notice but a greater chance that chemicals probably affecting only a small proportion of the population, yet possibly affecting thousands of people, will remain undetected. "A good working definition of a catastrophe," according to David Ozonoff, the head of the Environmental Health Section at the Boston University School of Public Health, "is an effect so large that even an epidemiological study can detect it."[22] It is easy to say that further study will eventually reveal mechanisms of cancer causation, so that much better controls can be devised. But what do we do in the meantime?

Asking the Right Questions about Animal Studies and Human Cancers

Regulation of exposure to chemicals, including the intermittent exposures to trace elements to which the general public is subject, is largely based on interpretations of animal cancer bioassays. If these tests are reasonably accurate in predicting the probability, sites, and severity of human cancers, then regulation of suspected carcinogenic chemicals is on firm ground. But if these animal tests are weak or worse, so that one cannot reasonably predict human cancers from them, then regulation rests on quicksand.

Whether or not rodent tests predict human cancers, they have many important uses. Research into cancer mechanisms or problems of the immune system, for instance, may be furthered by introducing novel genes into small animals, such as transgenic mice, to discover better how life systems work.[23] There is no doubt that models based on research with animals have increased our understanding of metastasis, a crucial stage in the spread of cancer.[24] None of the many invaluable uses of animal cancer tests, however, tells us whether they can come close enough with enough frequency to be a valid source of evidence in predicting human cancer.

In assessing how predictive of human cancers these animal tests are, we must ask, How reliable are the tests? If the tests were repeated on the same species, would we get nearly the same results? If they were repeated on different animal species, would we come up with similar results? If chemicals are carcinogenic in several animal species, is it more likely that they are carcinogenic to mammals in general, including human beings, than if they only cause cancer in a single species? We must distinguish rates and sites of cancer not only by species but also by age, because cancer is largely a disease of old age, and by sex, because males and females are affected differently.

As precisely as possible we wish to answer David A. Freedman and H. Zeisel's question, "Are chemicals that have been shown to be carcinogenic through experimental animals also carcinogenic to humans?"[25] The reason for their inclusion of the modifier "experimental" has to do with the particular conditions under which animals are tested. Therefore they also ask, "Do experimental animals (rodents, in particular) and humans have similar susceptibility to the carcinogenic effect of chemicals, or are rodents incomparably more susceptible than humans?"[26]

The answer to the first question is, "sometimes"; rodent cancers can appear in humans too, but we do not know when. The same is true of cancers in different types of rodents. The answer to the second question is, "yes, mostly, but not always." If we know that a single dose for dioxin ranges from 2,500 micrograms per kilogram (μg/kg) in guinea pigs to 5,000 in hamsters, a difference of 2,500, does that give us confidence about rodent to human transfers?

Suppose that we arrive at several different answers: one is that there is a 10 percent probability that a substance causing cancer in a mouse or rat at a given dose will do the same in a human being. From one point of view, nine times out of ten the extrapolation from mouse to human would be wrong; from another, why take chances with human health if the probability of getting cancer is that high? Were we to find, however, that the answers from the tests would tell us only within a factor of several hundred times to several hundred thousand times whether rodent cancers predict human cancers, that level of extrapolation might not be a reasonable approximation. The question is not only whether we can get an answer but what kind of answer we will get. To interpret results, we need to know something about human cancers.

Causes of Cancer

Cancer is not a single disease; it is a large collection of diseases, differing in their causes, in the types of cell in which they originate, in their locations in tissues and organs, and in prognoses for amelioration or cure. Common to all cancers is the uncontrolled growth of malignant cells that arise in some way from normal and healthy cells. Malignant cells can spread (or metastasize) to areas some distance from where they started, a usually fatal development.

Of the causes of cancer, of which too little is known, it will be useful for us to distinguish between genotoxic, meaning those harmful to genes (DNA), and nongenotoxic, those that do not attack DNA but might affect hormone levels or metabolism. A single substance can be both a carcinogen at one site in a human being and an anticarcinogen at another site. This indicates the need for a holistic analysis of the overall effects of a substance.[27] Hoping to do better in the future by learning more about cancer-causing mechanisms is not a popular approach compared with taking preventive measures now. Still, the founding director of the United Nations International Agency for Research on Cancer, John Higginson, argues that "apart from more intensified effort on certain already recognized causal factors, current research indicates that the eliminatory [regulatory] approach will have little impact on the cancer burden and that the mechanistic [research] approach, although difficult and slow, represents the most logical alternative. This will require long-term major investments and fundamental research and manpower."[28] Debate is over whether in the meantime regulatory programs should be intensified, diminished, or abandoned. Assessing the regulatory approach brings us right back to animal cancer tests.

The Process of Animal Cancer Testing

Around 1915 or 1916 scientists learned that they could induce cancer in animals by treating them with certain chemicals. The methods of giving animals cancer vary greatly. "Chemicals have been introduced into experimental animals by every orifice (orally, nasally, urethrally, vaginally, rectally), by various types of injections (intramuscular, intraperitoneal, intravenous, subcutaneous), by skin painting, by surgery, and by other methods."[29]

Approximately 30 percent of rodents get some form of cancer absent exposure to chemicals, though all 30 percent do not die of it. This is one reason why a control group is essential. Because a chemical's effects at high doses may not show up at low doses, it is necessary to subdivide the animals into different dose groups. Since sex plays an important role in cancer, a further subdivision occurs between male and female. Usually there are three dose groups (0, one-half or one-tenth the MTD, and the MTD) and two species. There are at least twelve groups of animals. By convention and by statistical necessity, there are usually fifty animals, most often rodents, in each group.

In three basic aspects animal testing differs from epidemiological studies: in its short time, its high cost, and its essentially statistical character. A great advantage of rodent testing is that these animals live only about two years; thus results are available relatively quickly. One can also test any chemical, including new chemicals, for which epidemiological evidence may not be available. But it is costly to keep animals under controlled conditions for up to two years. When each animal dies (or "is sacrificed") several pathologists must painstakingly examine around forty sites to search for tumors, some of which are so small they can be discerned only with high-powered microscopes.

That is why there is a team of pathologists who first work separately and then meet to resolve differences before their findings are accepted for further evaluation.[30] These pathologists consider whether the tumors or other abnormalities are actually induced by the chemical, an opinion based on what they know about the normal incidence of tumors and their experience. They ask themselves not only whether the incidence of tumors is higher but whether they are of a different size, shape, or color or contain any other signs that might show them to be similar to or different from naturally occurring lesions.[31]

To understand better whether the proper dose was administered, the animals are weighed to see whether they have lost appropriate amounts of weight and examined to see whether the dose was either so large as to threaten their lives from causes other than cancer or so small as to make its effects unnoticeable. When one multiplies the time these tests take, roughly three years, by the cost of keeping twelve groups of animals in controlled conditions, and then adds the cost of killing and dissecting them, as well as the cost of preparing and examining forty slides per animal and of reconciling differences, the substantial costs do not appear out of line.[32]

It is possible for a government regulator to conclude that the tests are inadequate or that the substance being tested is actually a carcinogen. But it is not possible under the rules to say that the substance is not, insofar as is known, a carcinogen. Instead, the most negative statement government scientists are allowed to make is that "the compound has not been shown to be carcinogenic."[33]

What do we actually know about the toxicity of different doses in different species?

Calculating Toxicity by the LD_{50} Test

In the field of pesticide regulation, lethality is calculated through the assignment of an LD_{50}, the lethal dose for one-half of the animals during the test period. The relevant number for aspirin is 730 mg/kg, signifying that 50 percent of the test animals died when exposed to 730 mg of aspirin per kg of their body weight.[34] The larger the LD_{50}, the more of a substance it takes to produce a toxic effect and the less harmful the chemical.

Among species most commonly used to carry out the LD$_{50}$ test are fish, birds, rabbits, mice, and rats; occasionally monkeys and dogs are tested. Generally about sixty animals of a particular species and a specific dosing method are used. The application is made by inserting a tube down the throat of the animal and forcing injection of vapors, or by spreading the substance on the skin.[35] The usual test lasts about two weeks; many animals die during that period, and the rest are killed off at the end. The usual symptoms are bleeding from the mouth or eyes, convulsions, diarrhea, and what are termed "unusual vocalizations." Rather than tolerate early death, according to the British Toxicological Society, "There is pressure on the toxicologists to allow the study to continue, even when the animals are in distress since their premature killing may alter the end-point of the study, and so possibly affect the classification of the material being tested."[36]

Animal rights advocates are not happy with this method. Whether one believes that the LD$_{50}$ test involves "a ritual mass execution of animals"[37] or that "the main information they give is an indication of the size of dose required to commit suicide,"[38] or even whether most experts consider "the modern toxicological routine procedure a wasteful endeavor in which scientific inventiveness and common sense have been replaced by a thoughtless completion of standard protocols,"[39] there is ample scientific doubt about the value of the LD$_{50}$ test for the purpose of predicting effects on humans.[40] One basic difficulty is that enormous differences between species are reported, with doses ranging from five to seventy-five times greater from one species to another, making findings suspect.[41]

Toxicity (Potency)

Unfortunately, the effects of acute (large, relatively instantaneous) exposures do not predict the effects of chronic (repeated, low-level concentration) exposures of a particular chemical. Large doses of vitamin D or sodium fluoride, for instance, can be deadly, but tiny amounts spread over time appear to be good for human health. The same is true for table salt and iodine and numerous other substances. But there is also evidence that some substances are quite harmless in one-time huge exposures while harmful in small doses given over long periods of time. Metallic mercury, for example, passes through the body without causing damage when taken in a single large dose, whereas a long period of small doses can be fatal.[42]

It may help put matters into perspective to adopt Andrew Rowan's sixfold classification scheme moving from supertoxic to practically nontoxic (Table 8.1).

But where are the nontoxic substances? These substances do not exist, because virtually everything can be toxic if the dose is large enough. E. M. Boyd and I. L. Goady show that even distilled water, entirely free of impurities,

Table 8.1   Grading of toxic substances

Grade (lethal dose per kilogram body weight)	Dose required to kill a person	Examples
Supertoxic (less than 5 mg/kg)	A few drops	A few pesticides, cyanide, nicotine
Extremely toxic (5–50 mg/kg)	A few drops to 1 tsp	Drain cleaners, morphine, amphetamine
Very toxic (50–500 mg/kg)	1 tsp to 1 oz	Disinfectants, degreasers, aspirin, boric acid
Moderately toxic (0.5–5 g/kg)	1 oz to 1 pt	Polishes, hairdyes, bleaches, kerosene, methanol
Slightly toxic (5–15 g/kg)	1 pt to 1 qt	Most cosmetics, soap products, ethyl alcohol, saccharin
Practically nontoxic	More than 1 qt	Foods, candles, kaolin, talc

Source: Adapted from *Encyclopedia Britannica,* 15th ed. (1983), "Poisons and Poisoning," vol. 14, p. 619; in Andrew N. Rowan, *Of Mice, Models, and Men* (Albany: State University of New York Press, 1984), p. 191.

can kill rats if they are forced to take enough of it, though, of course, the dose would be equivalent to an adult male drinking more than seven gallons of water at a rate of a gallon every twenty minutes. Rowan rightly concludes that "this demonstrates that, ultimately, the toxicity of a substance depends on the dose given and the manner in which it is given. As a result, chemicals are usually classified according to the dose required to produce a toxic or lethal effect."[43]

Many more agents or teratogens that cause physical defects in an embryo have been identified in rats than in people. Although one can find some examples of human teratogens that are also rat teratogens, there are many rat teratogens—from vitamins A and B to aspirin and insulin—that do not appear to act in the same way in human beings. Regretting the "lack of specificity in cause and effect," S. J. Yaffe concludes that findings from animal experiments "cannot be extrapolated from species to species or even from strain to strain within the same species, much less from animals to humans."[44] As Rowan puts it, "Part of the problem may be that rodent and rabbit embryos are not as well protected as human embryos, especially in the early stages of development."[45]

The best explanation to the public of the difference between very large and very small exposures to different kinds of species comes from reporter Richard Harris of National Public Radio, together with a number of cancer researchers and government officials. Their dialogue is instructive:

Penelope Fenner Crisp (EPA):   We're coming to discover that there are more differences between species than we had expected or, frankly, hoped that existed.

Harris: It turns out that a great many chemicals that can cause cancer in one species don't seem to do anything at all in another species . . . Now nobody's suggesting that these chemicals are harmless, but in some cases scientists believe that the standards may be vastly overstating the health risks. Again, this comes down to a necessary but flawed shortcut the EPA uses to size up a chemical. Scientists give a huge dose of chemicals to rats and then estimate the effects of that chemical at lower doses. By way of analogy, if you drop a bottle from 10 feet off the ground, it's pretty obvious what's going to happen.

(Sound of glass shattering)

Harris: This large drop is the equivalent of a large dose of a chemical, and it can be deadly. But what if, instead of taking one bottle and dropping it from 10 feet, you take 10 bottles and drop them from one foot? It's like giving many people a smaller dose of that toxic chemical. Here's what the EPA assumes will happen.

(Sound of several bottles hitting the ground and one of them shattering)

Harris: They figure one of the 10 bottles will break. The reasoning is that one-tenth the dose, or one-tenth the drop distance, will do one-tenth the damage. In reality, though, this is what happens.

(Sound of several bottles hitting the ground)

Harris: There is, in fact, a safe height you can drop a bottle from without breaking it, and John Doull from the University of Kansas says the same idea holds for toxic chemicals.

Doull: It is the dose, not the compound, that determines its adverse effects . . .

Harris: So, recently, researchers like Swenberg have started to dig deeper and ask why some chemicals trigger cancer in some animals . . . And Swenberg says one especially interesting example is unleaded gasoline. You may have seen the sticker at the pump warning that gasoline causes cancer in laboratory animals. Well, here's the story with gasoline.

Swenberg: It causes kidney cancer in male rats only, not in female rats and not in mice.

Harris: So what's going on? Swenberg decided to find out by studying those animals, and he discovered that a chemical in gasoline binds to a naturally-occurring protein that's only found in the kidneys of male rats.

Swenberg: And this results in a build-up of the protein and ultimately leads to the development of cancer. And since humans do not synthesize this protein, this is not likely to be a mechanism important to humans.

Harris: Swenberg says dozens of other chemicals besides gasoline cause this specific kidney cancer in male rats, including copy machine toner, a bathroom deodorizer and even a natural chemical called D-limonene.

Swenberg: It turns out that about two glasses of orange juice contain a carcinogenic amount of D-limonene for the male rat, but it has absolutely no effect on mice or on female rats, and I'm sure it has no effect on humans.

Harris: As a result of this research, the Environmental Protection Agency recently decided that if a chemical like gasoline only triggers this kind of kidney tumor in male rats and it doesn't do anything else bad, it's probably not going to cause cancer in people. So far there are just a handful of stories like this where scientists have actually figured out why a compound is causing tumors in certain animals. But there are a lot more studies in the works, including reassessments of dioxin, formaldehyde and certain PCBs.[46]

Knowledge of mechanisms yields far greater discriminatory power. With such knowledge scientists can determine whether there is a threshold below which no damage occurs or whether harm occurs in proportion to the dose. No mechanism, no dose-response relationship.

The War over the Dose-Response Threshold

A common statement about dose-response levels is that no one really understands what happens when people are exposed to very low levels of chemicals.[47] We have no difficulty finding substances, such as the heart medicine digitalis, that are helpful at low doses but can be fatal at large doses. But that does not answer the question of whether there are substances for which no threshold exists.[48] Given the considerable range of sensitivity among human beings, it can always be said that some hypersensitive people may be adversely affected; the traditional response has been to use a margin of safety to take care of the supersusceptible. Given also that chemicals may interact with each other to create cancers that neither substance would have alone, it cannot be said definitively that either is safe. By the same token, however, one chemical may render another harmless or less harmful.[49]

Going further into the "threshold controversy,"[50] it is instructive to read a semiofficial account by high-ranking EPA officials published in a major journal, *Risk Analysis*. The models the EPA uses attempt to establish an upper bound, marking the worst-case scenario, on the basis of a no-threshold linear response. Elizabeth Anderson and her colleagues at the EPA admit that "this recognition that the lower bound may . . . be indistinguishable from zero stems from the uncertainties associated with mechanisms of carcinogenesis including the possibility of detoxification and repair mechanisms, metabolic pathways, and the role of the agent in the cancer process."[51] In short, there may be no damage at low doses.

Furthermore, "most often there is no biological justification to support the choice of any one model to describe actual risk." Choosing a model would be easy if we had data on actual environmental risks to human beings, "in the absence of such data a variety of models can be used to fit the data in the observed range, but these models differ sharply [in the danger estimates they produce] at low doses." If the choice of model determines the results, because they "differ sharply at low doses," why bother with the experiment? Employing the justification that nevertheless these models are the best available, the EPA article continues, "It should be clear from the preceding discussion that the linear non-threshold model has been used by the EPA to place plausible upper bounds on risk, not to establish actual risk."[52]

But why? Use of the upper bound misleads people into thinking it is an actual estimate of hazard by an authoritative government agency when it is not. Use of worst-case scenarios makes no sense, moreover, when there is reason to believe the outcome may be zero.

If it is true, as Anderson and others say in their appendix, that "there is no really solid scientific basis for any mathematical extrapolation model relating carcinogen exposure to cancer risks at the extremely low levels of concentration that must be dealt with in evaluating environmental hazards,"[53] then why make one? The answer must be that without going from rodents to people most regulation of chemicals would lack a rationale that could be called scientific.

Multistage Models

Actually, extrapolation models combine two models in one: first, a model of the biology underlying cancer causation; second, the statistical approximation of that model. Getting accurate results depends on the predictive power of the biological model and also on whether the statistical approximation captures the causal structure of the model. If the model does not well describe cancer causation in human beings and if, on top of that, the statistical approximation does not well describe the model, the errors in both models multiply to give unsatisfactory results. The task is a daunting one.

We can take the Armitage-Doll model as representative of those used by governmental agencies in regulation. It seeks to describe the relationship between exposures to chemicals and the incidence of cancer at various ages for men and women.[54] The biological version portrays human cells as going through a number of stages that ultimately result in cancer. The hypothesis is that one or more cells receive an insult and then go through several changes that turn them into malignant cells, after which they proliferate. The times and the different stages are not specified. All these stages are probabilistic in that some cells under the same exposure will become cancerous and others will not.

"Put another way, with multistage models," Richard Peto explains, "when all the predisposing factors have been allowed for, luck has an essential role in determining who gets cancer and who does not."[55] Thus the stages in the models are essentially probabilities without the users knowing whether human cancer proceeds in those stages or according to those odds.[56] In the field of economics, these would be called Markov chain models, which means essentially that every present stage depends on results of previous stages. The time spent in the various stages is assumed to be proportionate to the exposure of the affected individual.

The basic difficulty with multistage models, as the reader may imagine, is that there is little reason to believe that they actually capture the biological process of cancer formation. At the same time, the statistical manipulations are very far from the causal requirements of the model so that there is no clear idea of what has been described when the result is cranked out.[57]

Thomas Fears, Robert Tarone, and Kenneth Chu have concluded that "there is danger in relying solely on the finding of statistical significance without incorporating biological knowledge and corroborative evidence such as the presence of a dose-response relationship for experimentally consistent results in different species or sexes."[58] But what if there is little or no biological knowledge?

To get accurate estimates of the probability that chemicals that cause cancer in animals also cause cancer in human beings, David Salsburg recommends applying "the bioassay to a number of innocuous substances. There have to be some compounds that are not human carcinogens, or the whole exercise of looking for carcinogens makes no sense." Yet, after examining the literature, he finds that "this was never done for the [rodent] lifetime feeding study."[59] His argument needs to be heard in full:

> Thus, it would appear that no attempt has ever been made to determine how well society can identify human carcinogens by feeding groups of 50 rats and mice, each, the suspect substance at maximum tolerated doses for their entire lives. Common scientific prudence would suggest that this assay be tried on a group of known human carcinogens and on a group of supposedly innocuous substances (such as sucrose or amino acids) before we either (1) believe that it provides some protection for society (sensitivity) or (2) believe it identifies mainly harmful substances (specificity). There is no substitute for such proper validation on any new bioassay.[60]

He believes that "we are confusing the effects of biological activity upon the old-age legions of rodents with the thing we fear, cancer."[61]

Mitogenesis, or, Is Cancer Caused by the Test and Not by the Chemical?

In addition, the tumors observed in animal cancer tests may be due to the huge doses delivered at the MTD. If true, this connection would doom the no-threshold idea, because the animal cancer tests would be measuring only what they themselves caused. Bruce Ames and Lois Gold, among others, claim that the chronic wounding caused by delivering heavy doses of a chemical promotes cancer by inducing cell division, a process called mitogenesis. As the animal is effectively injured or poisoned, it grows replacement cells, a process known to increase chances of mutation and hence of cancer.

The theory was prompted by findings that while cancer is thought to be accompanied by mutation, alteration, or damage of DNA, a large proportion of the chemicals that cause cancers in animal tests do not in fact damage genes examined in other tests. There are "too many rodent carcinogens," in view of our knowledge of cancer and our recognition that there would be a great deal more cancer around if half the chemicals in the world caused this terrible sickness.[62]

Samuel Cohen and Leon Ellwein confirm, "Chemicals that induce cancer at high doses in animal bioassays often fail to fit the traditional characterization of genotoxins. Many of these nongenotoxic compounds (such as sodium saccharin) have in common the property that they increase cell proliferation in the target organ." They argue that "the increase in cell proliferation can account for the carcinogenicity of the nongenotoxic compounds."[63] Similarly, Daniel Krewski finds a "fairly strong" correlation between carcinogenicity and toxicity, which one would expect when test animals are being wounded by being fed the maximum tolerated dose.[64] If the mitogenesis theory is right, then rodent tests run at mitogenic doses are invalid as predictors of human cancer from exposures below toxic levels. Ames states bluntly: "We think the current approach to cancer risk assessment is bankrupt."[65]

Obviously, many do not agree with the Ames-Gold thesis that the animal bioassay, by feeding animals the MTD, is itself causing excessive rates of cancer. Frederica Perera, for one, argues that a variety of international agencies, including those in the United States, have "adopted the general assumption of low-dose linearity for carcinogens—regardless of their presumed mechanism of action." In other words, there must be something to this assumption since so many agencies have gone in the same direction. Scientists usually do not give the "argument from authority" when they possess a theory they can validate with evidence. But here, according to Perera, there is a general lack of understanding of the mechanisms of cancer causation, a lack of agreement on a safe threshold level below which exposures would not be harmful to a diverse population, and "the desirability of preventing cancer through the use of testing and model systems, obviating the reliance on epidemiological data in humans."[66] The question is whether regulation should be undertaken as a replacement for the existing lack of knowledge. Vincent Cogliano and his colleagues have answered just this question: "In the meantime, EPA cannot ignore its responsibility to evaluate and control synthetic chemicals . . . since no one, including Ames and Gold . . . has yet devised an acceptable alternative."[67]

We disagree with this answer. The predicted cancer rate at 1-in-1-million (even 1 in 100,000 or 1 in 10,000) is so low it will never be detected by epidemiological studies or any other method unless we know a lot more about the mechanisms of cancer causation. Moreover, "The problem with . . . risk assessments . . . based on animal tests," the Office of Technology Assessment's Michael Gough tells students, "is that their theories cannot be tested."[68]

The Limitations of Statistical Models

The most important defect of animal cancer studies, therefore, is that the choice of statistical model overdetermines the results: it produces outcomes that vary by hundreds, thousands, tens of thousands, and occasionally millions of times. Yet, without knowledge of the biological mechanisms of cancer causation, there is no way of choosing among these models. In the study of the grain fumigant EDB, for instance, the probability that an individual would get cancer from eating food in which tiny amounts of EDB were present varies over a million times depending on the model.[69] Using the same animal data but different statistical models in regard to saccharin, to take a well known but extreme instance, led to differences of some 5 million.[70]

With estimates this far apart when different models are used, and without a plausible biological reason for preferring one model over another, the results are no better than guesswork. Indeed, we doubt that educated guessers would produce results so far apart. Would the reader accept a number in anything that mattered if it was between 100 and 40,000 times off and the reader did not know which way?

But if epidemiology is too insensitive and animal cancer tests are invalid, the question remains: How should the multitudes of chemicals be treated until we possess the knowledge to eliminate or restrict those that cause human cancers at low doses? Should our collective decision be to cut through the complexity by severe regulation? Would such a policy actually improve our health? What do we actually know about the sources of carcinogenic chemicals to which human beings are exposed? Two categories of interest are synthetic carcinogens produced by industry and natural carcinogens produced by plants to ward off their predators.

Natural versus Synthetic Carcinogens

It is possible to calculate the relative weight and potency of synthetic carcinogens as compared to natural carcinogens. By dint of rough-and-ready but serviceable calculation, Ames and Gold have determined that nature's chemicals, in weight and potency, exceed industry's chemicals by a ratio of 10,000 : 1. Synthetic chemicals ingested by humans amount to less than 0.01 percent of our daily diet.[71] How harmful can such a small proportion of synthetic carcinogens be?

Environmental groups, measured by the requirements they place on industry and regulatory agencies, think that tiny is still terrible. They point to the absolute, not relative, amount of synthetic chemicals. And they offer a number of counterarguments, including the assertion that as human eating patterns have developed alongside natural chemicals, the human body has grown accustomed to these chemicals but not to industrial chemicals. However, the evidence runs counter to the principal points in this argument. Most foods humans eat nowadays (potatoes, tomatoes, and so on) have been consumed widely only relatively recently, and human evolution works much too slowly to adapt to whatever insults to the body these foods might deliver. Nor is there necessarily a need for such speedy adaptation. The body contains numerous mechanisms for repair of small harms: antioxidants defend against radiation; skin sheds (the cells lining the mouth, skin, lungs, esophagus, colon, and other organs regularly are discarded, the damage they have suffered along with them);[72] T cells in healthy immune systems react against toxins; and more. Were the body's defenses too specific, it would be at a disadvantage in coping with new dangers.[73]

But, the counterargument goes, it takes several decades for cancers to come about. What about the thirty years that have already passed? Does anyone suppose that the United States was not an industrial power in 1963? Or in 1943, when the wartime economy was pumping up the chemical industry? Where, then, is the long-awaited rise in cancer rates or decline in general health rates?

As for the idea that the human body copes well with natural carcinogens but badly with synthetics, there is no supporting evidence. The body is a nondiscriminatory equal opportunity ingestor to which a chemical is a chemical is a chemical. Still, the counterargument continues, whatever comes from synthetic sources, however small, will add to whatever natural exposures there are; why should we accept this risk? Again, background levels (roughly, 99.99 percent) are very large compared with synthetic additions. Ames and Gold have shown that a cup of coffee contains approximately 15 g of potentially carcinogenic material, equivalent to the total amount of pesticide residues an average person would take into his or her body in a year,[74] so those who worry about synthetic chemicals can drink one less cup of coffee.

Is There a Cancer Epidemic?

In their study "Progress against Cancer" John Bailar and Elaine Smith assert that the best measure of the incidence of cancer is the age-adjusted mortality rate, which has increased. Hence they claim that "we are losing the war against cancer."[75] When we look at different kinds of cancer in the entire U.S. population, however, we see that colon and rectal cancers are down moderately, stomach cancer is down substantially, and breast and prostate cancers are moderately up. But these figures and conclusions leave lung cancer (largely due to smoking) out. There are apparently no national data separating smokers and nonsmokers.

But when Bailar and Smith do control for lung cancer, "the change in overall age-adjusted mortality from cancer since 1950 shifts from an eight percent increase to a 13 percent decrease."[76] Their point is that medical treatment does not help all that much, which is not our concern. If by an epidemic we mean that people can greatly increase their chances of getting cancer by smoking, then there is one. But if we mean that nonsmokers are experiencing an enhanced incidence of cancer, the evidence shows just the opposite.

An article in the July 1992 *American Journal of Public Health* unequivocally denies the existence of a cancer epidemic. Epidemiologist Richard Doll explains that when cigarette smoking is accounted for,

> the mortality from all other cancers, considered as a group, is decreasing in middle and old age. To detect an effect from new hazards of individual types of cancer we need to examine the trends in incidence and mortality in early middle age (that is, in adults under 50 years of age), as it is in the relatively young that the effect of new hazards is usually first seen. Examination of these trends shows that most of the important cancers are becoming less common in both Britain and the United States, the increase in the recorded incidence of the others is mostly an artifact of screening or the result of behavior that the individual can choose to avoid. Only the worldwide increase in cancer of the testis and the small increase in non-Hodgkin's lymphoma that is not attributable to immune deficiency remain unexplained and may possibly be due to new environmental hazards.[77]

Perhaps some comparison of causes of cancer and of the relative weight of potential carcinogens would help illuminate the size of the problem. A noted English statistician, Richard Peto, prepared his best educated guess about the causes of death from cancer and about remediation of those deaths in two tables reprinted as Tables 8.2 and 8.3.

Peto's pollution and industrial product categories, the only places where small-scale exposures might be found, are significantly below other causes. In the remedial context of cancer control, Peto estimates very low reductions in cancer rates from removing synthetic chemicals from food, water, and air.

The only comparative carcinogen database is Gold and Ames's HERP (Human Exposure Dose/Rodent Potency Dose). Table 8.4 is calculated from HERP and compares the estimated cancer hazards of DDT and EDB (a banned grain fumigant) with those of common foods. These comparisons reveal that regulation holds synthetic chemicals to limits far more stringent than the levels readily accepted for ordinary food and drink. The health benefits of such regulation, therefore, are bound to be minuscule or nil.

Table 8.2   Future perfect: Estimate of the proportions of cancer deaths that will be
found to be attributable to various factors

	Percentage of all U.S. cancer deaths	
Cause	*Best estimate*	*Range of acceptable estimates*
Tobacco	30	25–40
Alcohol	3	2–4
Diet	35	10–70
Food additives	<1	−5[a]–2
Sexual behavior	1	1
Yet to be discovered hormonal analogues of reproductive factors	−6	−12–0
Occupation	4	2–8
Pollution	2	1–5
Industrial products	<1	<1–2
Medicines and medical procedures	1	0.5–3
Geophysical factors (mostly natural background radiation and sunlight)	3	2–4
Infective processes	10?	1–?
Unknown	?[b]	?
Total	200% or more[b]	

Source: Richard Peto, "Epidemiological Reservations about Risk Assessment," in Avril D. Woodhead et al., eds., *Assessment of Risk from Low-Level Exposure to Radiation and Chemicals* (New York: Plenum Press, 1985), p. 6.

a. The next effects of food additives may be protective, e.g., against stomach cancer.

b. Since one cancer may have two or more causes, the grand total in such a table will probably, when more knowledge is available, greatly exceed 200%. (It is merely a coincidence that the suggested figures in the present table happen to add up to nearly 100%.)

The Continuing Debate

The scientific debate over how to regard the risks posed by small doses of animal carcinogens has not died down since the Delaney Clause was first passed. Every year professional journals publish numerous articles proposing newly refined statistical methods for characterizing carcinogenic risk and attacking the entire concept of trying to place a number on something we are still far from understanding.[78] Controversies rage back and forth over whether to divide the characterization of "carcinogenic chemicals" into overlapping subcategories of "initiators" and "promoters" or according to whether their carcinogenic activity is believed to operate through "mutagenic" or "epige-

Table 8.3   Present imperfect: Reliably established (as of 1981), practicable ways of avoiding the onset of life-threatening cancer

Method	Percentage of all U.S. cancer deaths known to be thus avoidable
Avoidance of tobacco smoke	30
Avoidance of alcoholic drinks or mouthwashes	3
Avoidance of obesity	2
Regular cervical screening and genital hygiene	1
Avoidance of inessential medical use of hormones or radiology	<1
Avoidance of unusual exposure to sunlight	<1
Avoidance of known effects of current levels of exposure to carcinogens (for which there is epidemiological evidence of human hazard) in:	
(i) occupational context	<1
(ii) food, water, or urban air	<1

Source: Peto, "Epidemiological Reservations about Risk Assessment," p. 6.
Note: Excludes prophylactic prostatectomy, mastectomy, hysterectomy, oophorectomy, artificial menopause, and pregnancy.

Table 8.4   Comparison of cancer hazards

Cancer hazard	Food (carcinogen)
0.0003	DDT, from residues remaining
0.0004	EDB, before ban took effect
0.001	Municipal tap water, four eight-ounce glasses (chloroform)
0.03	Peanut butter, one sandwich (aflatoxin)
0.03	Comfrey tea, one cup (symphytine)
0.07	Brown mustard, one teaspoon (allyl isothiocynanate)
0.1	Dried basil, one leaf (estragole)
0.1	Raw mushrooms, one (hydrazines)
2.8	Beer, twelve-ounce bottle (ethyl alcohol)
4.7	Wine, eight-ounce glass (ethyl alcohol)

Source: Kathleen McAuliffe et al., "How Safe Is Your Food?" *U.S. News and World Report,* Nov. 16, 1987, pp. 70–72.

netic" mechanisms.[79] Scientists argue over whether particular combinations of chemicals might act "synergistically" (exerting a joint toxic effect greater than the sum of their individual effects in isolation) or "antagonistically" (canceling each other out, or exerting a joint toxic effect less than their individual effects added together).

The major arguments on both sides of the debate over the Delaney Clause are outlined in the following chart.

Assumptions underlying Delaney Clause philosophy[80]	*Criticism in light of current knowledge*
High-dose carcinogens can be assumed to act by directly causing mutations or other DNA damage. Thus even 1 molecule of a high-dose carcinogen can cause cancer by producing a critical mutation.[81]	A limited number of high-dose carcinogens are directly mutagenic and may cause mutations even at much lower doses. Many other high-dose carcinogens are believed to form mutagenic compounds only after being metabolized. Others may increase tumor rates by causing cell death with associated proliferation of neighboring cells or by stimulating hormone activity. Small doses of these chemicals would not be expected to increase the natural mutation rate or cause cancer.
Nonoccupational exposure to synthetic chemicals is a major cause of cancer in industrialized countries. Reducing exposure to synthetic chemicals can significantly lower cancer rates.[82]	There has been no significant rise in cancer rates from the 1950s to the 1980s in the United States, except for lung cancer rates associated with smoking.[83] Most cancers are believed to occur as a result of a combination of factors, including smoking, diet, natural DNA damage that increases with aging, sunlight, hormonal activity, and viruses. Exposure to synthetic chemicals is unlikely to be a major cause of cancer, since our exposure to natural chemicals of equivalent potency is much greater than exposure to synthetic chemical residues.[84]
In an additive or multiplicative fashion, small, simultaneous exposures increase the total risk of cancer.	Some combinations of chemical exposures do have an additive carcinogenic effect. In other combinations, however, two chemicals may inhibit each other's effect, for example, by competing for

	cellular receptors or by stimulating the production of detoxifying enzymes. There is no way to predict in advance whether exposure to a combination of chemicals will have more or less carcinogenic potency than independent exposures.
Only a limited number of chemicals will be discovered to be high-dose carcinogens; thus people can be protected by removing these from commerce, without major economic consequences.	Nearly 50 percent of all synthetic and natural chemicals tested in animals have been found to increase tumor rates when administered at high enough doses. High-dose carcinogens are present in all common foods and thus cannot be eliminated. Removing synthetic carcinogens from commerce has proved to be far more expensive than anticipated in the 1950s, with the removal of chemical residues from waste sites alone expected to cost billions of dollars annually.[85]
High-dose carcinogens are inherently unhealthy at any dose and should thus be eliminated.	Some chemicals that are carcinogenic in animals at high doses are essential nutrients at lower doses, including selenium and fluoride.

Some Conclusions

What about human safety standards for chemicals? We can all agree that the less of a potentially toxic chemical people are exposed to, the less likely they will get sick from the chemical. This statement applies to chemicals that may cause cancer. But would it not be better if we talked about carcinogenic or toxic "doses" of chemicals, rather than calling the chemicals themselves "carcinogenic" or "toxic"?

If the only evidence about toxic effects in a chemical is from high doses, there is no good reason to apply the effects at lower doses to people or animals. Numerical extrapolations are statistical games and cannot provide insight as to real risks. These ideas hold true for all types of chemicals, carcinogenic and noncarcinogenic. There is no guarantee that any chemical dose will be "absolutely safe," but we can make a good guess that a particular dose will be insignificant compared with other potential disease factors.

Congress's attempts to ban chemical carcinogens in the nation's food supply seem to be based on beliefs that chemicals can easily be divided into those that "cause cancer" and those that do not, and that the public health benefits of eliminating "cancer-causing" chemicals entirely are greater than the expense. In fact, the categories of "carcinogen" and "noncarcinogen" are fundamentally flawed: many chemicals which may help cause cancer at very high doses will not cause cancer at lower doses. The cost of reducing chemical residues all the way down to nondetectable amounts has become increasingly expensive as technology for detecting ever-tinier amounts of chemicals improves. At the same time, the benefits of eliminating tiny amounts of synthetic chemical residues have been called into question given the presence of human defense systems that prevent damage from the low levels of natural chemical carcinogens people get every day in their diet. It is not the cost, however, that is the strongest argument against the criteria used to regulate chemical exposures in the United States today; the strongest argument is that there are no health benefits.

We propose that the U.S. government reject regulation based on weak causes and weaker effects. The government should instead use its resources and those in the private sector it regulates to enhance two approaches that show promise of developing a knowledge-based policy of cancer control: epidemiology and discovery of actual cancer-causing mechanisms in humans and other species.

Let us take another look at the major weakness of epidemiology: studies of human populations do not reveal harms from small doses of chemicals even if they exist. This weakness could be diminished by putting resources into doing larger studies and developing better statistics. As long as regulatory action is conducted at terribly small levels, however, the weakness would remain. But rodent cancer tests are no better; if they indicate a cause of cancer, that indication can only be random. Over time, mechanistic understanding is the only way of distinguishing those small insults that are harmful from those that the human body successfully defends against. What should be done in the meantime?

We should ignore small harms from small causes until we learn how to identify them reliably, until the damage done by generating so many false positives is exceeded by the health gains from discovering true positives. A desire to prevent cancers—even more, a desire to show the public that the government is trying to protect them—is not the same as actually providing protection. The pretense of protection, moreover, is expensive not only in its loss of money but in its loss of the very health and safety it is supposed to defend. The current method of risk assessment and regulation makes people sicker in the name of keeping them healthier.

The other shortcoming of epidemiology is that it takes a long time, since the latency period of cancers can be decades. But if a disease takes so long to occur, it cannot be striking people down at early ages or the evidence would already have shown up. We could argue over whether preventing deaths when people are already quite old should be part of governmental policy. There is no need to do that, however, because short-term harms perpetrated by preventive measures are palpable, while long-term gains, in the absence of knowledge about cancer causation, are dubious.

While the shortcoming of a long latency period pertains to occupational exposures, it is much less relevant to the general citizenry exposed to intermittent small doses. For the population at large, shorter-term evidence from epidemiological studies has value in that if workers exposed to comparatively large doses, even for only a few years or a decade, show no ill results, the probability is low that small and sporadic exposures will be harmful to the general public. In the same way, knowledge that symptoms decline or disappear when doses are reduced helps to assure us that the cause of the symptoms has been identified.

Perfection is not for this world. People in industrial democracies should accept the modest imperfections they know, while striving to do better, rather than the imperfections they can hardly imagine from invalid animal tests.

9

The Effects of Acid Rain on the United States (with an Excursion to Europe)

with Robert Owen Rye

Readers who treat the story of acid rain like a good murder mystery are on the right track. But they will find, as the mystery unravels, that most of the clues are misleading, the body is missing, and there is even strong doubt about whether a murder has been committed at all. Our objective is to evaluate the claims that have been made about the detrimental effects of acid rain and determine to what degree, if any, it has damaged the environment.

According to the National Research Council (NRC), an arm of the National Academy of Sciences, "Possible environmental consequences [of acid rain] include adverse effects on human health, acidification of surface waters with subsequent decreases in fish populations, the acidification of soils, reduced forest productivity, erosion and corrosion of engineering materials, degradation of cultural resources, and impaired visibility over much of the United States and Canada."[1]

There are hundreds of lakes in the northeastern United States, particularly in the Adirondack Mountains of New York, that show some evidence of recent acidification. Many of these lakes no longer have as many varieties of fish as they once did, and some are actually fishless now. Recent acidification, it is said, has led to the elimination of the fish in these lakes both directly through death due to acidity and indirectly through poisoning brought on by the leaching of aluminum from the soils around the lakes by the acidic water. Acid precipitation is also blamed for leaching other toxic metals and stripping nutrients from soils. Such stripping and leaching result in a reduced growth rate and the poisoning of whatever plant life is dependent on the affected soils. Thus, in agricultural areas there is said to be less food and in forested areas fewer and less healthy trees than before the advent of acid rain.

Acid rain has also been said to directly injure the leaves of plants, their source of food. A significant decrease in either forest or agricultural productivity would have profound economic implications, since we rely on agriculture for nearly all of our food and on forests for building materials, paper, and a myriad of other important products. Beyond the simple economic costs we must also consider the cost of lost beauty and wildlife if forests are severely damaged. In terms of other aesthetic and economic implications acid rain has been blamed for a great deal of damage worldwide to buildings, monuments, and statues of marble and limestone, highly soluble in acid. A general haze reducing visibility over the North American continent is also attributed at least in part to tiny particles known as aerosols that are associated with acid precipitation.

According to those who believe that acid rain is a problem of devastating proportions, the gaseous and particulate emissions of our power plants and our motor vehicles contain the seeds of destruction for much of the natural world. Sulfur and nitrogen oxides are belched forth from industrial sites and products in vast quantities. These compounds ride the winds. There they either react with other constituents of the air to form acids or fall out of the sky as particles. Regardless of the pathway, these waste products are deposited on the earth, where they interact with ecosystems, changing the chemistry of the regions where they land. These changes might be harmful. To protect the ecosystems that might be affected many have called for stringent controls on the emissions of the gases that are known to be acid precursors. But it is not enough to say that they might cause damage. We must answer the question of whether they actually do cause damage and, if so, how much of what kind of damage and at which level of exposure?

Defining Acid Rain

Acid rain is not a new term. R. A. Smith coined it in 1872 in a treatise on atmospheric chemistry in which he described three types of rain, the first two natural and the third more acidic.[2] From the beginning acid rain has meant rain that has been acidified by human activities. Its current association with wastes from smokestacks and cars implies that all rain that is acidic was made that way by human beings. This is simply not so. In fact, all rain, whether or not it has been exposed to anything man-made in the air, is acidic. Yet not all rain is "acid rain."

The acidity of a solution is determined by its proportion of hydrogen ions. Whether water is deemed acid or alkaline is determined by what is dissolved within. Thus, acid rain is any deposition of acids by wet precipitation or other means (dry deposition, fog) that is at least partially derived from man-made sources. Acidity is measured on the pH scale of 14, so that a neutral solution has a pH of 7.[3] As the pH decreases, the acidity increases. Since the scale is logarithmic, water with a pH of 5 is 10 times more acidic than water of pH 6.

Scientists once believed that pristine rainwater—that is, rainwater uninfluenced by human activities—had an average pH around the world of 5.6. Any long-term deviation from that value was thought to be indicative of human influence. As early as 1958 a number of scientists observed that the rain falling over much of the industrialized world was much more acidic than pH 5.6.[4] Ever since widespread analysis of rain chemistry started in the late 1970s annual average pH values of 4.0 to 4.5 have been found in much of the northeastern United States and southeastern Canada.[5] These low pH values represented an amount 10 to 30 times greater than expected acidity.[6] This large departure from expectations was blamed on industrial and vehicular emissions and was therefore dubbed acid rain. If these scientists were right and the deviations from "normal" acidity were entirely the result of American industry—primarily coal burning—then all that would be needed to restore the rain to normal acidity would be the elimination of coal burning. We could solve the problem with a wave of the legislator's pen. In reality, however, both the background and the human contribution are much more complex than this picture implies.

The assumptions that the average pH of pristine rainwater at each site is known and is 5.6 have proven to be incorrect. In fact, there is no single natural pH; the pH of rain is highly variable. The natural range can be inferred through three methods: ice core analysis, remote area precipitation analysis, and modeling. Data from ice cores dating from well before the Industrial Revolution show a minimum annual pH of about 5.0, not 5.6.[7] The snow in these ice cores was certainly not influenced by humans.

A 1982 study showed that pristine rains in remote areas had an average pH of 5.0.[8] More recently, the National Acid Precipitation Assessment Program (NAPAP) gathered together measurements of rain pH from areas all over the world where human influence was considered negligible. The average annual pH at those sites ranged from 4.4 at Camburito, Venezuela, to 5.8 in Calabozo, Venezuela.[9] NAPAP concluded that the average pH of rain in remote regions was closer to 5.0 than to 5.6.[10] Models of the contemporary atmosphere also indicate that the pH of pristine rainwater is not limited to a single value and that the average is lower than 5.6.[11]

Ancient data, contemporary data, and models all indicate that the pH of pristine rainwater can be highly variable. This variability precludes a simple model of human effects on the acid-base chemistry of rainwater.[12] It makes formulating an effective acid rain policy more complicated than simply setting the goal of restoring to 5.6 the pH of rain in the eastern United States. One cannot merely glance at a map of contemporary rain pH and know how much it has been affected by human activities.

Historical Data on Rain Acidity

If human industry has in fact changed the pH of rain, then it may be possible to establish a trend of increasing acidity from historical data from the northeastern United States. The presence of such a trend without a viable natural explanation would go a long way toward demonstrating that human beings have significantly altered the chemistry of rain. The absence of such a trend would indicate either that we have not altered the pH or that the change took place before measurements of rain pH were taken. It turns out that data from sites that have been continuously monitored since at least the early 1960s indicate that the pH of precipitation either remained the same or increased slightly.[13] None of the studies found rain getting more acidic. All other precipitation data from that era either point to the absence of a trend or to trends that can be linked to natural phenomena.[14] Unfortunately, there are no older reliable data.

What Sort of Acid Is Falling?

Instead of looking at how acidic the rain is, we need to look at what sort of acidity is in the rain. In most of the world it is carbonic acid; in the northeastern United States, however, it is sulfuric and nitric acids.[15] Of the two, sulfuric acid is generally the most prevalent.[16]

The approximate magnitude of chemical changes in precipitation due to man-made (anthropogenic) emissions can be inferred by comparing eastern North American precipitation with precipitation in remote areas.[17] Clearly the result of anthropogenic emissions, the deposition rate of sulfur in the eastern United States is eight to sixteen times the rate in distant regions of the earth.[18]

The natural sources of atmospheric sulfur include sea spray, bacterial decomposition of organic matter, reduction of sulfate in oxygen depleted water and soils, volcanoes, and forest fires.[19] There is no precise measure of background emissions, but NAPAP estimates that 1–5 percent of total sulfur deposition in the United States is derived from background sources, while the rest is man-made.[20] This estimate is consistent with all previous studies of sulfur deposition in the United States.[21]

Approximately 70 percent of anthropogenic sulfur emissions in the United States comes from electric utilities,[22] mostly from coal burning.[23] The concentration of sulfur in coal varies from region to region depending on the deposit from which it is mined, with western coal tending to be lower in sulfur than eastern or Appalachian coal.[24] Annual man-made emissions of sulfur dioxide in the United States peaked at 27 teragrams (Tg, equal to 10^{12} grams or 1 million metric tons) around 1970.[25] Although concern about acid rain has grown since then, anthropogenic sulfur emissions have actually declined by 27–29 percent as a result of the Clean Air Act of 1970.[26]

The Relationship between Emissions and Deposition

Legislators, environmentalists, and scientists all agree that if acid rain is a serious environmental problem worth addressing through regulatory actions, the rate of acid deposition should be a target for reduction. Since we have no control over depositional location or concentration once the acid precursor gases and particles have been emitted, all regulation must therefore focus on emissions. That the regulation does not bear directly on the effect we are concerned with—deposition—at first seems immaterial. What goes up must come down. If you decrease emissions by 25 percent, say, then you reduce deposition by the same amount. It seems intuitive. However, in atmospheric chemistry, as in many aspects of science, intuition does not always turn out to be correct.

What we are concerned about is acid falling on particular sites or ecosystems. Some of the emitted precursor gases are transformed into acids and fall on areas we wish to protect; others fall on places that are considered acid resistant. Some are not transformed until they are over the ocean, where they will do no harm at all. "To develop a meaningful assessment of the importance of these emissions it is necessary to develop atmospheric models capable of predicting the deposition patterns resulting from a distribution of sources."[27] We must therefore look at the available models that relate the sources or emitters to the receptors or areas that receive the acid precipitation if we wish to determine how much we need to change emissions.

The intuitive model (what goes up must come down) is the simplest source-receptor model and has proven to be popular. The *Washington Post* and *Science* both used that model in evaluating the NRC's call in 1981 for a 50 percent reduction in deposition to conclude that it had called for a 50 percent reduction in emissions.[28] But their conclusion about the particular emission reduction needed does not follow automatically given what scientists know about the complexities of atmospheric chemistry.[29]

Scientists know that the relationship between emissions and acid deposition is not necessarily linear—that it varies a great deal over time and space. As late as 1989 the models used to quantify this relationship were described as inadequate.[30] Although models have improved greatly over the past few years, scientists are still unable to quantify the source-receptor relationships of all aspects of acid rain.[31]

As a crude approximation, scientists have known for several years that about half of the sulfur emitted in the central and northeastern states is deposited in these states, 16 percent in Canada and the rest in the ocean, where it has no measurable effect on the chemistry or the biology of the system.[32] Yet if emission rates were reduced radically, it is not clear what effect that reduction would have on these fallout proportions.

Ideally, especially for purposes of regulation, scientists would like to be able to say with confidence that if the amount of sulfur dioxide released from smokestacks in a given plant in Cleveland is reduced by x amount, the amount of acid deposited on the Hubbard Brook area in New Hampshire would decrease by z. A model so precise is probably impossible, though there may be some aspects of the process that can be modeled fairly well.

The Models Evaluated

Between 1981 and 1986 the National Research Council published a four-volume review of the science of acid deposition. The third volume, *Acid Deposition: Atmospheric Processes in Eastern North America* (1983), examined the quality of the available models relating emissions to deposition. The survey found the models insufficient for informed policy decisions regarding emission control strategies: "We do not believe it is practical at this time to rely upon currently available models to distinguish among alternative strategies."[33] This conclusion was crucial, because it contradicted the widespread complaints that acid rain was already well enough understood to warrant stringent emissions regulations.

Without a reliable source-receptor model there is no way to tell which emission control strategy is best or even which ones will be helpful. For example, in one run of an early model the emission rate of nitrogen oxides was increased by 10 percent and all other factors were held constant. The increase in nitrogen oxides led to a 5 percent decrease in the sulfate (sulfuric acid) production rate,[34] which indicated that the sulfuric acid deposition was shifting farther away from the sources. This is an extremely important result, because it shows that it was plausible in the early 1980s for a phenomenon like an increase in nitrogen oxide emissions to affect the chemistry in the atmosphere in such a way as to slow the deposition of sulfate. According to this model, more of the sulfur would be deposited in Canada, where it would not be welcomed, and in the ocean, where it would do no harm.

The models available as of 1983 (and 1989 for that matter) were sensitive to small perturbations in their chemical and physical assumptions.[35] These assumptions were generally arbitrary, known to be only poor approximations of reality. Small perturbations were reasonable because there was no way of knowing if the changes would make the model more or less realistic. Such a condition made the results of model runs suspect and meant that given the distribution of emissions, the models were unable to give a general picture of the distribution of deposition. The NRC concluded that "results from such models should not be relied on in the development of control strategies for regional air quality problems in which chemical phenomena play a central role."[36] Acid rain certainly qualifies as a phenomenon with chemistry at its center.

Whereas the chemistry involved is best described by nonlinear equations, the assumptions about the chemistry were all linear in the early models of the source-receptor relationship.[37] The critical equation, known as the continuity equation, predicts that the relationship between the amount of sulfur put into the atmosphere and the amount of sulfate deposited in a given region downwind of that source will not be linear, because of numerous and diverse feedbacks. The models, however, generally assumed that no matter what amount of sulfur was released into the air, a fixed percentage of the sulfur dioxide would be converted into sulfate over any given time period.[38] With this assumption model results could be interpreted in such a way that a rollback strategy for emissions would lead to a proportional decrease in sulfate deposition.[39] This is not necessarily so, and in fact the continuity equation strongly indicates that it would be otherwise.[40]

It is an incontrovertible statement that the sulfur that goes up will eventually come down. Sulfur concentration may fluctuate over days, weeks, months, years, decades, centuries, millennia, but it cannot constantly increase, because if it did, all of the world's sulfur would eventually end up in the atmosphere. Nevertheless, we may not assume that all, or any particular fraction, of the sulfur will fall in the form of sulfate or sulfuric acid. As long as we cannot predict the magnitude of that fraction for any given atmospheric composition, we cannot really predict the effects of emissions regulations. Hence the NRC warning: "Given the state of knowledge of the physics and chemistry of the atmosphere in the context of long range transport in air pollution, *we advise caution* in using deterministic models to project changes in patterns of deposition on the basis of changes in patterns of emissions of precursor gases."[41]

Nonlinearity in NAPAP Conclusions

Having seen that determining the relationship between emissions and deposition is crucial to policy decisions and that a nonlinear relationship between them may lead to counterintuitive results from changes in emissions, we can now examine the most recent and accurate models available. These come from NAPAP research, the most ambitious effort to study acid rain, supported by large congressional appropriations.

The nonlinearity of the source-receptor relationship results from the chemistry involved in transforming gases into acids in the atmosphere. The primary reactions that convert sulfur dioxide, SO_2, to sulfate (sulfuric acid) require an oxidant, the most common being hydrogen peroxide, H_2O_2. If there is more SO_2 available than there is H_2O_2 to react with, only the amount that H_2O_2 can take care of will be transformed. A situation in which there is not enough H_2O_2 is described as oxidant-limited, which is what pertains in most areas in the eastern United States. Under these conditions the rate of transformation is independent of SO_2 concentration. That is, the rate will not change (or will change very little) if emissions are increased or if emissions are decreased, unless the decrease is enough to eliminate the limitation.[42]

Studies by NAPAP on the source-receptor relationship indicate that deposition reductions in one state will require emission reductions in several states upwind of it.[43] But decreased emissions in the United States would probably lead to disproportionately large decreases in deposition in eastern Canada, as much of the deposition there is due to long range transport and much of Canada is fairly distant from major sulfur sources.[44] In short, areas in the United States will, because of oxidant limits, be little affected by modest reduction in emissions.

Are drastic reductions in sulfur emissions worth achieving? In the long run (say, thirty years from now), this question will become moot, as the means of generating electricity turn from coal to newer types such as fluidized beds, which will emit many fewer acid precursors. There might also be other environmental objectives, like reducing global warming, should it occur, that would be harmed by reducing the amount of sulfur dioxide particles, which reflect heat upward into the atmosphere. Leaving these considerations aside, the question remains whether the enhanced acidization of rain due to human activities in the industrial democracies causes sufficient harm to be worth reducing at great cost.

The Effects of Acid Rain on Aquatic Ecosystems

The concern about acid rain and the environment is that levels of acidity higher than normal in rain will harm some or all of the plants and animals living in a given ecosystem. These suspected effects vary across the types of ecosystems, across regions, and across time. The impact on aquatic systems has received the most attention because the effects there appear to be dramatic. Those appearances have generally been misleading, however: the harm done is much less extensive than was once thought by scientists.[45]

In theory, aquatic environments (lakes, streams, swamps) should be affected by changes faster than terrestrial environments because the former have a smaller capacity to neutralize acidic inputs.[46] This hypothesis has been extensively tested and generally confirmed, but the damage has turned out to be much less extensive and severe than it was believed to be a decade ago. Widespread acidification of lakes and streams in both North America and Europe has been reported and has been blamed for the deaths of many aquatic environments. Reports have also indicated that lakes and streams that once held an abundance of fish are now devoid of them. These reports have generally been blown way out of proportion. There have been some documented cases of acidification and fish deaths, but the vast majority of lakes in the United States (even the Northeast, which has the most acidic rain) have not been damaged by acid rain.[47]

FISH LOSS

A wide variety of aquatic organisms are adversely affected by decreases in lake or stream pH (increases in acidity),[48] with the clearest evidence of acid intolerance being that gathered for fish. Many species of fish cannot survive in waters that are too acidic.

Fish loss has been documented at over 180 of the more than 2,800 lakes in the Adirondacks.[49] Something left the lakes fishless, and acid rain got the blame, for it was the most obvious change in the environment. But suspicion does not constitute proof. We know that some of those lakes teeming with trout were once stocked by the state of New York and are stocked no longer. The change has occurred not in the chemistry of the waters but in the policies of the people managing the lakes. Other lakes may actually have been acidified. We must turn to a systematic examination of the available chemistry data of supposedly acidified lakes if we are to determine whether acid rain is in fact at fault.

HOW HAS ACID RAIN AFFECTED LAKES?

The traditional view is that the addition of rain containing excess acid will lead to the chronic acidification of surface waters. Intuitively this seems obvious, yet most of the water in lakes has passed through soils before it reaches the lakes. Soils may neutralize the waters;[50] they may also acidify it far more than the rain ever could.[51]

The alternative hypothesis is that "acidification by 'acid rain' is superimposed upon natural processes of acidification."[52] In this view the dominant source of acidity in most surface water systems is the soil in the area.[53] The soil is a source of organic acids, such as formic and acetic acid, and with the addition of acid rain these organic acids are replaced by inorganic (sulfuric and nitric) acids. The pH is not radically affected one way or the other. Nevertheless, changes from organic to inorganic acidity, if extensive and severe enough, could adversely affect organisms in the water.

CORRELATIVE EVIDENCE IN SWEDEN AND NORWAY

The first studies of the effects of acid rain, conducted in Sweden, determined that changes in the pH of rainwater and of the lakes and streams in southern Sweden were related in time and space.[54] Studies in Norway also reached the same conclusions. An increase in the output of sulfur dioxide from the smokestacks of northern Europe and southern Scandinavia was said to be correlated with an increase in the acidity of the precipitation in the areas downwind of those smokestacks.[55] It was believed that rain from air masses that had passed through the emissions columns of the smokestacks contributed to the acidification of the lakes and streams. Acid rain was blamed for fish deaths, forest diebacks, and other environmental changes.

U.S. LAKES

Efforts to extend the conclusions from the Scandinavian observations to the northeastern United States and Canada have proven difficult and have raised doubts about the accuracy of the earlier Scandinavian conclusions. One significant barrier to showing a correlative relation, let alone a causal relation, between the increased acidity of precipitation and the observed modern acidity in northeastern surface waters, is a lack of historical data. There is a paucity of measurements of the pH of surface waters in North America from before the late 1970s.[56] Without accurate measurements of pH from the past it is impossible to estimate whether or not pH levels are lower now. The hypothesis that acid rain has lowered the pH and thereby killed fish cannot be tested without reliable data.

Until recently regional acidification was generally estimated by comparing data from surveys of lakes taken over the past several decades with data from recent surveys. Where there were measurements, flawed or not, the data generally indicated that many lakes had been acidified over the course of several decades. These lakes began to register an acid-neutralizing capacity ≤ 0; that is, they were unbuffered against the addition of acid from precipitation. Many other lakes were said to be approaching a state in which they would be unable to neutralize the acids added to them by acid rain.

The writers of one survey of New England lakes said that where historical data was available 64 percent of lakes had lower pH in the late 1970s than they had had before, and 70 percent had lower alkalinity (buffering capacity).[57] Other surveys also indicated that there either had been or was impending acidification of most lakes in the Northeast as a result of acid precipitation. The threat of an "aquatic silent spring"[58] loomed over residents and wildlife of much of the United States and Canada. However, we must look at the difficulties associated with the use of older data before drawing such a conclusion.

Many of the investigators who collected data before the 1970s (and even some since then) did not document the techniques they used in sampling or in analyzing their samples,[59] which means that their data cannot be compared with modern data. Where techniques were documented, many of them used in the past to determine the pH of lakes are not compatible with present techniques. This incompatibility was first demonstrated by M. H. Pfeiffer and P. J. Festa in 1980 in their study of high-altitude Adirondack lakes and ponds using both modern pH meter methods and old style colorimetric methods.[60]

The colorimetric data indicated that only 4 percent of the lakes tested in the 1930s had had pH below 5.0. The new meter method found that over 50 percent had less than 5.0 in the 1970s. A comparison of the data from the 1930s with the data from the 1970s showed that there had been a massive acidification of the surface waters of the Adirondacks. However, testing the waters with the old colorimetric techniques yielded a much different picture, showing only 9 percent of the lakes with pH less than 5.0, a percentage radically lower than the one yielded by the meter method. Given the uncertainty in the measurements, it was unclear whether the 9 percent measurement of the late 1970s was in fact significantly higher than the 4 percent measurement of the 1930s.[61] Thus the old colorimetric data could not be used to establish that acidification had taken place.[62]

The problem with using the historical data actually goes beyond incompatibility of techniques. Even if colorimetric techniques were to be used today, they might not be compatible with those of the past.[63] For example, some of the samples in the past were collected in soft-glass containers, a potential source of alkalinity because some of its constituents are soluble in water and basic in solution. Furthermore, we really do not know what the natural variability of acidity is over the course of a year or even a decade, so we cannot tell how representative the historical data are of the past condition of the lakes surveyed. It is possible that a number of lakes were sampled at a time of especially high or especially low acidity. "It is therefore difficult to draw quantitative conclusions regarding historical acidification based on these data."[64] All in all, the quality of the data from before 1970 is questionable.

Hubbard Brook, New Hampshire, is a site with clearly reliable data back to the 1950s. These indicate that there was no trend in the sulfate ion concentration, the pH, or the alkalinity of the water at this site. The lack of a trend does not really say much about whether or not the current conditions at the site have been influenced by acid rain, as there has also been no recent trend in sulfur deposition at Hubbard Brook.[65] If acid rain had acidified this site, it could have done so before the time for which we have data.

PALEOLIMNOLOGICAL STUDIES: A LOOK INTO THE FAR PAST

To get data from far enough back for comparison, we need to rely on the results of paleolimnological studies. These studies, performed today, can tell us what the pH of the water was decades or centuries ago.[66]

Any large changes, on the order of -0.5 to -1.0 units in pH or -50 to -100 meql^{-1} (microequivalents divided by liters) in alkalinity, that have taken place over the past century or so are almost certainly due to human activities. Paleolimnological data indicate that in the past it has generally taken hundreds to thousands of years to change a lake's chemistry significantly through natural processes.[67] If these changes are not only recent but also occur in lakes that are spatially related to man-made sulfur dioxide sources, then they are probably due to emissions from coal plants and cars. If changes on this scale have already taken place, acid rain is probably responsible for the deaths of many aquatic ecosystems and an immediate threat to many others. If, however, the changes are on the scale indicated by the lowest estimates from the colorimetric data, then the aquatic ecosystems are most likely much more resilient than has been assumed; that is, much less damage has been done than had been reported as of 1980. If the changes are this small, continued emissions at current levels are not likely to lead to significant damage in the future, as there would not be a significant degradation of either the pH or of the systems' capacity to neutralize new acid. Such a situation would indicate that as far as the acidity of aquatic ecosystems is concerned, there is no pressing need to reduce sulfur dioxide emissions.

The region of greatest interest is the Adirondack Mountains. The paleolimnological data indicate that 90 percent of all acidic lakes in that region were acidic before the Industrial Revolution. Thus most of the acidity in that region is not the result of human activities.[68] Most of the acidified lakes are at high elevations (greater than 600 meters) and in portions of the Adirondack Park where precipitation and sulfuric acid deposition are highest.[69] Over 25 percent of all lakes that are almost acidic—that is, that have a low acid-neutralizing capacity (ANC)—experienced either a decrease in acidity or no change. The 13 $meql^{-1}$ or smaller loss of ANC experienced by 50 percent of the lakes is tiny; it is not likely to render a lake uninhabitable for fish.[70]

For fish the most important measurement, in the short term, is the pH rather than the ANC. The most recent surveys indicate that 15 percent of all lakes in the Adirondacks larger than 4 hectares (where 1 hectare equals a 100 $\times$ 100 meter square) had a detectable pH decrease.[71] Such a decrease, if greater than 0.28 units, may be large enough to kill some species of trout and other fish if the lake started off barely able to support fish. These pH declines in the Adirondack lakes are generally due to acid rain; the change in the lakes' overall chemistry is not consistent with any of the other possible causes.[72] Scientists do not have enough data to determine the extent of acidification elsewhere, but preliminary results indicate that the Adirondack region is the only one that has experienced widespread acidification.

Almost one-quarter (24 percent) of the lakes in the Adirondacks have no fish in them.[73] Fish-response models indicate that half of these lakes may be fishless because of the presence of mineral acids in the water unrelated to acid rain. About one-sixth of all lakes that have fish have apparently lost at least one species because of acidification due to acid rain. These results are consistent with the available surveys, but they are by no means definitive.[74]

The outlook for other regions and for the future of the Adirondacks is fairly bright. Other regions have evidently been affected to a far smaller degree.[75] Furthermore, NAPAP research indicates that most lakes in the Northeast have reached a steady state with respect to sulfur. As long as acid deposition does not increase (it is currently decreasing), few if any lakes will be further acidified.[76]

Those lakes that have been acidified can be helped. Scientists have shown that the pH of a lake acidified by rain can be brought back up by adding alkalines. There are several proven techniques for restoring the pH, the best being liming—that is, adding limestone to the lakes—which generally has a predictable positive effect on aquatic ecosystems.[77] Many toxic effects due to acidification can be alleviated through liming, and fish can be successfully reintroduced.[78] All in all, we see that the lakes of the United States have mostly not been acidified and that there is a readily available and cheap technique for alleviating what acidification has occurred. Acid rain has not devastated our aquatic environments.

Why is it that early studies indicated things were so much worse than they seem today? Changes of over 0.5 pH units or 70 meql^{-1} ANC have given way to changes on the order of -7 meql^{-1} for the most vulnerable lakes and less for others. Although doubts about the quality of the data were available in 1980, indicating that there might not have been so much damage, it was widely believed that there had been. What made scientists assume that the pessimistic interpretation of the pH was the correct one, we must ask ourselves, even though they did not have good enough evidence to say so? What determined which interpretation of the data would be selected?

As is often the case, one interpretation was chosen over another because it fit an accepted theory and the other did not. What was known and believed about aquatic systems at that time indicated that there should have been significant changes due to acid rain. Thus, when scientists were presented with data that could indicate either a lot or a little acidification, they went with the assumption that there had been a lot because that is what they expected. This was good science, since to the best of their knowledge they were using a well-established understanding of acid-base chemistry to choose between two interpretations of data. All science involves the testing of hypotheses. If a hypothesis has served you well, then you believe it is true until it is proven otherwise. The data available were consistent with the hypothesis that acid rain would lead to acidification of certain areas.

Part of the traditional acidification hypothesis is that there are certain conditions under which surface waters are vulnerable to acidification. Sensitive surface waters have little to no capacity to neutralize acid and are thus vulnerable to acidification on time scales measured in years or a few decades. Hence, the sensitivity of surface waters is a crucial factor in determining policies on the emissions of sulfur dioxide and nitrogen oxides. The more surface waters there are that are sensitive, the more urgent it is to alleviate the acidification process through measures such as reduction of sulfur dioxide emissions, if we wish to preserve the life of our lakes and streams. The pessimistic interpretation of the limited historical data that was widely believed as of 1980 or so was in part based on a now obsolete set of assumptions about which conditions made surface waters vulnerable to acidification. These assumptions led scientists to believe that sensitive surface waters were much more common than they now appear to be.[79]

There are no direct measures of sensitivity. A number of chemical and geologic features of surface waters and their surroundings have been used over the years to estimate the likelihood that they would become acidified when exposed to acid rain. All of these methods have been based on the assumption that surface waters with low alkalinity are sensitive and those with high alkalinity are not sensitive. However, because there have not been alkalinity data available for much of the surface waters in the Northeast, scientists have had to use other quantities or qualities to predict what the alkalinity of many surface waters would be. In other words, they have had to use proxies for alkalinity, which is itself a proxy, to predict the sensitivity of most surface waters in the region. The evolution of these proxies, the models that they are a part of, and the results that they have presented to scientists and policymakers, have helped shape the debate over the importance of acid rain.

HISTORICAL DEVELOPMENT OF NOTIONS OF SENSITIVITY

Most of the early "efforts to determine patterns of surface water sensitivity to acidic deposition . . . relied on interpretations of bedrock distribution and chemistry."[80] This reliance makes intuitive sense to the geologist, since all of the soils in any given area are derived from the bedrock. The bedrock is the source of all the materials available to neutralize acid rain; it is the ultimate determining factor of alkalinity. This model has the advantage of relying on a fundamental property of the areas for which sensitivity is being predicted. Furthermore, it is easy to use.

To apply this theory about spatial patterns of sensitivity only two things are needed, both of which can be obtained without even going into the field. The first is a good geologic map that shows the chemistry and distribution of the bedrock in a region. These maps are available at just about any research library for every part of the country. The second requirement is an understanding of the chemistry involved, which also is available to just about anyone through basic geology and geochemistry texts.

Using this office-based approach, it was predicted that "the most well-known susceptible areas are those with a shallow overburden [soil] and quartzbearing bedrock, e.g. granites and gneisses."[81] Much of the northeastern United States and eastern Canada is covered with shallow soil and underlain by quartzbearing bedrock. Consequently, as of the early 1980s much of this region was labeled sensitive, because it was thought that the capacity to neutralize acid in much of its surface waters would either have been exhausted or be nearing exhaustion.

The problem with these predicted patterns of sensitivity is that they often conflicted with actual alkalinity data when such data were available. Scientists found that "there is a lack of spatial correlation between the patterns drawn by these efforts and the observed patterns of surface water alkalinity."[82] The alkalinity was still high in some areas where there should have been acidification, despite the generally accepted finding that acidification only takes place where alkalinity is low. The old sensitivity surveys were no good, yet they were all that was available as of the early 1980s.

Since then there have been more refined maps of alkalinity showing that there are fewer sensitive areas than once thought and that the sensitive areas have been affected less than was previously believed. In one such survey of the

Midwest, for example, less than 0.5 percent of the lake area in the region was deemed likely to be suffering long-term acidification. Lakes of 50–200 meql^{-1} are thought to be sensitive to periodic acidification.[83] This short-term number is the approximate amount of acid that would need to be neutralized during a severe acidic snowmelt. Only 6.2 percent of the lakes lie between 50 and 200 meql^{-1} and almost 83 percent of those are between 100 and 200 meql^{-1}. All told, only 6.6 percent by area of the surface water is sensitive to acidification, and the vast majority of that water is sensitive only to short-term acidification by snowmelt.[84] This acidification may or may not be sufficient to kill fish fry just as they are hatching, but if it is, the acid pulse would have to come at just the right time. In any case, the survey provides evidence that acid rain has not had and will not have a major impact on the pH of the lakes of the Midwest given current acid input rates.

ALUMINUM IN LAKES

Although changes in pH are not as radical as was once thought, there may still be significant effects on lakes arising from acid rain. These changes involve a shift from organic acids in the waters to sulfuric acids, with no change in pH. One of the most important effects of these possible changes is the concentration of free ionic aluminum.

Concentrations of aluminum that are toxic to fish can be reached at harmless pH levels.[85] Early studies indicated that survival of fish fry into adulthood was greatly impaired above 0.2 milligrams per liter (mg/l) of aluminum in the waters.[86] Such concentrations lead to severe necrosis of the gill epithelium, in which a high proportion of the cells on the surface gills die.[87] If the necrosis is severe enough, the fish itself will die. Not just any aluminum will have lethal effects, however; only the specific form free ionic, or dissolved inorganic, aluminum will.[88] The form of the aluminum is mostly determined by the soils that the waters flow through to get to the lake and the amount of dissolved organic carbon that is present in the lake.

Aluminum, one of the most common elements, is present in nearly all rocks and soils. Plant activity indirectly produces a great deal of organic acids in the soils, and these dissolve some of the aluminum that was left behind by the original weathering of the rocks. After dissolution, the aluminum is generally bound to the organic acid in a structure known as an organic aluminum complex. In alkaline soils nearly all of the aluminum that has been liberated from the soils and put into the soil water is organically complexed.

If there is a great deal of dissolved organic carbon in the waters, then the aluminum will probably be further complexed. If there is very little, then the complexes may dissolve, leaving the aluminum in its free or inorganic form. For fish, organically complexed aluminum is the least toxic form; in fact, "the ameliorating effects of organic complexation on aluminum toxicity suggests that lakes and streams with high organic carbon content may be suitable for successful fish production despite moderately low pH and high total aluminum levels."[89] However, those lakes and streams that do not have high amounts of organic carbon will most likely receive more than 0.2 mg/l of inorganic aluminum, an input toxic to many species of fish. Such doses are associated not with normal rain runoff but with acid precipitation.

Results of studies at high elevations as of 1979 indicated that acid precipitation causes leaching of soils leading to the dissolution of aluminum under inorganic conditions.[90] According to Johnson and colleagues, "In the Falls Brook watershed the hydrogen ion acidity of the ambient precipitation is largely converted into aluminum acidity by the time the precipitation makes its first appearance as stream water."[91] On the one hand, this replacement is good because it means that the aluminum acts as a buffer against the lowering of the pH of the surface waters by the runoff and soil water. Such buffering, which supplies inorganic aluminum to lakes, prevents the acidification and consequent fish damage that comes from the input of large amounts of hydrogen ions into those waters. On the other hand, fish do not care whether they are burned by acid or killed by aluminum. Dead is dead. Unfortunately, it is even more difficult to determine reliably if our actions have led to increased concentrations of labile aluminum than it is to determine the impact on acid-base chemistry.

There is a striking increase in the solubility of aluminum at about pH 5.5.[92] If the waters being provided to streams were to drop from above 5.5 to below that value, there would probably be a dramatic rise in the aluminum input to the streams.[93] Surveys of aluminum concentrations in U.S. lakes indicate that inorganic aluminum is generally associated with sulfuric acid in lakes with a pH less than 5.5.[94] The extent of lakes with elevated inorganic aluminum is not known.[95] For now this effect remains unquantified.

The lakes of the eastern United States have proved more resilient than scientists initially believed. Reports of acidification of over half of the lakes in the Adirondacks have been replaced by studies indicating that 15 percent have experienced some acidification. There are lakes that have lost their fish because of acid rain, but there are no massive fish deaths in the Northeast. Despite earlier concerns about sensitivity, lakes in other regions have been shown to be far less affected than those in the Adirondacks. All in all, our lakes do not appear to need immediate protection from acid rain.

Effects on Forests

Concerns about the effects of acid rain have extended to forests as well as lakes. Predictions of forest damage abound. If acid rain causes widespread forest damage, we have a compelling reason to reduce emissions of acid-producing gases. Forests are important to us: their beauty is a source of peace and relaxation for many, they are home to myriad animals, and they also provide much of our building materials and all of our paper. Fortunately, we are not destroying them.

The evidence available today indicates that acid rain is not a short-term health threat to U.S. forests. NAPAP found that the forests in the East are not experiencing a significant or widespread decline.[96] In fact, no one has documented a case of direct damage of vegetation by acid rain in North America.[97] Why then has there been such concern about the impact of acid deposition on forest health?

There was never strong evidence that acid rain had damaged forests, but before NAPAP's research there was reason to be concerned. In the 1970s and early 1980s unexplained local declines of some American forests, theoretical predictions that acid rain could damage forests, and reports of a dramatic decline of European forests all pointed to the possibility that acid rain could be a threat. Scientists did not know enough before NAPAP investigated to confirm or refute the existence of this threat.

As of the early 1980s a number of scientists had suggested ways that acid rain could adversely affect forest ecosystems. For instance, the rain could leach enough nutrients from the foliage of trees to leave the trees malnourished.[98] Another hypothesis was that if soils were acidified by rain, a number of indirect effects on trees would result.[99] And so on. None of these effects had been detected, but none had been ruled out either. There was no convincing evidence either way. Many of these damage hypotheses put forward in the 1970s and early 1980s were plausible given the state of knowledge at that time. All of them needed to be tested to establish dose-response relationships.

Scientists knew that large amounts of acid poured on any organism would kill it. They also knew that if enough acid was added to soils their capacity to neutralize acid would be overwhelmed. However, they did not know how much acid any particular soil type could neutralize over a given time period, nor how much acid any particular organism could tolerate in the short or long term. Thus they did not know what the impact of ambient levels of acid deposition would be.

SURVEYS

The first test of a damage hypothesis is whether or not forest decline is correlated, in time and space, with acid deposition rates.[100] If there is no correlation, then acid rain cannot be responsible for the decline. Before NAPAP investigated, there were few thorough studies of forest growth declines, while those studies available indicated that there were large variations in tree growth rates across the Northeast. There did not appear to be a general pattern of decline starting at a particular date, though surveys showed that a number of northeastern forested areas had experienced declines over the previous decades.[101]

The next step was to see if these declines could be correlated with acid deposition patterns in some way.

A 1977 study tested the hypothesis that forest declines were spatially correlated with acid deposition by looking at growth patterns in two stands of pine trees located in two different precipitation regimes, that is, trees receiving rain of significantly different pH. One stand was in the Bowl Natural Area of White Mountain National Forest in New Hampshire, the second in the Great Smoky Mountains National Park in Tennessee. "One site (the Bowl) was near the center of the acid precipitation region while the other site (Smoky) was at the periphery of the acid pattern."[102]

If acid rain was reducing the growth rate of the two stands, there should have been a marked difference in the amount that it reduced one as opposed to the other. There was not.[103] Such a negative result does not prove that acid rain at the given level of exposure has not harmed those forests, but it does suggest that other more important factors dominated their growth patterns. As the author concluded in 1977, "a correlation of forest growth and acid precipitation cannot be established at this time."[104]

A 1981 survey showed that the trees in the New Jersey Pinelands grew considerably more slowly than expected from 1955 to 1979. This twenty-five-year decline in growth was unprecedented in the tree-ring record for this forest.[105] Since none of the usual causes, such as pests or wildfire, were responsible for the slowed growth, the investigators looked for other factors that might correlate with growth rate. Stream pH was among several possible stresses that correlated with growth rate over that year span. As stream pH decreased, acidity increased and growth rate decreased. If stream pH was a good proxy for precipitation pH and the stream pH data were good, this correlation would indicate a correlation between acid rain and growth rate.

As we know, historical pH data is generally of questionable quality. Stream pH is not necessarily representative of precipitation pH, although the study group led by A. H. Johnson found that there was a coefficient of correlation of 0.709 between stream pH and the pH of rain that fell three months previously.[106] This correlation is good, but variations in stream and rain pH may both be a product of climatological factors that are also among the elements correlated with growth rate. The growth patterns may then turn out to be the product of the same force that causes fluctuations in pH rather than of the fluctuations themselves. Two things are certain though: something had slowed the growth of pitch, shortleaf, and loblolly pine in southern New Jersey in a way it had not been slowed before. And this slowdown took place in an era when the pH of the rain was believed to be more acidic than it had been before. These two observations led the scientists studying these forests to conclude that acid rain had to be considered as a possible factor in the decline.[107]

Several other studies in the late 1970s and early 1980s failed to find any obvious reasons for declines in various forests.[108] In many of these the only significant change noted in the environment was a change in the chemistry of precipitation. Given the lack of obvious alternatives, recommendations for investigation of acid rain's role were quite reasonable.

A crude sort of correlation was established in the mid 1980s. A critical assessment review paper by the Environmental Protection Agency in 1985 indicated that there were widespread tree growth reductions in areas that received high doses of acid rain.[109] Such a correlation is trivial, because the entire Northeast experienced rainfall of similar acidity. Most of the forests were fine. What was needed was an assessment of each forest that might be declining in order to see whether or not there were links to acid rain. Such surveys were performed by the National Acid Precipitation Assessment Group.

In 1986, at the midway point of its first term, NAPAP concluded: "No evidence confirms that acidic deposition of sulfur dioxide and nitrogen oxide gases at ambient levels are responsible for the observed changes in the Nation's forests."[110] This conclusion was based on many more surveys than the old predictions and vague correlations had been. The report went on to say: "No evidence exists that current atmospheric deposition will affect future forest productivity indirectly through impacts on soils."[111] This conclusion remained intact when the surveys were completed for the final report.

CONCERNS NAPAP INVESTIGATED

Concerns about acid-rain-induced forest health problems primarily focused on the high-elevation red spruce in the Appalachians, the sugar maple in Quebec, Ontario, Vermont, and Massachusetts, and the pines in the southeastern states.[112] As of 1991 none of these forested areas was demonstrably damaged by acid rain, although precipitation had not been ruled out as a contributing factor to health problems in all of them.

Growth rates of red spruce in the high-elevation stands of the northern Appalachians have decreased over the past three decades.[113] Severe weather, wind stress, insects, and fungal pathogens have all contributed to this decline.[114] High-elevation stands spend much of the year shrouded in fog, which is more acidic than rain at lower elevations with a typical pH of 3.5–3.7. The combination of high exposure times and highly acidic vapor means that the trees in these stands receive extremely high doses of acids. Some experiments have provided evidence that acid deposition at these levels may decrease the red spruce's ability to withstand severe winters.[115] However, scientists have not yet found evidence that the effects of high exposure levels found in experiments are present in the field.

Among the sugar maple trees in Quebec over half are thought to have experienced a slight decline and 4 percent a moderate to severe decline in the 1980s, according to estimates based on aerial surveys of foliation.[116] Acid deposition was one of the suggested culprits for this decline and for smaller, more isolated ones in Ontario, Massachusetts, and Vermont. But the evidence does not support this hypothesis.

Declines have been observed off and on since the early 1900s. The most severe one, near Quebec, occurred in the 1950s and 1960s, not the 1980s. Insect defoliation is known to be a significant factor in these declines, and lack of potassium—a nutrient naturally in low supply in much of the region's soils—is certainly a contributing factor. There is no spatial correlation between acid deposition rate and forest decline rate. Controlled experiments have not shown that the acid deposition rates found in the declining sugar maple stands will reduce the growth rate of these trees.[117] All of these factors indicate that if acid precipitation plays a role in the decline, it is a supporting role that cannot be determined at this time. NAPAP research indicates that more subtle effects may tie acid rain to these declines, but the subtle effects have not been sufficiently tested for, and until they are, they cannot serve as evidence one way or the other in the debate.

The growth rate of some pine stands in the Southeast has decreased over the past few decades. Some decrease is expected, for natural pine stands are maturing in this region. The maturity of the stands, combined with historical land use patterns, should lead to tree growth reductions. It is unclear if the reductions are larger than they would be as a result of these factors alone. There are visible symptoms indicative of acidic rain's involvement in the tree growth reduction. As for changes in soil chemistry, insufficient data are available to show whether such changes have been significant, and if so whether those changes have harmed the trees. The available evidence is not consistent with aluminum toxicity or nutrient depletion, the two primary mechanisms whereby acid deposition at high enough doses is thought to affect forests. As it stands, the scientific evidence available indicates that acid rain has not induced a decline in the health of the pines of the Southeast.[118]

There is no indication that forests other than those NAPAP singled out for mention in its final report are experiencing a widespread decline in growth rate or overall health. NAPAP's conclusion about acid rain's effect on U.S. forests is quite clear: "Compared to ozone and many nonpollutant stress factors, acidic deposition appears to be a relatively minor factor affecting the current health and productivity of most forests in the United States and Canada."[119] But what about the dreaded *Waldsterben,* the death of the forests in Europe?

EUROPEAN FORESTS

Although there has never been evidence of a widespread decline in American forests, there has long been concern that it is forthcoming, based mainly on the reported decline in Europe. The reports of massive deterioration in European forests in the 1980s were said to portend devastation for eastern U.S. forests, because the acidity of the rain in the two areas is similar. But there are two reasons why these reports do not predict a decline in the United States.

First, European forests are different. Most of the forests in Europe are managed, whereas only 20 percent of U.S. forests are plantations, the rest being either lightly managed or left entirely alone.[120] Almost all the soils in some countries in Europe, such as the Netherlands,[121] are acidic and considered vulnerable to further acidification. Most soils in the United States are not considered vulnerable.[122] As each forest will respond in its own way to acid rain, depending on its resistance to change, predictions based on European experience are not strong.

Second, and most important, there has in fact been no decline in European forests. On the contrary, there is strong evidence that they are growing rapidly. Recent surveys of forest biomass indicate that the growing stock of European forests increased by 25 percent from 1971 to 1990.[123] Nor is there indication that the rapid growth of forest resources will slow in the near future.[124] The findings of these growth surveys are consistent with forest studies done before 1980.[125] Why then were there reports of massive continentwide decline in the 1980s?

One reason was that a decline was expected. The most important study of forests conducted before 1980, the SNSF project (the Norwegian Interdisciplinary Research Program "Acid Precipitation—Effects on Forest and Fish"), indicated that although there was no evidence of a decline at that time, it could take place. No decreases in forest growth due to acid rain were found in this study, but the authors pointed out that harmful changes in soil chemistry could develop over the coming decades in the more susceptible forest regions—those with acid soils. In fact, they claimed that it was just a matter of time before such changes would take place and the forests would decline.[126]

The other reason for reports of forest damage was that the assessment method used during the 1980s, though flawed, indicated that there had been a decline. The damage SNSF predicted in 1980 seemed to appear during the 1980s. As of 1986, 23 percent of all trees in Europe were described by the United Nations Economic Commission for Europe (UNECE) as moderately or severely damaged by causes other than known natural stress factors.[127] Since then many of the damaged stands of trees have recovered.[128] If acid rain had been responsible for the damage, then without significant abatement in acidity the forests could not be expected to recover. When the forests did recover, despite continued acidity in the precipitation, some scientists began to question the damage thesis,[129] but in general reports of decline persisted throughout the 1980s.

"Forest decline" usually refers to a decrease in growth rate, not an easily measurable quantity on large scales. However, growth and foliage health are somewhat linked in that much of a tree's food comes from photosynthesis in the leaves. Because of this linkage, the UNECE chose to define forest decline in terms of a proxy—foliage loss and discoloration—more easily measurable than growth rate and indeed the most easily observed class of changes in forest conditions. Surveys of foliage changes are called ocular surveys because they can be done by visual observation, what Americans might call "eyeballing." These surveys were believed to indicate the overall health of each forest, but the assessment technique used was fatally flawed.

In an ocular survey, trees are generally placed in one of the five damage categories shown in Table 9.1. Healthy and slightly damaged trees are not considered to be declining, but reports of moderate damage—that is, observations of needle loss of 26 percent or greater and/or yellowing of more than 25 percent coupled with loss of more than 10 percent—were taken by the UNECE as clear evidence of damage and thus decline. Changes of that magnitude do not fall within the range of healthy or normal variations in a tree's foliage.[130] There is also little chance of human observational error with effects of that magnitude. Such a large loss of needles is obvious to the eye, and if more than a quarter of the needles are yellowed, a tree will clearly be discolored.

The reports of massive decline were based on continentwide ocular surveys which showed that at least 15 percent of the trees in each country in Europe lost a significant fraction of their leaves or needles during the 1980s.[131] There is no record of such ubiquitous damage occurring before 1980.[132] Clearly there had been a widespread and dramatic change of unprecedented nature, yet the change in European forests was evidently not a decline. The forests of Europe grew rapidly during the 1980s, and it was this rapid growth that accounted for the large number of defoliated trees.

Table 9.1 Forest damage on the basis of needle discoloration and needle loss

Damage class	Vitality	Needle loss (%)	Yellowing (%)
0	Healthy	0–10	0–25
1	Slightly damaged	11–25	26–60 at 0–10% nl[a]
			< 25 at 11–25% nl
2	Moderately damaged	26–60	26–60 at 11–25% nl
			< 25 at 26–60% nl
3	Severely damaged	61–99	26–60 at 26–60% nl
			0–100 at 61–99% nl
4	Dying	100	

Sources: Adapted from W. Uhlmann et al., "The Problem of Forest Decline and the Bavarian Forest Toxicology Research Group," in E. D. Schulze, O. L. Lange, and R. Oren, eds., *Forest Decline and Air Pollution: A Study of Spruce (Picea abies) on Acid Soils,* Ecological Studies 77 (New York: Springer-Verlag, 1989), p. 2; with corrections based on Sten Nilsson and Peter Duinker, "The Extent of Forest Decline in Europe: A Synthesis of Survey Results," *Environment,* 29, no. 9 (1987): 7. There were some illogical numbers in the "Yellowing" column in the original, which we assumed were in error.
    a. Needle loss.

As a forest matures it thins out. Large trees require more space than small ones. As the trees grow, some number may suffer defoliation as they fail to compete with their stronger neighbors for space and the resources that space provides. "A high rate of increase in the number of defoliated trees does not always indicate a declining stand but can be a sign of intensive growth."[133] Thus, the defoliation observed may actually be an indication that the forests of Europe have been thriving. This hypothesis is borne out by the growth surveys indicating that the total volume of wood has actually gone up by 25 percent during this time of thinning.[134]

Remarkably, the rapid increase in forest resources of Europe may be due to acid rain.[135] Acid rain provides nitrate, an essential nutrient that is in short supply in many forest soils. Thus, far from being a harmful agent in the forests, acid rain may actually be a boon.

SIGNS OF DAMAGE: THE FICHTELGEBIRGE

Each forest stand is unique. A general upsurge in the European forests does not preclude damage to individual forests. In fact, there is evidence that some European forested areas are being damaged by acid rain. As in the United States, the damage is occurring in high-elevation forests where the soils are naturally acidic and the exposure levels to acid rain are an order of magnitude greater than those at low elevations. For example, the Norway spruce forest in the Fichtelgebirge, a mountainous region in northeastern Bavaria, has clearly suffered damage over the past decade and a half, though there is no indication in the major study done on this spruce forest as to whether the growth stock has decreased or increased over this period.[136]

It should be noted that even in the Fichtelgebirge many stands of trees are unaffected by acid rain. A comparison of a healthy stand in the region to a failing one reveals that the declining one receives 50 percent more acid through rain and fog each year and has half the nutrient supply rate through rock weathering.[137] According to models, if conditions remain the same and no harvesting takes place, the healthy stand will not have a calcium deficiency for another 350 years and magnesium will take even longer to deplete—1,350 years. With harvesting, the calcium depletion could take place in about 100 years.[138] Lower-elevation forests, which receive much less acid than even the healthy Fichtelgebirge stands, should for the most part show less damage than the healthy stand just described.

Acid rain has not caused a dieback of the forests in the Western world. In fact, evidence available for the European continent strongly suggests that acid rain has actually helped forests by acting as a fertilizer. However, damage due to acid rain theoretically may still occur, particularly through the causal mechanism called foliar leaching.

FOLIAR LEACHING

When a leaf is exposed to acid deposition, some of its nutrients are dissolved, or leached, by the acid, making them no longer available to nourish the tree.[139] Acid-rain-enhanced foliar leaching was first hypothesized to damage trees in 1975.[140] Many scientists have speculated that great enough quantities of nutrients can be leached from the leaves of trees to kill those leaves or greatly reduce their photosynthetic capacity. A large enough loss of foliage will slow the growth of a tree or even kill it.

Foliar leaching is by no means a new thing. In fact, in humid regions such as the northeast United States it is a part of the normal nutrient cycle of all trees. Trees have always taken up nutrients from the soil and then lost some of them through their leaves as a result of leaching by precipitation. The problem is that acid precipitation may increase the loss rate while the rate at which the tree receives nutrients from the soil remains the same. In general, as the pH of precipitation decreases and acidity increases, the rate of leaching increases. If nutrients are leached too quickly, trees can suffer malnutrition. Potassium, calcium, and magnesium may be lost through this mechanism at a rate greater than the trees can compensate for.[141]

To date there are no documented cases in which foliar leaching has become a threat to the health of a U.S. forest.[142] Leaching by acid rain is a short-term concern. The leaves that are on deciduous trees today will be replaced next year. The needles that are on conifers if damaged will not recover, but new growth will continue to take place. There may be some long-term damage to trees suffering excessive foliar leaching, but the long-term viability of the forest is not threatened. If significant foliar damage is directly linked to acid deposition from human activities, then appropriate emissions regulations can be considered to alleviate the damage, but there is no need to take preemptive measures.

SOIL ACIDIFICATION

In the late 1970s scientists predicted that acid rain would permanently change the chemistry of some soils. Although soils are generally far better buffered than waters, their buffering capacity was known to be finite. To determine whether the buffering of soils was great enough to deal with the acid from acid rain, scientists developed models of soil responses. The changes they predicted through the use of these models included increase in acidity and loss of nutrients. Although a number of scientists have since concluded that changes such as acidification of soils are reversible,[143] the damage caused by them may not be readily so.

In one early (1977) model a hypothetical plot of land was subjected to rain with a pH of 4.0 for 100 years. The soil started out with a pH of 6.0 and had a fixed supply of nutrient or basic cations; that is, there was no source of nutrient ions to replenish those used in replacing hydrogen ions. This crude model indicated that at the end of 100 years the pH of the soil would drop to 5.2.[144] This drop makes chemical sense. If there was really a soil that had no source of basic cations, then it would get more acidic whenever acid was added. But since there is no such thing as a soil with no source of basic cations, the model does not describe reality. There are, as the authors clearly state, a number of false assumptions in the model, which the authors adopted to simplify the modeling process; each assumption led to an increase in acidification rate relative to reality.[145] Thus their model provided an upper limit for the rate of soil acidification that was far from realistic.

The results, rather than indicating that American forest soils were threatened, served to highlight the "resistance of most soil systems to pH change, [and] the small likelihood of rapid soil degradation due to acid precipitation."[146] Nevertheless, the authors also concluded that long-term acidification was of concern and would be very difficult to detect with short-term studies, given the slow rate of change in even the worst-case scenario. It is this awareness that has kept forest soil acidification a topic of discussion despite the total lack of data to support the hypothesis that it is occurring.[147]

Modeling of the southern forests indicate that they are the only forests in the United States that may experience soil acidification in the foreseeable future, according to NAPAP. Still, "whether such changes will (a) actually occur and (b) affect forest growth remain to be determined."[148] There may not be any damage as a result of acidification even if it occurs. In fact, beneficial effects such as fertilization may more than offset even these possible changes.

Broad generalizations about what acid rain will or will not do to low-elevation forest soils in the eastern states are not justifiable. There is significant variability from place to place in the properties that determine what impact if any there will be. This is true of both the deposition composition and loading but it is especially true of soil characteristics.[149]

As evidence regarding the various hypotheses of forest damage has been gathered, it becomes more and more clear that acid rain has not significantly damaged the forests of the United States. Continuing study is necessary, for we do not know what the long-term effects of acid precipitation will be. Some models indicate that soils in a few southern forests will be acidified within fifty years, though even there we do not know if the net effect will be negative. Like that of any other potential toxin, acid rain's effect is dependent on dosage. In the case of U.S. forests it appears that the dosage is not high enough to be toxic. Can the same be said of the conditions for growing food?

Agricultural Effects

Other than fish, which is less than 5 percent of our diet, and the occasional animal consumed as a result of game hunting, all of our food is derived from agriculture in one way or another.[150] It has been suggested that acid rain may significantly reduce the productivity of our agricultural lands by damaging the health of the plants themselves and by reducing the capacity of the soils to support them. If one or more key crops were adversely affected by acid rain, the economic costs could be severe.

Early studies, however, indicate that in some areas acid rain may actually be a boon to agricultural productivity. In England, for instance, where there has long been serious concern about the destruction of forests by acid rain, much of the soil used for agriculture is sulfur deficient, and the nutritional needs of ryegrass are often met by sulfur inputs both from fertilizers and from the atmosphere.[151] Thus it appears that the sulfur from acid rain actually increases the productivity of these lands.

To date there have been no documented cases of agricultural damage as a result of acid rain.[152] Farm soils are heavily managed. The acid loading from fertilizers is hundreds of times greater than the acid loading from acid rain for many sites.[153] Soils where acidification due to fertilization is a concern are limed in order to neutralize the acid effects,[154] and this liming will also take care of any acidification due to precipitation. Therefore, acid rain should not lead to any long-term acidification of agricultural soils.

Direct effects such as foliar leaching or burning have not been observed on farms. Controlled experiments indicate that most of the foliage of crops is not significantly affected by acid rain at the concentrations now found in the United States.[155]

In sum, it appears that the agricultural resources of the United States are not threatened by current acidity levels in the rain. As acid deposition has actually been declining over the past two decades, there is no reason to expect that farms will be damaged in the future either.

Human Health Effects

With our forests and farms safe for the time being, the logical question is, what about us? There are no direct human health effects of acid rain as rain. The gases that act as precursors, particularly sulfur dioxide at high enough concentrations, do contribute to respiratory difficulties for asthmatics. Studies have found that difficulty in breathing is experienced at sulfur dioxide concentrations above 0.4 parts per million in asthmatics.[156] The shortness of breath typical of a combination of exertion and exposure is rarely life threatening, except to extreme asthmatics, and does not appear to lead to long-term difficulties. Although the concentrations cited are found in many urban areas, they are generally found only near a major source of sulfur dioxide.[157]

Because acidic fog may pose a health risk, E. L. Avol and associates studied the effects of sulfuric acid fog on asthmatics, exposing the subjects to fog containing 0, 500, 1,000, and 2,000 milligrams per cubic meter of (mg/m^3) sulfuric acid and had them exercise intermittently for one hour. "The only significant effect observed was a decrease in peak expiratory flow after exposure to 2,000 mg/m^3."[158] In other words, the people tested were a bit short of breath as a result of the peak exposure level, which is higher than the peak exposure level of any city in the U.S. survey of sulfur dioxide and of all but one rural area.

Acidic fog is apparently a problem only in the Los Angeles basin, where the fog typically ranges from pH 1.7 to 4.0, which is far more acidic than fogs recorded elsewhere.[159] Even Los Angeles fogs probably do not have concentrations as high as 2,000 mg/m^3, that is, high enough to affect asthmatics.[160] Thus, acidic fog, as it exists in the United States does not appear to pose a health threat, given our limited understanding of it.

The Media's Treatment of Acid Rain
with Laura Evans

From the start, individuals questioning the significance of acid rain were given less coverage and, when their views were reported, were subjected to much greater scrutiny than those who claimed ecological catastrophe. Stories of impending disaster received high-profile coverage even when the basis of these claims was dubious. Many journalists were either unaware of the state of science on acid rain or chose not to communicate these developments to their audience.

To provide a suitable base of information from which to evaluate the media's treatment of acid rain, articles were collected from the *New York Times, Los Angeles Times, Washington Post, Wall Street Journal, Time, Newsweek,* and *U.S. News and World Report.*[161] Charts were compiled in which the slant toward or against harm from acid rain was assigned a negative, neutral, or positive value. Negative values denoted criticism of the acid rain thesis, with positive rankings indicating support and neutral rankings signifying balanced coverage. By assigning overall rankings to articles, ranging from −2 to +2, the picture of how rarely stories critical of the acid rain thesis appeared becomes clear:

Rank	−2	−1	0	+1	+2
Percent	3%	5%	35%	39%	18%

Antiregulatory ideas were discounted and downplayed through a variety of means. For the most part more negative articles tended to be placed farther back in the paper, diminishing the importance of a story and the chances that it would be read. Within an article the number of proregulatory paragraphs dwarfed the number of antiregulatory paragraphs, making it less likely that misgivings about acid rain harms could be adequately explained. Finally, a paragraph expressing skepticism was placed considerably later in the story than a paragraph warning of severe ecological damage.

It is not difficult to discover instances of the scrutiny to which skepticism about acid rain's impact was subjected. The most prominent example is the treatment of NAPAP findings that discounted claims that acid rain produced ecological devastation. By studying the *New York Times*'s coverage of NAPAP we can see how critically its views were treated, while environmentalists' opinions were seldom questioned.

The first article on NAPAP appeared in 1987 when the program made an interim report.[162] Journalist Philip Shabecoff carefully explained the study's findings and permitted the authors to defend their conclusions, but he also included condemnations. The findings are referred to as "bad science," a "startling misrepresentation," and even "political propaganda." The next article, appearing a few days later, described environmentalists' criticisms of NAPAP's study.[163] The actual findings are not presented until the eleventh paragraph. A third article, printed a few months later, focused on Canadian displeasure with NAPAP. No defense of NAPAP's conclusion, which is related in half a sentence, was included.[164]

When the program's final report was published, the first newspaper article on the subject implied that NAPAP had found damage to trees to be even greater than expected, which contradicts the actual conclusions.[165] The next article focused on NAPAP's studies of the acid-buffering capacities of American waterways. The headline proclaimed, "Study of Acid Rain Uncovers a Threat to Far Wider Area," because the report had pointed to the low buffering capacities of lakes and streams in certain regions.[166] Yet NAPAP's overall analysis was that further acidification was likely only in limited areas. The next article on NAPAP was fairer, quoting extensively and equally from its supporters and detractors.[167] However, the final story concentrated on reactions to the study, which supposedly had unleashed a "torrent of criticism." Only two paragraphs attempted briefly to explain what these findings were; all other discussions of the report were in the form of direct quotations from its critics.[168]

While the NAPAP report was thus treated harshly, environmentalist assertions were presented without question and often stated as fact. Stories declared impending disaster without evidence. A *Los Angeles Times* article stated that acid rain destroys freeways and galvanized steel, although the only study it referred to was one that had yet to be started.[169] One writer opened his attack on acid rain by attributing to it the following consequences: "Brook trout no longer crack the glassy surface of Big Moose Lake . . . They have vanished, along with crayfish and frogs, loons, kingfishers and most of the swallows. The chirping tree frogs . . . have largely disappeared or stilled their voices, and blights are killing the . . . trees. An unnaturally silent spring has fallen over sectors of this wilderness."[170] The author presented a sad and stirring image, but he did not give reasons for believing why acid rain was responsible for this loss of life. Likewise, in another article, acid rain had turned lakes into "aquatic death traps" while "forests . . . once lush and teeming with wildlife, were filling with dead and dying trees."[171] Another story asserted that acid rain "may cause heart and lung distress after long-term exposure."[172]

Reporters resisted NAPAP findings, much as they failed to report accurately other changes in scientific knowledge. More generally, as the following table illustrates, using the −2 to +2 ranking system, newspaper coverage of acid rain was not indicative of the developments in scientific opinion. In the 1970s and early 1980s the extraordinarily high proportion of articles supporting the thesis that acid rain is harmful at the level experienced in the United States and Europe was incongruous with the limited evidence at the time. With doubts of acid rain damage evolving over time one would expect the proportion of proregulatory articles to fall through the mid- and late 1980s and the early 1990s. While there was some fall-off in the mid-1980s, after this there was little movement; in the early 1990s proregulatory articles continued to outnumber articles downplaying the acid rain phenomenon:

	Rankings				
Years	−2	−1	0	+1	+2
1974–1982	4%	8%	16%	40%	32%
1982–1985	4%	1%	44%	35%	16%
1986–1988	2%	4%	31%	45%	17%

Conclusion

Acid rain is real; it does affect the ecosystems it falls on. The rain in the eastern United States has become more acidic as a result of sulfur dioxide and nitrogen oxide emissions from power plants, vehicles, and other manifestations of the industrial era. The sorts of acid that fall in rain have also been changed significantly by our actions. Both the increase in acidity and the change in acid type have had an effect on the environment. This impact was once thought by scientists and the lay public alike to be potentially destructive, but years of research have shown that devastation will not occur in lakes, forests, farms, or our bodies.

There has been some damage to lakes, but much less extensive and severe than was once thought. Initial assessments of our lake resources had indicated that over half the lakes in some areas had been acidified, and many were thought to have been rendered fishless by the effects of acid rain. These assessments were wrong. More recent studies indicate that only a small percentage of the lakes in the Northeast have become significantly more acidic as a result of acid rain, though the acidity has apparently killed the fish in a number of these lakes. Where there has been damage, however, alleviation through liming is possible. Furthermore, scientists believe that if the rain were to remain as acidic as it is today, no more northeastern lakes would be damaged. Therefore, immediate emissions reductions are not necessary to protect most lakes.

Forests were also once thought to be severely damaged or gravely threatened by acid rain. We now know that only a few high-elevation forests, representing less than 0.1 percent of our forest resources, have experienced damage attributable to acid rain. Few, if any, forests will be damaged in the next several decades even if emissions are not reduced. This finding of no or minimal damage is repeated with agriculture products and with human health.

If we look at titles of scientific articles about acid rain—"Acid Rain not Only to Blame,"[173] "Natural Diebacks in Forests: Groups of Neighboring Trees May Die in Response to Natural Phenomenon . . .,"[174] "New Perspectives on Forest Decline, Prophesies That 'New-Type Forest Damage' Would End in *Waldsterben* (Forest Death) Have Not Come to Pass: Greater Realism Now Characterizes Research on the Subject."[175]—the change in direction is apparent. The new realism is reflected in definitions of damage that pay attention to natural variation, to regional differences, and to observations of regeneration.

10

CFCs and Ozone Depletion: Are They as Bad as People Think?

with Robert Owen Rye

Global environmental issues are relative newcomers to the list of human concerns. We have only recently discovered that our actions in one corner of the globe may have harmful effects throughout the world. Nowhere has this concern been more evident than in the debate over the continued use of chlorofluorocarbons (CFCs), which are said to be thinning the earth's ozone layer. A sufficiently depleted ozone layer could let in enough ultraviolet radiation to cause widespread health problems. But there are a number of questions remaining to be answered. Thinning may not be occurring, or what thinning there is may not be serious. CFCs are economically important; the ozone layer is fundamentally important. Wisdom is clearly called for.

The governments of most of the world's nations have chosen to eliminate the use of CFCs by the year 2000. This policy will be costly. The roles that CFCs play, from refrigerants to foamants to cleaning agents and beyond, promote human welfare all over the world. Large-scale cheap refrigeration, one of the great inventions of the twentieth century, prevents disease and extends the shelf life of food, reducing spoilage and hence waste. Its wide use has clearly improved the health and nutrition of the industrialized world. As cleaning agents CFCs play a vital part in the electronics industry, allowing for hyperclean circuits. They are also used in air conditioning, which though not essential for human life has greatly improved productivity in the hotter and more humid parts of the industrialized world, not to mention provided great comfort. In all of these applications CFCs were chosen in part because they are nonflammable and nontoxic. Production has been in the billions of kilograms.[1] In short, they are extensively used and safe, which replacements may not be. Claims that there are substitutes as effective and no more costly, or even cheaper, are just that—claims.

Unfortunately, once they are made, synthetic CFCs leak into the atmosphere in large quantities. Once released, they rise into the stratosphere where their by-products destroy ozone. The 1987 Montreal Protocol on Substances That Deplete the Ozone Layer and the 1990 London revisions to the protocol together constitute the international community's response to the threat of CFC-induced ozone depletion. In these agreements future use of CFCs was first severely limited and then banned altogether as of the year 2000. The restrictions and ban were seen as necessary at the time they were passed to protect life on earth. But the conclusion that the evidence available justified a ban is open to question.

The purpose of this chapter is to determine whether or not the case against CFCs was good enough to warrant the steps taken in the Montreal Protocol when it was signed and whether or not the case had become sufficiently stronger between 1987 and 1990 to warrant the ban imposed in the London revisions. We expect that after reviewing the evidence you will agree that, with some reservations, we can say yes to Montreal and no to London.

There has been a significant change in scientists' understanding of the ozone layer since the London revisions were made. Although not strictly within our stated purpose, we will look briefly at this change at the end of our scientific assessment because it contains an important lesson about global environmental issues—we are still learning the fundamentals. The voyage of exploration has just begun. Surprises are bound to be waiting for us.

Concerns over CFCs

Concerns about CFCs can be traced to a single paper that appeared as a letter to *Nature* in June 1974 by Mario Molina and F. Sherwood Rowland. In this oft-cited letter the two scientists announced that CFCs put chlorine in the stratosphere, the layer of the atmosphere where most ozone is found, and that the chlorine destroys ozone.[2] They went on to say:

> It seems quite clear that the atmosphere has only a finite capacity for absorbing Cl [chlorine] atoms produced in the stratosphere, and that important consequences may result. This capacity is probably not sufficient in steady state even for the present rate of introduction of chlorofluoromethanes. More accurate estimates of this absorptive capacity need to be made in the immediate future in order to ascertain the levels of possible onset of environmental problems.[3]

They believed that continued emission of CFCs at the levels of their day would lead to depletion of the ozone layer. Since then there has been a great deal of research into the questions raised.

Before the famous letter, CFCs were universally recognized as safe chemicals with wondrous properties. When they were developed in the late 1920s, "To demonstrate their relative safety at ground level Thomas Midgeley, Jr., the research chemist who synthesized CFCs, inhaled vapors from a beaker of clear liquid and then exhaled to extinguish a candle."[4] Over forty years later CFCs still had a clean bill of health. In fact, J. E. Lovelock and colleagues, writing in *Nature* in 1973, said: "The presence of these compounds constitutes no conceivable hazard."[5]

Given the widespread belief in the safety of CFCs it is not surprising that Molina and Rowland's letter was largely ignored for several weeks. It was not until September 1974, when they first publicly discussed their theory at a meeting of the American Chemical Society, that they began to get some attention.[6] Clearly, they convinced some people, since the National Academy of Sciences created a task force in October to study the impact of CFCs on the ozone layer. That group predicted that the ozone layer would be depleted by 10 percent by the year 2050 or so if CFCs continued to be emitted at the rate they were then.[7] And so began a massive effort to understand the chemistry of the ozone layer.

A number of the better publicized studies have indicated that CFCs have already depleted the ozone layer or that they will deplete it in the future. Furthermore, the discovery in the mid-1980s of a phenomenon known as the "ozone hole" has fueled concerns about the CFC threat. The question is, why do we care? What is so important about ozone?

The Importance of Ozone to Life

Ozone is important because it absorbs potentially harmful solar ultraviolet (UV) radiation. If the concentration of ozone in the atmosphere were radically reduced, one would expect that the radiation in the band that ozone absorbs would reach the earth's surface in greater quantities than it does now. Although there is no consensus about how much ultraviolet radiation is acceptable, it is generally understood that without ozone in the atmosphere, life as we know it could not exist.

Ozone absorbs radiation of wavelengths from approximately 240 to 320 nanometers (nm). Radiation in the range of 240–290 nm is called UV-C, and even low levels of exposure to this radiation leads to the destruction of DNA and RNA, the basic building blocks of life. Fortunately, stratospheric ozone absorbs very close to 100 percent of the radiation in this range that reaches the upper atmosphere. Consequently, the actual exposure of living matter to this sort of radiation is trivial.[8] In fact, even with a significant decrease in the ozone concentration in the stratosphere it is unlikely that we would see marked increases in the amount of UV-C radiation that reaches the ground.[9]

The radiation that is of general concern lies in the range known as UV-B, which stands for biologically active ultraviolet radiation, 290–320 nm. This radiation is only partially absorbed by the ozone layer even when ozone concentrations are at a maximum, because the absorption efficiency of ozone in this band is much lower than it is in the UV-C band. It is the threat that humans and animals will be exposed to ever-increasing doses of UV-B that fuels the worry over the ozone layer, and it is the fact that chlorine from CFCs destroys ozone that prompts so many to call for the banning of these once universally useful chemicals.

The Accusation: CFCs Cause Cancer Indirectly

Scientists know that ultraviolet radiation causes skin cancers: both squamous cell and basal cell carcinomas have been firmly tied to exposure to ultraviolet radiation.[10] If CFC-induced ozone depletion leads to increased ultraviolet exposure, the incidence of these cancers will probably increase. Although these conditions are generally not fatal, melanoma, another form of skin cancer, often is.

According to the Environmental Protection Agency (EPA), 200,000 additional deaths due to melanoma will occur in the United States between now and the year 2050 because of CFC-induced ozone depletion. This prediction is based on an assumed 5 percent decrease in the earth's total ozone column and on an assumed 2 percent increase in UV-B at ground level for every 1 percent decrease in ozone.[11] There is an additional assumption of a 1 percent increase in the incidence of melanoma for every 1 percent increase in UV-B.[12] Thus, there will supposedly be a 10 percent increase in the incidence of melanoma by the year 2050, which will lead to these 200,000 deaths.

Numbers like these should cause one to pause and consider the strength of each of the assumptions the EPA has made. First of all, scientists have not yet shown that ultraviolet radiation does cause melanoma.[13] We will return to this point later. Second, the connection between CFC-induced ozone depletion and cancer cannot be tested directly. We must rely on evidence about the atmosphere from the scientific community to determine whether or not the fundamental assumption that CFC emissions will lead to more UV-B exposure is correct.

There are three main points in the ozone depletion hypothesis: there has been or will be a long-term (or secular) thinning of the ozone layer; CFCs are responsible for that thinning; human, animal, and plant health will be harmed by it. Each of these premises must be put to the test to determine whether the ban of CFCs was wise.

Basic Chemistry and Dynamics of the Ozone Layer

The term "ozone layer" refers to all of the ozone in the atmosphere. There is not a single layer of ozone at any point in the atmosphere, nor is there a constant amount of ozone over any one point or a single thickness over the whole earth at any given time.

About 90 percent of the ozone in the world is formed in the stratosphere, the layer of the atmosphere some five to thirty miles above the surface of the earth. Most of this ozone is formed in the equatorial regions. After formation it is transported toward the poles by the circulation cells (or winds) of the stratosphere. The higher latitudes are generally better insulated from ultraviolet radiation than the equator. Because the sun is nearly directly overhead in the equatorial regions, and because most of the ozone is transported away from them by winds, these regions receive about fifty times as much ultraviolet radiation as do the polar regions.[14]

Ozone is a gas created by the action of sunlight on oxygen. When a photon of ultraviolet radiation of wavelength less than 242.2 nm is absorbed by a molecule of oxygen, which contains two atoms (O_2), the oxygen molecule breaks apart or photodissociates into the two atoms. Each of these atoms may then attach themselves to molecules of oxygen to form two three-atom molecules of ozone (O_3). Ozone can in turn be destroyed when ultraviolet radiation splits off its extra atom.[15]

Formation and destruction of ozone take place constantly when the sun is up and to a lesser extent when it is down. There must be a net ozone production-destruction rate of zero on the time scale of centuries, otherwise all of the oxygen in the atmosphere would be in the form of ozone. Ultraviolet radiation forms more ozone than it destroys. Destruction through collisions with oxygen atoms takes care of about 20 percent of the excess,[16] while the remainder is removed by other reactions, the most important involving nitrogen oxides (NO_x), chlorine oxides (ClO_x), and hydrogen oxides (HO_x). Together these three gases account for the destruction of almost all of the rest of the excess ozone and thereby maintain the balance of destruction and creation.

All of these compounds react catalytically with ozone: that is, the ozone is consumed, but the other compounds are regenerated so that they may react with the next molecule of ozone they encounter. If any one of these gases reacted only with ozone, the ozone would rapidly be depleted. In fact, scientists have shown that during the early spring in the Antarctic stratosphere chlorine reacts almost exclusively with ozone, thereby generating the large-scale ozone hole.[17] Fortunately, under normal conditions, each of these gases reacts with one another as well. These reactions often lead to the formation of other gases that do not react with ozone.

There is a constantly shifting balance between the reactive gases that helps to determine how much ozone there will be in the atmosphere at any time. Any radical long-term increase in the amount of one of these gases in the stratosphere would most likely lead to a long-term decrease in the total amount of ozone there. But the depletion of ozone would not go on unchecked. The rate at which ozone is destroyed by these gases is a function of the amount of ozone available to be destroyed. As the concentration of ozone goes down so does the rate of destruction. Eventually, the destruction and production rates would again be balanced, but there would be less ozone than before. Such a scenario is possible given the addition of chlorine from CFCs.

If a CFC molecule absorbs radiation of the appropriate wavelength (primarily 185–215 nm), the molecule will photodissociate just as oxygen and ozone do. In the case of CFCs a highly reactive chlorine atom separates from the rest of the molecule, and when it comes into the presence of ozone it produces the following catalytic reaction.[18]

$$Cl + O_3 \rightarrow ClO + O_2$$

$$ClO + O \rightarrow Cl + O_2$$

$$Net: O_3 + O \rightarrow 2O_2$$

The net effect is to convert monoatomic oxygen and ozone into molecular oxygen.

Nearly all of the CFCs released will make it up into the stratosphere and release chlorine, for there are no known tropospheric processes that destroy a significant amount of CFCs.[19] The migration of CFCs can take decades, so that what we emit today will reach the stratosphere in the early twenty-first century.[20] A long-term trend of increasing stratospheric chlorine could change the balance of the ozone chemistry.

Natural Sources of Chlorine versus CFCs

There was chlorine in the stratosphere—from volcanoes, oceanic biota, and sea salt—long before mankind came on the scene.[21] CFCs are of concern because they may add chlorine to the normal pool and change the balance between the gases that help determine the concentration of ozone.

The current stratospheric chlorine concentration is over 3 parts per billion by volume (ppbv).[22] Scientists have estimated that the pre-CFC chlorine concentration was 0.6 ppbv.[23] This estimate, indicating a five- to sixfold increase since the late 1920s, is based on a model. Even if we do not accept the model, the available measurements indicate that chlorine has increased by a factor of 1.8 since 1977.[24] The measured rate of increase is consistent with the predicted rate of chlorine addition by CFC emissions. Furthermore, measurements indicate that fluorine concentrations have increased by a factor of 3.2, and scientists believe that CFCs provide almost all stratospheric fluorine.[25] A long-term increase in fluorine strongly indicates that CFCs are breaking down in the stratosphere and supports the notion that the observed increase in chlorine is a result of CFC emissions. Despite this strong indirect evidence, there are still questions about the background level of chlorine and its variability over the time period for which we have concentration data.

Volcanoes put out huge amounts of chlorine, but very little of it reaches the stratosphere. Most is washed out as hydrogen chloride (hydrochloric acid) within a day or so of emission. Some observers suggest that the amount of chlorine added to the atmosphere by volcanoes, which is thousands of times larger than that from CFCs, has a vastly greater impact on the ozone layer.[26] They fail to recognize that very little of the chlorine from volcanoes (and, we might add, swimming pools and chlorinated water) reaches the stratosphere and that nearly all of it from CFCs does. Only the chlorine that reaches the stratosphere matters here.

Exceedingly violent volcanic eruptions such as El Chichon in Mexico in 1982 do inject ill-quantified but certainly large amounts of chlorine into the stratosphere, perhaps 0.1–1.0 teragrams (Tg), or 10^{12} grams, per year into the atmosphere, of which a significant fraction reaches the stratosphere.[27] They may inject 17–36 percent as much chlorine into the stratosphere each year as CFCs.[28] Both of these estimates are much more tenuous than the estimates of annual CFC releases. Because volcanoes are not likely to contribute to long-term continuous increases in chlorine, they are not grounds for dismissing the role of CFCs, as has been suggested by some opponents of CFC regulation. Though volcanic contributions to stratospheric chlorine can be disregarded in the models of long-term ozone levels, eruptions may still have important short-term effects on stratospheric ozone.

Sea salt is another source of stratospheric chlorine that is not well-quantified. We do not know the mechanisms whereby chlorine from sea salt reaches the stratosphere. It is generally agreed, however, that the contribution from sea salt is not likely to vary a great deal over time. Unless it can be shown that the long term input from natural sources is highly variable and that there is compelling evidence for a recent influx of natural chlorine large enough to explain the recent increase in stratospheric chlorine, we may conclude that CFCs are the source of the added chlorine. The question then is what impact will the chlorine from CFCs have on the ozone layer?

The Evidence: Observations versus Models

Predictions of how much CFCs will deplete the ozone layer are made through computer-generated models of the atmosphere. Those who make these models have to deal with the constant change in our understanding of the stratosphere; what is realistic today is often not tomorrow. We who seek to comprehend the import of these models need to keep the evolution of this understanding in mind.

MODELS OF FUTURE OZONE LEVELS: THE QUALITY OF OUR CRYSTAL BALL

The first prediction of ozone depletion from CFCs was made by Rowland and Molina in 1975, less than a year after they had said that CFCs posed a potential threat to the ozone layer. They predicted that if CFCs 11 and 12, the most common CFCs, were emitted forever at the same rate they were emitted in

1973, there would be a steady-state decrease in the total amount of ozone in the earth's atmosphere of approximately 10 percent within fifty to seventy years.[29] Predictions have varied greatly from model to model and year to year since then (see Table 10.1). Yet for atmospheres in which the only added "pollutant" is CFCs, all models have shown a depletion of ozone over time. Many of these predictions have been accompanied by a confidence level of 68 to 95 percent. Still, there has been a great deal of disagreement over how much CFCs will affect the ozone layer. How can these two facts be reconciled?

The confidence levels do not tell the whole story of the uncertainties in the models. The NRC prediction of 6.0–7.5 percent, for instance, is actually best characterized by a range of 2–20 percent if all of the uncertainties in the assumptions that were then known are included.[30] This range is the true range of the models' predicted depletions, but it is not the one that is quoted anywhere other than in the report itself. The preparers of the report felt that "most informed persons would judge [2 percent] to be tolerable" and 20 percent to be intolerable.[31] Thus, they concluded, "What we know is sufficiently uncertain that continued release of CFMs [CFCs 11 and 12] at 1973 rates: 1) may be tolerable; 2) may be intolerable."[32] Such a conclusion is hardly a convincing condemnation of CFCs. Failure to report the extent of the uncertainty led to an unrealistic public perception.

Models have changed over time as scientists have learned more about the stratosphere. The chemistry involved has grown much more complex. When all of this started there were perhaps 50 reactions of interest; today there are some 200 that might be important in the stratosphere.[33] As they have added

Table 10.1 Early predictions of ozone depletion due to CFCs

Date	Agency	Emissions	Projection	Depletion	Timing
1976	NRC	CFC 11, 12	At 1973 rates	6.0–7.5%	50 years
1977	NASA[a]	1975 rates		10.8–16.5%	varies
1981	WMO	Constant		5.0–9.0%	~50 years
1984	NRC	Constant		2.0–4.0%	?

Sources: Compiled from data in the following publications. NRC (National Research Council), Panel on Atmospheric Chemistry, Assembly of Mathematical and Physical Sciences, *Halocarbons: Effects on Stratospheric Ozone* (Washington, D.C.: National Academy of Sciences, 1976), pp. 161–162. NASA (National Aeronautics and Space Agency), "Chlorofluoromethanes and the Stratosphere," ed. R. D. Hudson, publication no. 1010, 1977. WMO (World Meteorological Organization) and NASA, WMO Global Ozone Research and Monitoring Project, report no. 16, *Atmospheric Ozone, 1985: Assessment of Our Understanding of the Processes Controlling Its Present Distribution and Change* (Greenbelt, Md.: NASA Laboratory for Atmospheres, 1986), p. 728.

a. Nine models.

the effects of new reactions and improved their understanding of old ones, scientists have changed their predictions.

REACTION RATES

A model of atmospheric chemistry must include numbers that describe the rate at which each reaction will progress under a given set of conditions. As more laboratory work has shown that previous estimates were wrong, modelers have had to modify these numbers. These revisions have led to a change in the relative importance of various reactions in the models of the stratosphere and have modulated the degree to which CFCs are expected to deplete the ozone layer. Despite all the reaction rate changes, the model runs that had CFCs alone perturbing the atmosphere continued to show a significant decrease in the steady-state ozone column. That is not true of the model runs that ceased to assume that addition of CFCs is the only important source of change in the atmosphere.

MODELS SENSITIVE TO ADDITION OF GASES OTHER THAN CFCS

The focus on the role of CFCs has overshadowed the fact that ozone chemistry is complex and that catalytic destruction by chlorine is by no means the dominant process in that chemistry. There are a number of other trace gases whose concentrations in the stratosphere and upper troposphere are changing significantly. Some of these gases may affect the ozone column. A 1977 study indicates that increases in carbon dioxide would radically slow ozone depletion.[34] A 1983 study shows that if carbon dioxide, nitrous oxide, and methane increase at current rates, ozone might increase even with the added chlorine from CFCs.[35] In fact, "one-dimensional models predict that the magnitude and even the sign of the ozone-column changes due to increasing CFCs depend on the future trends of carbon dioxide, methane and nitrous oxide."[36] This statement is borne out in a number of studies showing that "the equilibrium concentration of ozone depends on the amount of catalytic agents which are present in the atmosphere,"[37] of which chlorine is only one. The most recent models include the changing concentrations of trace gases in the stratosphere.[38]

MODELS AT THE TIME OF THE MONTREAL PROTOCOL

The Montreal Protocol was based on the most up-to-date information on the ozone layer available in 1987. The most important relevant document produced shortly before that was the World Meteorological Organization (WMO) report *Atmospheric Ozone, 1985*. This publication, which spanned over a thousand pages in three volumes, represented the state of knowledge just before the drafting of the Montreal Protocol, and its predictions were crucial in determining what action should be taken.

A number of models are quoted in the WMO report. Each model was run several times, with different scenarios provided to determine the sensitivity of the results to modification of assumptions (Table 10.2).[39]

On the basis of these models scientists were certain that continued rapid growth in CFC use would lead to severe depletion of the ozone layer. They predicted that freezing emissions at 1980 levels would lead to a 5 percent or greater depletion in the long run. Cutting to 80 percent of 1980 emissions and allowing other trace gases to increase would lead to a fairly stable ozone layer, most models concluded, with a negligible loss if any.

In 1987 it was not reasonable to act as if emissions of CFCs would lead to depletion of the total ozone column. Some regulation of CFC emissions was in order, but the evidence did not support the elimination of CFCs. The models were unanimous, however, in their prediction that there would be a change in ozone distribution in response to CFCs and other trace gases. Was such a redistribution unacceptable?

All of the models predicted massive depletion, on the order of 60 percent for "CFC only" runs and 35 to 50 percent for realistic scenarios, in the altitude range of 40 kilometers (km), or 25 miles.[40] Although this depletion would be impressive, it would not lead to a significant depletion of the total ozone column for two reasons. First, most of the ozone is below 25 km and almost all is below 40 km. Thus huge decreases at high altitudes will not have a large effect on the total column thickness. Second, this depletion would be matched by increases in ozone in the lower stratosphere and upper troposphere. Some of this increase would be due to self-healing, a process by which the destruction of ozone at higher altitudes allows more ultraviolet to reach lower altitudes, where it is absorbed to produce more ozone. The rest would be made up by the production of ozone through the decomposition of methane.

In addition to the depletion at high altitudes, two-dimensional models indi-

Table 10.2   Model predictions at time of Montreal Protocol

Scenario	Depletion ranges	
	Six models[a]	Max Planck Inst. model
1980 emissions forever; no other changes	−4.9 to −9.4%	
Steady state 8 ppbv chlorine (80% of 1980 emissions); no other changes	−2.9 to −5.7%	−9.1%
80% of 1980 emissions with doubling of methane; 120% increase nitrous oxide	−2.3 to −3.4%	−6.0%
Same as above plus double carbon dioxide	+0.2 to −1.4%	−5.2%
3% annual growth in CFC emissions; 15 ppbv chlorine by 2040	All models show massive depletion of ozone layer	

a. See n. 39 for the sites of the first six models.

cated that some regions would undergo significant depletions during winter months, and the available data indicate that these depletions have already come to pass.[41] Nevertheless, winter ultraviolet levels are not as great a concern as summer levels, because people tend to be more heavily clothed and ultraviolet exposure is generally at a minimum in the winter. As long as the total column above any given point is not radically reduced in the summer, human health will not be at risk.

STATISTICAL ROBUSTNESS OF THE MODELS

The total uncertainty in the model predictions, given all of the uncertainties in the rate coefficients for those reactions that affect ozone concentrations, was a factor of 1.84.[42] All of the predictions in the models could be 1.84 times too great or too small. Thus, a prediction of a 5 percent decrease could be anything from a 2.7 to a 9.2 percent decrease. Scientists generally report predictions with much narrower uncertainties, because they assume that the reaction rate numbers are well known and the important reactions have all been accounted for in the model. Such confidence is not reflected in the comments in the general reports on ozone. A 1982 NRC report stated, "It is difficult to rule out the possibility of an important role for [chemical] species not now included in models, and, if history is a guide, there may well be future surprises in this area."[43] Such surprises are bound to come, as models cannot even accurately predict the current state of the atmosphere and, in fact, have come since the promulgation of the Montreal Protocol.

The models did not correctly reproduce the concentrations of the chlorine monoxide radical (ClO), nitric acid, and ozone in the upper stratosphere.[44] In fact, observed ozone levels at and above 35 km were 30–50 percent higher than the levels predicted by the models.[45] The discrepancies between reality and model led to speculations that (1) there was an unknown source of ozone (odd oxygen) in the upper stratosphere, (2) the destruction rate of ozone was overestimated at higher elevations, or (3) there were reactions of importance missing from the models.[46] There were approximately 200 gas-phase reactions accounted for in the models, and any of these could have been more (or less) important than was thought. Even if all of the reaction rates used were perfect, the list of reactions may not have been exhaustive.

If models fail to describe the current atmosphere, it is difficult to have confidence in their reproduction of future atmospheres. Nevertheless, models were all that scientists had to go by in predicting the effects of CFCs at the time of the Montreal Protocol. These models correctly reproduced the ozone concentrations at the levels at which it was most important—the middle and lower stratosphere. Errors in the upper stratosphere did not justify dismissing the models out of hand, but they did warrant questioning of the precision of their predictions. Furthermore, as "it is difficult to predict with confidence the future evolution of CH_4 and N_2O, whose atmospheric trends have only been noticed during this decade, and the reasons for which are not fully understood,"[47] the sensitivity tests with the trace gases had to be taken with a grain of salt.

We know that these trace gases are important in determining future levels of ozone, but we still do not know how important: "Even if numerically accurate models were complete in photochemistry and satisfactorily approximated those aspects of atmospheric motions and other physical processes that strongly affect ozone, the models still could not predict future ozone changes due to increasing chlorofluorocarbons unless they were supplied with the future trends of other trace species."[48]

The uncertainties in the models pointed to a clear need to test them against observations of the phenomena represented. The more uncertain one is of the validity of a model, the more important it is to temper confidence in the model's predictions with reliance on the data at hand. How reliable are past data?

Past Ozone: Observation versus Models

Scientists have conducted numerous studies of the ozone column data. The trends discerned have typically been based on data from no more than two and a half decades. A number of assessments of the available data were made prior to the writing of the Montreal Protocol and are reviewed in Table 10.3. All of these assessments indicated that there was no significant overall trend. As of 1987, when the Montreal Protocol was written, there was no evidence that ozone depletion had already started. Nor was there any evidence that CFCs were contributing to stratospheric chlorine. Only the models indicated there was cause for concern in the long term, and even there the concern was with unrestrained growth in emissions.

Model predictions after the writing of the Montreal Protocol indicated that

Table 10.3   Pre–Montreal Protocol trend analyses

Investigators	Time interval	Reported trend
Reinsel et al.	1970–1978	+0.28 ± 0.67%
St. John et al.	1970–1979	+1.5 ± 0.5%
Bloomfield et al.	1970–1979	+0.1 ± 0.55%
Reinsel et al.	1970–1984	−0.26 ± 0.92%

Sources: Compiled from data in the following publications. G. C. Reinsel et al., "Statistical Analysis of Ozone Data for the Detection of Trend," *Atmospheric Environment,* 15 (1981): 1569–1577. D. St. John et al., "Time Series Analysis of Stratospheric Ozone," *Communications in Statistics: Theory and Methods,* 11 (1982): 1213–1333. P. Bloomfield et al., "A Frequency Domain Analysis of Trends in Dobson Total Ozone Records," *Journal of Geophysical Research,* 88, no. C13 (Oct. 20, 1983): 8512–8522. G. C. Reinsel et al., "Statistical Analysis of Total Ozone and Stratospheric Umkehr Data for Trends and Solar Cycle Relationship," *Journal of Geophysical Research,* 92, no. D2 (Feb. 20, 1987): 2201.

even with the 50 percent reduction in emissions called for in that document, there would still be significant decreases in ozone during the mid-twenty-first century. A 6 percent loss of ozone in the winter in the middle latitudes—over Europe and North America—was predicted. Smaller losses in the tropics and greater losses in the polar regions were also projected, as well as smaller losses during the summer.[49] The projected depletion, reflecting an improved understanding of stratospheric processes, was much greater than that predicted before the Montreal Protocol. Did this mean we should have banned CFCs immediately? No.

The models also predicted substantial losses even with a ban on CFCs by the year 2000. That ban would mean that there would be a 2 percent smaller loss in ozone than if we continued to use CFCs at 50 percent of the rates of 1986.[50] If we phased out CFCs over the next few decades, there would be an even smaller difference in ozone loss between phasing out and banning CFCs.

In addition to the new model results, scientists also produced a revised analysis of all of the available data from ground-based ozone measurements at Dobson spectrophotometer stations. The report put out by the Ozone Trends Panel (OTP), an affiliate of NASA and several international scientific bodies, was a catalyst for the London revisions banning all CFCs by 2000. This report has been described as "a 16-month comprehensive scientific exercise involving more than 100 scientists from 10 countries, using new methods to analyze and recompute all previous air- and ground-based atmospheric trace gas measurements."[51] Attempting to correct for gross inconsistencies in the data and for calibration errors, the panel revised the data set.

The OTP analysis indicated that total ozone had decreased by 1.7–3.0 percent over any decreases that could be attributed to natural causes.[52] More than anything else, the OTP finding of a measurable and significant decline in total ozone led to the decision to ban the use of CFCs. The magnitude of the effort put into producing the report and the prominence of the people involved made this the definitive study of long-term ozone trends. Nevertheless, there are a number of grounds on which the report's conclusions have been questioned.

The OTP based its findings on data from 1965 to 1986, going from one solar minimum (approximately) to another, but the trends the panel reported are from 1969 to 1986, going from a solar maximum to a solar minimum. When the OTP's predictions are firmly negative, our reanalysis (formulated in Table 10.4), which takes into account solar variability, shows either no change or a very slight depletion in the 53–64° band.

It is not at all certain that the OTP correctly accounted for the effect of the eleven-year cycle of solar ultraviolet radiance, the sunspot cycle. The panel was not looking at a smooth decline in a constant value; rather, it found a long-term trend in a highly variable quantity. The variability of the ozone concentration over the eleven-year solar cycle due to solar variability was (and still is) thought to be between 1 and 3 percent.[53] The residual trend reported by OTP

Table 10.4   OTP ozone trends, 1969–1986

Latitude	Annual (%)	Winter (%)	Solar correction used (%)	Depletion with 3% variability due to solar	
				Annual (%)	Winter (%)
30–39°N	−1.7 ± 0.7	−2.3 ± 1.3	−0.1 ± 0.6	None	None
40–52°N	−3.0 ± 0.8	−4.7 ± 1.5	−0.8 ± 0.7	−0.8 ± 0.8	−2.5 ± 1.5
53–64°N	−2.3 ± 0.7	−6.2 ± 1.5	−1.8 ± 0.6	−1.1 ± 0.7	−5.0 ± 1.5

is 1.7–3.0 percent. These numbers are of similar magnitude. How then do we distinguish ozone depletion from natural variability?

NATURAL VARIABILITY

There are huge swings in the concentration of ozone over many time scales. As the NRC characterized all previous observations of the ozone layer in 1976, "The daily variations can be of the order of 50 percent; the seasonal mean variation may be 30 percent. The variability increases greatly with latitude and in higher latitudes varies with season."[54] Why does the concentration of ozone vary so widely?

The fundamental factor that determines the amount of ozone in the stratosphere is the amount of solar ultraviolet radiation that reaches it. When less ultraviolet radiation of crucial wavelengths (205 nm)[55] strikes the earth, less ozone is produced. The amount of UV radiation at these wavelengths varies over a number of time scales, but the only one that matters in the ozone trends is the eleven-year solar cycle. This is the amount of time between solar maxima, when sunspot activity is greatest, and minima. Astronomers have observed that as the number of sunspots increases, the intensity of radiation that reaches the earth (in most wavelength bands) increases, and that when the sunspot number decreases, so does the radiation intensity.

THE ELEVEN-YEAR CYCLE

Ideally, predictions of the effect of the variability of solar ultraviolet intensity over the eleven-year cycle would be based on well-understood and precise measurements of that variability. In the past few years satellite measurements have shown that there is much more variability in UV radiation than in the rest of the spectrum,[56] but the analysis of these measurements was not available at the time of the OTP report.

Satellite measurements of ultraviolet irradiance have been made since 1968, with the Nimbus 3 and subsequent satellites. Until recently these data were generally of dubious quality. It was difficult to deduce quantitative variation because of uncertainties in calibration, harshness of operating conditions, and so on.[57] The precision of the measurements made in the 1970s and early 1980s was no better than ±15 percent.[58] There was also a limited amount of data. The paucity and low quality of the data on solar irradiance forced astronomers to rely on inferred values of variations in ultraviolet irradiance, in other words, to use a proxy.

In choosing a proxy a scientist naturally seeks a phenomenon that is as similar to the one of interest as possible, and in this case variations in another wavelength were observed. The wavelength that has traditionally been adopted as the proxy for the ultraviolet band is the 10.7 centimeter (cm) band.[59] One study showed a .97 correlation coefficient between the variations in 10.7 cm solar activity and global mean total ozone.[60] Such a high correlation (1.00 is perfect) gave scientists confidence that the two variations were causally linked.

Scientists using this proxy found that ozone varied by 2–3 percent from solar maximum to solar minimum as a result of variability in ultraviolet intensity.[61] Other studies of ozone data came to similar conclusions.[62] This variability also agrees with predictions from contemporary models based on the available data and understanding at the time.[63] The 2–3 percent variability was the best estimate available at the time of the Montreal Protocol.[64]

IMPLICATIONS OF THE CYCLE FOR THE CONCLUSIONS OF THE OZONE TRENDS PANEL

The solar maximum of 1969 would be the time of a natural maximum for total ozone,[65] as the solar minimum of the mid-1980s would be the time of a natural minimum. Thus, the time period chosen by the OTP for its analysis, 1969–1986, maximized the potential effect that solar variability might have on the results.

Larger (but realistic) solar UV corrections than the OTP used eliminate some of the trends the panel found. The winter trends for the northernmost two sections remain substantial. Solar variability alone, then, may account for much of the observed changes in the total ozone between 1969 and 1986. The influence of variations in the sun's output stands as a clear alternate explanation, aside from chlorine chemistry, for some of the apparent long-term trends in total column ozone.

THE PANEL'S DATA

The global trend attributed to the OTP report is based on the data from Dobson spectrophotometers in the region from 30° to 64°N, because that is the only region where there are enough instruments to allow for the discernment of regional scale trends. However, "it is obvious from the distribution of sites in the present network that conventional ground-based observations cannot define the global total ozone, principally because of gaps in the network over the oceans but also because of limitations imposed by the instruments and the weather. No practicable extension of the network will change this conclusion. The only realistic approach will be to use satellites."[66] Why would OTP use data that could not yield an accurate picture of global ozone to determine global ozone trends? The reason is quite simple: there were no satellite data for the period being analyzed.

The OTP patterns that remain even with larger solar corrections occur in the winter. A small decrease in the northern mid-latitudes may be compensated for by a small increase in a region not covered by the Dobson network. Using the Dobson data alone one may plausibly suggest that some of the missing ozone may be redistributed rather than destroyed. Apparently "the circulation of the middle atmosphere exhibits strong interannual variability in the winter."[67] In fact, in discussing wintertime dips in ozone over Europe and North America in 1982–1983 and over the Far East in 1985, R. Bojkov and his colleagues state that "it is believed that these are mostly circulationally induced fluctuations."[68] But satellite data strongly suggest that circulation is not responsible for changes observed since 1979 in the Northern Hemisphere,[69] and there is no evidence of a radical change in circulation between 1969 and 1979. Thus the observed changes, if real, are probably due to destruction rather than redistribution.

The Bojkov group is well aware that determining long-term trends from data with periodic swings of large amplitude is not an exact science. They remind their readers that "over most of the stations considered ozone varies as much as 25–30% within a single year."[70] The large annual variations, however, show that small long-term decreases will not introduce us to unprecedentedly or even unusually high ultraviolet exposure. These variations are perhaps the best argument for skepticism about the reports by the OTP and by anyone else on the state of ozone depletion and its likely effects at this time.

MORE EVIDENCE AGAINST THE PANEL? DECREASING UV-B

A long-term study of the amount of potentially dangerous, erythemally weighted, radiation reaching the United States and Europe parallels the Dobson measurements of ozone. Erythemal weighting of ultraviolet assesses the intensity of radiation by measuring how effective a given wavelength is at damaging skin. It is not an absolute measure of the amount of ultraviolet that is reaching the earth but is rather an index of how much damage is likely to be done by whatever ultraviolet gets through. Analysis indicates that the amount of erythemally weighted ultraviolet reaching the ground at the one European and nine U.S. stations has decreased over the past decade and a half, roughly the same period as that of the OTP's reported decrease in ozone.[71] Thus increased ultraviolet exposure has not come to pass; in fact, the opposite has occurred. The downward trend in ultraviolet measurements does not contradict the observed ozone trends, but it does show that such trends do not necessarily lead to increased damaging ultraviolet.[72] One explanation for the decrease in ultraviolet at the ground is that as stratospheric ozone has been decreasing, tropospheric ozone has been increasing,[73] and low-level ozone is more efficient than stratospheric ozone at absorbing ultraviolet.[74] Thus, even if the total ozone has decreased, ultraviolet exposure over much of the Northern Hemisphere may have decreased.[75] Some scientists who model ultraviolet exposure predict that it has gone up globally.[76] They claim that measurements showing its decrease are the product of local pollution rather than a global trend.[77] Unfortunately, there is no global ultraviolet monitoring network to settle the dispute.

NEW DATA ANALYSES SINCE THE LONDON REVISIONS

In the years since diplomats wrote the London revisions, which called for a ban of CFC production by 2000, a great deal of scientific evidence bearing on the appropriateness of that ban has been amassed. We now have an analysis of all Dobson station data over three full solar cycles[78] and a full solar cycle of satellite data.[79] One pair of scientists has discovered that there may be a fundamental flaw in the ground-based measurements of ozone as they have been made to date.[80]

Analysis of the total ozone data over three cycles indicates that there has been significant ozone loss in all regions and in all seasons over the last two sunspot cycles (twenty-one years or more), as shown in Tables 10.5 and 10.6.

These trends are substantial. Since the analysis includes data from solar max-

Table 10.5   Global trends, 1970–1991

Period	Trend
Year round	−1.5% per decade
December–March	−2.1% per decade
May–August	−1.2% per decade
September–November	−0.8% per decade

Source: Richard Stolarski et al., "Measured Trends in Stratospheric Ozone," *Science,* 256 (Apr. 17, 1992): 344. Copyright 1992 by the AAAS.

Table 10.6   Regional trends

Region	Year round	Dec.–Mar.	May–Aug.	Sep.–Nov.
North America	−2.1 ± 0.3%	−3.2 ± 0.4%	−1.7 ± 0.4%	−1.1 ± 0.3%
Europe	−1.8 ± 0.3%	−2.9 ± 0.4%	−1.2 ± 0.3%	−1.2 ± 0.3%
Far East	−1.2 ± 0.4%	−1.8 ± 0.5%	−0.9 ± 0.5%	−0.4 ± 0.4%
26–64°N	−1.8 ± 0.2%	−2.7 ± 0.3%	−1.3 ± 0.2%	−1.0 ± 0.2%

Source: Stolarski et al., "Measured Trends in Stratospheric Ozone," p. 346.

imum to solar maximum, the investigators have minimized the chances that solar variability plays a role in the trend. Also, the data from the last solar cycle are supported by global data collected by satellites. Satellite data for the full solar cycle indicate that there has been a loss of 3 percent of all ozone between 65°N and 65°S between November 1978 and May 1990.[81] The satellite data indicate that the losses in the last twelve years have not been due to redistribution.[82] Because the loss is now observed in all seasons, we cannot simply attribute it to some unusual phenomenon in the winter.

Most of the loss has taken place in the lower stratosphere, where most of the ozone resides. There have also been substantial losses in the upper stratosphere, while tropospheric ozone has increased.[83] Losses in the stratosphere have supposedly accelerated to 5–15 percent per decade. Such a trend, if CFC induced and long term (or linear), would pose a serious problem.[84] But it was measured over only one solar cycle, over the last twelve years, which is not long enough to establish a linear trend. The inadequacy of one-cycle analyses is illustrated by the contradiction between the finding at one Dobson station of a significant downward trend over the past cycle and the results of a two-cycle analysis that shows no trend whatsoever.[85] Also, the concentrations were actually on the rise after 1986. Only time will tell if it is the beginning of a significant long-term loss of ozone.

A recent study of the Dobson spectrophotometer data at Uccle, Belgium, indicates that the trend in total ozone there may not be anywhere near what the OTP and others have said. An earlier study found a year-round trend of -3.53 ± 0.86 percent per decade.[86] De Muer and H. De Backer's new analysis indicates a trend of $+0.31 \pm 0.51$ percent.[87] In deriving this new trend the authors raise questions about the revision of the data and the measurements themselves. De Muer and De Backer checked the revisions done by the OTP, because "the question about the value of such an approximate revision of Dobson data is essential as to the reliability of the conclusions reached."[88] Their data revision eliminated much of the trend found by the OTP. The remainder of the trend was apparently an artifact of the measurement rather than a change in ozone concentrations. De Muer and De Backer corrected for changes in sulfur dioxide concentration and in so doing found that they were left with no trend.

Thirty years ago the inventor of the Dobson spectrophotometer figured out that sulfur dioxide could interfere with ozone measurements.[89] However, scientists had never corrected ozone data for SO_2 interference before last year, because they had believed that concentrations were too low to affect appreciably the measurements.[90] This is not true at all sites. Sulfur dioxide absorbs some ultraviolet radiation at the same wavelengths as ozone. In clean air there is not enough SO_2 to absorb a detectable amount of ultraviolet; in polluted air masses there is enough. Because satellite data are not susceptible to interference from SO_2, trends found with that data have not been called into question by

De Muer and De Backer's discovery.[91] It is interesting to note that Uccle is the most severely depleted site in the OTP and Bojkov group studies,[92] while the new analysis indicates there has been no depletion there at all.

The most up-to-date models as of 1991 indicated that continued CFC use at 50 percent of 1986 levels would lead to 2 percent ozone loss above that expected with a complete ban as of the year 2000. The available data indicate that global mean ozone may have decreased over the last twenty years, mainly in the last decade. The longer trend may not be real because the data may not be any good. Whether or not the decline is long term or short also remains an open question. The same cannot be said of the formation of the ozone hole. Scientists have shown that chlorine from CFCs is responsible for the hole, though they did not always know that for certain.

Periodic Antarctic Ozone Depletion: The "Ozone Hole"

The largest effect attributed to CFCs is the Antarctic "ozone hole." This inaccurate term describes a dip in total column ozone content over Antarctica during the first few weeks of the austral spring. There is never a time when there is no ozone in the column of air over any point in Antarctica or anywhere else. In that sense there is no hole in the ozone layer. Indeed, strictly speaking, there is no layer either.

THE IMPORTANCE OF INSTRUMENTATION

The periodic Antarctic ozone dip first came to the public's attention when J. C. Farman and associates observed in a letter to *Nature* that "the spring values of total O_3 in Antarctica have now fallen considerably."[93] The data they presented revealed that the total column ozone content over the British Antarctic Survey station in Halley Bay "was ~30% lower in the Antarctic spring seasons (October) of 1980–84 than in the springs of 1957–73."[94]

The drop in ozone content was observed only in the spring, in one place in Antarctica, and was much greater than had been observed at any other place over any time period. All these factors made the ozone dip a topic of interest for the atmospheric science community. It was an anomaly and anomalies cry out for explanation. No model in existence had predicted such a change. Even those scientists who were most convinced that global ozone was being depleted had not predicted a dramatic change in the ozone content of any part of the stratosphere in this century. The radicalism of Farman and his team's claim made it necessary for them to provide evidence that they were not working with bad data. They carefully documented their techniques and showed that their data were trustworthy.

Important questions remained about the phenomenon they had discovered, especially: (1) was the existence of the periodic ozone dip corroborated by observations made at other stations or by satellite? (2) was it a new and unique phenomenon, as it appeared to be?

In answer to the question of corroboration, it turns out that Farman's group was not the first to report a dip in the ozone column in the austral spring. Chubachi and colleagues had reported anomalously low column ozone measurements from Syowa Station in 1984 in a somewhat obscure publication and had been largely ignored.[95] The Halley Bay measurements were thus corroborated by at least one other station. But the satellite data, which offered the best coverage and most modern modes of measurement, did not corroborate their findings. The story of the satellite data, which in the end did confirm the ground-based observers, is a telling anecdote of the pitfalls of high technology.

There have been a handful of satellites in orbit around the earth that are or were equipped to measure atmospheric quantities such as the depth of the ozone column in the atmosphere. These satellites beam back the data to ground stations, where they are fed into computers. Because there are far more data coming in from satellites than scientists could ever hope to analyze themselves, the computers that store the data are programmed to filter and compile it. The compiling process is meant to put the data, which start off as enormous streams of numbers, into tables, graphs, and pictures or maps that give the scientists representations of the properties they are interested in studying. The filtering process is meant to eliminate data that arise from errors in measurement due to equipment failure or from conditions that render the observational equipment temporarily incapable of accurate measurement. Arbitrary limits, based on the experience of the scientists who program the computer, are set on various parameters in the computer, so that any datum falling outside those limits is either discarded or replaced with some "more reasonable" value.

In the entire history of ground-based ozone column measurements there had never been a confirmed or accepted measurement of significantly less than 250 Dobson units (DU) in Antarctica. (A Dobson unit is a measure of the thickness of the ozone column. A measurement of 1 DU is equivalent to a layer of ozone 1 micrometer, or 10^{-6} meters thick at standard temperature and pressure between the spectrophotometer and the background, or 2×10^{16} molecules per square centimeter in the column of air above the spectrophotometer. Thus, if all the ozone in the air above a spectrophotometer that was reading 250 DU could be compressed to a column of pure ozone, the column would be 250 millimeters thick.) Most, if not all, of the members of the atmospheric science community believed that any measurement of total ozone column content below 250 DU had to be in error. Thus, it was decided that all readings of less than 250 DU would be filtered out and replaced arbitrarily with the mean value for that point.[96] All of the raw data would be stored in the computer, but no human would ever see that there had been readings of less than the cutoff over any given spot unless the programming were changed.

When Farman's paper came out it was met with skepticism, because the satellite data did not bear out his story. After a time, however, scientists realized that the computer program used to process the data from the satellites automatically dismissed measurements as low as the one he had made, and to remedy this situation the filtering algorithm was removed and the data were reevaluated. Much to the delight of Farman and his team the scientists evaluating the unsanitized satellite data found that since 1979 the satellite had been faithfully observing the formation and the disappearance of this ozone dip every austral spring, while the computer had been just as faithfully deleting any record of the measurement. An added bonus of the removal of the filter was a full record of the extent and variability of the dip over a five-year period. There was already a huge amount of data available just as the phenomenon was being discovered.

The data from the Nimbus 4 satellite, which carried the Backscatter Ultraviolet (BUV) instrument, and from the Nimbus 7 satellite, which carries the Solar Backscatter Ultraviolet (SBUV) instrument and the Total Ozone Mapping Spectrometer (TOMS), were the most important data sets that were reevaluated. The Nimbus 4 provided data for the period from 1970 to 1972. Even without the editing algorithm there was no indication of a yearly spring depletion of ozone over the Antarctic. This finding corroborated Farman's claim that the dip was a new phenomenon, an undisputed but important observation. The Nimbus 7 had been in orbit since November 1978. It provided data for every October from 1979 to the present. The TOMS instrument on the Nimbus 7 was and is able to map total ozone over the entire globe on a daily basis. When the data from Nimbus 7 was retrieved and the filtering algorithm removed, scientists discovered that ozone had decreased over much of Antarctica during each September since 1980. The ozone level had reached a minimum value at some point in mid-October each year. The data indicated that "the low total ozone over Halley Bay . . . [was] part of a larger, elliptically shaped minimum region extending out to ~60–70°S."[97] The dip was not a local phenomenon or a pair of local phenomena; it was an anomaly of regional scale. And it had a definite starting date. "Antarctic total ozone in 1979 appears comparable to that observed in the early 1970s, indicating that most of the total ozone decrease has occurred since 1979, in agreement with the Halley Bay observations."[98]

All of the modes of measurement were now in agreement and there was an increasingly well-defined phenomenon to study. Those who were convinced of the dip's existence, a group that quickly grew to include most of the community of atmospheric scientists, set about the task of determining what had caused this apparent anomaly.

DETERMINING WHAT HAPPENED

The first step in finding a cause for an anomaly is to examine how it differs from expectations. In the case of the Antarctic ozone dip, expectations were derived from decades of ground-based observations and several years of discontinuous satellite observations. The ground-based observations, including spectrophotometric measurements of total column ozone content, as well as measurements obtained from instruments launched to various altitudes in the stratosphere, had provided scientists with a canonical—that is, lawlike, seriously believed to be certain—picture of the variability of ozone concentration with respect to time and space. This picture was derived from data over two decades and was based on consistent features of that data. Although this picture did not allow scientists to predict the precise concentrations of ozone at any particular time and place, there were a number of patterns they observed that came to be thought of as normal.

One of the most important was that the amount of ozone present in late September and early October, at the beginning of the austral spring, when the sun rose after the long winter night, was comparable to the amount present when the sun set in March. When Farman and his groups presented their case for the existence of an anomalous variation in the total ozone content over Halley Bay, one of their key exhibits was a graph that depicted what had come to be thought of as normal values in contrast with the apparently abnormal values the team had found. Ozone content varied from year to year during the period from 1957 to 1973, as it does now and as it always will. This variation was consistent with the variation seen in global ozone due to the variability of the solar ultraviolet output. There were maximum and minimum values that came to be accepted as the expected extremes.

The mean values for every October from 1980 to 1984 were lower than the lowest values recorded for October during the years 1957–1973. If for one year the October values had been lower than any recorded in that earlier period, then Farman and his colleagues should have written off those values as new extremes. Such a new extreme would have been no cause for alarm; it would merely have given them a greater range of expected variability than they had had previously. This was probably the authors' initial response, given that they took five years to publish their data. It was only the fact that the extremely low values recurred over a five-year period that led to the investigation of the phenomenon as an anomaly of interest. All of the observations for the springs from 1980 to 1984 were lower than could be explained by even the most extreme solar variability models. This pattern of extreme depletion has continued at least through 1992.[99]

Unlike the trend reported by the Ozone Trends Panel, this change was far greater than any existing explanation of the processes involved could account for. There had apparently been a fundamental change in the processes that determined the thickness of the ozone column over Antarctica during the austral spring. The question then was not whether or not ozone had been depleted but how it had been depleted.

Coping with Conflicting Hypotheses

In the year following the discovery of the dip a number of competing explanations were offered. The November 1986 supplement to *Geophysical Research Letters* contains forty-five letters that we take as representing the state of knowledge.

There are three basic categories of explanations for atmospheric phenomena such as sudden regional changes in composition: changes in the chemistry of the region, changes in the dynamics or circulation of the region, and some hybrid of the first two. The odd oxygen, O and O_3, that was over the Antarctic before the beginning of the austral spring had to go somewhere. Either all of the missing ozone had moved out of the region, all of the missing ozone had been broken up and its constituent parts had become parts of other compounds, or some of the oxygen atoms had moved out.

Farman's group believed that "the present-day atmosphere differs most prominently from that of previous decades in the higher concentrations of halocarbons."[100] Although blaming halocarbons now appears justified, the link was not always so clear. At that time the scientific community was divided on the question of what caused the ozone dip over Antarctica. There were two camps: those who held changes in circulation accountable and those who blamed CFCs. All of these scientists had the same data to work with and the same basic understanding of the way in which the atmosphere worked, yet they disagreed as to what caused the dip. The reason for their disagreement was that the right kind of data was not available to make a compelling argument for either one theory or the other.

THE DYNAMIC EXPLANATION: EVIDENCE AS OF 1986

According to the dynamic hypothesis, ozone-poor air from the troposphere welled up into the polar stratosphere driving upward the ozone-rich air that had been in the lower polar stratosphere. As the upwelling continued, the displaced ozone-rich air was forced away from the polar regions and into the temperate zones.[101] This hypothesis explained the apparent conservation of ozone between 44°S and the Pole and the decrease in the thickness of the ozone column over the Antarctic region. However, such an upwelling would be anomalous and therefore required some sort of change in underlying conditions.

The ozone in the lower stratosphere normally forms in the equatorial regions, rises there, travels poleward, and sinks at the poles. The action is like that of a conveyor belt. The cycle is completed when air from the troposphere, which is poor in ozone, rises into the equatorial stratosphere to replace the air that travels toward the poles.[102] If the circulation was reversed at the Antarctic, then the situation would be reversed: instead of the polar region receiving ozone from ozone-rich air, the Pole would become a net exporter of ozone for the duration of the reversed circulation. A plausible mechanism whereby a reverse cell could form that would fit the data on the ozone distribution in the Southern Hemisphere was offered in 1986.[103] Despite the apparent power of this redistribution argument, it lacked the one sort of data that would validate it—data that showed that the hypothetical transport was taking place.

THE CHEMICAL EXPLANATION: EVIDENCE AS OF 1986

In competition with this dynamic model, which was supported by a number of scientists,[104] was a handful of chemical explanations. Traditional photochemical models could not explain the Antarctic dip. The reactions involved in the old models were too slow and involved too many competing reactions to allow for the rapid decrease in ozone that was observed when the sun came up.[105] In the lower stratosphere, below 25 km, chlorine radicals are efficiently tied up in the reservoir species, chlorine nitrate and hydrochloric acid, through chemical reactions with nitrogen oxides and water, respectively. What was needed was a process resulting in the liberation of a large amount of chlorine. To allow for a much more rapid catalytic destruction of ozone by chlorine chemistry several scientists suggested heterogeneous processes[106] through which the chlorine could be freed to interact with the ozone without interference from nitrogen oxides and water: nitric acid crystals in clouds provide a surface on which the reactions that free the chlorine from the reservoir species—reactions that had only been observed in the laboratory as of 1986— could take place.[107] Over the six years since November 1986 and the landmark *Geophysical Research Letters* supplement, a compelling case for the roles of both dynamics and chlorine, and therefore CFCs, in the formation of the ozone dip has been built.

Guilty by Correlation and Association

The case against chlorine relies on dynamic features of the Antarctic winter and spring to set the stage for depletion. For chlorine to cause the ozone dip, nitrogen oxides have to be removed and chlorine has to be liberated. Apparently this process occurs in a dynamically isolated region.

Chlorine oxide (ClO) levels in the region of greatest depletion are two orders of magnitude higher than normal, which suggests that the depletion area is isolated from the surrounding area, since the ClO would otherwise rapidly diffuse into the surrounding air.[108] The concentration gradient is too high at that boundary for simple mixing to be taking place.[109]

If chlorine chemistry is responsible for the removal of most of the ozone in the ozone dip, then there should be an anticorrelation between the concentration of ClO and the concentration of ozone in the vortex. This is in fact the case: "Although the presence of anticorrelation between two variables does not in itself prove a causal relation, the data sequence tracking the evolution of O_3 depletion (in a contained region isolated from the atmosphere as a whole) from an initial condition in which there has been a marginal loss of ozone to a condition in which ozone has dropped precipitously spatially coincident with amplified levels of ClO is strong evidence of such a relation."[110] Scientists have found more than a general trend in which ozone and ClO decrease along a southward path. Rather, the jogs in the ClO concentration plot with respect to latitude match almost precisely the jogs in the other direction in the ozone plot with respect to latitude.[111]

The chlorine is freed by the heterogeneous process, when chlorine nitrate reacts with frozen nitric acid: the nitrogen is neatly tied up in frozen nitric acid and the chlorine is released. The chlorine gas (Cl_2) photodissociates, yielding two free chlorine atoms. Then the catalytic destruction of ozone by chlorine can take place without having a significant fraction of the chlorine tied up in reservoir species.

There are a few salient features that need to be examined in the argument implicating chlorine in the destruction of ozone and the formation of the ozone hole (we use the popular term here in the context of alarm). First of all, the inner polar vortex is defined by the region that is cold enough to freeze out nitrogen-oxygen compounds. Without that cold, there cannot be substantial loss of ozone. The only places sufficiently cold are over the two poles during the winter and early spring. Second, the outer vortex is characterized by strong barriers to mixing that prevent the influx of warm air. Thus the region can stay cold enough and sufficiently free of nitrogen to allow the rapid destruction of ozone by chlorine and chlorine monoxide radicals for months. There is no known mechanism for isolating the equatorial regions to the extent necessary, so there does not appear to be any reason to be concerned about the spread of this particular mechanism.

As for the polar stratosphere getting colder and colder and the chemically perturbed region growing, the size of the inner vortex appears to be permanently limited to the extent of the outer vortex. It cannot get any larger, or the air north of the boundary of the vortex (in the Southern Hemisphere) will continue to be heated by the sun when the sun goes down over the poles. Thus, the very thing that allows the chemically perturbed region to exist—its isolation—is also the very thing that prevents it from growing ad infinitum. Once the inner vortex, the chemically perturbed region, occupies the whole of the outer vortex, which is bound by winds, it cannot get any larger. Unless the wind fields, a phenomenon not controlled by the chemistry of the stratosphere and therefore not vulnerable to human tampering, were for some mysterious reason to change, the dip will continue to be confined to the polar regions.

In fact, the largest the dip can get may have already been established. In 1987 the Antarctic vortex was nearly in radiative equilibrium. There was virtually no transport of heat across the boundary of the vortex. There was also virtually no mixing between the air inside and outside the vortex. As the polar vortex could not have become much colder, the inner vortex—the region colder than

−78°C—could not have gotten much bigger. "It is therefore doubtful that the areal extent of the pool of cold air in the vortex could increase beyond that observed in 1987 through purely dynamical mechanisms."[112] In fact, in four of the last five years ozone depletions have reached that scope, with a dip of 14 million square kilometers each of those years.[113] The dip is not growing and, if scientists understand it correctly, cannot grow.

Yet the dip has been quite deep. Nearly all of the ozone from 14 to 22 km has been eliminated over the Antarctic continent in October of each of the four years with large dips. Typical October ozone levels over Antarctica were around 300 DU before the hole era; a value of 108 DU was measured in October 1991.[114] That is a depletion of nearly two-thirds of the ozone at that point in time and space. Fortunately, the dip exists almost entirely over unpopulated areas. Also, the perturbed air generally does not bring low enough ozone levels over these areas to make for higher ultraviolet exposure, because the sun is lower in the sky. The dip therefore poses little or no health threat to humans or their livestock and crops.

IS THE CASE AGAINST CFCS THE SAME AS THAT AGAINST CHLORINE?

Chlorine is responsible for the chemical component of the development of the ozone hole. Some claim that chlorine's role does not equal that of CFCs, but approximately 75 percent of the standing crop of chlorine in the stratosphere is from CFCs.[115] This percentage will increase as time goes on. It is therefore natural to blame CFCs for any effect that chlorine in the stratosphere has on the ozone in the stratosphere. However, in addition to the presence of chlorine, there are a number of other necessary conditions for the formation of the ozone hole. The temperature must be cold enough for the formation of the crystals. The vortex must stay coherent long enough for the chlorine and chlorine oxide to do their damage. Ozone depletion is tied to the strength and the temperature of the inner and outer polar vortex. The lack of an observed dip from 1957 to 1979 seems to implicate CFCs, but by the same token, it indicates that the conditions for the formation of the hole are not always present. Speculation in 1992 that an Arctic dip would finally emerge proved wrong, because the northern polar region warmed up, as usual, before there could be significant depletion.[116]

Would the chlorine that is present naturally have been sufficient to generate the dip?[117] It turns out that chlorine is crucial to the onset of the ozone depletion conditions: the rate of the primary chemical process involved in destroying ozone varies by the square of the concentration of chlorine.[118] We humans have quadrupled that concentration through CFC emissions, and consequently the rate of ozone destruction has increased sixteen fold. There might be years when the natural chlorine would cause further slight dips in the ozone in Antarctica, but the basic dip is clearly a product of CFC emissions. "The ozone hole has been associated with a fundamental change not just in the magnitude of

October ozone abundances, but in the character of the ozone seasonal cycle in Antarctica."[119]

Two radically opposed positions—that CFCs were responsible and thus the dip would grow way beyond its current bounds, and that the dip was merely a natural phenomenon and thus not a cause for alarm—were equally reasonable at first. As long as there was no preponderant evidence regarding the cause of the dip, there was reason to fear that it might spread, an apocalyptic symbol of the potential destructive force of humankind in tampering with nature's delicate systems. Now that we know better, now that we know that the dip is restricted to the polar regions, there is no longer reason to fear its consumption of the earth. The dip is over regions that are uninhabited by man or beast, and very few other living things are exposed to the ultraviolet radiation let through. That our chemicals could have such a profound effect on such a large region is a sobering thought, not a reason to panic. Nor is the existence of the ozone depletion necessarily a reason to ban CFCs entirely.

WHAT THE OZONE DIP TEACHES US ABOUT GLOBAL OZONE

Since Mount Pinatubo (in the Philippines) erupted in June 1991 global ozone has been unusually low.[120] The search for a cause of this drop of 2 to 3 percent below previous record lows has led to a new and deeper understanding of the processes controlling global ozone levels.

A number of scientists, in modeling the effects of Mount Pinatubo, suggested that aerosols—tiny particles or droplets suspended in the air—injected into the stratosphere by a violent eruption like Pinatubo's might provide sites for heterogeneous reactions similar in kind to those that take place on ice crystals in the polar region.[121] Laboratory experiments showed that such reactions could take place and indicated how fast they might take place.[122] Measurements made from airplanes revealed how much surface area the aerosols provided, how much NO_x was removed by the aerosols, and how much chlorine was freed.[123] These measurements revealed that the chlorine freed by the heterogeneous reactions taking place on the sulfate aerosols was most likely responsible for the record low ozone levels. They also revealed that "the observed changes in ClO in March [1992] are an upper limit for changes induced by increased aerosol SA [surface area]."[124] Beyond a certain point, an increase in aerosols does not result in more freed ClO. Thus, no more ozone could be destroyed by this process than was destroyed this time. There cannot be a dip comparable to the springtime dip over Antarctica as a result of this eruption. The drop that was recorded will be transient as the excess aerosols will be naturally removed from the stratosphere over the next couple of years.

There are always sulfate aerosols in the stratosphere. When scientists went back to their models and added in the effects of the normal aerosol levels on the chemistry of the ozone layer, they discovered something amazing—the whole picture of which processes destroyed what amount of ozone changed. NO_x was reduced to a limited role as models now indicated that active or odd hydrogen (HO_x) was responsible for 50 to 60 percent of ozone destruction.[125] Given that methane is the primary source of hydrogen in the stratosphere, it may turn out that changes in the rate of methane production and destruction in the lower atmosphere play a crucial role in the ozone layer. Nothing had changed in the stratosphere. Only our assumptions about what reactions were important and what amounts of the reactants were present.

The new estimates of the contribution of each catalytic cycle may be as shaky as the old ones. Our understanding of the stratosphere is only as good as the data we have available and the models we use to interpret that data. At this time there is little data on how much HO_x, ClO_x, and NO_x there is in the lower atmosphere. As scientists take new measurements there are sure to be more discoveries. The importance of chlorine has not been changed by this finding. The quest for understanding continues, but one thing is clear: CFCs are not the only thing affecting the ozone layer.

THE INDIRECT HEALTH THREAT OF CFC USE

The ozone-depleting properties of CFCs are touted as a serious health threat to many forms of life, including humans.[126] Melanomas are the only life-threatening health problem associated with the increased ambient ultraviolet radiation that is expected to arise. The EPA predicts that there will be a 10 percent increase in the incidence of melanoma by the year 2050 due to increased ultraviolet exposure from ozone depletion.[127] All we can say is that the relationship between ozone concentrations and ground-level ultraviolet exposure is not predictive at the sites where UV-B exposure has been measured. In other words, the connection between ozone depletion and UV-B increase has not been confirmed.

The EPA prediction is based on an assumption that there will be a 5 percent decrease in the amount of ozone globally by the year 2050 and that depletion will be uniform with respect to latitude and season. However, whatever depletion occurs is likely to be greater at higher latitudes and during winter—where and when people tend to stay indoors and to cover themselves when outdoors—than it will be at lower latitudes and during summer.

Although the EPA assumes that a 1 percent increase in UV will translate into a 1 percent increase in melanoma incidence, that incidence is not a simple function of UV-B exposure. If the ambient UV-B radiation level were the sole factor in determining melanoma incidence, then rates for that cancer would be higher nearer the equator because of the higher UV-B levels there. Epidemiological studies show that melanoma incidence actually decreases along a southward path in Europe.[128] Accordingly there is no simple, linear correlation between potential exposure to UV-B, as measured by ambient UV-B, and death as a result of melanoma.

Melanomas occur primarily in Caucasian people. Epidemiological work indicates that there is a strong positive correlation between the ease with which a person gets a sunburn (or, conversely, the difficulty of getting a suntan) and his or her chances of developing a melanoma.[129] Northern Europeans may develop melanomas more than southerners even though they get less UV-B, because they are fairer skinned and many of them take vacations in the south.[130] If the vacationing hypothesis is correct, then lifestyle would be much more important than ambient UV-B in determining melanoma incidence; the historical record confirms the importance of lifestyle.

If ambient UV-B determines the incidence of melanoma, then scientists should not have found any significant change in that incidence over the last several decades and certainly not before the 1960s. They did find, however, a 700 percent increase in melanomas in Connecticut since they started keeping records in 1935.[131] This rate is seventy times the increase the EPA predicts over the next sixty years due to ozone depletion. Jean-Pierre Cesarini states that "the incidence of melanoma doubles every 10 years, since 1930, in countries with an above average economic status."[132] This increase is very likely due to a change in lifestyle, as it has become fashionable to sunbathe and to spend a lot of time outdoors in shorts and short-sleeve tops. With such a rapid increase in the incidence of melanoma already underway, it would be impossible to detect the small increase that the EPA predicts. Furthermore, any reversed changes in lifestyle—toward greater protection against the sun by wearing more clothing, using sunblock, or spending less time in direct sunlight—would overwhelm the small increase in melanoma, provided the increase in UV-B is not enormous.

In general, atmospheric ozone is thickest at the poles and thinnest at the equator. Since the sun is also more directly overhead in the equatorial regions, the radiation has to pass through less atmosphere on its way to the earth. As a result of the structure of the atmospheric system, approximately fifty times as much UV-B reaches the ground in the equatorial regions as in the polar regions. Furthermore, there is less ozone over a site at 5,000 feet in elevation than there is over one at sea level at the same latitude. Although the gradient from pole to equator is not constant, a 1 percent decrease in the amount of ozone over any given place would be roughly equivalent to moving six miles toward the equator or 150 feet up in elevation.[133] We have never heard of someone deciding not to move from one place in the Northern Hemisphere to another further south because of concern about ultraviolet radiation. All in all, given the lack of causal linkage between UV-B and melanoma, the timing of the increased UV-B, and the current distribution of UV-B exposure, the threat to humans from CFC use, provided that use does not increase dramatically, is quite small.

THE THREAT TO OTHER LIVING CREATURES

Many agricultural plants are adversely affected by exposure to too much UV-B. "Ultraviolet-B radiation (280–320 nm), especially at high levels, could cause a variety of plant responses. These may be manifested either in subtle or dramatic changes in total biomass yield, or in damage to aerial plant parts."[134] Cucumber seedling growth and photosynthetic activity in radishes are impaired by increased exposure to UV-B.[135] Soybean production under conditions equivalent to a 16 percent reduction in ozone is severely impaired relative to production under current conditions.[136] There are myriad other examples. But these results are from laboratory-like conditions that do not allow other factors, such as the rest of the radiation spectrum, to vary as they would in nature. When this variation is allowed, plants typically respond with greater resistance to UV-B.[137] Nevertheless, at levels of UV-B equivalent to those produced by the reductions in ozone thickness projected for unrestrained growth of CFC emissions—that is, depletions on the order of 10 to 15 percent for the temperate regions of the earth—important plants like soybeans suffer detrimental effects. Wild animals and plants would most likely be adversely affected by large increases in the incident UV-B as well.

Given some restraints in emissions, however, the small increases expected in ultraviolet radiance should have little effect. Even if the increased exposure is larger than expected, common sense tells us that it will likely do little damage to agriculture or to wild animals, since the effect would be the same as moving closer to the equator.

Conclusions from the Scientific Evidence on the Threat of CFCs

Unrestrained growth in CFC emissions is clearly unwise. Over the years theories about the effects of continued emissions at controlled rates have varied in their predictions. The health threats that are expected to arise from ozone depletion and by inference from CFC emissions are ill quantified. The real knowledge gained, including its limitations, should be the grounds for regulation of CFC production and emission. Has this knowledge been used?

EARLY REGULATIONS

The United States banned the use of CFCs as propellants in aerosol spray cans in 1978. The media played an important role in the debate over this ban, giving widespread coverage to the battle between those scientists who believed there was a threat and the chemical companies who insisted there was no evidence of one. There was little if any discussion of the gaps in the models except by the chemical companies. This media attention was so effective that CFC use declined from peak levels in 1974 before the ban was even put in place four years later.

The U.S. ban was the first regulatory action taken as a result of the model predictions that CFCs would deplete the ozone layer. At the time no one claimed to have observed a thinning of the ozone layer, yet the models indicated that a significant depletion would result if CFCs were emitted at rates equal to or greater than the rates at which they were being emitted. These models constituted the best knowledge available at the time, and there were readily available alternative propellants, such as air and propane, to replace the CFCs. The hardship incurred in order to avert a potentially large threat seemed to be worth it.

At the time of the U.S. ban, the largest single application for CFCs was aerosol propellants. Thus the ban by the United States (and by many other countries in the years following) had a profound impact on total emissions. In fact, total CFC use was lower from 1978 to 1985 than it had been in 1974, the reference year for many of the predictions of the long-term effect of CFCs on the ozone layer. But the use of CFCs in other applications grew rapidly both in this country and abroad. By the mid-1980s it was clear that unilateral action by the United States would not be enough if there was to be a continued reduction in use. If CFC use was deemed a health threat that had to be met to protect the public, then more significant controls would have to be implemented.

THE MONTREAL PROTOCOL

The Montreal Protocol on Substances That Deplete the Ozone Layer is far and away the most important piece of regulatory legislation on any environmental issue. Initiated in September 1987, the Montreal Protocol is essentially an environmental treaty between the member states of the United Nations—over sixty nations had signed by mid-1990—that constitutes a commitment to control the use of CFCs 11, 12, 113, 114, and 115 as well as halons 1211, 1301, and 2402.

The parties agreed to limit their annual CFC consumption to their 1986 levels within six months, and to limit their use of halons to 1986 levels within three years. They agreed to limit annual CFC production and consumption to 80 percent of 1986 levels in any given country after June 30, 1993, and to reduce annual CFC production and consumption to 50 percent of 1986 levels after June 30, 1998. All of the countries also consented to exchange information on recycling, recovery, and destruction techniques for the controlled substances and on possible alternatives to those substances. Further provisions guarantee access to alternative technologies and substitute products for those countries that had not already developed the CFC technology for themselves, so that they would not be at an economic disadvantage.[138]

At the time the Montreal Protocol was written, no depletion in ozone had been observed, but there was a consensus among all those who model the future effects of CFCs that continued growth in the emission rate of CFCs (and halons) would lead to a significant thinning of the ozone layer. Scientists knew that a large decrease in the thickness of the layer would be unacceptable as too much ultraviolet radiation would impair agricultural activity and damage public health.

All of the earlier sections of this chapter that addressed the impact of CFCs on ozone apply as well to halons (bromofluorocarbons), which are used to put out fires. All halogens—chlorine, bromine, iodine—react catalytically with ozone. Halons contribute a growing percentage of the halogens in the stratosphere. Halons are brominated rather than chlorinated, and bromine is approximately ten times as efficient in destroying ozone as chlorine. Halon is used much less than CFCs, so the public eye is primarily on the latter. All of the models we have cited have addressed only CFCs, but if the use of halons continues to grow, they could eventually become more important than CFCs.

Although their selection was entirely arbitrary, ozone levels in 1986, which were not distinguishable from pre-CFC levels, were as good as any others as a choice for the ideal limit. All of the models available then indicated that a freeze at these levels of use would lead to a small but acceptable decrease in atmospheric levels on the order of 0 to 2 percent. The authors of the Montreal Protocol expected many countries to avoid signing or complying with the treaty, recognizing that the cuts might prove to be an unacceptable hardship. So they added the following caveat: "In order to satisfy the basic domestic needs of the Parties . . . and for the purposes of industrial rationalization between Parties, its calculated level of production may exceed that limit [50 percent of 1986 production] by up to fifteen per cent of its calculated level of production in 1986."[139] This provision is a strong point in the protocol. It allows for flexibility in meeting the needs of growing economies in the future.

Another of the strengths of the Montreal Protocol, or more properly the Vienna Convention, of which the Montreal Protocol is a product, is that it calls for the regular review of the scientific evidence in order to revise the regulations. It was this review process that produced the London revisions.

THE LONDON REVISIONS

The London revisions extend the scope of the regulations to cover a large number of CFCs and halons that were not previously marked as controlled substances. The revisions also offer potential alternatives to the controlled substances. Finally, and most significant, the diplomats who wrote the London revisions decided to ban the use of all of the controlled CFCs and halons by the year 2000.

The extension of the regulations makes perfect sense. The substances that have been added to the list all have atmospheric lifetimes comparable to those of the previously controlled substances. These newly controlled compounds also bear about the same load of halogens (chlorine in CFCs and bromine in halons) to the stratosphere as those covered by the Montreal Protocol. Thus they are expected to have equivalent ozone depletion potential (ODP), and their control is as justified as the control of the first batch.

The substitutes listed in the London revisions are all hydrochlorofluorocarbons (HCFCs). They and their companion alternative, hydrofluorocarbons (HFCs), are said to be better for the ozone layer. Switching from CFCs to

HCFCs "has an interesting ethical advantage: the damage to stratospheric ozone and hence to life on Earth would largely be confined to the lifetimes of those people under whose political control such release took place,"[140] provided there is any damage to the ozone layer or human life done by either CFCs or HCFCs.

Much of the world will have to rely on these new substances for economically essential tasks. For example, the value of the equipment in the United States that is dependent on CFCs is approximately $135 billion. This equipment should last twenty to forty years on average,[141] so it is not economically feasible to replace it all at once. The authors of the revisions have tried to allow for continued economic growth while plans are developed to control emissions of all CFCs (and halons). But alternative substances compatible with existing equipment have generally not been achieved to date.

Despite the great potential costs of such an action, the complete ban of CFCs is to take effect as of the year 2000. Although the original protocol already called for a limit of 15 percent of 1986 production,[142] this provision was not stringent enough to encourage rapid development of safe HCFCs and HFCs. Moreover, few HCFCs or HFCs had been tested to determine whether or not they are health hazards. Even after the testing it will take years to find out whether or not they are as safe as CFCs, which they almost certainly will not be. Still the decision was made to bank a goodly portion of our economic future on the development of these chemicals. What is the justification for that?

The London revisions appear to have been motivated by concern that the depletion of the ozone layer was much more rapid than was previously suspected. The revisions' authors perceived not just a little dip but a big hole. Coverage of the Ozone Trends Panel report stirred up fears. In an article in the science section of *Newsweek,* for instance, the OTP report was the impetus behind the headline "More Bad News for the Planet: A Grim Report on Ozone."[143] The authors reported on all the largest trend numbers with no reference to uncertainties or to previous contradictory assessments, gave the EPA projections for deaths as a function of the degree of depletion of ozone, and then declared that environmentalists had seized on the report as proof that a ban is needed. The reader was told by Rafe Pomerane of the World Resources Institute that " 'if this set of facts had been on the table [at Montreal] . . . we would have had a complete phase-out' of CFC production. 'Now we have to catch up with the damage that has already been done.' "[144]

It was not only environmentalists who saw doom in delaying the ban on CFCs. In responding to a statement that the Montreal Protocol was "a crucial first step," Sherwood Rowland, one of the discoverers of the relationship between CFCs and ozone, said, "it ought to be a first step in a sprint, not leisurely steps 10 years at a time."[145] Clearly he agreed that a ban was in order. Both science and environmentalism had spoken, and in light of this confluence of opinion the authors of the *Newsweek* article concluded, "Indeed, the very survival of the planet could depend on hastening that pace."[146]

Such a fearsome conclusion was unwarranted. A 2 percent ozone loss over the expected loss after a ban hardly deserved an apocalyptic prediction. Reports in popular magazines, newspapers, and television continued to give the projections without any discussion of the possibility that they could be wrong. Reading the news about atmospheric ozone was like riding a rollercoaster. The headline in *Science News* shouted, "Ozone Layer Shows Record Thinning"; the leader of the team that made the measurements, James F. Gleason of the NASA Goddard Space Flight Center in Greenbelt, Maryland, stated, "these are the lowest values we've ever seen."[147] Gleason's own report in *Science* bore the heading "Record Low Global Ozone in 1992."[148]

The rhetoric culminated in a press conference held by members of the impressively named Airborne Arctic Stratospheric Expedition on February 3, 1992. The scientists involved reported that exceptionally large amounts of chlorine and aerosols in the Northern Hemisphere stratosphere created conditions for significant loss of ozone not only over the Antarctic but, as a NASA spokesperson put it, over "very populated regions." No doubt this sparked then-Senator Al Gore of Tennessee (now vice-president) to speak to his colleagues about the specter of an "ozone hole over Kennebunkport [Maine]," then the summer home of President George Bush. Predictably, major newspapers editorialized about moving more rapidly against the menace, and the Senate voted unanimously to speed up the phaseout of the CFCs agreed to in the Montreal Protocol.[149]

Our readers are aware, however, that other conditions must be met to ensure ozone depletion, including extremely cold temperatures. As things turned out, temperatures rose and depletion leveled off at 10 percent, not a new record and not over populated areas. This episode left the scientists concerned about ozone depletion and the public officials who relied on them with egg on their collective face. It resulted in "The Ozone Backlash," a headline not in a supermarket tabloid but in staid old *Science*.[150]

Professor James Anderson of Harvard University saw nothing wrong with what he and his colleagues had done. Afterward he said that "the discovery of the extremely high chlorine monoxide levels over the Arctic was, from a scientific point of view, a very serious one. We felt it was a straightforward matter of releasing the information and discussing what we had seen." Atmospheric scientist Richard Stolarski got closer to the heart of the matter: "At the time of the press conference they qualified everything properly. But the tone that came across was that this was an unmitigated disaster and we're all going to die, which in a sense just gives fuel to the Limbaughs, who think it's all hogwash."[151] Had all the qualifications been mentioned at the press conference or added afterward to counter the banner headlines, the whole episode would have been aborted.

A phenomenon himself, Rush Limbaugh, a popular conservative talk-show host, varied in describing the ozone depletion as a "scam," "balderdash," and "poppy cock." Building on doubts raised by a variety of scientists, especially those in a book called *The Holes in the Ozone Scare*—most of them arguing that in view of the massive amounts of chlorine from volcanoes, biomass, and sea salt, the addition of CFCs was trivial—Limbaugh blasted the theory in his best-selling book *The Way Things Ought to Be*.[152]

Gary Taubes writes in "The Ozone Backlash" that "while evidence for the role of chlorofluorocarbons in ozone depletion grows stronger, researchers have recently been subjected to vocal public criticism of their theories—and their motives."[153] The reference to motives alludes to allegations by Limbaugh and others that the purpose of the press conference was to increase federal government appropriations for ozone research. Taubes suggests that as the evidence supporting proponents' theories gets better, they are being treated worse in public. Our analysis should help the interested citizen appraise the validity of this statement: yes, there is increasingly persuasive evidence that chlorofluorocarbons do deplete ozone; but, no, the entire theory, of which CFCs are an important but not the only part, is not necessarily true. For one thing, the theory does not yet predict depletion at the right atmospheric level. In the upper stratosphere depletion is only about half of that predicted. For another, the expected increase in UV-B radiation has not yet appeared. And the natural variation in ozone levels is still about as great as the predicted depletion.

Worse still from the standpoint of the proponents, some scientists who approved of the regulatory efforts have modified their conclusions. "The current and projected levels of ozone depletion do not appear to represent a catastrophe," stated Michael Oppenheimer, of the Environmental Defense Fund. "It's a serious concern," Stolarski said, "but we can't show that anything really catastrophic has happened yet, or that anything catastrophic will happen in the future." Summing it all up, reporter Boyce Rensberger of the *Washington Post* asserted: "In fact, researchers say, the problem appears to be heading toward solution before they can find any solid evidence that serious harm was or is being done."[154]

On a see-saw subject, the equivocal conclusions of atmospheric scientists M. D. Schwarzkopf and V. Ramaswamy should stand for the time being: "The present results suggest uncertainty in both the magnitude and the sign of the decadal total ozone forcing."[155]

Summation: A Call for More Observations

The evidence that CFCs stay in the atmosphere, go up into the stratosphere, release chlorine, and destroy ozone is strong, but there is no clear evidence of global ozone depletion. Mitigating circumstances, such as concurrent emissions of methane, carbon dioxide, and nitrous oxide, as well as large natural variations in ozone, make the detection and prediction of a long-term thinning of the ozone layer problematic. The chemistry of the stratosphere is certainly undergoing rapid change, and as our understanding of the subtleties of that change improve we may be in a position to say that unacceptable changes are taking place. But we are not in that position now.

The question of whether or not the emission of CFCs and the consequent addition of chlorine to the stratosphere will eventually lead to a lower steady-state concentration of ozone remains open. As Fred Singer writes: "The science of stratospheric ozone is at an interesting crossroads. The CFC-theory is not yet good enough to explain observations; and the observations are not yet good enough to confirm the theory. The policy question is whether drastic worldwide controls should be instituted immediately or whether one should wait for a better scientific understanding."[156]

The world's leaders have concluded that we should institute controls now. The evidence only partly supports this position. We will have to live with that fact, and if someday we find out that CFCs really are not all that bad, we will look back and wonder why we put ourselves through the pain of banning them. Then again, if they are bad, we will look back and think that we saved ourselves from a catastrophe. But if we are allowed to use all the knowledge in our common possession, a different conclusion stares us in the face: from the very same body of new evidence that enables us to conclude firmly that CFCs are responsible for the bulk of ozone depletion comes strong indications that the harm from such depletion will vary from little to none. Ozone depletion should be effectively monitored; but it does not deserve to be ranked as a global disaster toward whose alleviation huge resources should be devoted.

11

Who's on First? A Global Warming Scorecard

Were the stakes not so incredibly high—countering global warming by withdrawing carbon from the economy would require hundreds of billions of dollars and promises to "radically change the most vital functions of human economies"[1]—the question of whether future temperatures will be a lot hotter or colder than they are now would be a fascinating intellectual puzzle. It touches everything, the skies, the land, the oceans, the clouds, the heavens, the winds, and volcanoes, glaciers, floods, and droughts as well. It also features conflicts between well-defined personalities. To keep a global warming scorecard, it is necessary to know the players as well as the rules of the game.

One way to make sense of the debate over the existence and extent of global warming is to view it as a conflict between two major theories, one called positive feedback, in which every undesirable departure from the norm amplifies deviant behavior, and the other called negative feedback, in which processes operate so as to reduce deviations, thereby keeping temperature pretty close to where it started.[2] Negative feedback is exemplified by the spirit of James Lovelock's Gaia hypothesis, in which the natural world is so constructed as to keep its essential relationships intact. Positive feedback would be exemplified by an arms race in which country X reacts to country Y's armament by increasing its own level of weaponry, which in turn leads Y to increase its arms. If it is true that cold brings back warmth and warmth induces cold to return, then negative feedback is occurring and fears of global warming must be exaggerated. But if warmth breeds ever greater heat in an endless upward spiral, then positive feedback applies and even the worst scare talk underestimates the gravity of the situation.

Feedbacks from what? The concern of those who predict significant global warming is that as the amount of carbon dioxide in the atmosphere increases, largely as a result of industrialization, the enhanced warming it causes will lead to a series of positive feedbacks that warm the earth more than humans can stand. These feedbacks involve not only other atmospheric gases—water vapor, methane, ozone, nitrous oxide—but also clouds, oceans, land masses, and much more. The critics of the global warming thesis depend on most of the feedback being strongly negative. Because of the huge part played by water vapor, moreover, any significant decline in its positive feedback would be sufficient to undermine the warming thesis.

The basic difficulty in determining whether and to what degree there will be global warming is the same as bedevils us in regard to the projected thinning of atmospheric ozone: the effects we are concerned with lie within the realm of natural variation. That is why there has been no unambiguous detection of a warming signal, no "fingerprint" or indicators that would lead scientists to believe that warming was, at long last, on its way.

Each theory entails certain predictions that enable us to tell if it is on the right track or not. If the earth has not warmed while carbon dioxide in the atmosphere has risen substantially, something else must be going on to mitigate this effect. To judge the adequacy of competing views, we can use a scorecard of the extent to which predicted evils do or do not come to pass. The global warming theory predicts greater warming in winter, for example, with hotter temperatures during the day. So the proponents of the thesis, as we shall see, get credit for winter warming but lose credit because most warming has taken place at night.

When as a boy in Brooklyn in the late 1930s and early 1940s I sat in the bleachers at Ebbets Field, one of many constant hawker cries was, "Scorecards, scorecards, you can't tell the players without a scorecard." With the use of a scorecard, I believe that interested citizens not only can determine who is on what side but also can decide who has the better arguments.

Because so much importance is attributed to the amount of carbon dioxide (CO_2) in the atmosphere and the hypothesized rapidity of its accumulation, a peruser of the global warming scorecard needs to understand its role in terms of runs, hits, and errors. The larger the amount of CO_2 in parts per million and the quicker it accumulates, the worse its hypothesized warming effects. It turns out, however, that the amount and speed of accumulation depend on other factors that are not as well known outside the climatological community as they should be but are nevertheless deemed crucial by virtually everyone. One is the uptake, that is, the degree to which CO_2 is reabsorbed into the land or oceans. Places and processes that absorb carbon are typically called carbon sinks. When I refer to the case of the missing CO_2, I mean that a substantial portion of the carbon produced by natural processes, such as excretion by termites, upwelling from soda springs, and man-made sources, especially burning coal and oil, remains unaccounted for.

What is called the residence time of carbon dioxide in the atmosphere, before it is reabsorbed in a terrestrial or ocean sink, is also crucial. The oceans, which are now believed to be the biggest carbon sink, turn over approximately every thousand years, which places a limit on how much CO_2 oceans can absorb. Simple arithmetic dictates that if CO_2 grows exponentially while absorption of CO_2 on earth grows only numerically, or not at all, there is going to be a lot of CO_2 left with effects so extreme they would likely overcome any negative feedbacks of which we are aware. But if the life of CO_2 is significantly shorter than has been assumed, it will grow a lot slower in the atmosphere, thereby giving us a longer period of time to figure out what is going on.

I am not sure how many readers remember the hilarious routine by Abbott and Costello called "Who's on first?" I have used their theme as the title of this chapter, because it is easy for everyone concerned, the lay public and scientists alike, to be confused by the many causal factors—from sunspots to ocean plankton to higher and lower clouds of varying sizes—which are considered to be the most powerful players at work in determining weather.

Under the Greenhouse Blanket

Analysis of the air bubbles inside the Antarctic and Greenland ice sheets reveals that before the Industrial Revolution the concentration of carbon dioxide in the atmosphere was approximately 280 parts per million (ppm). Scientists agree that the trend in measurement, based in significant part on observations from Mauna Loa in Hawaii, grew to 315 ppm in 1958 and is now somewhere over 350 ppm.[3]

The greenhouse effect might better be termed the greenhouse blanket. The various greenhouse gases, preeminently water vapor but also methane, chlorofluorocarbons, nitrous oxide, and ozone, let light through to heat the earth but then trap that heat, keeping the earth warmer than it otherwise would be. By predicting that the gases making up the greenhouse blanket would grow to a doubling of the carbon dioxide equivalent,[4] those who devise models of global circulation set off a vast train of concern.

On June 23, 1988, James Hansen testified before a Senate committee that there was a 99 percent chance that a significant global warming already under way would in fact occur by the middle of the next century unless a significant reduction in emissions of carbon dioxide were achieved. The uproar this testimony caused can be seen from an article in the Santa Rosa, California, *Press Democrat* of November 22, 1988:

> In only 60 years or less, all life on earth will be dramatically touched from a planetary heat-rise such as the world has not undergone for 10,000 years, environmental scientists predicted Monday.
>
> California, of all the nation's states, stands to suffer among the most severe effects from a projected 5-degree (Fahrenheit) median heat rise on earth by 2050, said Peter H. Gleick, a director of the Pacific Institute for Studies in Development, Environment and Security in Berkeley . . .

"The fuse is lit," Gleick warned. "And we're waiting for the bang . . . climatic changes are coming. Now is the time to prepare for them . . ."

Warnings from Gleick and others to civic, scientific, educational, business and industry leaders, included expectations that earth's agriculture will change; some animal species will become extinct; ocean levels will rise to drown sea level structures, and snow levels will rise . . .

Runoff from snow pack will be drastically cut because only 2 degrees of rise (Fahrenheit) will mean a 1,000-foot rise in the low edge of the pack. About 20 percent of [Pacific Gas and Electric's] power comes from runoff water.

Only a 3.3-foot rise of the Pacific Ocean by polar melts would triple the volume of delta water. It would drown at least 30 percent of levee-protected land, causing a "hundred year" flood that would shove salt far into normal freshwater ecosystem.

Such an ocean rise, unless protective measures are taken, could wipe out San Francisco and Oakland airports with damage in the billions.[5]

Self-described as a newsletter "Providing News and Resources to the Movement for Environmental Climatological Justice," *Rachel's Hazardous Waste News* proclaimed, "global warming will be the greatest challenge that humans have ever faced."[6]

In a 1981 article in *Science,* Hansen and his colleagues at the NASA Institute for Space Studies in New York claimed that "potential effects on climate in the 21st century include the creation of drought-prone regions in North America and central Asia as part of a shifting of climatic zones, erosion of the West Antarctic ice sheet with a consequent worldwide rise in sea level, and opening of the fabled Northwest Passage."[7] A few scientists then voiced doubts. But, as Republican Representative Claudine Schneider from Rhode Island declared at a global warming symposium at Tufts University, "Scientists may disagree, but we can hear Mother Earth and she is crying."[8]

Why the Public Is Rightfully Confused

Aware that many of his scientific colleagues felt that the signal of greenhouse warming could not be detected amid the range of natural fluctuation, Hansen told various interviewers that he felt he had to speak out. "A couple of weeks before the 1988 testimony," he stated, "I weighed the costs of being wrong versus the costs of not talking; the costs of not talking seemed much heavier."[9] He believed that summers would get hotter and that the probabilities for very high heat were increasing.[10] He has said, we should be "very careful about how hard we are pushing the climate system, because we just don't know when it might respond in a very non-linear way."[11] Mother Nature, he implies, will take revenge unless industrial societies clean up their act.

Instead of seeing a policy of massive carbon withdrawal as interference with the way things are, Hansen depicts the production of CO_2 by industry as a dangerous experiment. "Computer simulations, supported by paleoclimate studies," he has told the press, "suggest that the potential greenhouse climate change within a century could rival the difference between today's climate and the last great ice age of 20,000 years ago." Although he and his colleague Andrew Lacis limit their claims, pointing out that there are other mechanisms that change climate, such as solar variability or the reflection of heat into space by aerosols, they term these possible influences "speculative."[12] In the midst of these self-confessed uncertainties, they go on to state: "It is clearly desirable to reduce the ultimate magnitude of the 'experiments' that man is carrying out on the earth . . . Otherwise Americans will be soon finding ourselves . . . up the proverbial creek without a paddle."[13]

Should evidence surface that a drought or heatwave was caused by phenomena other than CO_2-induced warming, Hansen's prime public supporter, climatologist Stephen Schneider, has already prepared a riposte: in testimony before the Senate committee he explained that greenhouse warming "may tilt the balance such that conditions for drought and heatwaves are more likely, but cannot be blamed for individual drought."[14] What evidence, then, would count against the warming thesis?

Where others see the failure of a greenhouse signal to manifest itself as a reason to doubt the hypothesis of significant warming, Hansen plays it exactly the other way. To him the delay in the response time ("it is clear that a large part of the eventual greenhouse warming owing to anthropogenic gases already in the atmosphere has not yet occurred") is even more ominous, "because the positive climate feedbacks associated with a high sensitivity only come into play in response to the warming." A wait-and-see policy is intolerable, he concludes, "because the magnitude of this climate time bomb will grow if greenhouse-gas emissions continue to increase."[15] If we see evidence of greenhouse warming, that is bad; if evidence does not appear, that is worse.

There is no doubt, stated Michael Oppenheimer, scientist at the Environmental Defense Fund, that Hansen's Senate testimony "shook up a lot of people." Never had Oppenheimer "seen an environmental issue mature so quickly, shifting from science to the policy realm almost overnight."[16]

"Do you think," a reporter asked Schneider, "that too many scientists today, like yourself, are crying wolf when there may be no such animal threatening us?" Schneider has written that "knowledge of the probability that wolves do lurk in the forest should be sufficient information for deciding whether to take preventive action."[17] As it is with all other environmental and safety issues of which I am aware, the ultimate argument, the card that trumps all others, is the precautionary principle—we should not take chances with safety.

When Hansen predicted a "high degree of cause and effect" between man-made carbon dioxide and hot weather,[18] his testimony led to visions in which, as Robert Balling put it, "an assortment of interrelated changes devastate the ecosystem of the planet."[19] The Intergovernmental Panel on Climate Change (IPCC), sponsored by the United Nations World Meteorological Organization and mostly but not entirely made up of scientists working for governments, generally supported theories of increased global warming,[20] although the bulky report expressed considerable qualifications, whereas the short policymakers' summary was more directly alarmist. Given two opinions within the same document, advocates of the global warming thesis claimed there was a scientific consensus while opponents denied that any such consensus existed.

Around seven hundred of what *Science* called "heavy hitters," including roughly half the members of the National Academy of Sciences and forty-nine Nobel Prize winners, signed a petition urging President George Bush to pay serious attention to global warming, an appeal orchestrated by the Union of Concerned Scientists.[21] However, a survey by Greenpeace, the environmental advocacy group, found that while 13 percent of the climatologists surveyed thought an unstoppable greenhouse effect was already under way and 33 percent thought it possible, 47 percent of respondents were doubtful. A survey of climatologists conducted by the Gallup organization for the Science and Environmental Policy Center (of which I am a board member) revealed a big split among scientists on whether greenhouse warming had actually occurred and if so whether it was serious.[22] Fred Singer produced a list of scientific luminaries who denied everything, especially the contention that global climate models reflect reality. With Richard Allee of Pennsylvania State University, an authority on the character of ancient climates, stating that "yet by the time we can with confidence say that the greenhouse is here, it will be too late," and Singer observing that all sorts of interest groups use the greenhouse effect to further their own agendas ("My nuclear friends are happy to promote the greenhouse effect. My natural gas friends are happy . . . a lot of scientists promote the greenhouse effect because of increased funding"), all could agree with Schneider that "the public is rightly confused."[23] I hope a scorecard will help.

If Global Warming Is Guilty, Where Are Its Fingerprints?

"It is shown," Hansen and his colleagues wrote in 1981, "that the anthropogenic carbon dioxide warming should emerge from the noise level of natural climate variability by the end of the century."[24] By my own count, a considerable majority of climatologists who write, lecture, and talk to reporters believe that there has been a 0.5–1.0°C warming in the last century, mostly by 1940.[25] Yet I cannot find sufficient evidence in the climate record, or in the analyses thereof, to support this contention. Part of the difficulty may lie in the haphazard and unreliable ways in which temperatures were kept over the past couple of hundred years. Measurements made by pulling up water in leather buckets and using uncalibrated thermometers do not inspire confidence. Neither do those made after running the water through engines before its temperature is taken. There is no reason why these admittedly unsatisfactory measures should have led to conclusions of systematic warming rather than randomness or cooling.

The exception is what is called the "heat-island" effect, created as cities grow up around measuring posts and register systematic increases in warming, though not quite systematic enough to explain the temperature record.[26] Basically, as vegetation thins out and the ground is covered over in cities, surface moisture declines and thus the sun's heat goes into warming the ground.[27] Yet, in answering the question in the title of their article "Are Atmospheric 'Greenhouse' Effects Apparent in the Climate Record of the Contiguous United States (1895–1987)?" Kirby Hanson, George Maul, and Thomas Karl of the National Climatic Data Center, using readings from over six hundred stations and comparing the years 1895–1969 to 1970–1987, conclude, "Test results indicate that overall trends are near zero."[28]

A glance at Karl and associates' data on the U.S. temperature record since 1900 (Figure 11.1) reveals the difficulties. On a worldwide level, the great bulk of warming occurred between 1880 and 1940, a period during which carbon dioxide emissions grew only slowly. Growth of emissions has been substantially greater since then, but the record stubbornly refuses to reveal net warming. The warming that was supposed to take place at the poles has not occurred, and hemispheric warming is out of phase with predictions. More-

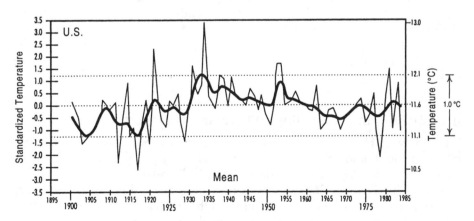

Figure 11.1   U.S. temperature record since 1900. *Note:* This figure shows the annual average and smoothed temperatures, weighted and corrected for "urbanization" effects. *Source:* S. Fred Singer, "Global Environmental Policies: Drastic Actions Based on Shaky Science Make for Bad Policies," *In Depth: A Journal for Values and Public Policy,* Spring 1992, p. 31, based on data from T. R. Karl et al., *Historical Climatology Series,* 4–5 (Asheville, N.C.: NOAA National Climate Data Center, 1988).

over, according to climatologist Andrew Solow, "The historic warming is consistent with a natural recovery from the Little Ice Age" that began around 1300.[29] As the Marshall Institute critics of the warming thesis note, *"a compilation of temperature records for the U.S. reveals no statistically significant warming trend over the last 50 years."*[30]

The former director of the Scripps Institution of Oceanography, William Nierenberg, reported that in a gathering of 150 climatologists, "it was clear that the overwhelming majority could see no increase in the average global temperature in the past 100 years."[31] The difference was solely between those who saw no warming trend over the past 100 years and those who saw none over the last 50 years. *Science* reporter Richard Kerr, attending a conference on global warming, recorded the uniform opinion that the validation of "fingerprint" approaches comparing recent climate trends with estimates of warming are a good decade away. "Assuming the model predictions are correct," he wrote, "studies presented at the workshop show that any intensification of the greenhouse is not yet detectable above the background of natural climatic noise."[32]

Comparing temperatures in the United States, the former USSR, and China since 1951, George Kukla and Thomas Karl discovered that "a range between the mean maximum and the mean minimum temperatures decreased in all countries in all seasons." They are not the only ones to have concluded that "we lack adequate understanding of the causes of the different changes of the maximum and minimum daily temperatures."[33]

Studies by climatologists Robert Balling and Sherwood Idso in Phoenix, Arizona, and elsewhere, revealed that while moderately high temperatures are going up, extremely high temperatures are not. In the United States in general they found the greatest warming occurring in the winter. This raises minimum rather than maximum temperatures.[34] Their findings have been bolstered by Karl and associates, whose research demonstrated that temperatures in the Northern Hemisphere are warmer in the evenings and show higher minimums during the days but not higher maximums.[35] In a study lasting over three decades and employing weather balloons in the former Soviet Union, Greenland, Scandinavia, sixty Arctic locations, and the United States—the largest study of its kind—"more than one million temperature measurements recorded in the Arctic over a 30-year period show no evidence of global warming."[36] Roy W. Spencer of the Marshall Spaceflight Center in Huntsville, Alabama, and John Christy of the University of Alabama collected microwave data through the Tiros-N series of weather satellites for 1979–1988. Since the Southern Hemisphere's temperature went down and the Northern Hemisphere's went up during this decade, the two effects canceled each other out. "We found that the earth's atmosphere goes through fairly large year-to-year changes in temperature and over that 10-year period we saw no long-term warming or cooling trend."[37]

Perhaps the most powerful argument against what most climatologists feel is an increase in global temperature over the last century and a half is statistical. People are interested in getting at the substance of stories, not in analyzing the statistical background by which important components of the argument are calculated. Yet, as Stephen Schneider and Norman Rosenberg inform their readers, "Climate prediction, like most other forecasts of complex systems, is essentially a process of extrapolation from the past and present to the future employing statistical methods."[38]

In a seminal article, "Testing for Trends," Wayne Woodward and Henry Gray take the standard model for assessing whether a trend exists and note that it estimates its values from past data and then uses variables to predict temperature in the future. The basic assumption is that "if there is a trend in the past, there will be a similar trend in the future." However, they ask, "*if conditions remain the same, should we predict the temperature to increase in the future for an extended period of time?* then the validity of the previously discussed tests [for determining whether a trend exists] must be questioned."[39] If they are right, as I think they are, the whole question of temperature is up for grabs.

Those who attempt to determine trends in temperature change use what are called ARMA (Auto Regressive-Moving Average) models, which basically hold that the value of a process at time T is a linear combination of what the process was in the past plus a linear combination of noise or error terms. "In our research," Woodward and Gray continue, "we fit ARMA models to the warming trend data sets, and it should be noted that the best forecasts based on these fitted ARMA models do not predict continued warming."[40] They conclude that "the correlation structure for the model is not nearly strong enough for the ARMA-based forecast to predict an increase over any extended period of time."[41]

No wonder A. H. Gordon of the Institute for Atmospheric and Marine Science, at the Flinders University of South Australia, titles his paper "Global Warming as a Manifestation of a Random Walk." The series he uses are the sea surface and annual land temperatures for the years 1861–1988 expressed as departures from the 1950–1979 mean for the Northern and Southern Hemispheres. Instead of working forward to arrive at expected temperatures for the present and future, Gordon works backward from the assumption that the surface global warming has been 0.5°C in the last 100 years. He does this by reconstituting the temperatures into a series of changes from a present to a past year. To show a trend toward warming, there must be more pluses than minuses or the means of the pluses from year to year must be greater than the minuses or both. Yet,

an analysis of 128 years of anomalies of surface temperature of the globe and of the separate hemisphere shows that there is no significant difference between the means of the positive and negative values, and furthermore that the frequencies of the positive and negative values are those to be expected by chance with a probability not far removed from 0.5 . . . The resulting path closely resembles the kind of random walk that occurs during a coin-tossing game.[42]

The stock market, the reader might wish to know, is often described as operating like a random walk, which is why no one consistently outguesses it over the long term.

What would have happened, Wallace Broecker asks in his "Global Warming on Trial," if there had not been any atmospheric (greenhouse) gases more extensive than those prevalent in 1950? "The answer by both sides would have to be that instead of remaining the same as it was in 1850, the planet's temperature would have undergone natural fluctuation, which could have been as large as the changes measured over the last 100 years."[43] As Karl, Heim, and Gray ask, in regard to the fingerprint manifestation of global warming, "If not now, when?"[44]

Reactions to the inability or even unlikelihood of detecting a greenhouse warming signal vary. Stephen Schneider, who would "be surprised if it doesn't happen," expects it will take one or two decades before a clear signal can be detected. "But," he wonders, "how do you assign a probability to something when you have no objective means of doing so? You base it on physical intuition and then state your assumptions. By my intuitive reasoning, a greenhouse signal has been detected at an 80% probability." What Schneider calls his "faith," he continues, "is based on the principle of heat trapping by greenhouse gases and the billions of observations that support it. All that objective stuff rests on assumptions. The future is not based on statistics, it's based on physics. Objectivity is overplayed."[45] If by "objectivity" Schneider means impartiality, I disagree. If he means that statistical manipulations are no substitute for theoretical expectation, I concur. But is the theory right?

Hansen's basic argument is that since surprises are possible, we should guard against them. Climatologist Richard Lindzen, who argues that "the data provide no evidence of man-induced global warming," replies that "this is indeed an unanswerable argument. The point, quite simply, is that ignorance is ignorance. It is certainly not a basis for action. If it were, the only logical response would be total paralysis."[46]

Since the predictions of future warming come from global circulation models, the place to seek enlightenment is through consideration of the strengths and weaknesses of these computer-generated devices. The problem is not whether they have weaknesses—all models do—but whether these weaknesses are severe enough to restrict their use. It may be, as M. C. MacCracken and F. M. Luther assert, that "the only applicable method for projecting future climate is the construction of mathematical models based on the full set of fundamental physical principles governing the climate system."[47] If this were an intellectual exercise, the point would be well taken. Ultimately, however, the global warming scenario is aimed at influencing public policy. Supposing, then, that one does not have "the full set"; what follows?

How Accurate Are Global Climate Models?

The basic purpose of global climate models is to simulate the weather based on a series of simultaneous equations that can be solved through calculations performed on high-speed computers. The difference between incoming and outgoing energy, which we call temperature, is affected by many variables, including heat-absorbing and -reflecting capacities of the oceans. The earth is cooled by air currents, which involve winds and clouds. Whether heat is reflected out into space (causing cooling) or back to earth by the atmospheric blanket (causing warming) depends on concentrations of atmospheric gases, the largest of which is water vapor. Thus models of global climate must encompass the swirl of atmospheric currents as well as their interactions with land masses and oceans that both absorb carbon and send up water vapor. How well do these general circulation models do? How good, in general, is weather forecasting?

In a recent twenty-three-year period the forecasters of the National Weather Service achieved an average success rate of 8 percent in predicting whether temperature would be above or below normal in a ninety-day period. Whereas chance would allow them to be right 33 out of 99 times, they came close to the mark 38 times. Their success rate in predicting rain was 4 percent.[48] Evidently weather prediction, even in the short term, is no easy task, not to mention predictions decades and centuries ahead. One can only sympathize with climatologist G. L. Stephens when he says that "there is hardly a single aspect in dealing with climate change that can yet be described with confidence."[49]

To accomplish the enormous task of weather prediction, two things are necessary: vast amounts of data about all the variables, which require extremely large calculating capacities, and representation of the physical and chemical processes in the form of theories or models to process this data so as to produce approximately correct results. Processing tasks have become remotely feasible only because of the development of high-speed computers, fast though not quite fast enough yet, and a community of computer modelers willing to try their considerable scientific talents and entrepreneurial skills in bringing these models to fruition. In the trade they are called general circulation models, or GCMs.

All large-scale models are based on bets that the vast complexity of the natural world can be represented in simplified form. If one follows R. A. Reck in the belief that models must "describe all pertinent atmospheric processes" to make "model sensitivity to parameter perturbation [change in basic variables] . . . equivalent to that observed in the real world," including such "pertinent coupling schemes between parameters" as cloud-ocean-atmosphere interaction,[50] then current GCMs are far from satisfactory. In this regard I would accept the defense of the global modelers who believe that improving their models is the best way to get accurate predictions in the future. But that does not answer the question of whether these models are sufficiently developed now to accept their conclusion of substantial global warming by the middle of the next century.

Models require compromises. When modelers learn that they are not able to generate the average temperature of the earth and come close enough to the distribution of temperature between the poles and the equator, for instance, they make "arbitrary adjustments referred to as 'tuning.' "[51] Because their basic theory tells them that without the intervention of other factors an increase of carbon dioxide in the atmosphere should warm the earth, and perhaps because they wish to use variables that human beings might control (or be held responsible for controlling for policy purposes), they adjust the models to come closer to actual measurements. "Thus," Richard Sanford claims, "their method is a perfect example of the fallacy of context-dropping," in which other variables that might affect the results are given a secondary place.[52] The response to such criticisms is that all models must omit some variables or they would be too unwieldy to use, and that other variables are being included to the extent modelers can understand their effects and devise programs capable of dealing with them. When variables such as oceanic heat flux are acknowledged to be important but their dimensions are unknown, they are for the time being ignored. The value of the solar constant is not known as precisely as the modelers would wish, so they use approximations.

Such devices lead major modelers M. E. Schlesinger and J. F. Mitchell to wonder whether "there is an inherent limitation in our ability to validate the accuracy of GCM perturbation simulations [running the models as if certain variables were much smaller or larger], which thereby affects our confidence in the accuracy of the GCM simulations of CO_2-induced climate change."[53] When the equations of processes that affect temperature on earth or in the atmosphere require too much tuning or take up too much computer time, as is the case with heat flows in the soil and the constitution and reconstitution of sea ice, these variables may be "parameterized," that is, given arbitrary values.[54]

To their credit, modelers try to introduce these processes as soon as they are able: the result is greater accuracy but also greater complexity. The more new variables that are related ("coupled") to each other, the better the model represents the real world, but the greater the difficulty the modeler finds in putting the parts together. Once modelers try to synchronize atmospheric and oceanic models, Solow observes, the difference in time scale—atmospheric time is counted in days, oceanic time in years or millennia—makes it difficult to parameterize one of the variables to come closer to observed values without destroying its relationship to the others. "As a consequence, coupled models typically drift into an unrealistic state. To avoid this drift, flux correction terms are introduced."[55] But if the adjustments ("flux correction terms") make the model, what are we getting?

There are significant differences of opinion about the underlying physical processes that ought to be taken into account in attempting to predict global climates. Solar radiation is not the only source of heating or cooling; there is also convection, through which heat is moved by air motion. In a number of articles climatologist Hugh Ellsaesser has argued against "the inherent assumption that all energy transport within our atmosphere is by radiative transport." He, along with Lindzen, believes that convective cooling exerts a large effect.[56] Towers of water vapor rise, forming clouds that reflect sunlight into space. As the earth heats, they argue, winds increase convective movement away from earth, thus constituting a negative feedback cooling the planet.

Predictions of large degrees of global warming critically depend on an indirect effect under which the warming initiated by CO_2 causes surface evaporation of water into the atmosphere, thereby increasing the effect threefold. Ellsaesser and Lindzen, by contrast, claim that deep convection from fast-rising towers of water vapor creates greater precipitation and more rapid condensation, so that the upper atmosphere will thin out and dry. "In the lower latitudes," Ellsaesser argues, "the rise in CO_2 emissions will produce a 3 to 1 rise in greenhouse blanket thinning due to condensation. That's exactly the opposite of what the models predict." Thus convection produces an effect comparable to that of carbon dioxide but, as Sir James Lovelock says, "in opposition to it." "We will eventually discover how naive we have been," Karl told columnist Warren Brookes, "in not considering CO_2's effects on cloud cover and convection. As CO_2 speeds up the hydrological cycle, more convection creates more clouds and more cooling. So, the greenhouse effect could turn out to be minimal, or even benign."[57]

Any number of global warming critics, including some modelers themselves, object to the lack of spatial resolution. The use of 500 kilometer (km) spaces leaves out of the equation a lot of detail of interest to people who live in much smaller areas. Someone interested in thunderstorms or how regional climate varies with changes in the atmosphere will not find helpful general circulation models as presently constituted.[58]

Equally important, GCMs as we have seen depend on historical time series, the future depending on a succession of pasts. But historical data are lacking in many variables from ice cover to rainfall to ocean currents to thickness of clouds. No wonder a climate modeler at the National Center for Atmospheric Research, Warren M. Washington, says, "we are starved for measurements."[59]

To these complaints, many of which are acknowledged, the response is that the climate modelers can and will probably do better. But right now we must ask whether the models produce results close enough to actual weather to be believable and therefore usable in making public policy. The basic determinations of the 1990 Intergovernmental Panel on Climate Change, the most authoritative to be found among the global modelers, predict mean global warming of 4.2°C around the mid-twenty-first century, with the North Pole regions reaching as high as 18°C.[60] Let us follow general GCM predictions in more detail.

Given the already calculated increase in all atmospheric gases in the past hundred years or so, the Northern Hemisphere, containing the largest land mass, should be warming by 1°C and the Southern Hemisphere by less. As the Marshall Institute notes, "However, the observed temperatures show no significant difference in temperature trends in the two hemispheres."[61] According to Balling, "something was either retarding warming in the Northern Hemisphere or stimulating warming in the Southern Hemisphere." He is undoubtedly right when he says that "any comprehensive vision of climate change into the next century must be able to account for this hemispheric difference in temperature trend."[62]

In winter, should there be warming, snow will melt, less heat will be reflected, and more will dissipate. Thus GCMs predict there will be more warming in the winter and less in the summer. Balling's test produced warming winter trends in some parts of the country though not in others.[63] Good enough. Score one for the warming thesis.

Hansen and his colleagues, though by no means all global modelers, predict that there will be an increasing frequency of extremely high temperatures.[64] To test this prediction, Balling and Idso found areas in the United States that had significant warming or cooling in the past forty or so years. They found that the mean summer temperature from 1948 to 1987 had actually cooled somewhat. Thus they concluded that "there is no sound observational basis for predicting an increase in the frequency of occurrence of extreme high summer temperatures in response to greenhouse warming."[65] But Hansen bet that the early 1990s would be among the highest recorded, and he was right.[66] I shall score this as a draw.

One thing good models should do is predict the past. Yet GCMs retrodict twice as much warming for the past couple of hundred years than appears in the climate record. Most modelers would agree with O. Thiele and R. A. Schiffer that "even the most comprehensive global climate models greatly oversimplify or misrepresent key climatic processes."[67]

Ellsaesser contends that without a fingerprint of global warming we have no grounds for believing that it is occurring. The difference between the warming he observes during the past century (0.5°C) and that predicted by GCMs (4.2°C or more) is nearly ten times. "How much bigger does it have to get," Ellsaesser asks, "before we are willing to accept that the models might be wrong?"[68] The bottom line belongs again to Solow, who expresses my view when he states: "When natural variability is so great, it is dangerous to attribute any particular change to any particular cause."[69]

Thus far our discussion of global warming has been quite general. We have asked whether and to what degree GCMs follow the temperature record. But we have not gone into the particular variables that are at issue—from carbon dioxide to clouds to oceans and the various feedback processes that determine whether what starts as warm will become warmer still or cooler.

The Residence Time of CO_2 in the Atmosphere

Life is carbon and carbon is life. Not for nothing do science fiction writers refer to life on earth as carbon-based or human beings as carbon creatures. Carbonates in the crust of the earth and in the sea, organic compounds in the soil and rocks and their interplay with oxygen, hydrogen, phosphorus, nitrogen, and various metals have created life as we know it. Gases containing carbon, primarily carbon and methane, bubbled out of the earth during ages past to form carbon dioxide in the air.[70] Even today, a group of Japanese scholars has observed fluid bubbles containing 86 percent CO_2 emerging from the sea floor at a depth of 1,335–1,550 feet near Okinawa. Immediately gas hydrates developed on the bubble surfaces and coalesced to form pipes interacting with volcanic sediments.[71] The formation of the earth is not something that happened only once; it continues to happen. Table 11.1 tells us how rich

Table 11.1   Carbon sources and amounts

Form	Carbon mass (10^{18} gm)	Relative to life
Calcium carbonate (mostly in sedimentary rocks)	35,000.00	62,500.0
Ca-Mg carbonate (mostly in sedimentary rocks)	25,000.00	44,600.0
Sedimentary organic matter (as kerogen)	15,000.00	26,800.0
Oceanic dissolved bicarbonate and carbonate	42.00	75.0
Recoverable fossil fuels (coal and oil)	4.00	7.1
Dead surficial carbon (humus, caliche, etc.)	3.00	5.4
Atmospheric carbon dioxide	0.72	1.3
All life (plants and animals)	0.56	1.0

Source: Robert A. Berner and Antonio C. Lasaga, "Modeling the Geochemical Carbon Cycle," *Scientific American,* 260, no. 3 (Mar. 1989): 78. Copyright © 1989 by Scientific American, Inc. All rights reserved.

Note: Amount of carbon found on the earth in various forms is listed both in units of 10^{18} grams and amounts relative to that found in all life. Far more carbon is stored in carbonates (the fossil remains of animal skeletons) and kerogen (the remains of soft animal tissue) than in living or recently dead organic matter, indicating that the geochemical carbon cycle is ultimately responsible for regulating atmospheric carbon dioxide over geologic time scales, which are measured in millions of years.

we are in carbon and where in general it is stored and in what proportional amounts.

In school we learn about the biological cycle through which carbon in the atmosphere is absorbed by plants and transformed by photosynthesis into organic material, which is then recovered through bacterial decomposition and respiration. The larger geochemical cycle transfers carbon to sedimentary rocks, oceans, and the atmosphere. Atmospheric carbon is only one part of this cycle. Insofar as is known, most carbon is stored in rocks representing remains from soft tissues as well as skeletons of ancient animals and plants, mostly of the sea but possibly from the land as well. Over long cycles of time the carbon from decomposing rocks combines with oxygen and forms carbon dioxide, which eventually finds its way into the atmosphere.[72] Over eons CO_2 in the earth's atmosphere has risen and declined with enormous variation. From gas bubbles on the ocean floor, from geysers and erupting volcanoes, there have been for the longest periods much more atmospheric CO_2 than there is today, during the Cretaceous period perhaps eight times as much. Modern concerns were initiated by Swedish chemist Svant August Arrhenius, who recognized the existence of the carbon trap or blanket in the atmosphere and predicted that if other factors did not intervene, the more carbon trapped there, the greater the earth's warming trend.

Because the amount of CO_2 in the atmosphere is very small compared with the amounts entering and leaving, emitted and absorbed by the various processes discussed, even small changes in the flux make a big difference. Some scientists have come to the conclusion that weather in the past has been largely controlled by these fluxes,[73] but others disagree. Another basic difference is that even those who think CO_2 controls climate differ over whether it is industrial emissions that control the controls, so to speak, or whether it is the rate of degassing in the earth (liberating carbon to flow into the atmosphere) that is the causal factor. Thus Robert Berner and Antonio Lasaga conclude: "We are therefore in accord with Fisher that atmospheric carbon dioxide in the world climate in general is mainly controlled by tectonism [the movement of huge terrestrial plates on the ocean floor] processes taking place deep within the earth."[74]

Since the Industrial Revolution began in the mid-eighteenth century, and with increasing speed in recent times, the amount of atmospheric CO_2 has increased by about 25 percent, and various estimates hold that it will grow a lot faster. Taken together, all atmospheric gases have risen by perhaps 50 percent since that earlier period. There may have been warming but not warming to the degree expected from global climate models.[75]

Much has been made of the Vostok ice core, which facilitates measurement of CO_2 trapped in ice bubbles over hundreds of thousands of years. By correlating these measurements with temperature, the claim has been made that the level of CO_2 by itself is largely responsible for variations in climate.[76] A glance

at the famous ice core graph (Figure 11.2) strongly suggests that the two phenomena—atmospheric CO_2 and variation in temperature—closely track each
other. But which came first? Generally two assertions are made: one is that the
Vostok data provide powerful confirmation of the global warming theory; the
other is that although the data confirm the parallel, no one knows what drove
up CO_2 to much higher levels in the more distant past.[77] Not surprisingly, some
scientists have argued that over long periods of time climate change determines
the amount of CO_2 and not the other way around.[78] Others believe that variations in the earth's orbit around the sun, called orbital forcing, explain certain
temperature patterns,[79] which result in the release of large amounts of CO_2
into the atmosphere. No one knows which came first, the chicken of carbon
dioxide or the egg of higher earth temperatures.

   "I was shocked to discover, when I went to Oak Ridge in 1975," Freeman
Dyson recalled, "that nobody knew what happened to half the carbon that we
were burning." Roughly 6 or 7 gigatons of carbon (1,000 million metric tons)
is being poured into the atmosphere, but something less than half, 43 percent,
is somewhere else, though nobody knows where. "This," Dyson continues, "is
the mystery of the missing carbon. It is preposterous to claim any ability to

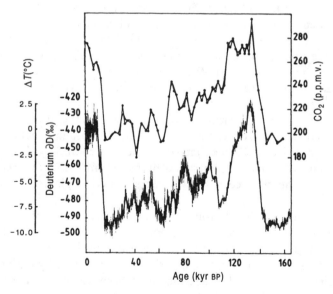

Figure 11.2   Long-term variations in global temperature and atmospheric carbon
dioxide. *Note:* Observe how one tracks the other. It seems highly unlikely that future
patterns will be different. *Source:* J. M. Barnola et al., "Vostok Ice Core Provides
160,000 Year Record of Atmospheric CO_2," *Nature,* 329 (Oct. 1, 1987): 410. Reprinted with permission from *Nature,* copyright 1987 Macmillan Magazines Limited.

predict the future of the carbon cycle so long as we lack a rudimentary under-standing of what has happened to the carbon in the past."[80]

Everyone is concerned about the case of the missing CO_2. Roughly 6 billion tons of carbon are added to the atmosphere each year through burning fuel, cutting down trees, and eroding of soil. Around 3.5 billion tons remain in the atmosphere. What happens to the rest of the carbon? Pieter Tans and his colleagues at the National Oceanic and Atmospheric Administration assert that 1 or 2 billion tons of CO_2 are somehow taken into trees and plants in land masses, the terrestrial sink, mostly in the Northern Hemisphere.[81] The great mystery is what there is on land capable of storing that much carbon. If there is an available terrestrial sink, of course, then fears about global warming are exaggerated. But if the land is at or near its total absorptive capacity, which is hard to say considering how little is known, it might start releasing a great deal more into the atmosphere, and that would be even worse.[82] Other scientists hold that ocean kelp beds are storing a great deal more than currently realized. Tans and his colleagues conclude that "understanding the role of the land in the sea [carbon] budget must include a reanalysis of the contribution of mid-latitude reforestation as well as studies of the feedbacks between ecosystem functioning, climate, and atmospheric composition"[83]—which is tantamount to saying that this whole field of study should begin again, hardly a confidence-building statement. But if one reads articles stating that "land biota is most probably a net source of carbon for the atmosphere"[84] and a report headlined "Fugitive Carbon Dioxide: It's Not Hiding in the Ocean,"[85] the mystery deepens.

All of the models discussed above are premised on the supposition that CO_2 will remain in the atmosphere for 200 to 300 years. This long residence time explains why global warming supporters fear that a speedy buildup, along with a slow exit, will create extreme warming. In a path-breaking article, however, Chauncey Starr and Scott Smith "show from historical global fossil energy use that the ocean/biosphere/atmosphere system exhibits a characteristic CO_2 residence half-life of about 37 years for anthropogenic sources. That is, if man were to stop emissions immediately, after 37 years approximately half of man's total contribution would be removed from the atmosphere."[86] This model does not depend on our poor understanding of interactions among oceans, atmo-sphere, and biosphere but rather looks carefully at data connecting energy use to atmospheric measurements of CO_2.[87] On the one hand they develop good reasons for believing that large-scale reductions in the use of fossil fuels around the world, even with substantial conservation measures, are quite unlikely; on the other hand a much shorter residence time than commonly assumed means that while CO_2 would continue to accumulate, it would do so at a much slower rate. Their model contrasts with the swift timetable designed by Hansen and Lacis: the latter drastically reduces fossil fuel emissions on the grounds that sulfate aerosols, produced by the same combustion, have hidden a large global warming effect that, because of the long residence time of CO_2, will grow much worse much faster over time.[88]

There may well be more burning of fossil fuel into the next century, but that does not necessarily mean there will be that much more CO_2. In the time that it took the United States to double its production from the 1950s, for instance, the amount of CO_2 produced increased by only 7 percent. Furthermore, a rerun of the "business-as-usual" scenario of the IPCC revealed that its reading of last century's record was faulty. M. Heimann and his colleagues at the Max Planck Gesellschaft contend that the increase in CO_2, which the IPCC expected to double by 2060, may not reach that level until 2100.[89] And now we know that doubling will take even longer.

Clouds: The Wild Card

"The *cloud* feedback is probably the most complex and uncertain geophysical feedback in the climate's system," asserts climatologist Daniel Lashof.[90] "All clouds have a split personality," scientist-reporter Richard Monastersky notes; "They feature some characteristics that cool the Earth and others that warm it."[91] The president's Office of Science and Technology Policy called clouds the top scientific priority in the study of global climate change.[92] This is not a contentious statement. "Clouds," greenhouse warming critics Robert Jastrow, William Nierenberg, and Frederich Seitz claim, "are the largest single source of error in the greenhouse forecasts."[93] Climatologist Robert Cess adds: "Clouds have been the Achilles heel of climate modelers."[94] Clearly, clouds are a subject worth discussing.

It is important to note that the consequences of cloud movements are much greater than the consequences of CO_2 doubling. The reason is that doubling CO_2 reflects a heating force of 4 watts per square meter (watts/m^2), whereas clouds reflect upward or downward around 75 watts/m^2.[95]

The trouble is that some cloud formations cool while others warm. High-altitude cirrus clouds warm. Marine stratocumulus clouds, by contrast, cool the atmosphere. The droplets of water that constitute them radiate sunlight deep into outer space. Since cirrus clouds cover only 16 percent of the earth, whereas stratocumulus cover around 34 percent, their cooling effect dominates.[96] Anyone who has tried to deal with difficult problems can sympathize with modeler Anthony Slingo of the National Center for Atmospheric Research, who reflects that "the trouble is the lack of a very firm theoretical foundation for the way we treat clouds in a general circulation model. When we get a result on climate change, we don't really know what confidence to give it."[97]

How much do clouds, or rather their treatment in models, matter? A change in the way clouds were modeled by adjusting the amount of water they contained led modelers of the British Meteorological Office to reduce their 5°C warning, after CO_2 doubled, to 2°C.[98] In general, in the United States and throughout the world, the twentieth century has been a time of increased cloud cover.[99]

In a positive feedback model relating clouds to climate, a warmer earth heats the air above it, leading to the formation of fewer droplets and hence less water in clouds. Consequently there are fewer and less dense clouds, more sunlight reaches the earth, and the initial atmospheric blanket effect is magnified. The negative feedback model holds that, as the earth warms, the consequent increase in water vapor leads to the formation of more clouds, which in turn radiate heat outward, thereby cooling the earth.

Most modelers in the past, like Hansen, have adopted the positive feedback view.[100] But climatologists V. Ramanathan and W. Collins, both from the Scripps Institute of Oceanography, have argued that there is substantial negative feedback; that is, that the earth regulates itself. "I cannot see," Ramanathan asserted, "how the planet can have a runaway greenhouse effect." He and his colleagues studied what happened in the Pacific during the 1987 El Niño, when the sea surface temperature increased by a few degrees. As the surface of the sea grew warmer and warmer, they discovered, clouds kept forming until they virtually shut out the sun, thus initiating cooling, which caused the clouds to dissipate.[101] This conclusion was immediately attacked by Rong Fu and associates, who claim that the effects found in the area near the El Niño are not true of the Pacific as a whole.[102] Ramanathan and his colleagues answered with the results of a multisatellite experiment that measures temperature more directly, thereby demonstrating that "clouds cool the planet more than they heat it."[103] By recording changes in reflected light from clouds the experimenters determined that more heat was radiated away than was absorbed, thus resulting in cooling.[104]

The stakes are high. In making calculations for temperature predictions Hansen and his colleagues determined that the optical properties of clouds led to a positive feedback of 0.22. The succeeding literature varies from predictions of similar and slightly larger positive feedback on the one hand to suggestions that feedback might be negative on the other.[105] To provide an idea of the immense difference in models, Ramanathan and his team concluded that "the size of the observed net cloud forcing [that is, impact on earth's climate] is about four times as large as the expected value of radiative forcing from a doubling of CO_2" and that this feedback is negative.[106] Ramanathan showed that differences in the way various GCMs treat clouds and water vapor led to a difference of 6°, larger "than the entire temperature increase predicted by most of the models."[107]

Is it true, as Monastersky wrote in *Science News*, that "at present, neither theorists nor large-scale computers can predict with accuracy whether cloud systems will help or hurt a warming globe"?[108] And how would we know? Patrick Michaels and David Stooksbury developed a list of effects they expect to follow from greater low-level cloudiness brought on by sending various industrial products into the atmosphere. By determining whether their expectations-cum-predictions are met or not, a scorecard can be kept for future use:

1. Night warming from both the increase in greenhouse gases as well as the increase in cloudiness,

2. Reduced daytime warming because of cloud albedo [the proportion of solar radiation reflected from its surface],

2a. A consequent decrease in the daily temperature range,

3. The greatest warming (night effect) of clouds should occur on (long) winter nights,

4. The greatest cooling (day effect) should occur on (long) summer days,

5. The least warming (night effect) should occur on (short) summer nights,

6. The least cooling (day effect) should occur on (short) winter days,

7. Cloudiness should be enhanced near . . . North America and Eurasia, and

8. These effects should be concentrated in the industrial (northern) hemisphere.[109]

The attacks and counterattacks on the effects of clouds will continue for some time. Because what happens to clouds is connected to what oceans do, I shall pursue our quest for feedbacks into the briny deep.

The Mysterious Oceans

"Most climatologists agree," Robert Balling informs us, "that the key to understanding global climate change lies in the understanding of ocean-atmosphere interaction."[110] William Nierenberg adds that "the rate of carbon exchange between the atmosphere and the upper ocean because of photosynthesis respiration and nonbiological chemical and physical processes is 20 times greater than the emissions due to human activity."[111] It was generally believed (and is now contested) that 40 or 50 percent of atmospheric CO_2 is picked up by the ocean. The ocean also takes in more than half of the incoming solar radiation and then stores it for redistribution by currents. All sorts of other processes move water up and down faster or slower depending on circumstances. In general the warmer the ocean, the more CO_2 it releases to the atmosphere and vice versa. The implications for global climate models are stated by Andrew Solow: "Climate models that assume a passive ocean typically exhibit more rapid warming and greater polar amplification than models that assume some oceanic response."[112]

The case of the missing carbon was resolved in the past by assuming that oceans took in 2 or 3 billion tons and terrestrial sources the rest. That was before Pieter Tans, Inez Fund, and Taro Takahashi, affiliated with the Lamont-Dougherty Geological Survey, argued that the oceans could not be taking up more than a billion tons and hence there must be a large as yet undiscovered terrestrial sink. Other scientists then began to argue that ocean winds affected growth of ocean plankton, which in turn affected their ability to absorb carbon.[113] Along came oceanographer William Dillon of the U.S. Geological Survey, who observed that vast amounts of hydrates in the Arctic permafrost and in deep sediments contained immense amounts of methane and carbon dioxide. These hydrates, he contends, hold about twice the carbon that all the gas, oil, and coal reserves on land hold—approximately 10,000 billion metric tons. This calculation explains the growing popularity of research into naturally occurring gas hydrates.[114]

Though there is strong evidence that the North Atlantic absorbs considerable carbon dioxide (which is to say that it takes in more than it puts out), there are differences of opinion about whether Northern Hemisphere oceans absorb all the carbon or whether other oceans play a part as well. Does this matter? Well, the atmosphere contains about 20 percent of the amount of carbon dioxide in the oceans. Should as small a proportion as 1 percent of ocean CO_2, not to mention what is locked in hydrates, be released into the atmosphere, CO_2 levels would grow rapidly. Unable as yet to discover how man-made and natural sources of CO_2 get divided among land, ocean, and atmospheric spheres, we cannot predict how the oceans would react to a great deal more CO_2 production than now exists.[115]

There are many proposals and actual research projects under way to determine what is happening to carbon. P. D. Quay, B. Tilbrook, and C. S. Wong measured the inorganic carbon that was dissolved in the surface waters of the Pacific and discovered that between 1970 and 1990 it had decreased by about 0.4 ppm. Thus they calculated the absorption of CO_2, its uptake, to be 2.1 billion tons (or gigatons) of carbon per year. As they say, "this amount implies that the ocean is the dominant net sink for anthropogenically produced CO_2 and that there has been no significant net CO_2 released from the biosphere during the last 20 years."[116]

How much anthropogenic CO_2 is going into the ocean, students of global warming want to know, and how much is being reabsorbed by photosynthesis? To answer these questions, we must know how much carbon is stored and the rate at which soil and humus are changing. The common belief among scientists is that a combination of deforestation and agriculture has led to less absorption of CO_2, thereby allowing more to rise into the atmosphere. But CO_2 is also a significant fertilizer. Therefore, some have concluded that carbon storage in terrestrial reservoirs has increased despite the toll taken by foresters and farmers. They point to a global greening of the earth's natural vegetation and soils driven by increased atmospheric CO_2 content and by the fallout of anthropogenic NO_3 and NH_4 nutrients.[117]

In an ingenious study based on the atmosphere's oxygen content, Ralph Fee-lince and Stephen R. Shertz of the National Center for Atmospheric Research were able to compare the atmospheric level of oxygen with the amount required for burning coal, oil, and gas. Their conclusion is aptly summed up by Wallace Broecker and Jeffrey Severinghaus: "Global greening appears to be compensating for losses by afforestation and agriculture."[118]

Although the oceans absorb a good deal of whatever heat exists, they warm up only slowly.[119] Some scientists suggest that this could go on for decades or even centuries. What does it portend? Hansen and other global modelers suggest that ocean thermal inertia could be storing up a great deal of heat, which would then respond to increased warming with a devastating release of much larger amounts of CO_2 into the atmosphere. Suppose, however, that as the ocean heats up, water vapor rises and forms clouds that reflect sunlight into space. Whether oceans react to heat by cooling or warming, they depend, like all living things, on the sun for life.

The Solar Cycle: Is Nothing Else Important?

"The Earth's climate," Solow proclaims, "is driven by solar radiation." By the time it reaches the earth's surface it has an impact of about 240 watts/m² of energy. The earth absorbs part of this heat and sends the rest back into the atmosphere. The shortwave radiation from the sun is remitted from earth as longwave infrared radiation into the atmosphere, so that eventually the outgoing and incoming energy come into balance.[120]

Only since 1978 has solar radiance been accurately measured. It is known to fluctuate with the eleven-year, somewhat irregular, sunspot cycle. Changes over this cycle on the order of 0.1 percent (or 0.25 watts/m²) are equivalent to atmospheric effects on climate during the time period. Because this degree of solar impact (or, as climatologists say, "forcing") on the earth's climate changes periodically, it has not been thought to have long-term effects. Until recently, therefore, changes in the sun's power had not figured large in the global climate debate.[121] Enter an intriguing proposal by Danish researchers that led *Business Week* to headline its story "Could a Hotter Sun Be the Culprit in Global Warming?"[122]

In 1991 E. Friis-Christensen and K. Lassen proposed that variations in the length of the solar cycle, and hence the warmth of the sun, exert a strong influence on global warming. Basically, the sun warms up, the earth gets warmer, and when it cools down, temperatures decline. The problem they faced was the lack of a good measure of the total energy output of the sun. The solution they chose was remarkable in its simplicity: namely, the length of the solar cycle. The power of their explanation lies in a remarkably tight correlation between temperature and the sunspot cycle, reaching a near-perfect level of 95 percent.[123]

"The fit is so good," atmospheric scientist J. A. Eddy stated, that "the burden of proof that something's wrong almost rests with any detractors."[124] Indeed, Keith Shine at the University of Redding mused that "if it's correct, we have to change our view of climate fundamentally. It's an incredible correlation; it would imply that almost nothing else [besides solar variation] is important in the climate system."[125] Nevertheless, skepticism remains, as Friis-Christensen and Lassen predicted in their article, because of the lack of a physical mechanism that ties the length of the sunspot cycle to the output of the sun.[126]

The following year, 1992, brought two papers casting grave doubt on the sunspot thesis—but not, I think, in a way that validated large-scale warming. Proposing an energy-balance model covering the period between 1776 and 1985, P. M. Kelly and T. M. L. Wigley demonstrated that man-made effects have been stronger in influencing climate over this period than has the solar cycle.[127] This finding was apparently reinforced by another historical investigation by M. E. Schlesinger and N. Ramankutty, who used their climate-ocean model to evaluate the relative effects of solar forcing, aerosols, and greenhouse gases in the period from 1630 to 1985. On the one hand, they concluded that greenhouse gases will continue to be larger than other forces in influencing climate. On the other, they noted that the influence of the solar cycle is sufficiently strong to reduce the predicted changes in the solar temperature linked to hypothetical doubling of carbon dioxide by almost one half.[128] Now, patient reader, which is the more important conclusion: that the solar cycle does not overwhelm everything else or that the predicted and feared global warming, due largely to man-made CO_2 emissions, is reduced by nearly half?

There is one last factor we should be aware of: a field of paleoclimatology exists in which scientists believe that many of the recent climate oscillations are driven by cyclic variations in the angle of earth's spin axis and the shape of its orbit around the sun. They find climate cycles of several hundred thousand and even millions of years buried deep in the earth's past.[129]

Feedbacks: More Positive or Negative?

"The sensitivity of the climate system to anthropogenic perturbations over the next century," climatologist Daniel Lashof begins a review of the subject, "will be determined by a combination of feedbacks that amplify or damp the direct radiative effects of increasing concentrations of greenhouse gases."[130] Having looked at clouds and oceans and the sun, we can see that these alone or in combination may overwhelm all other effects. But there are other feedbacks, especially those coming from atmospheric trace gases, that have been the focus of attention. And for good reason. Carbon dioxide alone, as predicted in most global climate models, would not produce sufficiently high temperatures, when faced with negative feedback from other sources, to produce the feared effects. Rather than leave the matter at the level of rival claims, as if there were nothing more to be said, I shall review feedbacks, in the spirit of this book, to see whether as knowledge grows these feedbacks are estimated to be positive or negative. I will concentrate not only on the existence of forces affecting climate but on the size of these effects.

Let us start with atmospheric trace gases. The most sizable of these by far is water (H_2O), whose atmospheric concentration is about 1 percent. Compared with water, the concentration of CO_2 is around 0.04 percent. The other main trace gases are methane (CH_4), chlorofluorocarbons (CFC-11 and CFC-12), nitrous oxide (N_2O), and ozone (O_3). Methane now amounts to around 20 percent of atmospheric trace gases, but its future rate of growth is in doubt. Water vapor warms the earth by blocking the heat reflected upward. Should the vapor form certain types of clouds, however, they may reflect heat heavenward, thereby cooling the earth. As the earth warms, evaporated surface water rises, cools as it drifts higher, and forms droplets of water, which both block out heat and reflect it back to earth. Which mechanism is dominant is the question.

Methane enters the atmosphere from numerous sources.[131] For a long time methane production was thought to grow about 1 percent a year as a result of increased cattle raising, growth of rice, and reliance on natural gas.[132] Whereas around 1800 methane was approximated at 0.75 ppm in the atmosphere, in our time it has risen to around 1.70 ppm.[133] Stephen Schneider has expressed the fear that as the earth warms, the extensive permafrost in the Arctic and Siberia will warm, thereby releasing a great deal more methane with substantial effects.[134] Studies of fossilized dung suggest to some paleoclimatologists that dinosaur flatulence may have been responsible for the considerably larger warming experienced in remote times.[135] In the spring of 1992, however, well-researched articles appeared demonstrating that the buildup of methane was, for a variety of reasons (better recovery from oil and gas wells, fewer rice paddies and cattle), far less than feared, dropping from 13.3 to 9.5 parts per billion (ppb) per year.[136]

Unlike methane, which has a residence time of only a decade, CFCs are believed to last in the atmosphere for centuries. Despite the fact that they are measured in a few parts per trillion, they exert powerful effects. Because CFCs are produced almost entirely by human forces and because the Montreal Protocol has greatly limited their production, they do not hold the threat of distant warming. But they should still be important into the twenty-first century.[137]

We covered in the previous chapter the depletion of ozone in the lowest stratosphere. "Our results," say V. Ramaswamy, M. D. Schwarzkopf, and K. P. Shine, of the Atmospheric and Oceanic Sciences Program at Princeton University, "indicate that a significant negative radiative forcing results from ozone losses in middle to high latitudes in contrast to the positive forcing at all latitudes caused by the CFCs and other gases." Their "results suggest that the . . . contribution of CFCs to the greenhouse climate forcing is substantially less than previously estimated."[138]

One of the major critics of the global warming thesis, climatologist Richard Lindzen, claims that "there is compelling evidence that all the known destabilizing feedbacks in the models may actually be stabilizing (negative) feedbacks."[139] To make the point more directly, another critic, Sherwood Idso, asserts:

> It is difficult to avoid the conclusion that the Earth's climatic machinery is programmed, so to speak, to maintain the status quo. That is, it does indeed appear that the planetary climate system actively counteracts both heating and cooling perturbations; although these negative feedbacks produce readily detectable short-term deviations from the Earth's mean climate state, they clearly have a tendency to perpetuate the long-term climatic stability of the planet.[140]

This is tough talk.

Hansen and Lacis respond to Lindzen's claim that even if CO_2 doubled, it would lead to less than a 1.5°C increase in the earth's temperature: "His argument, buttressed by a metaphysical presumption of a need of the planet for stability, is based primarily on the hypothesis that the net impact of increasing moist convection (expected to accompany greenhouse heating) is a drying of the atmosphere, resulting in a negative water vapour feedback." But, they argue, "Lindzen's hypothesis of a negative water-vapour feedback cannot be reconciled with real world data."[141] Let us take a further look. Another group of climatologists, from the Goddard Space Center, argue that satellite data confirm the existence and strength of positive feedback from water vapor.[142]

In general circulation models, the upper tropospheric water vapor—which is the largest greenhouse gas, ranging from 3 to 12 km above the earth—increases along with the surface temperature. "Without this feedback," Lindzen asserts, "no current model would predict warming in excess of 1.7 degrees Centigrade—regardless of any other feedback."[143] Satellite studies purporting to demonstrate positive feedback, he believes, are defective in one essential respect: they do not attend to the decline of humidity in the upper troposphere. This drying implies a decrease in warming from any increasing carbon dioxide.[144]

An examination of the snowline record around the equator in past eons led Lindzen and De-Zheng Sun to contend that the increased humidity in the middle and upper troposphere would cause strong negative feedbacks.[145] Lindzen and his colleague claim that as the sea surface warms, clouds will grow and become more buoyant, thus reaching the upper levels of the troposphere where the weather is colder, so that they will hold a lesser amount of water vapor. The less the water vapor, the less the positive feedback, the less the warming. My guess is that climatologists will concur with Idso that "the great

CO_2-induced water vapor feedback effect that the GCMs predict for the Earth is most probably a model artifact and not a characteristic of the real world."[146] Until Lindzen's critics have a chance to get back at him and other scientists enter the fray, I will score this proposition as neither proven nor unproven.

Aerosols

"Other than greenhouse gases," say Hansen and Lacis, "the larger known global climate forcing is that due to changing atmospheric aerosols." Indeed, "aerosols are the source of our greatest uncertainty about climate forcing."[147] Hansen and Lacis consider the effects of these aerosols potentially significant but smaller than the greenhouse effect itself. Other scientists disagree. Gaseous sulfur compounds rise into the atmosphere primarily from the combustion of fossil fuels, but also from volcanoes and the burning of biomass on earth. Sulfur oxide forms water droplets, creating sulfuric acid (H_2SO_4), which then acts to brighten existing clouds.[148] The prevailing theory is that a larger number of brighter clouds should reflect heat back into space and thereby add a cooling factor to the climate equation.

In 1990 R. J. Charlson and his colleagues demonstrated that even without cloud formation, sulfate particles backscatter radiation from the sun, thereby cooling the surface of the earth. According to Balling, who cites numerous authorities, "a scientific consensus is growing that the aerosol sulfates are acting to retard, to some degree, the expected warming from the buildup of greenhouse gases."[149] Charlson and six of his colleagues produced a consensus statement holding that the influence of sulfates in the lower atmosphere could range from 0.5 to 2.0 watts/m^2.[150] How much is that? "One-third of the global temperature trend of the past 100 years disappears when the stratospheric aerosol index is considered!"[151] Yet an anomaly remains to be explored.

Most fossil fuel is burned in the Northern Hemisphere; it therefore emits about one and a half times the amount of sulfur aerosol as does the Southern Hemisphere. Yet the temperature in both hemispheres matches. "We are left with a puzzle," according to Steven Schwartz of the Brookhaven National Laboratory. "Why isn't the earth responding to the difference in the emissions?" Some scientists think there might be a plankton-climate connection, and others believe that the distribution of sulfate aerosols is so patchy that it is hard to say exactly what it does in which place.[152] If warming increases the growth of plankton, as has been hypothesized, then plankton must emit a gas called dimethylsulfide, which acts like sulfuric acid to form aerosols with the same cooling effect. "Global climate is quite sensitive to the population of these particles," Charlson asserts. "I believe that we have identified an important missing piece of the overall climate puzzle."[153] Monastersky cites Charlson as asserting that the aerosols' "combined direct and indirect influences may be just about as strong as the greenhouse gases."[154]

Arguing that "anthropogenic SO_2 [sulfur dioxide] emissions may exert a significant cooling effect on climate in the Northern Hemisphere through backscattering of solar radiation by sulfate particles," Jos Lelieveld and Jost Heintzenberg isolate a mechanism—oxidation of man-made SO_2 droplets within clouds into sulfates—that makes sulfur dioxide emissions an efficient backscatterer of SO_2 emissions on earth.[155]

Proponents of the global warming thesis sometimes claim that the cooling effect of sulfur emissions, man-made as well as natural, may have masked underlying warming. To the contrary climatologist Thomas Karl reports that these emissions have actually decreased substantially in the United States.[156] "Therefore," the Marshall Institute's report concludes, "they could not have masked the expected greenhouse temperature increase in the U.S."[157]

Behind this critical exchange scientists continue to argue the validity of the concept of man-made carbon dioxide as culprit. Advocates of the global warming thesis are interested not only in demonstrating that the earth is likely to warm significantly in the next century but also in showing that this will result from anthropogenic emissions, which can be controlled. Their preferred policies cannot be separated from their science, because the one, carbon withdrawal, cannot be separated from the other, man-made carbon as the great cause. Similarly, if the opponents of the global warming thesis can show either that the earth is not warming or that if it is, the increase in man-made carbon dioxide is not at fault, they will leave the proponents without a policy and return this matter to the locus of fascinating scientific debate. Because sulfur dioxide is produced along with carbon dioxide by industrial emissions, if the one substantially offsets the other, then humankind is not doing harm to itself, and the whole tone of the debate shifts.

Qualifying their conclusions by saying that these might be off by a factor of two or four, Joyce Penner, Robert Dickinson, and Christine O'Neill estimate that smoke particles from the burning of biomass might move the global radiation balance in a cooling direction. Like sulfur particles, these smoke particles increase cloud reflectivity and enhance cloud condensation. "Together these effects," they reason, "although uncertain, may add up globally to a cooling effect as large as 2 watts per square meter, comparable to the estimated contribution of sulfate aerosols. Anthropogenic increases of smoke emission thus may have helped weaken the net greenhouse warming from anthropogenic trace gases."[158] The burning of the forests, whatever its other effects, may be keeping the rest of the earth from heating as much as it otherwise would.

Warning that temporary cooling produced by sulfur aerosols may provide a false sense of security, just when intervention against the greenhouse effect is best undertaken, Hansen states that "we may have a larger warming in store than the present trend would suggest." Compare this view with Balling and Idso's contention that it is "SO_2-modulated energy balance perturbations rather then the CO_2 greenhouse effect [that is] . . . the agent responsible for the decreasing diurnal temperature range of the past three decades."[159]

Volcanic Eruptions

"The modeling of the 'Little Ice Age' and other features [of the earth's climate history] suggest that most of the century-to-millennia events are driven by variations in volcanicity."[160] Few climatologists go as far as Reid Bryson does in this statement, though the thesis has its advocates. It is widely agreed that the earth cools when volcanic eruptions scatter radiation from the sun.[161]

The eruption first of El Chichon (in Mexico) and then of Mount Pinatubo (in the Philippines), the latter the biggest bang in the twentieth century, produced a haze of dust particles of light-reflecting gas and ash some fifteen miles above the equator. Scientists at the National Oceanic and Atmospheric Administration, making use of their polar-orbiting weather satellites, estimated that temperatures in the tropics have been lowered by an average of some 4°F.[162] While the ash drops out of the atmosphere after a few weeks or a month, the sulfur dioxide gas turns into the aerosol drops just discussed and stays in the stratosphere from one to three years, thereby blocking a good deal of the sun's solar energy. Data from the Nimbus 7 satellite support an estimate of 15 million tons of sulfur dioxide, plus or minus 5 million tons, ejected into the atmosphere.[163] Hansen commented, "either we're going to see substantial cooling over the next two years or our global climate models are wrong."[164] Students of climate note that volcanic eruptions are but one of a number of phenomena that lead to cycles of heating and warming in relatively short periods of time.[165]

Is the World's Ice Melting?

Melting of the world's ice, including glaciers, is seen by some as a greenhouse signal. But, as usual, there are differences of opinion over whether there is more or less ice. Part of the problem is that only about 1 percent of the glaciers in the world are actually observed. On the basis of data provided by the Permanent Service on the Fluctuations of Glaciers (renamed the World Glacier Monitoring Service), Fred B. Wood of the U.S. Office of Technology Assessment reported that "advancing glaciers are shown to have increased from about six percent of observed glaciers to 55 percent." Some are retreating—that is, growing smaller—but most appear to be expanding. He concluded with the usual plea for more observation.[166]

There have also been reports that polar sea ice has diminished in size by some 6 percent.[167] At roughly the same time (1988–1990) an article described that glaciers in Norway as "growing bigger in apparent defiance of global warming." According to Olav Orheim, director of the Antarctic Section at the Norwegian Polar Research Institute, global warming theories have missed the main show, which is that as temperatures rise more snow falls. Consequently, "in Western Norway, all small glaciers, which respond most quickly to climate changes, are now advancing."[168]

As time has passed, predictions by the International Panel on Climate Change, and other global climate modelers, of precipitous rises in sea levels due to thermal expansion of water following global warming have become less frequent. At the South Pole, observed glaciologist Charles R. Bentley of the University of Wisconsin, slight warming led to more snow that pulled water in from the sea.[169] By 1992 the direction of change had reversed, so that now the expectation is that sea levels might drop a foot or two in the next century. By examining the geological record over the past 130,000 years, the indication is that a warmer climate leads to the growth of ice sheets rather than rising sea levels.[170] Mark F. Meir, director of the Institute of Arctic and Alpine Research at the University of Colorado, told *Science News*, "We have revised rather drastically our best estimates of how much global sea level will rise due to greenhouse warming,"[171] adding that "our understanding of the system is not very good at the moment." Again we see that although water expands when warmed, negative feedback from clouds to snow can lead to a different result.[172]

Is the Earth Warming or Cooling?

Among the intriguing suggestions for treating the earth as if it had a natural thermostat (that is, as if it were capable of large negative feedbacks) is that as the planetary surface warms, the ocean's phytoplankton increases, thereby metabolizing dimethylsulfide (DMS), which diffuses into the atmosphere, thus increasing cloud reflectivity. As the earth cools, there is less DMS produced, thereby allowing more of the sun's radiation to hit the earth, enhancing warmth.[173]

Carbon dioxide is a strong fertilizer, whose higher concentrations should greatly increase growth of vegetation, which would send up more DMS and thus cool the earth. For this reason, Idso wants to encourage research on how soil microbes enhance DMS production. "Clearly," he writes, "the lowly soil microbe is a force to be reckoned with in the great unfolding drama of global change."[174] In a 1984 study Hansen and others predicted that warming would decrease vegetation and snow, thus decreasing the earth's reflectivity. This would be a positive feedback.[175] The evidence, however, suggests that warmth increases vegetation reflectivity, thereby providing modest negative feedback. By lowering transpiration (the water lost in plants that goes into the atmosphere), thus closing the pores in plants that let the water escape, CO_2 fertilization reduces the need for water. Less transpiration also means that plants will do better in arid areas. With some qualifications, doubling the amount of CO_2 available to a plant reduces its need for water by about a third.[176]

How good can it get? Hugh Ellsaesser wonders whether "we should be thinking of adding carbon dioxide to the atmosphere" to save fossil fuels for the future, as a contribution to energy sustainability, and to ward off what has long been thought of as a possible return to the lower temperatures of the little ice age. Why would anyone, including a noted climatologist, think that increasing the atmosphere's CO_2 content might be beneficial?

In 1975 there appeared in the *New York Times* two headlines. One read "Scientists Ponder Why the World's Climate Is Changing: Major Cooling Widely Considered to Be Inevitable"; the second, "Theory of 'Greenhouse Effect' of Sun's Rays Is Challenged." Two years earlier, in 1973, *Science Digest* readers learned that "at this point, the world's climatologists are agreed on only two things: that we do not have the comfortable distance of tens of thousands of years to prepare for the next ice age . . . once the freeze starts, it will be too late."[177] In those days panels of scientists predicted that a drop of 1.6–2.0°C in the world's climate—parallel, one might say, to the proposed increase in warming—"would lead to an unstable condition in which continental snow cover would advance to the Equator . . . [so that] the oceans would eventually freeze."[178] A book published in 1976 called *The Cooling* held out a future of famine and devastation, "world chaos, and probably world war, and this could all come by the year 2000."[179] In the same year Stephen Schneider wrote *The Genesis Strategy,* advocating drastic action to prevent or mitigate the coming global cooling. Why would climatologists then and even now think that the world might be in for a new period of advancing glaciers and lower world temperatures?

Leaving out blips on the weather screen, like a series of cold winters, those who study climate over eons—say, the last 850,000 to 1 million years—observe that climate today is a lot warmer than it has been. These warm periods, lasting something like 11,000–12,000 years, are given the name interglacials, suggesting that glaciation was the norm. Of course, for those paleoclimatologists who studied the far distant past, the preceding 40 million years during which dinosaurs lived were a lot warmer than today.[180] The indicators on which scientists based their forecasts of a coming global cooling were a doubling not of CO_2 but of temperature minima since 1930, unfavorable monsoons from 1920 to 1965, and excellent growing seasons in the U.S. wheat belt from the 1950s through the 1970s, which apparently suggested that this warmth could not last.[181]

Predictions of Drought in the Central United States

All climate models, as far as I know, predict rises in temperature along with increases in CO_2 and increases in global average precipitation. They also make other micropredictions that need to be considered: for example, the incidence of drought is expected to increase in the central United States from 5 to 50 percent by the mid-twenty-first century.[182] The predictions of the general circulation models are based on a rise in temperature followed by an increased rate of evapotranspiration, especially during the summer, so that a deficit would develop in the moisture content of the soil, thus increasing the frequency of drought.

But the evidence is against it. We have seen that temperatures are either not rising at all or not as quickly as predicted, and thus the rate of evapotranspiration need not go up. For the models to predict correctly, there would have to be an increase in rainfall during the winter and a decrease in summer, but that has not taken place.[183] In fact, there is more rainfall over the United States as a whole, and the levels of moisture in the soil are rising rather than declining.[184] Indeed, six of the nine states showing the greatest increase in soil moisture content are located in the central part of the country.[185] As Kirby Hanson and his colleagues conclude, "Neither is there evidence of change in winter or summer precipitation on the northern plains during that [1895–1987] period."[186]

Increasing CO_2 in the Atmosphere

"Will man's production of CO_2 prove the salvation of life on our planet?" Sherwood Idso asks. "Or is the rising CO_2 content of the air the very essence of the problem?" "This," he rightly states, "is *the* environmental issue of our time."[187]

A growing number of scientists argue that increased carbon dioxide at, say, the 600 ppm predicted for the middle of the next century, is likely to enhance rather than detract from life on earth. For one thing, there is apt to be about 10 percent more rain. For another, evidence shows that most warming is taking place during the winter and at night, which should lengthen growing seasons. Although global temperatures might rise as much as 1°C, evaporation rates would remain stable and soil moisture content the same. Moreover, atmospheric CO_2 sucked up by plants should greatly increase their growth rates while substantially decreasing their need for water.[188] Supporters of the global warming thesis generally agree that enhanced amounts of CO_2 benefit vegetation. But if warming goes way up, they believe that plants will die of heat prostration before they can take advantage of these possibilities. The positive feedback they envisage arises when warming speeds up plant respiration, thereby accelerating the decay of organic matter, which would release more CO_2.[189]

A Seventh-Inning Scorecard

The global warming game is still being played. We are in the middle of what might well be an extra-inning contest. Thus our scorecard cannot be complete.

But as we try to collect our wits during the seventh-inning stretch, a glance at the scorecard (Table 11.2) will prove revealing: true claims are outweighed by the false, and negative feedbacks dominate the positive.

We were already aware of decreases in positive feedbacks and increases in negative feedbacks that limit predictions of global warming. Now James Hansen agrees with Patrick Michaels that most of the warming that has taken place occurs at night. Although Hansen still expects daytime heat to go up over time, thus validating the greenhouse thesis, science reporter Boyce Rensberger concludes that "Mr. Hansen's forecast, however, is not as dire as it used to be."[190] Adding to the optimistic account is a paper by Jonathan DeCall of the University of Wisconsin at Milwaukee to the effect that in the past four decades there have appeared no signs of warming whatsoever in the Arctic.[191]

To top it all off, there is increasing evidence in the news, to use the headline in the *New York Times,* that "A Forest Absorbs More Carbon Dioxide Than Was Predicted." In the study of Harvard Forest in Petersham, Massachusetts, investigators discovered that the forest absorbed approximately one-third more carbon than was assumed in the computer models that generate predictions of substantial global warming.[192] As for the concern that Arctic temperatures have been increasing in intensity and depth for some time, exacerbating global warming, it now appears that Arctic pollution has substantially declined and that warming from that source is "roughly half" what was feared in the past.[193]

The Media and Scientists on Global Warming

In a study of media coverage of the greenhouse debate from January 1985 through August 1992, investigators at *Media Monitor* studied 67 opinion articles and 387 news stories in *U.S. News and World Report, Newsweek, Time,* the *Wall Street Journal,* the *New York Times,* and the *Washington Post,* as well as 73 television stories broadcast in the evening on NBC, CBS, and ABC. In deciding whether the greenhouse warming theory was justified, the proponents outnumbered the opponents in all these stories 9 to 1. It is interesting that of the news stories 61 percent affirmed the theory while only 5 percent rejected it; whereas in editorials and op-ed pieces, in which scientists had a greater say, 50 percent affirmed and 22 percent objected. Almost half the stories contained suggestions of gloom, doom, and disaster.[194] In a Gallup poll of experts reporting on the same issue, 60 percent agreed that global temperatures had indeed risen but only 19 percent attributed the increase to human activities. Scientists displayed considerable disagreement, with half of those polled saying that there might be significant global warming in fifty to one hundred years and half stating otherwise.[195] This deep division should be contrasted with the often expressed opinion that the experts agree that global warming is upon us.

A professor of atmospheric science at Colorado State University, Roger A. Pielke, reported that at a department reunion of some two hundred professors,

Table 11.2 A global warming scorecard

Prediction	True	Maybe	False	Positive feedback	Negative feedback
Greenhouse signals are detected above natural variability			X		
Statistics used to extrapolate temperature are valid			X		
Expected 1°C global warming materializes			X		
Northern hemisphere warms more than Southern			X		
Winter warms more than summer	X				
Extreme high temperatures appear with increasing frequency			X		
Prediction of past temperature is valid			X		
Residence time of CO_2 is long, i.e., nearly 200 years			X		
CO_2				X	
CO_2 is increasing in atmosphere	X				
CO_2 is increasing fast, according to GCM scenario			X		
Clouds			X		X
Glaciers are retreating			X		
Sea level is rising because of global warming			X		
Solar radiation, not CO_2, drives climate		X			
Water vapor by convection is nearly as large as in GCM				X, decreasing	
Methane				X, decreasing	
Methane is growing fast according to GCM scenario			X		
CFCs				X	
Aerosols					X
Ozone loss					X
Smoke, temporary					X
Volcano, temporary					X
Phytoplankton					X
Vegetation reflectivity					X
Droughts are increasing			X		
Soil moisture is declining			X		

students, and alumni, not one answered affirmatively to the question of whether human activities had led to global warming. Why, then, he asked rhetorically, had atmospheric scientists not spoken out? "The main reason," in Pielke's opinion, "is that most of us have not been asked whether we think that policy should be made predicated on the belief that global warming is, in fact, occurring." He was asked to comment on the IPCC report, but he believes that his views were "completely ignored" because they did not comport with those of the sponsors. His judgment is that no one knows how to make climate predictions.[196] A review of the scientific literature has persuaded me that this is true. When one thinks of the alternative hypotheses—that climate is controlled by volcanic eruptions, oceans, clouds, phytoplankton, vegetation, the sun, and aerosols—placing immense bets on a single basis for predictions is unwise.

12

Reporting Environmental Science

with Brendon Swedlow

Citizens usually first encounter claims of chemical harm to health and the environment by opening the morning paper or tuning in to an evening newscast. Consequently, the task of citizenship in science, of informed participation in decisions that require some understanding of science, begins with the evaluation of the mediated science found on our doorsteps and in our living rooms. In asking whether the claims of harm are true, we must therefore also ask whether the expertise we are being offered to evaluate those claims is a true representation of scientific views of the hazards. If government officials seem to overreact and to overregulate, it is in part because they must respond to a citizenry that gets its scientific knowledge of harm largely through newspapers, newsmagazines, and television.

Assuming Rather Than Assessing Danger

A basic question is whether scientific information is even used to assess claims of harm reported by the media. Those studying New Jersey newspapers came up empty-handed when investigators asked a panel composed of a journalist, scientist, industry representative, and environmental activist to evaluate media treatment of man-made environmental health risks.[1] Science did not enter into these stories in support of risk assessments because claims of harm were often not even explicitly made. Harm was simply assumed to flow from the presence of a particular chemical in a place where it was not supposed to be.

"Risk is so casually mentioned in the articles that it's like people have read five stories in depth [on the topic]," commented the environmental activist.[2] "So many of the articles make the assumption that readers will translate the event into danger . . . They assume that if they report a fire, or an odor, or smoke particulate, that means it's bad. And frankly it is; the question is how bad."[3] "What I decided must be going on is that the reporters have made assumptions that the reader knows what the . . . risks are."[4] The industry representative concurred. "They're accepting the fact that dioxin is dangerous, that takes care of it, put it in the lead, I don't have to worry about it any more, let's get on with the indictment."[5] "I've read some of these articles in the past and said 'right on' as an activist," the environmental advocate concluded. But "looking at them in a slightly academic manner, I thought, 'So what?' They are catchy and get citizens involved but don't give any content."[6]

Environmental reporting in New Jersey is not easily distinguishable from reporting in the rest of the country. Some of the same investigators did a content analysis of 564 network news reports (appearing from January 1984 through February 1986) on "man-made chemical, biological, and physical agents that create risk in the indoor, outdoor, and occupational environments."[7] They found that while expert and advocate sources were more likely to present information on environmental risk, and hence science, than government sources, the experts tended to be legal or medical professionals rather than scientists and neither they nor the advocates spent much time evaluating risk. "Attorneys and judges described judicial proceedings, not their scientific bases; local health inspectors or consultants detected the presence of a substance, not the dangers of exposure to it; and family physicians described the symptoms, not their possible connection to chemical exposure. Typical advocates represented special interest groups like a local citizens' organization demonstrating against a facility or veterans' groups requesting benefits for victims of Agent Orange rather than national organizations capable of placing sophisticated risk information in an environmental advocacy perspective."[8]

A national random sample, taken in 1993, of 512 reporters who had covered at least one environmental story at a television station or newspaper also reveals that many journalists choose to report on reputed environmental hazards in a way that omits scientific information and does not provide any evidentiary basis for evaluating claims of harm.[9] When reporters were offered two environmental story scenarios and three possible themes for reporting them, barely a majority in one case and only a plurality in the other chose risk assessment themes (see Tables 12.1 and 12.2). Almost half of the respondents in both cases consequently would have reported the story in one of two ways that did not require them to identify or delineate the risk raised by the scenario. Scientific views of the environmental dangers were therefore necessarily omitted. These results are particularly significant since responses to the scenarios were weighted to reflect more heavily the views of reporters at larger newspapers, which are generally thought to be better equipped to report scientific elements of an environmental hazard story.

Experts Say Media Overstate Man-Made Causes of Cancer

Beyond the basic question of whether science enters into reporting claims of harm to health or the environment is the question of how science and scientists

Table 12.1   Covering a controversy over power

Suppose a coalition of parents and environmentalists claims that high-voltage power lines near a school put their children at risk of getting cancer. It wants the school closed. The power company says there is no danger. If you had to choose one theme for your feature story, which would it be?

	Reporters		
Theme	*All (262)*	*Print (206)*	*TV (40)*
Evidence to support or allay concerns that power lines cause cancer	52%	48%	66%
Claims and counterclaims of the coalition and the power company	32	35	24
Means of reducing the adverse health effects of living or working close to high voltage power lines	13	14	8
None of these	3	4	2

Source: American Opinion Research, Inc., *The Press and the Environment—How Journalists Evaluate Environmental Reporting,* conducted for the Foundation for American Communications (Los Angeles: Foundation for American Communications, 1993), p. 59.

Table 12.2   Covering a chemical spill

A derailed tank car of metham sodium spilled into a river this week. Fish and vegetation are dying. If you had to choose one theme for your feature story, which of the following would it be?

	Reporters		
Theme	*All (262)*	*Print (206)*	*TV (40)*
The nature of metham sodium, its uses, and its effects on plants, animals, and people	43%	45%	36%
Response to the spill by the railroad company and government agencies	31	30	34
The railroad company's responsibility for the accident	21	21	20
None of these/refused	5	4	10

Source: American Opinion Research, Inc., *The Press and the Environment,* p. 60.

are used in these stories when they do make an appearance. One way to assess media distortion of environmental science is to compare a random selection of expert opinions on cancer with the opinions voiced by journalists and their sources—some experts, some not. In one study, six of seven media sources agreed that the United States faces a "cancer epidemic."[10] Among experts, fewer than one in three shared this view. Similarly, while two of three media sources believed that cancer-causing agents are unsafe at any dose, more than two of three experts believed that safety is related to dose. Finally, while half of media sources said that human cancer risks can be extrapolated from animal tests, roughly a quarter of surveyed experts believed this (see Table 12.3).

S. Robert Lichter and Stanley Rothman also studied the relative importance cancer specialists, media sources, and reporters assign to various environmental (as distinguished from inherited) causes of cancer.[11] Cancer is widely feared and its suspected causes often reported on, but journalists and their sources mention some causes much more than others. Randomly selected cancer experts also believe that some causes are much more important than others, but their priority list looks quite different from that in media accounts.

Lichter and Rothman measured the importance experts gave environmental causes of cancer by asking them to rate each of thirteen causes on an eleven-point scale, where 0 meant an environmental factor was not a cause and 10 indicated the factor was a very important contributor to cancer. A sample of 401 cancer scientists was randomly selected from current members of the American Association for Cancer Research specializing in carcinogenesis or epidemiology, and those selected were interviewed over the phone by the Roper Center for Public Opinion Research in January and February 1993.[12]

Lichter and Rothman measured the importance the media attribute to an environmental cause of cancer by tallying the number of times reporters or their sources stated that a particular environmental factor was a suspected, probable, or known cause of cancer. Major causes by this measure were those most frequently associated with cancer from 1972 through 1992 in *Time,*

Table 12.3   Comparing media sources and expert opinion on cancer

	Percentage agreeing with claim	
Cancer claim	*Media sources*	*Cancer experts*
U.S. faces cancer epidemic	85	31
Cancer-causing agents unsafe at any dose	66	28
Can base human cancer risks on animal tests	50	27

Source: Adapted from "Is Cancer News a Health Hazard?" *Media Monitor,* 7, no. 8 (Nov./ Dec. 1993): 4.

Newsweek, U.S. News and World Report; the ABC, CBS, and NBC evening newscasts; and prominent articles from the *New York Times, Washington Post,* and *Wall Street Journal.*[13]

Almost all of the cancer specialists considered tobacco smoke a major environmental cause of cancer (rating it 7 to 10 on an eleven-point scale), while more than half thought the same of diet and exposure to sunlight (see Table 12.4). Secondhand tobacco smoke and workplace chemicals were thought to be major causes by roughly two in five experts. Finally, sexually transmitted disease, herbicides and pesticides, and air and water pollution were thought to be major causes of cancer by little more than one in four of the specialists.

Tobacco, including chewing tobacco and snuff, was also associated with cancer more often than any other *single* environmental cause by reporters and their sources (see Table 12.5). But beyond this concordance of reporting and expert opinion media accounts did not closely mirror scientific thinking regarding environmental causes of cancer. While journalists linked food addi-

Table 12.4   Causes of cancer—scientist survey

Aspect of environment	Mean	Percentage rating cause as:			
		Major	Moderate	Minor	DK
Tobacco smoke	9.21	**96**	3	1	0
Diet	6.38	**52**	34	11	4
Sunlight	6.33	**54**	32	13	2
Environmental smoke	5.74	**42**	35	22	1
Chemicals in workplace	5.44	**37**	27	25	3
Sexually transmitted disease	4.74	28	34	34	4
Pesticides and herbicides	4.72	26	**36**	34	4
Air and water pollution	4.70	26	**37**	34	3
Infectious diseases	3.96	14	36	**43**	6
Drugs (medical)	3.64	13	27	**54**	6
Food additives and preservatives	3.27	10	29	**57**	4
Chemicals in the home	3.05	7	27	**62**	4
Radiation (medical/dental)	3.00	10	22	**67**	1

Source: Adapted from S. Robert Lichter and Stanley Rothman, "Scientific Opinion vs. Media Coverage of Environmental Cancer: A Report on Research in Progress" (Washington, D.C.: Center for Media and Public Affairs, Center for Science, Technology, and Media; Storrs, Conn.: University of Connecticut, Roper Center for Public Opinion Research; Northampton, Mass.: Smith College, Center for the Study of Social Change, 1993), table 1.

Note: Level of concern measured on an eleven-point scale, with "major" rated 7–10, "moderate" 4–6, and "minor" 0–3; "DK" used for those replying "Don't Know." Boldface numbers reflect modal rating (plurality).

Table 12.5 Cancer causes cited in news stories, 1972–1992

Cause	Affirmed	Suspected	Total
Manmade chemicals	304	194	498
Tobacco	213	79	292
(Secondhand smoke)	23	15	38
Food additives	165	108	273
Inheritance	155	117	272
Hormones as drugs	115	153	268
Pollution	127	95	222
Man-made radiation	109	103	212
(Medical/dental)	15	29	44
Infection	121	79	200
Pesticides	117	77	194
Asbestos	104	59	163
Man-made food contaminants	70	69	139
Dietary choices	67	69	136
Sunlight	92	42	134
Other drugs	70	59	129
Lack of natural defenses	28	58	86
Natural chemicals	52	32	84
Natural food contaminants	55	27	82
Alcohol	40	39	79
Lifestyle	34	36	70
Clothing additives	32	34	66
Cosmetics	35	24	59
Dioxins	29	19	48
Radon	29	15	44
Hormone imbalances	13	28	41
Aging	2	7	9

Source: Lichter and Rothman, "Scientific Opinion vs. Media Coverage," table 6.

tives and hormones used as drugs to cancer almost as often as tobacco, roughly only one in ten experts considered food additives and any kind of drug major causes of cancer. Moreover, dietary choices and exposure to sunlight—major causes for more than half the scientists—ranked low on the list of causes cited by journalists and were reported as affirmed or suspected cancer causes about half as much as tobacco, food additives, and hormones used as drugs.

On the topic of man-made chemicals as environmental causes of cancer media references and expert opinion diverge most noticeably. The Roper survey asked scientists about chemicals at work and in the home as causes of cancer. Less than two in five considered workplace chemicals major causes and less than one in ten thought likewise of chemicals used in the home. The survey did not ask about outdoor man-made chemicals. But it did ask about pesticides and herbicides and about air and water pollution. Barely more than one in four experts considered these major sources of cancer. In media accounts of cancer causation, however, pollution, pesticides, and man-made, nonmedical radiation were each linked to cancer about two times for every three times cancer and tobacco were associated.

Moreover, as Table 12.5 shows, even when pollution and pesticides and a host of other man-made chemical causes of cancer were separately considered, there remained a huge residual category of man-made chemicals that reporters or their sources said cause cancer. Sometimes these chemicals were identified, but most of the time unspecified man-made chemicals were linked to cancer. In media accounts, man-made chemicals were associated with cancer about five times for every three times that tobacco and cancer were linked.

Lichter and Rothman also asked their cancer specialists to rate media coverage of eleven alleged causes of environmental cancer (see Table 12.6). Pluralities found the risk to be overstated for five factors, stated fairly for six, and understated for none.

Finally, the experts were asked to rate, on a ten-point scale, various specific news outlets for the reliability of their reporting on environmental cancer (see Table 12.7). Network news, most people's source of news, was considered highly reliable (rated 7 to 10) by hardly more than one in twenty experts. Weekly newsmagazines were trusted by fewer than one in ten experts to report accurately on cancer. The lack of reliability of network reporting is perhaps not surprising given the constraints of time and the visual format of television. But one would assume newsmagazines do a better job, since they have more time to put stories together and more flexibility in the way they present them. What is really surprising, however, is how poorly the *New York Times* fared in this survey. Only slightly more than one in five experts considered the nation's flagship daily a highly reliable source of information about cancer, while almost one in three considered it unreliable (rating it 3 or less on the scale). The only way citizens can get an even chance of acquiring trustworthy information about cancer is to read *Scientific American,* and the only way to improve on those odds is to go to the nation's top medical journals and read real science, exactly what this book has emphasized. Even those sources, however, were considered accurate by less than three in four experts, which underscores the need to evaluate critically all sources and scientists in the way elaborated in Chapter 14.

We can only conclude with Lichter and Rothman that their findings pose a true challenge to citizenship in science, particularly where the citizen is unwilling or unable to go beyond media accounts of environmental cancer causation: "A regular news consumer would get the impression that many man-made substances pose greater risks than cancer researchers believe, while failing

Table 12.6 How scientists rate media portrayals of cancer risk

Source of risk	Overstated	Understated	Stated fairly	Don't know	Ratio (over to understated)
Nuclear plants	61%	9%	23%	6%	6:1
Pollution	54	13	25	8	4:1
Food additives	53	13	23	10	4:1
Radon	36	8	50	6	4:1
Pesticides	42	19	32	8	2:1
Chemicals in home	34	18	35	13	2:1
Chemicals in food (naturally occurring)	39	31	22	9	1:1
Dietary choices	29	26	39	6	1:1
Chemicals at work	28	26	42	5	1:1
Sunlight	6	31	60	3	1:5
Tobacco	4	36	59	2	1:9

Source: Lichter and Rothman, "Scientific Opinion vs. Media Coverage," table 4.

Table 12.7   Cancer experts rate reliability of cancer reporting

Outlet	Percentage of experts considering outlet highly reliable source of cancer information
New England Journal of Medicine	72
Journal of the American Medical Association	55
Scientific American	54
New York Times	22
Weekly newsmagazines	9
TV network news	6

Source: Adapted from "Is Cancer News a Health Hazard?" *Media Monitor,* 7, no. 8 (Nov./ Dec. 1993): 5.

Note: Highly reliable sources rated 7–10 on a ten-point scale.

to appreciate the magnitude of risk that experts assign to other factors. Further, even a careful news reader would be misinformed about much of the scientific debate over cancer causation."

Developing Standards for Reporting Environmental Science

Before chronicling a similar distortion of scientific opinion in media coverage of global warming, it may be useful to pause for a moment and examine the criteria that journalists use and that critics have suggested for reporting environmental science. Journalists have traditionally been taught to strive for balanced reporting, but exactly what this means in the context of complex, multifaceted environmental issues has not been as clear as in some other, simpler areas. Historian John C. Burnham, for example, claims that "journalistic treatments of environmental matters have gutted science of content by reducing it not to a method for finding truth but to a conflict of authorities and even personalities quoted in the media."[14] He believes global warming reporting "has lent itself particularly well to trivialization by juxtaposition of competing authorities," a belief shared by at least a couple of former reporters. For example, Teya Ryan, now senior producer of Turner Broadcasting's *Network Earth* and vice president of the Society of Environmental Journalists, complains that "balance" is "often artificial, a matter of giving equal air time or newshole space to dissenting views of questionable merit."[15] She suggests that global warming journalism has been "most muddied" by this practice. Conceding that there is "rarely 100 percent certainty in science," she nevertheless thinks that although "an overwhelming majority of scientists believe that the theory of global warming is essentially correct," because of the norm of balanced reporting "journalists have felt compelled to seek out contrary points of view, in some cases calling on experts of doubtful expertise and motive." So "who is right?" she wonders. "With a balanced report the audience is left with more questions than answers." A former reporter for the *Nashville Tennessean* and now the vice president of the United States, Albert Gore, Jr., shares this critique of balanced reporting, also in the context of global warming journalism. "Climate change is measured over years, not days or moments," he noted while a U.S. senator. "And, for most observers who lack the advanced training of a Ph.D. scientist, the research experience of the experts, or the time to study the science and the evidence, it merely becomes a case of reporting both sides. On even the hottest summer day we can't say we're witnessing global warming, so if scientist A says one thing, search high and low for scientist B who will disagree." He draws attention to the 700 members of the National Academy of Sciences who wrote then President George Bush urging action on global warming and then criticizes the way the media reported this appeal. "Six or seven members take the other side of the argument, but they are given equal billing with the 700," he laments. "Our most eminent scientists are telling us we face a problem of unforeseen proportions. Yet the media, compelled to search high and low for 'the other side,' find those who say there is no problem and amplify those voices so that soon the protests of the few are as powerful as the concerns of many."[16] Sociologist Dorothy Nelkin believes that this striving for balance has its historical roots in a striving for objectivity, itself the result of journalists' admiration and emulation of scientists' success in separating facts from values. "Ironically," Nelkin observes, "this notion of objectivity is meaningless in the scientific community, where the values of 'fairness,' 'balance,' or 'equal time' are not relevant to the understanding of nature, where standards of objectivity require, not balance, but empirical verification of opposing views."[17]

Gore's criticism points toward one kind of solution, which is to sort out scientific hypotheses and evidence according to their "scientific standing."[18] If a majority or consensus view can be identified, then it can be reported as "scientists say global warming has arrived," with minimal acknowledgment, if any, of dissenters. Another, more radical, solution—called advocacy journalism— is suggested by Ryan and others: "This is what I saw as a reporter. This is who I talked to. This is my perspective, and here are my suggestions for change. If you want another point of view, find it from another broadcaster or newspaper."[19] *Network Earth,* she says, is "a program that will always come out on the side of the environment," a program that reports "that global warming is a reality and then offer[s] possible solutions."[20] All this seems clear enough and certainly does not represent a fringe set of views; a senior *Time* magazine editor, Charles Alexander, told participants at a Smithsonian Institution conference on the environment in 1989 that he "freely admit[ted] that on this issue we have crossed the boundary from news reporting to advocacy."[21] Yet both Ryan and Alexander have simultaneously hedged their position in various ways, seemingly caught between an old set of journalistic values and the new ones they are advocating along with environmentalism. "Your facts must be secure, and you must be ready to defend them," counsels Ryan, who then, in an Orwellian turn, insists on "not simply reporting the facts, but telling the truth."[22] In admitting that he "still [has] to be a journalist and ask hard questions of the environmentalists,"[23] Alexander appears to support Ryan's concession that "advocacy does not mean that Greenpeace is always right and the oil companies are always wrong."[24] Yet these statements muddy their earlier positions and leave us with an unclear journalistic standard. Both seem to believe that most of the truth essentially lies with the environmental movement. "I have three children," Alexander said before promising to get tough on environmentalists, "and I don't want them left on an overheated planet," thereby accepting as true exactly the claims that are in question.[25]

A mechanically applied balancing approach to reporting environmental science is clearly problematic, as is the formula of only reporting those views with "scientific" standing. These approaches are shortcuts and surrogates that do not really give the reader an idea of what scientific views exist and how scientists discriminate among them; nor do they allow the reader to assess the empirical bases of rival views. After all, dissenters may have a better understanding of how nature functions than the majority does. Advocacy journalism is not the solution to these problems either. Instead of striving toward the rarely realized ideal of accurately reporting scientific knowledge, advocacy journalism veers into politics, where the object is to select from all scientific views those that best support a position. Harnessing science to sloganeering and propagandizing does not seem likely to help anyone understand what scientists really think. In fact, this approach suggests we do not really care what they think, which represents a regression to a time when superstition and religious convictions determined human understandings of nature.

A better approach is simply to do a better job of reporting the scientific elements of a story, though this is obviously far from simple. Ryan and Gore themselves make comments that point toward this solution: Gore has said that his foregoing statements about global warming journalism are "not an argument for one-sided reporting, but rather for reporting that recognizes the subtleties involved in this issue and that accurately weighs opinions and research";[26] Ryan implies that "the dueling perspectives approach," when "responsibly done and extremely well done," is superior to advocacy journalism because it does more to empower people, "and empowering people is, in part, what covering the environment is all about."[27] We can only agree: reporters who facilitate citizens' understanding of scientists' views about environmental hazards are helping democracy function. Journalists should try "to analyze and write to allow for the best possible public evaluation of the arguments and data being used on all sides," as journalism professor Phillip J. Tichenor recommends in an oft-cited article about science reporting.[28]

As Bob Engelman, an environmental reporter for the Scripps Howard News Service put it: "The challenge is the same as for any other [issue] that is complex, important and likely to remain in the public eye: Learn the issue. Maintain skepticism. Seek out all viewpoints. Ask probing questions. And report the story as accurately and fairly as you can."[29] Readers must be given the materials to evaluate scientific understandings on their own terms. They should certainly be told what are the majority and minority views and what the affiliations of the disputants are, as well as the social and political contextual factors, but all of this information should not be substituted for reporting the scientific elements of a controversy. The reporter should try to take readers on a full tour of such debates on all their levels. But this requires the journalist to make the trip first. The primary obstacle to this approach is the unwillingness of news organizations to direct resources toward environmental reporting, toward furthering the scientific and environmental education of their reporters and editors. As Amal Kumar Naj, who writes on technology and environmental issues for the *Wall Street Journal,* noted, "When Dan Quayle appeared in his vice presidential debate and got to the 'greenhouse effect,' some editors looking for slips had to look up clips or call their beat reporters to check if the gaffe-prone candidate had gotten the facts right."[30]

Scientific Consensus on Global Warming a Media Artifact

Getting the facts right does not happen too often with environmental science issues, if global warming is any indication. Many advocates, including the now prominent former journalists cited above, have tied their hopes for governmental action on the environment to the global warming issue because they believe that scientific evidence of harm is stronger in this case than in others. At the same time, they claim that coverage of global warming misrepresents scientific evidence by overreporting the views of skeptics. Remarkably, the opposite is true: journalists have overstated the nature and extent of convergence among global warming experts. In fact, media accounts provide the only instances of anything approaching the claimed consensus among scientists.[31] If global warming constitutes environmentalists' best example of journalistic distortions, one has to wonder what they are talking about when they claim journalists have botched reporting on other environmental issues.

When climatologist James Hansen testified at a congressional hearing in 1986 that within the next century the greenhouse effect could cause temperatures to rise to "the warmest the earth has been in the last 100,000 years" but added that he "would like to understand the problem better before I order any dramatic actions," the *New York Times* and *Washington Post* led with the more extreme statement. "This was dramatic language, spoken with the authority of science, and it got front-page coverage in the *Washington Post,*" notes Stephen Klaidman. "Never mind that Hansen's basic plea was for better understanding of the problem; his testimony contained headline-grabbing language."[32]

Hansen again testified before Congress in 1988, this time stating that he was "99 percent" confident that the atmosphere had gotten warmer and that he had "a high degree of confidence" that there was "a cause and effect relationship between the greenhouse effect and the observed warming." These words became the basis for inaccurate reports, made by Philip Shabecoff of the *New York Times* among others, that Hansen had said "it was 99 percent certain that the warming trend . . . was caused by a buildup of carbon dioxide and other artificial gases in the atmosphere." The misrepresentation, as Klaidman observes, comes from the desire to write a hard lead, a first paragraph without significant qualifications.[33]

Fellow climatologist Stephen H. Schneider pointed to the complicity of some scientists in producing distorted media reporting when he admitted that

> On the one hand, as scientists we are ethically bound to the scientific method, in effect promising to tell the truth, the whole truth and nothing but—which means that we must include all the doubts, the caveats, the ifs, ands, and buts. On the other hand, we are not just scientists but human beings as well. And like most people we'd like to see the world a better place, which in this context translates into our working to reduce the risk of potentially disastrous climatic change. To do that we need to get some broad-based support, to capture the public's imagination. That, of course, means getting loads of media coverage. So we have to offer up scary scenarios, make simplified, dramatic statements, and make little mention of any doubts we might have.[34]

Such deficiencies in public climatology and its reporting were only amplified by the amount of attention these views received. In a recent seven-year period analyzed by Robert Lichter, director of the Center for Media and Public Affairs, Hansen was cited or quoted more than twice as often as any other scientific source (52 times). The next most frequently used scientific source was Michael Oppenheimer, of the Environmental Defense Fund (25 times). Only two of the ten most often cited scientists in global warming coverage over this period can be described as critics of the theory. These findings are based on Lichter's content analysis of the *Washington Post, New York Times, Wall Street Journal, Time, Newsweek, U.S. News and World Report,* and ABC, CBS, and NBC coverage of global warming from January 1985 through June 1991.[35]

Not surprisingly, then, four out of five scientists cited in the media maintained that scientific understanding of the greenhouse effect was sufficient to begin regulating human activity seen to enhance the effect. Of the various types of media sources, scientists were the least likely to express doubts about global warming models and scenarios. Whereas "proponents of global warming theory outnumbered opponents by a margin of nearly nine-to-one," among scientists in the media this ratio was 33 to 1. In other words, for every 100 scientists used as sources only 3 could be described as critics of global warming. Reporters expressing an opinion were comparatively more reserved, with a ratio of theory supporters to theory opponents of little more than 8 to 1. By these standards, administration officials were extremely reserved, with supporters outnumbering opponents by roughly 3 to 1.

Claims that the greenhouse effect was real reached their apogee in 1989, when nearly three-quarters of sources held this view and none contested it. As Lichter notes, the "realness" of global warming was the key question for the media, though coverage "subsumed several distinct scientific controversies, from the validity of historic temperature data to the role of industrial activity in global climate change," which, depending on the issue, are "real" to experts in varying degrees.[36] By 1992 the level of support expressed in 1989 had dropped by more than half and opposition had grown to the point where disbelievers were almost half as numerous as believers (see Table 12.8). This decline resulted not from media reports including more skeptical scientists but from increased numbers of Bush administration officials contesting claims of the reality of global warming.[37]

Opposition to global warming theory did not enter the media through straight reporting as much as through editorials, op-eds, and signed columns. Throughout the period 1985–1992, 61 percent of sources in straight reports argued for the reality of global warming and only 5 percent denied it, while in opinion pieces 50 percent of sources voiced support and 22 percent denied it.[38] Not only did the media portray a virtual scientific consensus on global warming, but they gave this apparent consensus a prominent place in their coverage. Scientists were quoted or cited twice as much as government officials,

Table 12.8 Media sources and the "reality" of global warming: Is global warming real?

Reporter and source	Yes	No	Maybe
Overall, 1985–1992	60%	7%	33%
Over time, 1985–1987	82	8	10
1988	69	3	28
1989	74	0	26
1990	47	9	44
1991	48	10	42
1992	35	14	51
By source type, 1985–1992			
Scientists	66	2	32
Reporters	59	7	34
Administration	42	13	45

Source: Adapted from Center for Media and Public Affairs, "The Great Greenhouse Debate," *Media Monitor,* 6, no. 10 (Dec. 1992): 2.

the second most relied on source, followed by environmental groups. Nearly 40 percent of all sources were scientists.

Given the strong support scientists used as media sources showed for the "reality" of global warming, it is difficult to comprehend environmentalists' criticism that the media overreported opposition to global warming theory in the interest of journalistic balance. When we look behind the scenes and ask, as the Gallup organization did, a random sample of scientists what they think about global warming—and, moreover, do not merge all global warming issues into one question about "the reality" of global warming—a far more variegated portrait of scientific opinion emerges (see Table 12.9). The 1991 Gallup survey of 400 randomly selected members of the American Meteorological Society and the American Geophysical Union reveals much more doubt and uncertainty among climate, atmospheric, and oceanographic scientists regarding global warming than is found among scientists used as media sources.[39] Three-fifths of surveyed scientists say they believe in a historic increase in temperatures, but only one in three of these attribute the increase to human activities, which translates into less than one in five (19 percent) of all surveyed experts believing in past human-induced global warming. Moreover, almost half (49 percent) of those believing in a historical increase think it could have had natural, nonhuman causes, meaning that almost one in three of these experts think past global warming is within the range of the earth's natural temperature fluctuations. Finally, more than half (53 percent) of those directly

Table 12.9   Scientific opinion on "reality" of global warming

Claim	Percentage agreeing (N = 400)
Global temperatures have increased in past	60
Human activity caused increase	19
Increase could be natural fluctuation	29
Warming is presently occurring	66
Evidence supports this view	41
Evidence does not support this view	21
Warming is not or may not be occurring now	34
Don't know if warming is occurring	24
There is no current warming	10
Temperature will increase at least 2°C over next 50–100 years	47
Scientists understand climate change well[a]	51
Media inform public well[a]	30

Source: Adapted from Center for Media and Public Affairs, "The Great Greenhouse Debate," p. 6.

a. "Well" refers to rating of 7–10 on a ten-point scale.

involved in research on global warming do not believe in a historical increase of any kind.

Two-thirds of the randomly selected experts believe that human-induced global warming is presently occurring, which equals the two-thirds of media experts who think global warming is a "reality." But here again there is a significant caveat missed by the media: little more than two in five surveyed experts (41 percent) think scientific evidence supports the theory of current, human-induced global warming, and more than one in five (21 percent) believe that the evidence does not support the theory. This gap between evidence and belief should be systematically explored by the media; instead, it is systematically ignored. Moreover, we should hear more from the critics and skeptics: 15 percent do not believe temperatures have increased, 10 percent do not think they are currently on the rise, and almost 25 percent are unsure on both counts. If one of four experts do not believe in past or present temperature increases, and if another one in four are uncertain about these increases, and if, further, almost one of three who believes in current warming does not think the evidence supports this belief, why don't we hear about it from journalists? All this disbelief, doubt, and uncertainty should add up to more clear criticism of global warming than the 1 in 33 sources who can be identified as critics in media accounts.

Some Conclusions about Media Distortions of Environmental Science

What kinds of omissions and distortions can we identify in the reporting of environmental science? First, it is clear from the panel evaluation of New Jersey newspapers, the content analysis of network news, and the survey of national and local media that claims of harm to health and the environment are often not supported by any references to scientific evidence. In fact, sometimes harm is not even explicitly claimed but is simply implied by the mere mention of a particular chemical, like dioxin, or by the reporting of a leak, fire, or spill.

Second, only about half of all reporters choose to address issues of harm to health or the environment in a way that relies on scientists or scientific information to determine the extent and nature of that harm. The reader who would like to exercise some citizenship in environmental science should be especially vigilant when it comes to these kinds of stories, always asking for evidence of harm, even where that issue is skirted entirely. Environmental hazards need to be established and assessed, not simply assumed, before proceeding to assign responsibility for them or asking what should be done.

Third, the comparison of reported scientific opinion and randomly selected scientific opinion regarding the causes of environmental cancer and the "reality" of global warming shows that when journalists do utilize scientists and scientific information in assessing claims of harm they do so in a way that misrepresents scientists' assessments. Beyond accurately reporting that cancer specialists consider tobacco a major cause of cancer, journalists emphasize cancer risks (by repeatedly linking cancer with the supposed cause) that few scientists consider important, while deemphasizing (by mentioning less often) risks that many scientists say are important contributors to cancer. Reporters focus on man-made causes, especially chemicals, while scientists are much more concerned about natural causes, like diet and exposure to sunlight. The comparison of twenty years of reporting on the causes of cancer and a two-month "snapshot" of scientific opinion would in itself not be definitive—since the media's cumulative emphasis on man-made causes could simply have reflected an earlier scientific emphasis on these causes—but when this information is combined with the ratings these scientists gave reporting on cancer, it is safe to conclude that significant distortions actually exist. Since pluralities of cancer experts think that the risk was overstated for five factors, stated fairly for six, and not understated for any, we know that the media currently are not doing a very good job of communicating scientific views on this issue. More important, the cancer causes that experts see as most overrated in media accounts are exactly those that journalists reported most frequently from 1972 to 1992.

The distortions of global warming science that occur in media reports are of two types. One is the failure to disaggregate the question about the "reality" of global warming into its constituent elements. Had the media really explored scientific opinion they would have realized that the "reality" of that opinion encompasses questions of historical versus present versus projected increases in global temperature and that the scientific community has different views on each of these issues. Journalists might have realized that a large proportion of experts think there have been no historical changes and that even among those considering these changes real, many think they were not caused by humans and are within the range of natural fluctuation. Even among believers in a present increase, the media might have pointed out, many also concede that current evidence does not substantiate their views, and that, further, the proportion of those who do not believe in these increases is much higher than the 3 critics out of every 100 experts who appear in media accounts. As for projected increases, the media have failed to report that the experts are about evenly divided on whether there will be a significant temperature increase in the next 50–100 years. This failure to distinguish, to disaggregate global warming issues, produces one type of distortion in reporting, which is closely tied to another: as with the reporting of cancer causation, scientists' views are misrepresented to the point that a false center of scientific gravity is created. With cancer reporting, the media inverted the relative importance of various causes. With global warming reporting, the media neglected to air the doubt, dissent, and uncertainty that actually existed in the community, creating a scientific consensus that did not really exist. Whatever the "reality" of global warming, that is the reality of global warming reporting.

These are only a few of the omissions and distortions found in mediated science.[40] The citizen who wants to evaluate claims of harm to health and the environment would do well to recall how the media conveyed or failed to report scientific aspects of environmental stories. The reader should be on the lookout for recurring scientific omissions. In reporting on the cranberry scare (involving the chemical 3-AT), PCBs, and dioxin at Times Beach, journalists consistently failed to point out that any potential health consequences of exposure depend on the extent of the chemical dose received. In the 3-AT and PCB stories, reporters also failed to note that animal tests have only limited if any predictive power of health effects in humans. Reports on the cranberry scare made much of a synthetic carcinogen without mentioning surrounding natural carcinogens. And at Love Canal, informal survey results were published without discussion of their limitations, while disease incidence in the community was detailed without background rates for those same diseases being presented.

The reader should be aware that the media can misstate scientific knowledge. On the first television program on Agent Orange, sources were allowed to talk about the exposure of veterans and birth defects in their offspring when there was and remains no plausible mechanism for male-mediated birth defects from such exposure. Meanwhile, in reporting on acid rain, the *New York Times* attributed to the National Acid Precipitation Assessment Program (NAPAP) statements that were in fact contrary to its conclusions.

The reader must also watch for overgeneralizing, as in the case of ozone depletion, where big trend numbers were reported without the many caveats that really determine their plausibility. Another problem is substituting elaborate imagery of harm for a demonstration of its link to the claimed cause, as when one reporter evoked Rachel Carson's *Silent Spring* in lamenting a lifeless lake—without ever showing that its state had anything to do with acid rain. Media reports engage in "guilt by association," as when Alar is described as a pesticide, its legal definition, when it in fact is a growth regulator, or when DDT is discussed in the context of farmworkers getting sick from pesticides, when it was organophosphate insecticides that made them ill.

Still another distortion that citizens who want to understand environmental science by reading media accounts of it should take into account is the silence— or outrage—accompanying the rehabilitation of a chemical of ill repute. With DDT, there was no reporting of its coming to the rescue of a substitute that could not control the tussock moth; with dioxin, its demotion to a "weak carcinogen" was not covered nearly as widely as earlier claims of its deadliness. The media vilified the NAPAP report for its evidence of low acid rain damage, and the reintroduction of apparently harmless nitrites provoked a great deal of media criticism having little scientific basis.

"Consider the source." This rule can be a way to dismiss unpleasant facts, but it can also provide clues about the kind and direction of distortions in environmental reporting. Beware unidentified scientists, as in Love Canal reporting. In that episode citizen's tales figured prominently, as did citizen and government sources in Times Beach journalism. This can, and did, have distorting effects, especially in television coverage, as Times Beach again demonstrates. Injured people make for strong images but misleading attributions of blame. Government officials can get stampeded into admitting harms and doing things to fix them that are not scientifically defensible. Get your science from scientists whenever possible. Also watch out for manipulation of the press by interested parties. The case of the Natural Resources Defense Council and Fenton Communications releasing their Alar report on *60 Minutes* is probably the most egregious example. But environmental groups also took part in the publication of a leaked list of dioxin sites in Missouri; leaks were the cause of a premature release of a not-yet-peer-reviewed pilot study of chromosome damage in Love Canal. Finally, an Agent Orange trial attorney baldly admitted that plaintiffs' counsel tried to get a favorable jury by publicizing dubious claims about the herbicide's harmfulness.

A more systematic consideration of sources and reasons for the particular distortions observable in all these cases is beyond the scope of this chapter. However, it may be worth noting that the 1993 survey of 512 journalists (mentioned above) showed some clear patterns of where environmental reporters go for story ideas, information, and data (see Tables 12.10, 12.11). About half rely most on government officials, press releases, or reports and a third on environmental and consumer groups. One in six reporters rely most on academics, universities, and professional journals for ideas, while one in twelve most frequently turn to these sources for information and data. Only one

Table 12.10   Devising story ideas

Which of the following sources do you use for story ideas about the environment?

	Reporters		
Source	*All* (244)	*Print* (197)	*TV* (46)
Government officials, press releases, or reports	40%	45%	22%
Environmental activist groups	22	20	28
Academics, universities, and professional journals	17	16	20
Consumer groups	7	6	15
Business or industry executives, press releases	3	2	5
Company or industry publications	1	1	2
None of these	10	10	9

Source: American Opinion Research, Inc., *The Press and the Environment*, p. 56.
Note: Asked of reporters who covered at least one environmental story.

Table 12.11 Obtaining information for stories

Which of the following sources do you use most for data and information about the environment?

Source	Reporters		
	All (244)	Print (197)	TV (46)
Government officials, press releases, or reports	51%	53%	44%
Environmental activist groups	25	23	33
Academics, universities, and professional journals	8	8	7
Consumer groups	4	3	9
Business or industry executives, press releases	3	3	2
Company or industry publications	1	2	0
None of these	9	10	7

Source: American Opinion Research, Inc., *The Press and the Environment*, p. 57.
Note: Asked of reporters who covered at least one environmental story.

reporter in twenty-four rely mostly on business or industry executives, or company and industry publications and press releases. Thus, the relative influence of sources on the ideas and arguments presented by journalists points in the direction of the distortion observed in environmental reporting. Of those sources with predictable biases—environmentalists and consumers on the one hand and business and industry on the other—the former are relied on by the media eight times more than the latter.[41] To push the analysis a step further, this disproportionately high use of environmentalist and consumer sources occurs because reporters have their own biases. More than twice as many journalists think that "many" journalists are biased against business as think that "many" are pro-business.[42] Citizens who want to understand science well enough to make informed decisions on environmental issues certainly have their work cut out for them.

13

Citizenship In Science

Doubt requires justification just as much as belief does.
  —Hilary Putnam

It has been said that democracy requires a scientifically literate population. When we consider what this lofty view demands, our hearts might well sink. But if democracy requires, rather, the capacity of intelligent people to uncover the information they need and to make sense of it, our hopes may rise. Of course, this attainment comes at a cost in time and effort. For the run-of-the-mill environmental problem, usually involving low-level chemical exposure, an expenditure of some 100–200 hours should be sufficient. This is an effort that can be made every once in a while, an effort that, if exerted in collaboration with friends or neighbors, can lead to growing understanding and feelings of confidence. Were a group to address something like five to eight subjects a year, dividing them among an equivalent number of members, in a five-year period members would have a substantial grasp of twenty-five to forty issues. The club would be a powerhouse.

As things stand, a stroll down the aisle of the environmental section of a bookstore reveals an array of apocalyptic titles. Then voices are heard saying it is not so. How is the citizen to judge, to plumb the truth? Our society is in fact so polarized over these issues that widespread citizen understanding and participation appear to me the only hope of narrowing the gap between knowledge and action. If that is so, the standards for understanding these matters must now include firsthand acquaintance with the original research reports around which the controversies continue to swirl.

Experience persuades me that by following simple rules individuals of ordinary intelligence can learn what they need to know by reading original scientific studies rather than books devoted either to reassurance or to alarm. By preparing themselves to interrogate experts—indeed, by going back and forth between experts with rival views—citizens can sharpen their understanding of why there are disagreements. And by talking to friends, neighbors, and colleagues citizens can develop their own preferences.

Once upon a time, it was thought necessary to protect democracy against the machinations of the Dr. Strangeloves of the world who would hide their moral obtuseness or viciousness behind a command of esoteric lore. Now all sides have their scientists. The problem is not only how to choose between rival views based on access to the same body of knowledge but also how to know enough to make sense of the subject. The scientific literature is more specialized than ever, but the controversies now give us, as citizens, a better chance to make up our own minds.

How to Read Real Science

In the mid-1950s, when I first started graduate school, it was not thought necessary for social scientists to learn much of anything about scientific subjects. Institutions, processes, norms, these were our meat, not parts per trillion or biomagnification or whether ninety-seven types of squirrels constitute one species or many.

When the Graduate School of Public Policy at Berkeley began in 1969, I did not think it necessary for students to learn science along with the many other things they had to know. The matter was discussed, as I recall, but it was soon decided that making students into pseudoscientists would not be as desirable as encouraging them to become competent policy analysts. I have since come to believe that although it is far easier for a scientifically trained person to learn how to be a policy analyst than vice versa, nonscientists without prior training can learn to read the scientific literature and make sense of it for policy purposes.

There are two basic directions that must be followed in moving toward this goal. First, one must believe that it can be done. Without this conviction, novices are apt to consider the task impossible, at least for them. Second, one must not abandon the enterprise when one comes upon words, phrases, sentences, or paragraphs that one cannot understand. Rather, read only what you do understand and read that slowly and carefully.

The rule works well because articles in scientific journals are blessedly short. Therefore the lay reader is not dependent on a single report but rather may hope to learn from twenty to forty articles averaging no more than two or three pages. Moreover, controversies are in many ways easier to understand than basic science, for authors desire to make their points effectively and to refute the opposition. If the reader takes notes about differences, the points at issue will gradually become clarified.

Wise lay readers understand that they are not scientists and have no pretense of becoming such. What they want is a pretty good idea of what is generally agreed upon and what is not. There is almost no disagreement, for instance, that the danger is exceedingly remote of something going wrong with the release into the environment of genetically engineered organisms. It helps in this issue to understand that what disagreement there is concerns whether this remote and improbable danger constitutes a threat, not whether it is at all likely to occur.[1] No one reading a variety of pieces about global warming, as another example, would believe those who peddle the line that there is consensus among scientists on this subject. The lay reader soon discovers that scientists do agree that more carbon dioxide in the atmosphere means greater warming, but their dispute revolves around whether other factors, like oceans and winds and solar flux, might not alter or even reverse this one effect.[2] Then the claim of consensus can be seen for what it is, a tactic in the struggle over whose views will be accepted and become the basis for public policy.

There are a few things that the interested nonscientist needs to know. One is how to use computer search routines to locate the scientific literature, and another is how to follow leads from one footnote to another. Dictionaries of scientific terms are helpful, and they usually provide some idea of the nomenclature and dimensions used in a particular field. It is also useful to know what kind of specialist to consult. In regard to chemicals, for instance, toxicologists are the right people. In my experience they are willing to be helpful if the lay analyst has taken the trouble to become informed generally about the literature in the field.

Because the very idea of a journey into the scientific heartland may be forbidding, a more detailed itinerary is in order. Begin by undertaking a computer search or obtaining an index to scientific periodicals in which articles relating to your subject—say, toluene or PCBs—are found. Look for an article containing a review of the literature. If you do not find one, begin with any article and follow the footnotes to other articles. This is called, appropriately, the cobweb method. As you read, circle only those sentences or phrases you can understand, however few. Read these over slowly two or three times. Keep reading and rereading.

By the tenth article you should know what a few of the disagreements are about but perhaps not yet be cognizant of the big picture. As you proceed, make lists of arguments pro and con from the various papers. As you read further, consider only the information that bears on these arguments. Somewhere around the twentieth article you should begin to get the warm and welcome glow of déjà vu. There probably are only thirty-two (a mythical but cozy number) things to know about a subject, and when you know these the rest gets repetitious pretty fast.

By this time you know the state of play: what the arguments and the differences are about. Then you can call or write the scientists involved or contact scientists nearby (ecologists, toxicologists, biochemists, immunologists) and ask them questions designed to clarify your remaining doubts. Going back and forth between contending positions and the people who hold them is instructive and, my students found, fun.

Remember always that you are a citizen and not a scientist. Your task is to use the knowledge you have accumulated to improve your and your fellow citizens' judgments about what ought and ought not to be done.[3] Is the potential danger being discussed substantial and immediate or is it trivial and overblown? Would a little cleanup and alleviation suffice or would only a massive effort protect human health or prevent severe environmental degradation? Your reading and discussion will prove sufficient to educate your policy preferences. And that is all a citizen can ask.

The task is formidable, but it is by no means hopeless. As a citizen, moreover, the investigator has credibility among other citizens. Indeed, if you imagine that your effort is part of a group endeavor in which others are investigating various potential risks, the task can become an exchange of information among peers who have a lot to contribute to one another's continuing education. When experts disagree—or when it is difficult to tell whether they disagree and about what—it is up to us, the lay public, to find out for ourselves.

Understanding what these disputes are about in the sense of locating them within existing knowledge enables lay investigators to ask much better questions. If one wishes to see what genuine empowerment looks like, observe students or citizens who demonstrate that they have understood the scientific issues in a controversy and come to a reasoned conclusion.

Would Rational People Wish to Learn about Real Risks?

There is a flaw in my advocacy of citizens investigating risks for themselves that stems from the difference between individual and collective rationality. When individuals hear that a product may be harmful to their family's health, they are not disposed to stop and undertake a year-long study. It is easier to take one of two actions: forget it or avoid it. The "forget it" response is based on the belief that these alarms are likely to be too trivial to be worth bothering with. Over time individuals who ignore the scare talk about Alar on apples or arsenic on Chilean grapes or cancer from saccharin may feel that their lives will run more smoothly.

But they are not likely to be in a majority or even a substantial plurality. Who knows, most people feel, whether there is truth to this? By "avoiding it," citizens satisfy themselves that they are protecting themselves and their loved ones, whereas those who "forget it" may well develop a bad conscience. "I would never forgive myself," one can almost hear a woman saying to herself, "if anything happened to . . ." Suppose, furthermore, that a tragedy later strikes her family; a child develops leukemia or a husband, a stroke. How would our citizen know whether this might not have occurred had she avoided the cursed chemical?

To put the citizen's problem more poignantly, suppose that she knew what proportion of the warnings were true, how many people would be affected, and how badly. She would still have trouble deciding what to do, for she might believe that even if only one out of many turned out to be affected, it would still be worth acting as if all were affected. Back we are to two themes, pro and con, that have repeatedly surfaced in this book. If it is wise to follow the precautionary principle by always seeking to avoid possible dangers, then the individual is right in avoiding all possible harms, even if she believes that almost all are probably false or grossly exaggerated. If we believe that health and safety are functions of a society's standard of living, however, then the losses in life that accompany a lower standard of living have to be taken into account. Were that done, most claims of harm afflicting people would have to be true to outweigh the health and safety losses due to the costs of many false alarms.

Observe that I used the phrase "if *we* believe." There lies the crux of the matter. Even if one person believed that safety is a collective product, a function of the standard of living in the society as a whole, there is no way for that individual to appropriate the benefit of that belief. If everyone else follows the "avoid it" strategy, the individual who adopts the "forget it" strategy will pay the price if something goes wrong but will get none of the benefits, inasmuch as the actions of others will impose their costs on society as a whole.

Scientific Jabberwocky

The importance of distinguishing scientific jabberwocky from accepted understandings within scientific and professional subfields is nowhere better illustrated than in the Alcolac decision. In this case the Missouri Court of Appeals upheld a decision against a chemical company involving large sums of money allowed to plaintiffs who claimed that their immune systems had been severely damaged by certain chemicals. The most revealing aspect of the case involves the testimony of certain witnesses with degrees in a field called clinical ecology, whose testimony was far outside any known understanding of immunology but still was accepted by this court as valid, though other courts have ruled the evidence inadmissible. I wish everyone could read the critique of the court's decision by Richard S. Cornfeld, an attorney, and Stuart F. Schlossman, a doctor and immunologist, but we will have to be satisfied with a brief focus on rules for evidence accepted by immunologists and violated by the so-called expert testimony.[4] The citizen will want to ask where we would be if such claims were routinely accepted as valid.

In the field of immunology the most important part of evaluating a patient's immune system is his or her history of infectious diseases. In addition, a diagnosis of a suppressed immune system, as the medical terminology has it, must be based on laboratory tests with established diagnostic uses. In evaluating such tests small deviations from the "reference range"—the readings found in the general population—are not considered sufficient evidence by immunologists. Finally and foremost, even if test results are outside the reference range, they must be repeated under the same conditions and with the same results before they can be believed. All of these conditions were violated in the evidence that led to the Alcolac decision.

The clinical ecologists testified that their clients had suffered "such terrors as chemically induced AIDS" and "chemically induced immune disregulation," conditions which have no meaning among immunologists. Yet they could show no illness in any of their clients consistent with a suppressed immune system: no infections, no immune disease.

Following the court presentations carefully, one soon finds inconsistencies in addition to unscientific methodology. The citizen evaluator needs to ask what each test, however strange or peculiar its name, is supposed to measure. For instance, a plaintiff whose claim of damage is based on having an elevated OKTD-3 test and a depressed OKT-111 test should be reevaluated because these tests measure the same thing.

There is one aspect of the illuminating critique, however, with which I do not agree entirely: "Because the jury is ill-equipped to evaluate expert testimony that purports to be based on research on the frontier of science," Cornfeld and Schlossman state, "the court must exercise control over the type of testimony a jury can consider."[5] I do agree that the judges should have informed themselves about this matter, perhaps by calling on their own expert testimony, or even reading an introductory text on immunology, and I agree that "the jury is ill-equipped." But the thought ought not to end there. I am ill equipped and so are my colleagues only if we do not take the trouble to inform ourselves. There is nothing about a juror's intelligence that would prohibit him or her from understanding this issue once given appropriate reference materials. Without reading about immunology, however, everyone is in deep trouble. It might be helpful if the professional society of immunologists prepared an encyclopedia-type article including the kinds of tests acceptable for evidence, as Cornfeld and Schlossman propose, a document that could be given to judges and juries.

Popular Epidemiology

It has become commonplace to denigrate the accepted understandings and protocols within scientific and technical fields as if they were mere power plays or the results of professional prejudice. If the experts are the enemy, using their knowledge to put ordinary people down, perhaps citizens should hire their own experts or, better still, do themselves the scientific work that impinges on their lives. This movement is called popular epidemiology.

According to sociologist Phil Brown, "Popular epidemiology is the process by which laypersons gather scientific data and other information, and also direct and marshal the knowledge and resources of experts in order to understand the epidemiology of disease."[6] Of the several papers I have seen on this subject, Brown's is best at bringing out the operational components of this novel set of methods seeking to combine acquisition of knowledge with citizen activism.

The first claim of popular epidemiology, Brown informs us, is that "professionals generally concern themselves with disease processes, while laypeople focus on the personal experience of illness." Specifically, Brown claims that "many people who live at risk of toxic hazards have access to data otherwise inaccessible to scientists. Their experiential knowledge usually precedes official and scientific awareness, largely because it is so tangible." The knowledge he refers to is that "people hypothesize that a higher than expected incidence of disease is due to toxics."[7] This, of course, is exactly what is at issue; if one begins with a conclusion—"disease is due to toxics"—what is the purpose of research?

A second claim of popular epidemiology is that "scientific knowledge is shaped by social forces," such as economic and political interests, and limited in the resources at its disposal. True. But where does this "social construction" approach lead? Is it wrong, as Brown suggests, for "scientists and officials [to] focus on problems such as inadequate history of the site, lack of clarity about the contaminants' route, determination of appropriate water sampling locations, small numbers of cases, bias in self-reporting of symptoms, obtaining of appropriate control groups, lack of knowledge about characteristics and effects of certain chemicals, and unknown latency periods for carcinogens"?[8] Whoever does scientific work, lay people or professionals, must be concerned about such matters as water sampling or the whole undertaking becomes nonsense.

Brown observes correctly that "traditional approaches also tend to look askance at innovative perspectives favored by activists, such as the importance of genetic mutations, immune disregulation markers, and nonfatal and nonserious health effects (e.g., rashes, persistent respiratory problems)."[9] Someone might tell the worried parties that no popular, unpopular, or any other kind of epidemiology is going to help solve the mystery of rashes. Never mind. Genes mutate all the time; if it is alleged that certain chemicals cause cumulative and hence harmful mutations, that suggestion requires a specific investigation. But "immune disregulation," whatever that is, let alone its markers, is another matter. The term itself comes from the clinical ecologists already mentioned and is not accepted science.

Brown argues that popular epidemiology is good science on four grounds:

1. Lay involvement identifies the many cases of "bad science," e.g., poor studies, secret investigations, failure to inform local health officials.

2. Lay involvement points out that "normal science" has drawbacks, e.g., opposing lay participation in health surveys, demanding standards of proof

that may be unobtainable or inappropriate, being slow to accept new concepts of toxic causality.

3. The combination of the above two points leads to a general public distrust of official science, thus pushing laypeople to seek alternative routes of information and analysis.

4. Popular epidemiology yields valuable data that often would be unavailable to scientists. If scientists and government fail to solicit such data, and especially if they consciously oppose and devalue it, such data may be lost.[10]

The first point is a species of conspiracy theory; it directs citizen anger against the authorities without increasing knowledge of either effects or causes. The second point speaks well of standard science in that it refuses to accept vague and unsupported notions of causality. This admirable protection against falsehood is also converted into a form of system blame. The third is ad hominem; worse, citizen distrust is directed against either a fictitious entity ("official science") or the only real knowledge, the scientific literature. The whole thing is an exercise in producing anger, not knowledge. About the fourth point, "valuable data," something more may be said. To popular epidemiology Brown attributes the discovery of a number of "hazards and diseases," including two studied here—Agent Orange and asbestos—about which there is little or no truth.

I note with pleasure that some of the citizen activists Brown studied in Woburn, Massachusetts (where chemicals alleged to cause cancer in children had entered water wells), have come to like science. They could, I think, learn how to conduct worthwhile studies. But I also believe that the criteria of good scientific procedure Brown portrays as denigrating are the main protections citizens have against falsehood. Charlatanry is worse when you do it to yourself.

Cancer Clusters

It would require several tomes the size of telephone books to describe and evaluate the hundreds, perhaps thousands, of episodes all over the United States in which environmental groups and citizens complain about clusters of cancer in their localities that they believe are abnormal and linked to chemical emissions. These episodes perhaps more than any other have angered people and led to the widespread feeling that government and industry conspire to harm residents' health and that chemical contamination must be stopped. I have met several people involved in these incidents whose fury makes conversation difficult. Their feelings are not in doubt. The question I wish to raise is the usual one: are these charges justified by the evidence?

In largest part, my answer is no. The most comprehensive analysis is a survey of state health departments in which Stephanie Warner of the University of

Miami School of Medicine and Timothy Aldrich of the Oak Ridge National Laboratory report "that cancer cluster investigations undertaken by state health departments have been largely unproductive." Only a few states recorded positive associations and many of those come from occupational exposures.[11]

Why do the overwhelming proportion of studies of cancer causation by chemical contamination come up empty? There is a good reason: these claims are false. The amount of chemical involved is too tiny to cause the alleged harm. The effects are missed because the causes are weak. Weak causes are unlikely to produce the strong effects (cancers, physical deformities) about which complaints are made.

A sympathetic and discerning survey of cancer clusters has been made by Lawrence Garfinkel, vice-president for epidemiology and statistics and director of cancer prevention at the American Cancer Society, who establishes four criteria for assessing purported clusters:

1. If the cluster includes cancers of many different sites, it is almost certainly not a true cluster, and it cannot be attributed to a particular environmental agent . . .

2. The number of cases reported in a cancer cluster generally will be more than the number expected in a geographic area. Although the difference may be statistically significant, it could be a chance phenomenon; it does not necessarily mean that the cancers are caused by or even associated with some environmental agent.

3. Many community investigations are limited by our knowledge of the environmental agents that can cause cancer . . . This is a very small percentage of all the agents that produce the fumes, particles, and dusts to which we are exposed.

4. Before it can be maintained that a cause-and-effect relationship exists, an investigation must show that the people who developed cancer had been exposed to suspected environmental agents and had greater exposure than a series of controls.[12]

In a letter to *Science* Martyn Smith of the Department of Biomedial and Environmental Health Sciences at the University of California at Berkeley refers to the episode in Woburn, Massachusetts, as "an excellent example of irrationality in an environmental health issue." His argument is grounded in the finding that the chemicals in the Woburn wells had no more carcinogenic material than a glass of chlorinated tap water (containing chloroform, a "hazard index" of 9.1, compared with 11.6 for tap water).[13] Hence Woburn fails the fourth test.

Part of the Woburn problem is the belief of local residents that the harms they experience, like childhood leukemias and miscarriages, are way above what might be expected by chance. Such judgments require understanding probability, not relying on intuition. Thus an excursion into randomness and coincidence is necessary.

Random Behavior and Coincidence

"With a large enough sample, almost any outrageous thing is likely to happen."

People mislead themselves when they do not appreciate how large a number random occurrence may entail.[14] I call the belief that randomness comes in a thin homogeneous layer the Cuisinart theory of randomness, whereas actually randomness includes a considerable degree of clumping. Cancer clusters, it follows, are not unusual but common, what we should expect.

What most of us do not know about statistics undoubtedly would fill volumes, yet many of us have learned how to use computers, and for these there are inexpensive randomization programs. Plug in the numbers: so many cancers of a certain kind, a certain number expected in a population of a given size versus so many that actually occur. Run them through the program and see the real range of randomness. Often the actual number at a specific site falls within random variation. If not, the case for harm is stronger.

Once the search is on for small clusters, the possibilities are virtually infinite. Consider the case of video display terminals (VDTs) and clusters of miscarriages. Part of the problem is that around one in five pregnancies ends in a spontaneous abortion; the exact proportion depends on how early one starts counting. Among the 10 million or so women who use VDTs, pure randomness is bound to result in many clusters of miscarriages.

What Carol Mock and Herbert Weisberg say about political scientists is true of all of us as citizens: "there is not enough respect for the role of coincidences . . . Thus, political scientists often systematically underestimate (and underreport) the role of chance and the actual probability of the improbable in their observations."[15] We must make the crucial distinction between possibility and probability. When there are binomial distributions—that is, when we ask whether chemical X does or does not induce cancer at a given exposure—"there is a high probability of obtaining some significant results by chance when multiple tests are undertaken even when the probability of obtaining a particular significant result by chance alone is set low."[16] Moreover, the chance of finding a significant result rises quickly with repeated trials, so that fourteen independent tests yield a 50-50 chance of one being significant.[17] In Mock and Weisberg's words, "pattern is also a characteristic of chance."[18]

The importance of this relationship—that repeated tests produce positive results, in other words, evidence of harm—cannot be overemphasized. The only way to avoid false positive results is to follow good rules of evidence. A result is not meaningful unless it fits within a prior expectation generated by a theory and comes out the same way on another trial.

The Wrong and the Right Conditions for Caution

I am not arguing that regulation or advice on public health should require incontrovertible evidence. Indeed, throughout this book my coauthors and I have argued in favor of abiding by preponderant (not conclusive) evidence. The stringency of the proof required, I think, should vary with the purpose to which the knowledge is put. To consider this question I begin with the best example I found of the opposing view.

In an excellent paper M. G. Marmot of the Department of Community Medicine at University College, London, argues that "this lust for absolute 'proof' represents a view of science that is mistaken, dangerously so, since it interferes with two types of endeavour: translating scientific evidence into public health policy and pursuing research into the social causation of ill-health."[19] Bias in favor of one's theory, he believes, provides motivation to defend it in case it should prove true. "Such creative prejudice should be distinguished from the stubborn post-hocism that clings to a theory in the face of mounting evidence to the contrary."[20] Scientists whose work impinges on public policy, Marmot continues, should give their best judgment even while research continues, for "the argument that we should wait for certainty is an argument for never taking action—we shall never be certain."[21]

Though each of Marmot's propositions is true, I believe that he draws the wrong conclusions by failing to take into account the kind of decision on which a scientific recommendation is being made. There is a big difference between proving a negative (this chemical at that dose will not cause cancer) and proving a positive (this chemical is likely to cause cancer at a certain dose). There is a bigger difference between the standard of evidence required to subject healthy people to fundamental dietary changes than there would be to alter the diet of those who show symptoms of debilitating disease.

Placing oneself in the position of a member of a committee advising a government about dietary recommendations puts the problem in perspective. If the government waited for certainty or for total agreement, it would be paralyzed, for there are always exceptions, contrary findings, gaps in knowledge. Should the expert who knows what there is to know, including the contradictions, say he does not know enough? Failure to support preponderant knowledge may lead to much avoidable harm. What should be done?

Here we have to introduce conditions. If the recommendation is just that, a recommendation, I would go with preponderant knowledge. But if it is more like a requirement, I would not. Suggestions are one thing, mandatory action is another. A good case in point is the recommendation to avoid foods containing certain kinds of cholesterol. A massive campaign is under way in which doctors direct patients to undergo rigorous and permanent dietary change to bring their cholesterol levels down. I think this is inappropriate. Advice likely to be taken by a large part of the population should not be given without greater confidence than is now warranted.

In regard to cholesterol, I suspect that evidence will sway one way and then another, which is normal; but what will the reaction be when the evidence points in the opposite direction? Those who see their task as pointing out possible harms may simply move on to the next. But public officials, who must husband public trust, should not follow the fashions. Science writer Gina Kolata reports that there is emerging "sometimes grudging" agreement that very low cholesterol levels make death from other causes more likely.[22]

Consider in this context the concerns of a scientist who has been among those pushing people into a low cholesterol diet. This is not a small shove; doctors are broadly advising patients who show no symptoms of disease to reduce radically their intake of foods containing cholesterol (or, at least, the wrong kind of cholesterol). Without going into the evidence, I wish to associate myself on a more general level with the reflections of Dr. Antonio Gotto, who is described as a leading figure in the effort to change what Americans eat so as to reduce cholesterol: "We are moving to a position of policy where we would be lowering the cholesterol levels of millions of asymptomatic people and keeping their levels low for a lifetime . . . It behooves us therefore to be sure we are not doing any harm."[23]

With the goal of doing as little harm as possible, I have been advocating making more use of preponderant scientific evidence. Others involved in the debates over the health hazards and benefits of modern technology wish to avoid harm by weakening the forces in society they believe are marketing harm.

Progressive Science

In the journal *Science for the People,* which its sponsors correctly describe as progressive, Rick Hester, a pseudonym of an employee of a state public health agency, analyzes the environmental cancer debate. He considers the views of Richard Doll, Richard Peto, and Bruce Ames on one side and on the other Sam Epstein and Joel Schwartz, whose rebuttal to Ames was "cosigned by a list which comprises a virtual 'who's who' of progressive governmental scientists in the U.S." Unlike earlier analysts, however, Hester does not claim there is a cancer epidemic. "My reading of the cancer epidemiologic data is that there is still no convincing evidence for a chemically-induced cancer *epidemic* even though there are certain types that may be increasing in some areas because of historical exposures."[24]

Hester then turns to the area "where the popular perception of cancer is most at odds with the mainstream scientific opinion." In Woburn, Massachusetts, organized opinion was directed not against the much greater carcinogenic potency found in a mushroom or in wine, Hester notes, but against "the possibility that recklessly handled industrial solvents may have caused leukemia deaths in children." Whether Bruce Ames and his colleagues or Richard Doll and his supporters in the scientific establishment like it or not, "people are moved to act on what they perceive as 'outrageous misconduct' by people who have put profit before the health and safety of their neighbors (or their workers)."[25] Here Hester helps us understand the difference between risk perception and risk consequences: perception of misconduct replaces evidence of actual harm.

Hester concludes: "The overall import of our work has been to advance criticisms that expose the corporate greed at the root of many outrageous instances of excess cancer among workers or in communities." In his last sentence, he makes it clear that he will "continue to critique the diversions and obfuscations of the scientific apologists for the multinational companies."[26]

But, as we have seen, many if not most recent environmental and safety alarms are false, largely false, or unproven. If we do not want environmental ethics to be compromised by false claims based on bad science, we must separate social policy from science and return to truth telling.

An excellent example of the truth telling I envisage is Michael Castleman's "Who Dunnit," in the magazine of the Sierra Club, which provides an instructive guide to the perils of interpreting epidemiology in regard to small scares.[27] With miscarriage rates so high, for instance, in a company that employed 1,000 women using video display terminals, 150 women could miscarry and still fit within the usual rate of spontaneous abortions. I could not express my view better than Castleman's last paragraph does: "So before jumping to the easy conclusion about any cluster, consider its size, specificity, background rate, and your own political assumptions. Environmentalists compromise their credibility when they automatically blame pollution for a cluster that might be caused by something else—or that quite possibly is just a fluke."[28] Is Castleman's caution justified, one might ask, or is he just kowtowing to experts? Why should citizens adopt the criteria of proof used by scientists who may not care about their health?

The Knowledge Is in the Science

Discussion of the relationship between experts and citizens is commonly marred by a fundamental misunderstanding.[29] The error lies in confusing the science with the scientists. Being human, experts are error prone; they may be biased or ill informed. The progress of science depends not on the infallibility of scientists but on the processes of competition among them. Knowledge rests not with particular scientists but in the science itself.

Suppose that citizens feel that any amount of a chemical, however tiny, in their food or water or air is too much. Perhaps these citizens should search out their own experts who will tell them what they want to hear—that parts per million or billion can hurt them or that miscarriages, deformities, and leukemias can come from tiny amounts of the chemicals they fear. Let us further assume what is usually the case: that though these scientists have credentials, their views are at variance with the scientific literature on the subject. What have these citizens gained? Before, they were ignorant in the face of industrial and governmental scientists; now they are ignorant in the face of their own scientists. If their experts are mistaken, the citizens involved may worry needlessly or seek treatment for nonexistent illnesses or direct their anger wrongly at those they believe have harmed them, thereby adding discord to their other ills.

The only sure way to know what we want to know is through the science itself. Citizens who train themselves to read and understand the primary sources, the original scientific studies, can participate meaningfully; those who do not, cannot. Of what, to cite the title of this chapter, does citizenship in science consist?

The Competent Citizen

In "The Problem of Civil Competence," Robert Dahl addresses the standards citizens should meet to be considered competent. Though these standards vary in their severity, it is widely agreed that most citizens fail to meet them. Instead of proposing an apparently unobtainable ideal, Dahl seeks to reduce the requirements by asking, "What is a good enough citizen?" For most people, he argues, "it would be more accurate to say that they are *occasional, intermittent, or part-time* citizens." Fair enough. Yet he fears that even at this level few "possess sufficiently strong incentives to gain a modicum of knowledge of their own interests and of the political choices most likely to advance them, as well as sufficiently strong incentives to act on behalf of these choices." What might be done, he asks, about this "cognitive difficulty?"[30]

Dahl recites the well-known limits of experts, from their narrow vision to their overconfidence. He is especially concerned about a problem that lies at the heart of the relationship between knowledge and action as it relates to citizenship:

> The sheer complexity of public affairs means that experts are generally no more competent over a range of policies than ordinary citizens, and may even be less competent. Ironically, many of the basic qualities of experts conspire to impair their broader judgment: their narrowness of outlook, understanding, and values; their idiosyncratic biases reflecting their own field of specialization; their unwarranted confidence in the applicability of a model or method that is at best valid only under certain very limited conditions; their all-too-human tendency to force highly complex phenomena onto the Procrustean bed of theory; their blindness to the impact of their own ideology on their analysis; their unwillingness or inability

to recognize the limits of their own expertise. Alas, while noting that experts are often woefully incompetent outside their own narrow specialty may provide the rest of us with a certain wry amusement, it does nothing to solve our problem.[31]

Exactly so.

At this point Dahl reaches the question of this section, which he calls appropriately "the problem of trustworthy surrogates." All of us, he rightly claims, must rely on others for most of our judgments. But who and how? "How best to organize this task presents formidable problems, of course, not least of which is to give citizens easy access to expert views while at the same time encouraging them to examine claims to superior knowledge or infallibility with considerable skepticism."[32]

I advocate neither submission to nor denigration of experts. Instead, I propose a combination of citizen labor and development of citizen skills to study the scientific and technical literature so as to understand what is known and, with that understanding, to make intelligent use of experts in order to reach informed judgments. Becoming knowledgeable is the right way to exercise judgment, and it is also the best guarantee that citizen preferences will shape public policy. One does not become powerful by indulging in willful ignorance. Play the game, I say, but do not throw out the rule book.

14

Detecting Errors in Environmental and Safety Studies

with Robert Owen Rye

Like most innocents, we thought that stories about the ozone hole growing larger were based on direct measurements, that there was an instrument (or set of instruments) that somehow measured higher and lower levels of atmospheric ozone.

Looking at the title of a paper by Richard Stolarski and his colleagues—"Measured Trends in Stratospheric Ozone"—one might well imagine that ozone is measured, and in a way it is, but not directly. The measurements are but part of a statistical model. What Stolarski and others actually did was take various offsetting factors, such as atmospheric nuclear tests and seasonal and solar variations, and adjust the ozone time series accordingly.

Kenneth Towe, of the National Museum of Natural History, in a letter to *Science* advised that these adjustments do "not mean an actual ozone-related increase in the ultraviolet flux to the earth's surface any more than adjusting salaries downward to remove the effects of inflation will mean an actual decrease in the income taxes due on them. Both are adjustments made 'on paper.' "[1] Agreeing that "perhaps a better title would have been, 'Trends in stratospheric ozone deduced from measurements,'" Stolarski argued that nevertheless the trend data were correct and that "the solar ultraviolet should respond to the trend in the ozone. Unfortunately, our database on trends in ground-level ultraviolet radiation is not yet robust enough for us to be able to make any quantitative statements."[2] From measured to deduced is quite a distance. The moral of this story is, go beyond titles to read and then compare what authors say with what they actually do. And, of course, read letters to the editor.

How easy it is to be misled, not least when one becomes overconfident. Citizens evaluating potential dangers should be alert to the numerous ways in which clues apparently point in one direction only to end in blind alleys or to lead in another. That is what makes science fun and citizens despair.

Nihilism is not the point. Distrusting everyone and everything, especially one's own judgment, is self-destructive. Instead, the citizen should learn to recognize patterns of representation so as to avoid being controlled by them. Each citizen is his or her own private eye. All of us would-be risk detectives need experience in being misled and still, in the end, catching the real villain or, as the case may be, the false accuser.

To be a risk detective the citizen has to be alert to the typical errors and defects in environmental studies. Scientists and governmental experts are apt to make mistakes in experimental design, measurement, and inference. To make these flaws clear, we will look at a number of studies linking different chemicals to human health problems or environmental damage. We find a number of methodological flaws that appear repeatedly in the scientific justifications for regulatory practice. From these flaws and insights from previous chapters, we will derive rules to guide the citizen risk detective.

Scientific studies of the effects of chemicals should do four basic things: define an exposed population (or environment) in contrast to an unexposed one, evaluate the extent of exposure to a chemical agent, determine whether there is a significant difference in health between the exposed and the unexposed populations (or environments), and apply that understanding to the question of what effects the suspect substance will have on a broader exposed population (or environment). If scientists know how much exposure it took to produce the differences between the exposed and the unexposed, then they can hope to predict that such differences will arise in similar exposure circumstances. Scientists, however, must distinguish between random variation and differences induced by exposure. Only those effects that can be shown to occur at a higher rate than mere chance can confidently be attributed to the chemical. The flaws we will discuss are not niceties; their existence weakens studies to such an extent as to invalidate conclusions regulators draw from them.

The Need for Controls

How does a person find out if lakes or trees are declining or even dying? Simply eyeballing will not do. We must follow *Rule One: Use appropriate controls*. Looking does not necessarily tell the investigator about the history of the subject in question—which is necessary to know before appropriate inferences can be made. If you do not find fish in a lake, that does not tell you whether it ever had fish or, if it did, whether the government or a private club has stopped stocking it. Looking must be supplemented by studies with controls, and in many studies this means selecting appropriate control groups or establishing baselines for comparison. One great difficulty in understanding claims about global warming and ozone depletion, for example, is that the effects lie within the range of natural variation; therefore some control for this variation is necessary to substantiate claims and policy inferences.

A control group should be as similar as possible to the study group except for exposure to the chemical whose effects are being observed. Obviously scientists must make choices as to what factors they are going to make a point of controlling. In an epidemiological study designed to determine whether or not exposure to a chemical causes lung cancer, for instance, the investigators should make certain that the smoking habits of the control group are as similar as possible to those of the study group. Any result indicating that the exposed group had both a higher incidence rate of lung cancer and a higher rate of smoking than the control group would naturally be suspect.

This example may be translated into a general principle: if the study group has a property that the control group does not have and that could cause the postulated effect of the suspect chemical agent, then a false-positive result may be derived from the experiment. The term "false positive" describes an erroneous correlation between exposure and a health effect, which leads the analyst to the wrong conclusion. The proper selection of control groups helps to eliminate false-positive findings. If mice are naturally subject to liver tumors, for example, their expected rate of tumor formation has to be controlled for to get an accurate result.

CONTROL FOR AGE

With each disorder there are a number of factors that are known to be strongly correlated with its incidence rate. One of the most commonly related factors is age. For instance, it has been established that a pregnant woman's age is an important factor in determining the risk of stillbirths or miscarriages.

The wives of workers exposed to high concentrations of vinyl chloride at a plant in Pennsylvania reportedly had a statistically significantly higher rate of stillbirths and miscarriages than the wives of workers who were not highly exposed. But no effort was made by the investigators to find out the ages of the women in either group at the time of conception.[3] It is every bit as plausible that age was responsible for the differences in stillbirths and miscarriages as it is that vinyl chloride was responsible. Without such data there is no way to rule out age as the main causal factor.

CONTROL FOR SMOKING

Scientists have conducted a number of occupational studies of the effects of exposure to radon, finding that there are significantly more lung cancers among uranium miners, who are exposed to high doses of radon, than their control group counterparts. In all but one of these studies the smoking history of the miners was unknown,[4] which meant that the investigators did not know what sort of smoking habits the control group should have. This sort of control is critical, because scientists have found that 85 percent of all lung cancers are due to smoking. They also know that on average smoking is much more common among miners than it is among the general population, but they do not know what the rate among these particular miners is.[5]

A Canadian study of the effects of saccharin polled all those who had developed bladder cancer over a twenty-two-month period to find out whether they had used artificial sweeteners and if so how much they had used.[6] The investigators found that bladder cancer was strongly correlated with sweetener usage. They matched each case with a person of the same sex, age, and location, but they failed to inquire into the smoking habits of either the subjects or the controls. Given the prevalence of smoking in society and its well-established carcinogenicity, this failure gravely weakens the conclusions drawn in the study.

CONTROL FOR GENDER

Men and women have different health vulnerabilities, one example being the higher incidence of breast cancer in women. There are many cases in the animal research record of sex differences in response to chemical agents. Thus saccharin was much more of a problem at high doses for male rats than for female rats in both of the major animal saccharin studies performed in the early 1970s.[7] By contrast, female rhesus monkeys, when fed 2.5 parts per million Aroclor 1248, suffered from reproductive difficulties that were not exhibited in the control group, while the males in the study experienced no decline in fertility as a result of exposure.[8]

A recent study found 3 cases of soft-tissue cancer among 1,520 workers who were potentially exposed to dioxin for at least one year and had died at least twenty years after exposure.[9] The investigators compared the rate of soft-tissue cancer in this group with the general population and found that at a 95 percent confidence level the rate of soft-tissue cancer was twice that found in the general population. All of the workers studied were men. The segment of the general population used for comparison was an undifferentiated group of men and women. The differences, if any, between men's vulnerability to this type of cancer and women's is unknown. If men are much more susceptible, the rate found in the study might be no higher than expected without exposure to dioxin. If men are much less susceptible than women, then the results of the study actually underestimate the risk posed by comparable exposure to the general population. If there is no difference between sexes, then the control group is appropriately representative. Given a lack of data, each of these hypothetical situations is equally plausible.

Most of what we know about the effects of radon on humans comes from studies of miners. Since mining is strongly dominated by males, a study of miners should be compared with a control group that is also mostly men. By relying on the miner studies for our assessment of the risks of radon in the general population, we may underestimate or overestimate the risks to women. That is, predictions of the effects of radon may be skewed in ways that cannot be corrected with the data available.[10]

A Food and Drug Administration (FDA) task force criticized a study of the effects of nitrites on rats for not distinguishing between male and female rats.[11] This failure was one of a number of flaws that led to the dismissal of this study as a valid test of the safety status of nitrites.

CONTROL FOR PREEXISTING CONDITIONS

Scientists should control for conditions that existed or may have existed before the introduction of the chemical agent under scrutiny. This principle applies for both exposed population studies and assessments of putative environmental damage.

Scientists have frequently used average annual rain pH as an indicator of human influence on the chemistry of rain. On the assumption that pure rainwater has a pH of 5.6, they have taken measurements of less than 5.6 as a sign that human activities have acidified the water.[12] Over the course of the 1980s, however, a number of studies showed that pristine rain has a pH ranging from 4.4 to 5.8.[13] This vast natural variability makes it impossible to say with credibility that rain with pH less than 5.6 results from human activities. A measurement of pH is therefore not a measurement of the atmosphere's absorption of man-made acids.

In a study of workers exposed to formaldehyde two subjects developed a rare form of cancer known as nasal carcinoma. Investigators associated the disease with formaldehyde because 50 percent of 206 rats exposed to high doses for prolonged periods of time developed the same sort of tumor (inhaling air containing 14.3 parts per million of formaldehyde for six hours a day, five days a week, for a lifetime). None of the control animals developed this sort of cancer.[14] The two cases of nasal carcinoma in the worker study were considered a significant finding, yet it turned out that both subjects had had severe sinus problems that required surgery.[15] It is thus impossible to know whether the cancers were purely a result of exposure to formaldehyde or whether the workers' preexisting nasal problems made them susceptible to this particular cancer.

To determine whether a chemical has caused excess cases of a particular health problem in a given population one must first know what the incidence rate would have been without exposure. This rate is generally determined through the use of a control group, the most basic being the general population, whose incidence rate is known as the background incidence rate.

Birth defects and miscarriages are fairly frequent in the general population: the background incidence rate of birth defects is 2–4 percent nationwide;[16] the rate of miscarriages and spontaneous abortions is 15–20 percent.[17] Miscarriages and birth defects at Love Canal were associated with chemicals seeping through swale beds, but this association was not based on a quantitative assessment that demonstrated more occurrences than normal, as determined by national and regional rates and comparison with a place nearby with the same sort of people but no chemical-laden canal. Rather, it was based on a survey that indicated there had been a number of these defects near the swale beds. In fact, the biologist who concluded that migrating hazardous wastes were responsible did not know whether or not the rate of defects was higher than that expected for a normal population.[18] Later tests comparing evidence of harm to people at Love Canal with the health of those nearby showed no difference in miscarriages or birth defects.[19]

A popular piece on the effects of electromagnetic fields on children indicated that clusters of cancers could be attributed to proximity to power stations,[20] but this *New Yorker* article included no data on the nationwide incidence rates of the cancers for comparison. The Connecticut and North Carolina health departments subsequently showed that the incidence rate in those supposed clusters was not significantly higher than the background rate. The California Department of Health Services found that exposure levels in the area described were not higher than in most other areas and concluded that electromagnetic fields were not a likely contributing factor to the cancers found there.[21] The article confused proximity with causation.

If we have adequate controls for preexisting conditions, then we have something that our data can be compared to. Thus we have two more rules for the risk detective.

Rule Two: Establish the baseline. Increases and decreases do not come from nowhere. They are selected partly on the basis of existing data and convenience and partly to make whatever point the user has in mind. By knowing where the base begins, we learn interesting things: How long, for instance, have measurements been made?

Rule Three: Vary the baseline to determine whether the conclusion is robust. Would it matter for conclusions if a different baseline were chosen? Going from the winter to the summer, the earth is warming; going from summer to winter, the earth is cooling. Choosing the Cretaceous era as a baseline, the earth today appears starved of carbon; choosing the beginning of the Industrial Revolution of the nineteenth century, the earth appears to be carbon rich or at least richer. If the conclusion is the same no matter what the baseline, it is on firmer ground.

The Need for Correct Measurements

It should go without saying that efforts to measure the effects of environmental agents should not themselves change the properties they are meant to measure, other than in predictable and well-quantified ways. In studies of both low-level chemical exposure and large-scale environmental phenomena, however, scientists sometimes introduce confounding factors in the very act of measurement.

In the 1930s scientists frequently collected lake-water samples in soft-glass containers prior to analysis. It happens that soft glass contains materials that are soluble in water and tend to make the water less acidic; as a result, pH measurements consistently indicated that the lake was more alkaline than it actually was.[22] Since the magnitude of the pH changes from the time of collection to the time of analysis is unknown, the results of these analyses are dubious.

Impurities, even carcinogens, have been accidentally introduced into the food supply in a number of animal feeding studies. The conclusion of the Environmental Protection Agency in 1972 that DDT caused cancer in laboratory animals relied heavily on a study in which the animals may have been fed food that was contaminated with aflatoxin, a known carcinogen.[23] There is no way to know if any of the cancers found in the study group were caused by DDT or if all of them were in fact a result of exposure to aflatoxin. At the time of the ban (1972), no other study had found DDT to be an animal carcinogen at the dose (3 parts per million) administered in this study.

One of the first large-scale studies of the effects of saccharin ran into similar contamination problems. The saccharin fed to mice contained o-toluene-sulfonamide, a chemical that may contribute to the formation of tumors in the bladder—precisely the health problem linked to saccharin by the FDA study in question.[24] The contamination was a possible confounding factor that rendered the results of the study questionable.

The assertion that forests were dying all over Europe provides another instance of the importance of correct measurement. Like the Black Plague that was said to have destroyed one-third of the European population, the death of the Black Forest in Germany, the horrendous *Waldsterben,* was portrayed as the next development in the continuing saga of humankind destroying its own habitat. This observation, when direct measurements were made, turned out not to be true: "The fertilization effects of pollutants override the adverse effects at least for the time being."[25] What can the risk detective learn from the spread everywhere of false beliefs about forest decline? Be skeptical and whenever possible follow these rules.

Rule Four: Remember that parts are not necessarily wholes. Thus leaves are not necessarily trees; nor trees, forests; nor individual forests, regional forests. Reports of a massive dieback of the forests in Europe were based on surveys of damage to foliage, attributed by a number of scientists to acid rain.[26] The foliage was indeed in the condition they said it was in, but the total wood volume of the forests increased rapidly over the course of the so-called decline.[27] What the scientists were interested in was the health of the trees, but what they looked at was the health of the leaves. By identifying a harm other than the one they were looking for, European forest researchers reached an incorrect conclusion about the state of the health of their forests.

Rule Five: Count what counts. If it is trees, do not count leaves; if it is forests in total, do not just count one or two, take a sample. Recall, in regard to ozone, that if it is people's health that matters, we must measure ultraviolet radiation, the cause of skin cancers, not only ozone loss. We cannot always count what counts—indirect measures are also an important part of science—but when we can we should.

Rule Six: Follow trends. Looking for trends decreases dependence on appearances. Waiting may be frustrating, but it is also essential. If trends in forest growth or decline had been monitored, we would not have been misled for so long. Moreover, if we know the trends we can use *Rule Seven: Establish the normal range for the phenomenon in question as a standpoint from which to judge the trends.* Should the public be alarmed if rain becomes more acid? Because much depends on how the "normal" composition of rain is established, the risk detective wants to know if this has been done in a reliable way, such as by measuring acidity in ice cores dating to before the Industrial Revolution. Knowing that all rain is (and has been) naturally acid means that a greater increase must occur before we can blame human activity.

Measures have to be not only accurate and correct but also compatible. To measure a change in a physical property, such as acidity, the before and after measurements must be compatible. Comparisons of pH data from the 1930s and the 1970s indicated that there had been a massive drop in the pH (increase in the acidity) in most lakes in New England.[28] The old data indicated that only 4 percent of the lakes had a pH of less than 5 in the 1930s, while the new data showed over 50 percent with such a low pH in the 1970s.

When the old techniques of measurement were used to check whether the pH change in the 1970s was real, researchers found that only 9 percent of the lakes had a pH less then 5.[29] The change was one of technique not pH. It is not that one type of measurement was better than the other but rather that their inconsistent use led to considerably exaggerated results.

Because scientists must be able to calibrate across time in order to make certain of compatibility, they must document what they do in making their measurements. This is not always done. Many of the studies of lake-water pH performed in the past did not include such documentation,[30] which means that scientists of today cannot compare measurements in a meaningful way.

The lack of compatibility suggests another rule for our risk detective. *Rule Eight: Use the same types of measurement consistently.* It is not surprising that the acidity of rain was measured in two different ways, nor is it surprising that mixing up measures led to confusion or even upward bias. What requires explanation, though not in this book,[31] is that this blatant fault, while gradually corrected in the scientific literature, has been ignored in the public media. Had the public known of the confusion in the pH measures, it would not be surprised now to learn that the original fears of vast harm from acid rain were greatly exaggerated. Risk detectives who read the scientific literature would not be so surprised.

The Need to Identify and Quantify Exposure

There are three basic techniques for assessing chemical exposure: measurements, models, and proxies. A measured and timed dose is the ideal index of exposure. All other indices are good only to the extent to which they approximate this ideal. Again and again when measurements replace assumptions—as with the decline of European forests—results change radically.

A dose-response relation cannot be established without quantification of exposure. An apparent association between exposure and the incidence rate of a health problem tells us nothing other than that at some unknown level this agent may be hazardous to health. What is critical is what dose is hazardous under what conditions, which means that investigators should document exposure amount, duration, and mode or path as completely as possible.

Some studies offer no quantification of exposure at all. In a set of Swedish studies of the effects of dioxin exposure, subjects were asked whether they had been exposed to dioxin on at least one day in the past five to ten years, with no differentiation made between those who had been exposed to a small amount on one occasion for a few hours and those who had been exposed for days on end to large amounts.[32] There was certainly no effort at rough quantification of each individual's exposure. Even though those who were labeled exposed were more likely to be sick than those who were not, without quantification it is impossible to determine what dose posed what risk.

When scientists do not have sufficient data on exposure, they use models with estimates based on their understanding of the factors controlling the level of exposure. It often happens that those controlling factors are not well understood. Models are less effective than measurements because they are only estimates, but they can serve if well designed.

Proxies are the least-favored assessment technique, for they are indirect indices. A proxy is some quantity or quality that is presumed to be indicative of the level of exposure, for example, an electric meter outside homes for actual exposures inside homes. This technique leaves considerable doubt about the actual exposure level.

Unfortunately, there are few data on low-level exposure for many environmental agents. Scientists, for example, have not been able to determine what exposure levels are involved in the genesis of the cases of leukemia associated with benzene.[33] Most who study benzene are convinced that exposure to high doses does indeed cause leukemia, and some suspect that lower, but not necessarily the lowest, doses may cause leukemia as well. But as long as they do not have a good measure of the exposure levels in those cases where benzene is associated with leukemia, they cannot determine the dose-response relationship.

Scientists sometimes rely on models of exposure that are of unknown quality. In waste site studies, for instance, investigators generally have good measures of concentrations of chemicals at particular points at or near the sites, but often few or no data on how much of those chemicals reach the populace. In many instances default values, which are sheer assumptions about matters that influence ultimate exposure, are used in place of actual measurements. The Environmental Protection Agency (EPA) acknowledges that exposure may be significantly overestimated by such assumptions.[34]

Measurements beat assumptions. In assessing the likely by-products of waste incineration, EPA investigators published a report that indicated about 33 percent of the waste stream in the United States was paper and that items such as polystyrene, fast-food packaging, and diapers constituted nearly 25 percent.[35] Their estimate was based on extrapolation from assumptions about how much of these substances were produced, used, and thrown away. Actual measurements of material from landfills tell a much different story. They show that polystyrene is only 0.33 percent of landfill volume, fast-food packaging is 0.25 percent, and disposable diapers are 1.9 percent. Paper, by contrast, is about 50 percent, rather than 33 percent, of the volume.[36] Such a radical difference between the predicted composition and the measured one points to the need to rely as little as possible on extrapolation and as much as possible on actual observations.

Hence our super sleuth must follow *Rule Nine: Prefer measurements to estimates.* Accuracy and reliability depend on keeping in touch with actual measurements. Only then can it be discovered whether it is the measurements or the models that demonstrate danger.

Under ideal laboratory conditions, scientists know the exact chemical composition and size of a dose, the precise way it is introduced into the body, the time it is introduced, and the exact duration of exposure. Yet in human studies exposure is not easily quantified.

Many studies must rely on the recall of exposed subjects or remaining family members. The investigators in such studies do not generally seek an actual measure of exposure. Instead they start from a group of people who, say, have cancer and ask how any of them have been exposed to the chemical in question and to what extent they have been exposed. Since people have a tendency to look for causes when they see effects, these cancer victims are more likely to remember exposure than those who do not have cancer. This effect is called recall bias.

A Swedish study found that farmers exposed to dioxin were six to seven times more likely to have soft-tissue cancer than those who were unexposed.[37] This strong correlation—found in no other dioxin study—may be an artifact of the assessment method used rather than an indicator of the hazards of exposure. The investigators asked subjects or their relatives (when the subjects had died) to recall whether or not they had been exposed to dioxin on one or more occasions at least five to ten years earlier.[38] This technique obviously makes the measurement of exposure highly subjective, and there is no way to know whether those interviewed recalled accurately. Evidence from other studies with better measures of exposure show far lesser correlation or none at all.

A study of the effects of low-frequency electromagnetic fields on children indicated an increased risk of childhood leukemia associated with the use of some electrical appliances. But this conclusion was based on parental responses to a questionnaire.[39] Whether the responses accurately reflected the children's exposure, not to mention whether the use of appliances is an accurate indicator of exposure, is anyone's guess.

In studies of the effects of waste sites, investigators commonly use questionnaires to determine to what extent the community has been exposed.[40] A comprehensive review of the studies done at twenty-one of the National Priority List sites (set through the mechanisms of the Superfund) indicated that studies at only seven of them had involved measurements of exposure.[41] At those sites where questionnaires were relied on there is clearly a high potential for recall bias. After all, "everyone knows" that these sites are dangerous; anyone who is sick and lives near a site may associate the sickness with the site. Without actual measurements of exposure, such as "body burden" measures that show how much of the chemicals derived from the site are present in a person's body, it is impossible to know whether there has actually been any exposure and hence whether there is any link between exposure and illness.

A major Canadian study of the effects of saccharin relied on recall surveys. People who suffered from bladder cancer were asked if they had ever used a sugar substitute and if so how much and how often,[42] while a control group was then asked the same questions. Here, as with other studies that rely on recall, there is the question of whether the survey gets at the exposure levels of the people involved or just at the likelihood they will remember consuming what they have been told is a potential carcinogen.

The Love Canal Homeowners' Association attempted to determine whether or not the residents had an unusually high illness rate. Their hypothesis was that underground swale beds were carrying toxic chemicals from the disposal site to houses. Their study was based entirely on the memories of the residents, who knew that scientists had said there was a problem with toxic waste in the town. All recorded illnesses were self-reported; no effort was made to verify the existence or nature of the health problems with physician's records.[43] These unverified illnesses were plotted on a map constructed from the memories of "old-time residents," which was meant to indicate the locations of swales carrying the toxic chemicals from the canal to people's homes.[44] Both the predicted locations of exposure and its purported effects were derived from human recall. The importance of recall to this study seriously undermined its explanatory power. Therefore, it makes sense for the risk detective to follow *Rule Ten: Be aware of recall bias in assessing exposure.*

A final consideration in gauging chemical exposure is duration. Most studies of the health effects of an environmental agent involve chronic exposure, with results generally applicable only to repeated and long-term exposure. Yet regulators often forget or ignore the duration factor when they set exposure standards.

There is a huge difference between daily exposures of five and seven hours to a chemical. There is an even larger difference between a one-time exposure and daily seven-hour exposures. Nevertheless, the EPA used results from animal studies, which showed that a lifetime diet of less than 500 parts per million (ppm) of PCBs might adversely affect some animals, to justify requiring less than 50 ppm of PCBs in all outdoor capacitors.[45] A spill from one of those capacitors could not have exposed any animal, human or otherwise, to 50 ppm of PCBs unless it had lapped up the fluid. Furthermore, the exposure duration would certainly have been short for any human involved. Most people do not live on capacitors, and the PCBs would have rapidly volatilized or bonded with the soil. Either way, one should not compare chronic ingestion of 50 ppm PCBs to one-time exposure to a spill of a fluid containing 50 ppm PCBs. Instead one should follow *Rule Eleven: Consider the duration of exposure.*

The Need to Assess Effects

A principle fundamental to all studies is that scientists identify what they are testing for and make certain that what they find is what they claim it is. Yet massive misdiagnosis, incomplete diagnosis, or otherwise incorrect assessment of a condition is not an infrequent part of risk regulation.

When determining whether or not an environmental agent has done damage and if so, how much, scientists must identify the expected effect and measure that. In studies of acid rain, as we have seen, by looking at pH measurements scientists did not get at the desired effect. Since humans add sulfur dioxide and nitrogen oxides to the environment, which form sulfuric acid and nitric acid, measurements of these acid concentrations rather than of pH would have been more appropriate. Eight to sixteen times as much sulfur falls in the United States as in remote regions of the earth.[46] Regional sulfur enhancement clearly distinguishes the rain over the United States from the background rain, and thus using this increase as a measure allows for a meaningful discussion of the human impact on the atmosphere. Using pH alone does not.

Love Canal raised great concern over disease clusters. A study of the effects of the chemicals from the waste site on the local population found that there were heavy clusters of disease around swale beds. The investigators included miscarriages, crib deaths, nervous breakdowns, hyperactivity, epilepsy, and urinary disease in their analysis.[47] These clusters looked quite impressive, but did they really tell us about the dangers of exposure to the chemicals at Love Canal?

Science cannot establish a causal link between an ill-defined exposure to chemicals and a cluster of various disorders. Every area may have clusters of "illness," which are random or have a common cause. Each disorder must have its own cause, but whether several causes are related can be established only· by looking at the individual disorder and determining each cause separately. The chances that a cluster is random go up dramatically as one increases the number of disorders included. The presence of the Love Canal clusters, which appear to include whatever disorders the investigators found in the area, is not readily separable from chance. Hence *Rule Twelve: Evaluate separate effects to determine if there is really something to worry about.*

Finally, the risk detective should heed *Rule Thirteen: Be aware of the extrapolation of effects.* Since it is easier for scientists to establish the results of high doses than of low ones, they often do not directly examine the effects of low-level exposure. This is understandable, because a high dose generally has more of an effect than a low one. In many instances the effects of chemical agents on human beings have been quantified only for doses that are orders of magnitude higher than those to which the general public is exposed.

A number of officials and commentators have taken the position that there is no safe dose, however low, for a chemical that has been shown to be a hazard at however high a dose. The FDA banned the use of aminotriazole (3-AT) as a weedkiller in cranberry culture in 1959 on the basis of a study in which the lowest exposure level displaying a potential effect on rats was about 1,500 times higher than that which a 150 pound person would have received by eating 1 pound of cranberries with a 1 ppm residue of 3-AT every day for life. The 1 ppm residue figure was what the manufacturers requested as a maximum allowable level for cranberry products sold in the United States, but the FDA decided to ban it entirely because it claimed there was no way to establish a safe dosage.[48] This is like saying that a highly improbable harm is the same as a highly probable harm.

The saccharin debate also brought up the no-safe-dose argument. Here a professor of environmental sciences stated that the Delaney Clause "wisely allows no human discretion based on dosage . . . since there is no valid scientific basis for such discretion."[49] The opposite is more nearly true. There is no scientific basis for assuming that a chemical is dangerous at low doses just because it is dangerous at high levels. Aspirin will kill you if you take too much, yet it is a valuable drug as a painkiller and as an anticoagulant.

The Need to Connect Cause with Effect

A cause must precede its effect. This precedence is axiomatic in all discussions of causation, yet there are a number of instances in investigations of the effects of environmental agents where this axiom has been overlooked.

Whenever possible a comparison should be made between rates of health problems in a study population before exposure and after exposure. It would be wrong to assume that the obvious is always done, as exemplified by a study in which investigators blamed vinyl chloride plants for causing birth defects in neighboring areas. This study, unfortunately, did not include data on what the birth-defect rates had been in the areas where the plants were located before the construction of the plants.[50]

In another instance where questions of timing have been raised, DDT was blamed for a decline in the population of a number of bird species in areas where the pesticide was used extensively. One of the most publicized of these declines was the drop in the bald eagle population. However, a number of sources have indicated that the bald eagle population was already decreasing in many areas before the advent of DDT spraying.[51]

Determining whether or not recovery has occurred after exposure has ended is also important. If DDT was responsible for the bald eagle decline, its removal should allow for recovery. In fact, there is apparently a correlation between the recovery of bald eagle populations and the halt of DDT spraying in a number of areas.[52] Furthermore, there are indications that the decrease in the bald eagle population may have accelerated after spraying started.[53]

But scientists also have contradictory temporal evidence about bald eagles. In some areas the eagle population appeared to increase during the time that DDT spraying was going on.[54] Such contradictory conclusions resulting from temporal correlations point to the need for other criteria for evaluating a chemical's role in affecting health. When there is neither quantification of effect (the extent of decline before and during spraying and the recovery after spraying) nor of the cause (the doses of DDT involved in each area), one cannot reach correct conclusions.

The need to connect cause and effect means that the risk detective should follow *Rule Fourteen: Seek the mechanism.* Though it is by no means true that when you seek you will always find a causal mechanism, looking for one helps the observer understand what it might look like, what it would have to explain, its data requirements, and the past difficulties in coming up with an appropriate cause. Knowing that a plausible mechanism has not yet been identified or tried out provokes necessary skepticism. Asking what determines the weather and coming up either empty-handed or with a hodgepodge of variables should leave the investigator wondering how we can be so certain of the future temperature of the earth. When seeking causation it is also a good idea to keep in mind *Rule Fifteen: Establish conditions of applicability.* Rather than addressing, for example, the stultifying question of whether clouds in general add to or detract from warming, we are better off establishing what sort of clouds under which conditions cool or heat and (if known) to which degree. One danger when causation is not clear is the quick acceptance of residual explanations. The risk detective should be wary of this lazy reasoning and should follow *Rule Sixteen: Do not accept residual explanations.* In other words, do not accept the reasoning that if it is not Y or X (too little rain or some blight), then acid rain must be the culprit.

The Need to Replicate

Is one positive result enough to warrant a ban or other regulatory action even if there are several negative ones?

The most frequently cited study of the effects of low-level lead exposure on children indicated that there is a strong correlation between exposure level and impaired mental function.[55] The same investigators also found a correlation between umbilical cord lead levels and congenital abnormalities.[56] The results reached in these studies have never been replicated, despite the many similar studies conducted by other scientists. If the lead was truly responsible for the defects, then studies of similar exposure levels should consistently show the same results.

In 1959 the results of experiments of the effects on rats of aminotriazole (3-AT), the weedkiller, were released. The study found that four of twenty-six rats fed diets of 100 ppm 3-AT developed cancerous thyroid tumors.[57] Although further studies have failed to confirm findings (rats and hamsters fed daily doses of 100 ppm for their lifetimes did not develop tumors at a rate greater than the controls),[58] the FDA decided to ban the use of 3-AT on the basis of the one positive result.

A study published in 1968 revealed elevated DDT concentrations in the blood fat of liver cancer and leukemia victims compared with people who had died as a result of an accident or a gunshot.[59] No other study conducted before or since has found a correlation between DDT and cancer.[60] Even this one may indicate only that as cancer victims lose weight the concentration of DDT in their bodies goes up, because the rate of DDT elimination is not proportional to the rate of weight loss.

Where the overwhelming majority of the studies available indicate that a chemical is not a carcinogen but one study finds that it is, the negative results should be conclusive. If the negative results were reached in poorly designed or executed studies and the positive result was based on a well designed and executed study, then the positive result should be taken more seriously. Even then, scientists must be able to replicate the results before they can confirm or deny the causal implications of a study that shows a correlation between exposure and illness.

A study of the effects of an environmental agent should be replicable. The work done should be carefully documented to allow other scientists to check the conclusions and to determine whether or not they are warranted by the data. Replication of a study also helps to reduce the chances that the conclusions are drawn at random. All the more reason for risk detectives to heed *Rule Seventeen: Don't draw final conclusions from one study;* rather, look to other studies for confirming or contradicting results.

During the interval between publication of the original study suggesting harm and the replication study adverse health effects might continue. If so, that is a loss. However, if the first indication of harm generates regulatory action, there are also negative effects, though not such evident ones: lives are disrupted; jobs are lost; standards of living are lowered; all of which in turn reduces health rates.[61] Worst of all, as standards of evidence weaken, society pays the costs of unsubstantiated allegations of all kinds.

The Role of Skepticism

The flaws we have studied are quite disparate: failing to control for age and smoking is different from recall bias or failing to compare background rates. Yet there are strong similarities. Consider first how obvious they are! Except possibly for the mix-up over different methods of measuring the pH of lakes, the defects are of the most elementary character: failing to control for sex, relying on studies that have not been replicated, omitting relevant comparison groups. These faults are not esoteric, the kind that nonscientists would not understand. What is amazing is that bona fide scientists commit them and even more that they are used for governmental regulation policy.

A second common feature is that most of the defects that invalidate the results of the studies also bias those studies toward finding harm. We have found that the flaws either increase the probability that a positive result will be found or, less commonly, have a neutral effect. Only one error in one study produced a clear bias toward a negative result.

There is no need to invent conspiracies to explain this tendency. Proponents of the existing regulatory system (many of whom want even stricter regulation of chemicals) tell us proudly that they do not want to take chances with public health. The experimental defects recorded here are tolerated and the results from the studies in which they occur are used to set regulatory standards on the assumption that possibility of harm is much more important than probability of no harm. This better-safe-than-sorry precept, the environmentalists' precautionary principle, is the single most important decision rule in the environmental and safety field.

The rule of rules is that whether the claims are of excessive harm or of no harm at all, they should be approached on a "show me" basis. In other words, follow *Rule Eighteen: Be skeptical.* It is essential that citizens not just rely on scientific studies but "lean" on them as well. Then the risk detective soon becomes a specialist in puncturing inflated claims. Warned by the now-known-to-be-false cancer epidemic and asbestos scare as well as the claims of the death of forests when they are more alive than ever, our amateur investigator is no longer so easily fooled.

Suppose our sleuth is overwhelmed: too many clues pointing in too many directions. What to do? *Rule Nineteen: Keep score.* This approach raises awareness both of what constitutes evidence and of what the conflicting opinions are. Looking at masses of data and listening to rival arguments can muddle the mind. First, separate out the arguments, so that evidence pro and con can be accumulated. Keeping a scorecard enables a sort of summation. Second, if the count is pretty evenly divided, without good reason for assigning heavier weights to one factor or another, there may be no way to decide without learning more. But if, as in the case of global warming, the points for one side grow larger and larger, a conclusion is indicated: not merely the Scotch verdict "unproven" but the reasoned verdict "likely not serious."

In September 1992 news reports attributed hurricanes and floods to global warming. Are there in fact more such natural disasters than there have been? Are they more severe? The answer to the first question is "no" to the second, "maybe." Is there a causal path linking severity with warming? Doubtful. The risk detective knows that global warming is the only criminal who keeps revisiting the scene of the crime but stubbornly refuses to leave fingerprints.

The greenhouse gumshoe with a sense of humor knows that if everything is most important, nothing is. What better example than global warming? No doubt proponents still believe that man-made carbon dioxide is the most important causal force. There are scientists who argue that clouds are the determining force, while others think that oceans control the weather. Within these camps there are champions of ocean-bottom hydrates, phytoplankton, and tectonic plates. There we have it: the prize for the great climate mover and shaker goes to—the land, the oceans, clouds, deep earth, and, of course, human beings. The advantage of knowing who the suspects are is that one can keep an eye on them. Even more important perhaps is understanding that more than one force may well be at work.

Finally, there is *Rule Twenty: Seek diversity, not uniformity of opinion.* We must welcome dissent, because it will bring new clues. It is by comparing the evidence supporting and opposing rival views, after all, that such truths as may be found are likely to emerge.

Answers to the great question—are claims of harm to life and nature from technology mostly true or mostly false or is the picture mixed?—depend not so much on tentative conclusions about one issue but on the trends of conclusions to a considerable number of issues. What citizens need to know can be determined by citizen sleuths consulting one another. These consultations based on individual detective work can then be amplified by discussions with scientists, but only after the citizens know what to ask and how to make sense of what they hear.

Conclusion:
Rejecting the Precautionary Principle

The precautionary principle says that to avoid irreparable harm to the environment and to human health, precautionary action should be taken: Wherever it is acknowledged that a practice (or substance) *could* cause harm, even without conclusive scientific proof that it has caused harm or does cause harm, the practice (or emissions of the substance) should be prevented and eliminated.

— *Rachel's Hazardous Waste News*, 1993

Let us ask what appears, for the moment, to be a completely irrelevant question: why did no one (except Senator Daniel Patrick Moynihan) predict the imminent collapse of the former Soviet Union? Conservatives, I would say, had a stake in believing in a dangerous enemy. Liberals had a stake in not denigrating a socialist system lest that reflect badly on their hopes for a welfare state. But why did the intelligence apparatus, which is supposed to be more dispassionate, not come closer to the right result? One reason, no doubt, is that the Soviet authorities themselves, though they had intimations, did not realize they were on the brink. They had fudged their own data so thoroughly they did not know where they were. But our intelligence people are supposed to be professional analysts not taken in by wish fulfillment.

I believe that two sources of distortion at work among the intelligence community are relevant to environmental and safety as well as to foreign policy issues. One, well known, is that maintaining a secure career means not straying too far from current views. If you stick your neck out by suggesting something radically different, you are subject to attack as ideological or peculiar by those whose favorite presumptions are being upset. Whereas, if for comfort and cover, you stick to prevailing views, plus or minus a little to show your independence, your safety is pretty much assured. And, if you are mistaken, you will have so much company that it will be difficult to single you out; others, more powerful than you, will have a stake in avoiding retribution.

Another attitude that thwarts accurate predictions is the adoption of worst-case scenarios, defended as acts of prudence. If the United States overestimates the threat from the Soviet Union, this defensive doctrine goes, it will have spent too much money, which it has in abundance, but the nation will in any case have been defended. If the intelligence authorities underestimate the threat, however, less money may be spent on military defense, but the nation is exposed to destruction and defeat. Therefore, it is safest to assume the worst. Thus if the Soviet or East German economies could have been growing at 8 percent a year or declining at 4 percent, the safe strategy is to assume 8 percent growth and prepare defense systems accordingly. Would not any prudent person want to protect the nation by making "conservative" estimates? What could be wrong with bounding uncertainties in this way to preserve the nation?

Everything. This formula diverts attention from the best available knowledge to the hedging mechanism. Before you know it, the formula itself predetermines the results. By this time, the elected and appointed may have forgotten, if they ever knew, that they are taking out what the formula has put in. If we take an overly aggressive view of the threat, we will take overly aggressive measures against it. Since the results of doing too much can be as disastrous as doing too little, what should be done? Provide the most accurate estimates and let the decision makers determine what discount to give them.

Why bring defense policy into this? Because exactly the same thing is being done in regard to environmental and safety policy. The only difference I can see is that conservatives used worst-case safety-first rules for defense policy and now liberals are using these rules for environmental and safety policies. But how can actions designed to secure safety actually increase harm?

The precautionary principle is a marvelous piece of rhetoric. It places the speaker on the side of the citizen—I am acting for your health—and portrays opponents of the contemplated ban or regulation as indifferent or hostile to the public's health. The rhetoric works in part because it assumes what actually should be proved, namely, that the health effects of the actions in view will be superior to the alternative. And this comparison is made favorable in the only possible way—by assuming also that there are no health detriments from the proposed regulation. The rhetoric seems to present a choice between health and money or even suggest health with no loss whatsoever, for a tangential presumption is that industry will find a better and a cheaper as well as safe way. Something (health) is gained with nothing lost (no adverse health effects from the bans or regulations).

Two propositions need to be examined, the empirical and the moral. The empirical question is whether the health gains from the regulation of the substances involved are greater or lesser than the health costs of the regulation. One should not assume; one has to demonstrate, to supply evidence and argument, to justify that "take no chances" is empirically valid in protecting health. A responsible person cannot just believe but must have reasons derived from evidence. And preponderant evidence is against the proposition that health would be improved or maintained by regulating minuscule amounts of chemicals or withdrawing huge amounts of carbon dioxide from our economy.

The moral issue is, what norm states that health is the only value or even the dominant value? Emphasizing a single value, to which all others must be subordinated, is a sign of fanaticism. Whatever happened to other values? How much is a marginal gain in health worth compared with losses in other values such as freedom, justice, and excellence?

My main objection, however, is not that small gains in health are coming at the expense of other valued qualities. My great objection is that overall there are no health benefits from regulation of small, intermittent exposures to chemicals.

The Monumental Reversal

Life is full of uncertainties. We know something but not everything about the probabilities of certain events occurring or about the consequences of a given act or circumstance. We may not even be entirely confident of our own preferences, especially as these may depend on a less than complete understanding of the effects of pursuing our initial desires. Nevertheless, we are not usually paralyzed. We make use of whatever understanding we have, self-understanding as well as understanding of how others will react. At best, we observe the consequence of our actions, modify them accordingly, and proceed. In academic literature this is called incrementalism, the method of successive limited approximations. Sometimes it is called trial and error, meaning not that every conceivable trial is attempted but that trials not expected to be excessively harmful may be attempted. The rationale behind this "science of muddling through," as C. E. Lindblom termed it, is that small moves facilitate action while reducing the demand for knowledge.[1] By contrast, comprehensive decision making requires the abrogation of uncertainty by demanding full knowledge of consequences before action is taken.

Incrementalism has weaknesses. It may not be possible, for instance, in the midst of vast numbers of interactions, to observe the consequences of small moves. Successive limited approximations may be too cautious a strategy, too small and too slow to succeed, whereas bolder and faster moves might be successful. Nevertheless, incrementalism as a trial-and-error mechanism for decision making under the usual conditions of limited knowledge, bolstered by Herbert Simon's work on the cognitive limits of rationality, won the day in the 1960s both as a description of how choices are made and as a prescription of how they ought to be made.[2] The intellectual collapse of comprehensiveness as unattainable, given human limits, led to the demise or weakening of maximization postulates. Unless complete knowledge exists—agreement on preferences, understanding of all moves and their consequences—it is not possible to maximize objectives. It seemed then as if the battle were over.

In our time, however, comprehensiveness has emerged in a new form, one that is more difficult to recognize. In effect, the argument has been radically reversed: a lack of comprehensiveness now requires preventive action by government. If all is not known, that is, if a product or practice is not proved safe, then it must be banned or restricted. There is no need for opponents to demonstrate harm. It is up to the substance's supporters to prove a negative—namely, that the substance could not cause cancer—a task that requires comprehensive knowledge of all consequences. I recall Harvey Brooks writing that the only proof of a negative is an impossibility theorem demonstrating that the contemplated action or reaction is contrary to the laws of nature.

The immensity of the change requires reemphasis: private action requires proof of the absence of harm; governmental action requires no proof of harm. The relative role of the citizen and the state have been reversed. In the past it was the citizen who was entitled to act and the state that had to justify its intervention; now it is the state that intervenes by right and the citizen who has to give reasons for acting. The reversal of the usual course of action has profound implications. The question is whether this monumental reversal is justified.

The pattern—as we have painstakingly described in this book—is distressingly familiar. A study or observation spurs scare stories. A researcher finds a correlation between a suspect chemical and cancer in rodents. Horror-struck reporters and citizens demand stringent regulation of the chemical. Later on, articles containing criticisms of the original studies appear in scientific journals; the original authors retreat a bit but not much. New studies are undertaken, and the original charges are rebutted or the results are seen to be inconclusive. But the new studies do not keep pace with the publicity. The chemical has acquired a bad reputation. The chemical company becomes involved in costly lawsuits, while its product is likely to be withdrawn from the market. Production stops; lives are disrupted. There are no observable public health benefits, yet a victory for public health and the natural environment is declared.

Is the Precautionary Principle Good for Our Health?

The traditional guideline for toxic substances—reduce exposures by 10 or 100 or 1,000—has in the case of cancer been replaced by risk assessment, which makes worst-case calculations at each step of the process. This multiplication of improbabilities is justified in a blindingly simple way: it would be morally impermissible to do less than might be done. The only safe assumption is that the worst in cancer creation might happen at every step and therefore safety levels must be increased accordingly. Though imagining the worst is not usually recommended as desirable for a person's mental health, this stance enables the assessors to argue that they are being careful with human life. But is this true? Is human health conserved by mandating ever-lower exposures? Everyone will agree that there has to be a stopping point short of infinity. The trouble is that the argument from infinity does not answer the retort, "So what would be wrong with even more protection?"

There are analysts who want to rescue risk assessment from the multiplication of worst cases by arguing for analytic purity. By assuming the worst, with uneven definitions of how bad that can be, these purists claim, risk assessment becomes confused with risk management. They recommend that the least-biased risk assessment be performed and then the degree of caution desired be applied by administrative managers or elected political officials. I agree that this separation of science from politics, though it cannot be entirely achieved, is worth striving for. After all, if it is all politics, why bring in science? We would like to improve and not harm health and, without knowledge, we are at risk of doing harm. As John Graham, Laura Greene, and Marc Roberts put it pithily, "Without science there is no basis for regulation."[3]

But this argument, though accepted by a few, has failed to be broadly persuasive. What public officials want to be accused of not protecting their constituents' health to the fullest possible extent? This line of thought and action was evident in the statements from Arthur Flemming, Secretary of Health, Education, and Welfare during the cranberry scare of 1959. Until it could be proved absolutely that a chemical or mineral would not ever cause cancer, he had to ban or severely regulate the offending substance.

Analytical purists were not in good rhetorical shape then. If asked why they should not calculate conservatively on behalf of human health, they had no answer except "yes, but." Yes, they wanted to protect society, but they thought the obligation to place that desire in their calculations belonged to someone else. To make their point effectively they needed to but did not argue that their principle of accurate risk assessment was better for health than a bias toward maximizing possible hazards.

A number of scientists, including Bruce Ames and Lois Gold, tried two other tacks: (1) the substances in question were far too small a portion of the whole cancer-causing picture and too puny compared with the much larger natural background to make them important; and (2) the administration of the maximum tolerated doses in rodent tests were more likely to increase cancer rates than the test agents themselves. They proposed a theory—mitogenesis increases mutagenesis and therefore carcinogenesis in laboratory animals—to explain why so large a proportion of the test chemicals were causing cancer in rodent tests.[4] In other words, it was the test that produced the cancer. This scientific attack has shaken but not stopped regulatory science. Why not?

I do not disregard the importance of beliefs about scientific evidence. As scientific opinion steadily moves against the validity of high-dose rodent tests as currently conducted to predict human cancers at low doses, it will be more difficult for regulatory scientists to hold up their heads among their peers. No one wants to be looked down on as a practitioner of discredited science. Usually practices that lack justification are gradually abandoned, but in regulatory science the demand for justification is directed elsewhere: show that the substances involved are safe.

In regulatory science "might possibly cause" is good enough to justify considering every substance suspect, thereby justifying animal cancer studies. To break the stranglehold that rodent lifetime feeding studies have on regulatory science, it is necessary to show that they are valueless for predicting cancer in human beings. Though many if not most of the scientists concerned would agree that rodent tests are worth little, they will not say "worthless" until the evidence is overwhelming. "The most commonly used method to determine whether something is dangerous over long periods and low doses," science reporter Matthew Wald wrote in the *New York Times*, "is little trusted even by those who use it."[5] Yet any scientist can think of "maybes" that justify the practice.

Whether the proponents of basing regulation on animal cancer tests are right (tiny amounts do cause significant numbers of human cancers, so the tests help prevent cancer) or wrong (tiny amounts do not cause cancer), they win by being seen as defenders of health. Something is awry with the "heads I win, tails you lose" argument. What is wrong is assigning only health benefits, and no health costs, to regulation.

In *Searching for Safety* I sought to place the risk debate on another plane by asking and answering a quite different question. Instead of assuming that modern industrial society is detrimental to human well-being, I looked at health and accident rates in the United States and other Western democracies. Instead of inquiring into a nonexistent state of affairs (why is health declining in democratic industrial societies?), I sought to find out why these societies have not only the richest but also the healthiest people in world history. Why had morbidity decreased decade by decade, often year by year, for over a hundred years? By investigating the sources of improvement in health, I argued, we can follow an important criterion of decision: while attempting to protect health, do not weaken the forces responsible for unparalleled improvements.[6]

In general, health is a product of standard of living. Genetic inheritance matters, though we do not know how much. Health habits (eating, drinking, smoking, sleeping) matter a good deal. This said, everywhere, among and within nations, richer people are healthier than poorer people. Wealth buys not only better medical care but, more important, better organization, communication, and knowledge, those capacities that enable us to create what we need when we need it. The reverse is also true. Better organization and knowledge create wealth. Any way you want to put it, the idea that a lower standard of living reduces health rates is one of the best supported propositions involving human behavior.[7]

The rhetorical strength of "conservatism"—of "erring on the side of safety"—is that it appears to be unalloyed good. But if the substantial resources needed to clean up carcinogens lead to a lowered standard of living, public health becomes poorer than before. Instead of being an unalloyed good, stringent and therefore expensive regulation can worsen as well as improve health.

A noted decision analyst, Ralph Keeney, has created a model tried against data. He calculates that a cost of some $7.5 million per hypothetical life saved will cause equivalent harm from lowering living standards.[8] The important thing is not the exact numbers but the view that expenditures of national wealth on regulating minute quantities of chemicals actually undermine any health improvements achieved. From any gains in health due to regulation must be subtracted accidents and illness contracted during efforts to remove the offending substance and losses in health due to lowering the standard of living. Environmentalists are fond of saying that "no act does just one thing," wisdom they should apply to regulatory measures.

Conservatism in regulation would be appropriate if there were a limited area of uncertainty that needed to be bounded. Then we could decide how far to hedge against the bad news being worse than we had thought. When there is an ocean of ignorance, as produced by rodent cancer tests, there is no way of hedging, because we do not know where we are. How far are we from knowing where animal rodent cancer tests leave us? We may be 200 to 200,000 to 2,000,000 times from the truth, which is nowhere.

What, then, would be an appropriate strategy for ignorance, for not knowing probabilities of harm? The answer lies in what ecologists call resilience, whereby robust species adapt to and surmount newly arising adversities. Resilience in human societies requires growth in knowledge, communication, wealth, and organizational capacity, the resources that enable us to craft what we need when we need it, even though we previously had no idea we would need it.

What would we do, I ask, if we acted on the basis of the knowledge we have, imperfect as it is? We would be acting on experience with low-level, intermittent exposures. We would assume dose-response relationships to have low-end thresholds—points below which exposures are innocuous—unless there was evidence to the contrary. This step alone would rule out most claims of harm or make the resulting regulations far less drastic. We would beef up epidemiological studies and pursue research on mechanisms of cancer causation. From such studies would come slow but steady progress in understanding and controlling cancer. True, small effects might be missed, but rodent cancer studies would also miss them. Far fewer false leads would be followed.

Has Greater Knowledge Increased or Diminished Evidence of Harm from Technology?

How would the nation's health have fared if we applied the lessons learned from the case studies in this book? The cranberry scare would not have occurred, both because the amounts of the chemical would have been perceived as too tiny to be important and because existing evidence already suggested a thyroid mechanism that required much larger amounts of the suspected substance. The results: far fewer frightened consumers and no undeserved losses by anxious farmers, let alone losses to the manufacturers and their employees.

The episode of PCBs represents interesting mixed results. Regulators had good reasons for limiting exposures of PCBs to fish and animals through leakage and washing into streams. They could also make the case for getting rid of PCB transformers in buildings on the grounds that if the transformers caught fire, they would create a substance decidedly dangerous to people nearby, including firefighters. But there was no good reason for requiring the destruction of expensive electrical equipment containing trace concentrations of PCBs. Not only was the nation saddled with over $700 million in unnecessary spending, but the workers who had to replace electrical equipment amid high voltage lines were subject to unnecessary deadly danger. As for the PCB-contaminated building in Binghamton, New York, it was the policy of keeping it sealed tight for years that was harmful, not the building's contents. After the initial fire and the cleanup, had the building been vented skyward so trace elements of PCBs could be released into the open air with no danger to anyone, the building could have been opened quickly. I understand that public panic prevented this safe and sensible action. So much the worse for those who generated this panic.

Had DDT been allowed to live out its useful life, it would have lasted in weaker form for limited uses until equivalent or better substitutes were available, and people killed by substitutes for DDT would be alive. Harm to animals could have been mitigated and would, in any event, have diminished as use kept decreasing.

As far as anyone knows, there is not a single substantiated incident of harm flowing from Love Canal. The best use of that episode is as a compendium of how not to do research. Moving people out of their homes is a disruptive and harmful act; it should not be done without strong reason. Although it made sense to evacuate those who had very high levels of chemicals measured in their homes (a small minority), moving the others was done in the absence of any known mechanism of transmission or any study to see if demographically similar communities not near Love Canal suffered similar symptoms. When this study was eventually done, the results were negative. The rule of replication had not been followed. Here, again, we see that precautionary behavior—when in doubt evacuate—was bad for the well-being of those concerned. We also see that securing experts to tell us what we want to hear is not a good way to help ourselves.

Times Beach, Missouri, with its Environmental Protection Agency workers in moon suits and bewildered and despairing citizens, and Agent Orange, with its vision of poisoned American soldiers—both episodes invoking visions of retribution against the immoral actions of industry and government—are actually examples of self-flagellation. If the handlers of chemicals, who had by far the largest exposures, showed few ill effects, why should soldiers who had only trace exposures to Agent Orange become deadly ill? If the highest exposure to dioxin in Missouri (at Quail Run) produced minimal harm, why should the smaller amount found on dirt roads in Times Beach do significant damage? Finally, if among all occupationally exposed Americans only those with the highest exposure show any increase in cancer, and those with comparatively low exposure show no rise in cancer, why should people with much lesser environmental dioxin contact develop cancer? The dioxins that bind these two cases together did wreak havoc among rodents. But once more conservatism proved misleading, as the difference in the physiology of mice compared with guinea pigs, hamsters, and people ruled out the possibility of the same sinister results. Instead of helping, the cleanup harmed health as a result of increasing worry and expense. Meanwhile Agent Orange and Times Beach left a wreckage of broken hearts that ought to enter into someone's cost-benefit calculus.

The more that was learned about low levels of arsenic, the less harmful it appeared. Alar was not harmful to begin with, a finding strengthened by subsequent research. The claim that early fall of apples without Alar leads to greater insect infestation and hence greater use of pesticides has not yet been evaluated. Chrysotile asbestos, the kind used in schoolrooms and buildings, was not harmful in small, intermittent doses, and everything learned since confirms that view. It is a serious matter, in my opinion, to impose billions of dollars in costs that help no one and might harm those who do the work. Of nitrites, the less said the better. From the start, there were regulatory errors best understood as straining to find harm. Let us look more closely at how time has treated a few of the more celebrated episodes.

SUPERFUND

There have been four main ways in which knowledge pertinent to the alleged public health threat from inactive chemical waste sites has evolved since the passage of the Superfund law in 1980.

First, we have learned that the phenomenon of chemicals migrating from waste sites into the environment is far more widespread than originally understood. The number of identified sites has soared beyond expectations, and whole new categories of sites, such as underground gasoline storage tanks, are coming into view.

Second, fifteen years of epidemiological investigations have failed to turn up persuasive evidence of insults to human health in communities near inactive waste sites. At the inception of the Superfund program few such studies had been conducted, and there were dark expectations about what an accelerated program of epidemiology would turn up. But these fears have not been realized.

Third, new doubts have been cast on the validity of high-dose animal testing as a basis from which to project risks from low-dose human environmental exposures. Since the projections of threats to public health from inactive chemical waste sites rest almost exclusively on animal testing, a lessening of confidence in these tests erodes the very foundation of the Superfund program.

Fourth, we have learned that cleaning up sites to highly stringent standards is even more difficult than had originally been thought. Incineration has proved politically controversial and of debatable effectiveness, groundwater-containing structures plunged into the earth have leaked, and attempts to pump and treat aquifers to drinking-water purity have proven futile.

It is one thing to say what virtually everyone agrees on: Superfund is too slow and too expensive and lawyers take far too much of the insurance fund. It is another to claim, as I do, that Superfund, for all the billions spent on it, provides no health benefits. Indeed, I would go further and argue the likelihood that Superfund (except for its valuable emergency actions) damages health. The basic facts are that the amount of potentially harmful material is too minute and too far from people to do much harm. When one learns, in addition, that current technology is not able to clean up this muck to bring it within desired standards, it becomes clear that we are going through cleanup for its own sake.

ACID RAIN OVER TIME

Greater knowledge has certainly led to estimates of lesser damage from acid rain. When acid rain first came to public attention scientists were concerned about the degree to which humans had changed the chemistry of rain and the effects that change would have on lakes, streams, forests, and humans. They still are, but their concern has greatly diminished in magnitude.

When scientists first attempted to quantify the amount humans had acidified rain, they estimated that rain had become ten to thirty times more acidic in the eastern United States than it had been before the Industrial Revolution.[9] Recent estimates indicate that the acidity is actually within the range of the nonindustrial norm, or at most two times as acidic as the most acidic pristine rain.[10] Humans have changed the chemistry but not increased the acidity anywhere near as much as the claims of harm would indicate.[11]

The injury to forests was also substantially overestimated. Early concerns included soil damage leading to malnutrition of trees.[12] There was widespread concern that substantial harm had been done, and many local declines were blamed on acid rain. But a massive survey of U.S. forests found only one area showing evidence of detrimental effects of acid rain: high-elevation spruce stands. Other than this one group, which represents less than 0.1 percent of the forested area in the United States, there has been no forest damage due to acid rain. The results of this survey led the National Acid Precipitation Assessment Program to conclude that "compared to ozone and many nonpollutant stress factors, acidic deposition appears to be a relatively minor factor affecting the current health and productivity of most forests in the United States and Canada."[13]

OZONE HISTORY

Estimates of the long-term damage to the ozone layer have gone up and down since the first report that CFCs would lead to ozone destruction.[14] All models have indicated that continued growth in CFC emissions would lead to significant ozone depletion, yet there is no simple trend in the predicted magnitude of that effect. Still, research conducted since the Montreal Protocol clearly predicts greater depletion than was expected earlier and is based on a significantly better understanding of the processes involved. Scientists initially believed that thinning of the ozone layer would be undetectable for decades, but it has already been observed and strengthened the hypothesis that thinning is due to CFCs. The initial objections that the decrease could merely be a result of the solar cycle have been met, and the decrease is now actually larger than it had been,[15] making the threat posed by CFCs probably greater than originally predicted. I say "probably" because evidence of substantially increased ultraviolet radiation, which would do the harm, is scant, and the atmospheric conditions required to maintain large-scale depletion over time are hard to meet. Because ozone thinning is still within the range of natural variation, it could still turn out to be temporary, but evidence over time has tended to strengthen rather than weaken the thesis.

GLOBAL WARMING THEN AND NOW

The stories of catastrophe from global warming have hardly abated over the years. But both theory and evidence have shifted increasingly to somewhere between no warming and low warming. New findings about the residence time of carbon dioxide in the atmosphere have revealed that it will accumulate much more slowly, giving us time to learn more before taking drastic action. That many of the feedbacks are now either negative or less positive than they were once thought to be, provides further reassurance that such warming as has occurred has taken place at night and in the winter, which lengthens the time and temperature it would take to generate harmful consequences.

Looking back at the array of environmental and safety issues, many of which, like Love Canal and global warming, have become imprinted on the public consciousness, we can discern a clear pattern: the more that is known, the less reason there is to fear the worrisome object and the weaker the rationale for preventive measures. The one partial exception is CFCs leading to ozone depletion.

Of all the subjects I have studied or read about in over three decades as a social scientist, environmental and safety issues are the most extraordinary in that there is so little truth in them. Perhaps where there is smoke there are smoke makers. Lots of people, it turns out, for want of evidence of harms incurred or averted, hold the view that environmental programs reflect motives other than their stated one.

Are Environmental and Safety Issues about Something Else?

Atmospheric scientist Hugh W. Ellsaesser, in an article titled "The Politicizing of Climate Science," tells us that "the real reason for the move to protect the ozone layer by banning freons is to build the legal machinery and the constituency for the attack on carbon dioxide [blamed for alleged global warming]."[16] Of course, with a different perspective the buildup of carbon dioxide could be a force for good. To him it is "remarkable that increased carbon dioxide is not suggested as just what is needed to prevent or delay the onset of the next glacial—which, if anything, is already overdue."[17]

In agreeing with fundamental criticisms of Superfund as "slow," "inefficient," "self-contradictory," "unnecessarily difficult to manage," and, worst of all, not providing "environmental benefit that nearly matches the social sacrifice from the expenditures made," policy analyst Milton Russell, instead of condemning the program, suggests that by standards other than public health the same outcomes would be judged successful.[18] Indeed, these noncleanup, nonhealth objectives, in Russell's estimation, dominate the political process to the extent that the admitted flaws would be considered unfortunate but necessary adjuncts to those other purposes. And there is no doubt that Russell is right when he observes that Superfund has been supported consistently by large bipartisan majorities in Congress.

Some purists think that Superfund is about cleaning up old hazardous waste sites. How naive! Its defenders, Russell tells us, see Superfund as an important tool in restructuring the way the nation deals with the production of hazardous waste. In this regard Superfund is supposed to draw attention to the long-term consequences of current actions. It molds private (and government) incentives toward waste minimization so that future generations will not be harmed by current disposal practices. It encourages proactive search to identify problems before they otherwise would become evident. The environmental reach of Superfund, in this view, is substantially broader than the sites it touches.[19]

From this perspective Superfund is a consciousness-raising device. Evidently unsure of the real reason for Superfund, Russell continues:

> Another possibility is that Superfund is the setting for a modern morality play which teaches lessons even as it provides good theater. Evil is personified in those who presided over depositing the wastes. Injured innocence in those who live near the sites. Good triumphs over evil as those responsible for the wastes are forced to make restitution and suffer punishment at the hands of the forces for good, which arrive just in time to save the innocent and mete out justice. The appeal to those who can ally themselves with these forces for good is obvious . . .

Another desired function of Superfund may be to create "whipping boys" in the true medieval meaning of the term. Bad things happened; punishing a convenient party (whether directly responsible or not) satisfies the drive for retribution; it provides some restitution to those harmed and presumably changes future behavior by others. In this case, it is the fact of punishment, restitution, and changed behavior, not its narrow fairness in apportioning blame or its instrumental efficiency, that really matters.[20]

What bad things? If there is no or low probability of harm from abandoned sites, why are they bad? How can there be desirable outcomes from Superfund if its activities do not improve health?

The theme of good guys versus bad guys is a perennially popular one. In earlier times, however, morality plays were about the courage of entrepreneurs who overcame those who ignorantly stood in the way of industrial and technological progress. Why is business the bad guy when it used to be the good guy? It is not obvious. The American population is among the healthiest in the world; mortality has been decreasing decade by decade for over a hundred years, including the decades before the enactment and implementation of Superfund. And since when have Americans been unconcerned with "narrow fairness"? How can injustice today make for a better tomorrow?

Not to be denied, Russell wonders whether, in addition to being a vehicle for people who can become prominent and powerful through promoting a Superfund approach, it enables people "to respond effectively against the 'faceless others' who are seen to harm them."[21] Should Superfund provide scapegoats for worsening crime, failing education, rising home prices, whatever bothers people?

Before responding to this line of argument, I would like to deepen it by referring to an op-ed piece titled "Alar and Moral Outrage; or, Why Consumers Are So Upset about a Little Bit of a Chemical in Apples," by Dr. Edward Groth III of the Consumers Union and Professor Peter Sandman of Rutgers University. Their critique is worth considering because it brings to the surface elements of reasoning that are often submerged. They feel that outrage over Alar is warranted, despite their judgment that the scientific evidence does not allow a determination of whether Alar is really dangerous. Experts from different sides make different assumptions leading to different conclusions. So "the Public sees conflict and confusion. And that upsets people." I agree that the public is upset; I disagree with the other conclusions they draw. In their words:

Actually, it's mostly ethical and value dimensions of a risk like *Alar* that determine its capacity to provoke public outrage. And many of those aspects are, at heart, moral issues. For instance:

It's not fair. Children consume comparatively huge amounts of apples and apple products, and so bear much higher theoretical risks from *Alar* than adults do.

That's unfair. Growers reap most of the benefits of using *Alar,* while everyone shares the risks. That, too, offends our sense of justice.

It's involuntary. People can choose whether to smoke or to go sky-diving. Someone else decides whether there will be *Alar* in the apples and apple products we eat. Consumers reasonably may ask, "Who gave *them* the right to put my child at risk?"

We can't control our risk. Driving a car, one has a sense of being in control of a potentially risky situation. Not so with *Alar.* We can't see, taste, or smell it; we don't know if it's there; if it is, we can't remove it. People who once felt perfectly competent to choose foods for their children now must wonder if anything is really safe. They feel a loss of control over their own lives.

Someone's responsible. Biochemist Bruce Ames has said that "naturally occurring pesticides"—substances that plants produce to discourage animals from eating them—pose cancer risks thousands of times greater than the hazards of synthetic pesticides and other agricultural chemicals. Even if he's right (and it's debatable), his argument is beside the point. Nature may not be benign, but She's blameless. She's not making a business decision to sell or spray *Alar,* while someone else bears the risk. It's no surprise that people who shrug at the hazards of radon in their basement or invite skin cancer at the beach get hopping mad about *Alar.*

It's unnecessary. If *Alar* were an inherent trait of apples, we'd shrug it off, too. But it isn't, and we have choices—each with its own benefits, costs, and risks. If consumers don't want *Alar* in apples and apple products, it needn't be there.[22]

At the outset I should say that I am in fundamental agreement with the "risk is about morality" position, though I think it is misapplied. If the question is "who perceives dangers as strong or weak?" all of the hypotheses are appropriate candidates for testing. With anthropologist Mary Douglas I wrote a book on risk perception called *Risk and Culture,* whose thesis is that individuals choose what to fear to support their way of life. Insofar as perception of the dangers of modern technology are concerned, the lead against technology is taken by people of egalitarian beliefs who wish to weaken corporate capitalism as the source of the inequalities they abhor. When Karl Dake and I tested a variety of theories (knowledge of danger, risk-averse personalities, and so on), the cultural thesis won hands down. Hierarchists fear social deviance, individualists fear regulation, and egalitarians fear technology.[23]

These fears are moral. Now I wish to argue that although moral-cultural theories explain risk perception and may well determine governmental policy, these theories (mine as well as theirs) are out of place in determining risk consequences. The presence or absence of physical harms is not the same as perception of harm. Yes, risk from technology is about other things, but it should also be about health and safety. There is something very wrong when the reason given in public is not the real reason. That is why the argument is usually put the way Groth and Sandman present it. The truth is not known, the evidence is equally weak or strong on both sides, so the bewildered citizen who wishes the experts would agree might just as well decide by visceral reaction: get rid of the stuff whatever the cost.

Conclusive proof is rare, especially proof of a negative (no possible harm). But it is likely that no child has been harmed by drinking apple juice with trace elements of an Alar metabolite. If, as Groth and Sandman say, an Alar-treated apple is better for a child than a candy bar, how could the Alar scare be justified? Small is usually insignificant. It is highly improbable, for instance, that low-level electromagnetic fields would cause harm or that tiny residues of PCBs in electrical transformers would hurt people or that very small amounts of arsenic in water do damage; hacking asbestos out of schoolrooms is idiotic, there is no other term for it. The only actual harm to human health we can prove is that suffered by evacuees from Love Canal and Times Beach and by people killed by substitutes for DDT.

There is no big secret in discovering why most people are worried about getting cancer from chemicals. They are told to worry by environmentalist organizations, hear the claims reiterated endlessly in the media, observe them reinforced and ratified by high-level public officials. If the public is confused, I say "stop confusing them." We the citizenry do not have to be confused, which is why this book aims at encouraging citizenship in science.

There are nearly a thousand chemicals in a cup of coffee, of which fewer than thirty have been tested for cancer in rodents.[24] Are coffee drinkers involuntarily subject to the remaining hundreds? Or is human life full of unrecognized goods and bads? Are the workers and orchard owners who grow apples and manufacture apple juice the only ones who benefit? Such a question leaves out the taste and nutrition benefits of apples and apple juice. It suggests, in addition, that the benefits of industry in a capitalist society go only to the owners of capital, a theory long discredited. Ask the people in the former communist countries whether an economy in which profit is eliminated is better for their health.

A major point among those imputing motives in discussions of risks from technology is that people are outraged because they believe they have been misled or cheated by government officials. They are entitled to know about the dangers they face, the argument goes, and if not, vital information has deliberately been withheld from them. What is demanded is mind-boggling. If we were actually informed about every substance or process or product that might harm us, we would not be able to do anything else but absorb that information. The list of possible everyday harms is immense. Want to avoid getting murdered? Don't live in or start a family: most murders are committed within families. Try demanding perfect reassurance from a loved one at every moment, about everything, and then see how long that relationship lasts. The demand for perfect foresight—worse, a demand to be warned of all possible harms—cannot be met.

Ah, but it is argued, most of the exposures to harm are voluntary acts where we know what we are doing and accept the risks involved. Everything about this argument is wrong, including the idea that most people know the dangers that arise from ordinary activities. As Mary Douglas and I contend in *Risk and Culture,* the assertion proceeds as if voluntariness were self-evident, fixed and immutable. On the contrary, where the classification is the decision, to declare an act involuntary is to condemn it.[25]

Who is kidding whom? Would people not make different decisions if they knew that Superfund cleanups had no health benefits or did more harm than good? I am waiting for the day when the president of the United States, or a cabinet member, or a congressman says that despite the fact that no improvement in health can come from a regulation, there are real reasons for having it.

The Preventive Society

What would social life be like, we may ask, if it were geared to preventing dangers? What would our society be like if we collectively took out insurance against any calamities that might occur? Fortunately, we do not have far to look. In *The Genesis Strategy,* written in the 1970s on the expectation of global cooling, atmospheric scientist Stephen Schneider described a series of institutions that would serve to prevent such dangers from occurring or would mitigate their effects.

If Schneider thought global cooling likely then and considers warming more likely now, that is because he believes the evidence has changed. Here I wish to make use of Schneider's Genesis Strategy (named after Joseph the Provider, who saved the known world from starvation by storing food in Egypt against a coming famine)[26] to show what a strategy of prevention—"maintaining large margins of safety to secure our means of survival"[27]—would be like. Back in the mid-1970s, facing the possibility of global freezing, which would greatly reduce food production, Schneider asked whether it was right to be unprepared for "a recurrence of known types of climatic fluctuations," for "to ignore such possibilities is tantamount to gambling with the lives of those most heavily dependent on the weather."[28] How would such precautionary preparation, a Genesis Strategy, be carried out?

One element would be the promotion of a variety of energy technologies so that if something happened to one, whether it was accidental or due to sabotage, a natural calamity, or an act of terror, "the chances of economic collapse would be neglibly small." Aware that diversifying energy sources might mean less economic efficiency, Schneider urges that "these costs should be viewed as insurance against the even costlier and sudden demise that would follow the malfunctions or sabotage of one of a few highly efficient power production systems."[29]

To guard against "the worst possibility," which "is at least as likely to occur as the best possibility,"[30] Schneider advocates the establishment of a number of institutes with special functions. The task of the Institute of Imminent Disasters would be to determine "the probable costs of avoiding any and all perceived disasters impending, and the costs of recovery after their occurrence."[31] What would constitute sufficient evidence of impending disaster? For Schneider "the crucial question . . . is not whether we can *prove conclusively* that disaster really lurks ahead, but rather whether we can afford to be unprepared for its not unlikely occurrence."[32] While not "predicting famine (or feast) with *certainty*," Schneider warns against being "browbeaten by Pollyannas opposing a food reserve following the good news of one or two fine harvests."[33]

To buttress these warnings, the Institute of Resource Availability would provide an independent source of information on the availability of food supplies. When the Institute of Imminent Disasters issued its warnings, these would be conveyed to still another research center, the Institute of Alternative Technologies, which would fund research on technologies to supplement or replace existing ones so as to avoid or mitigate expected disasters. Data and discovery from the three aforementioned institutes would be supplemented by a fourth, the Institute of Policy Options, whose task would be to consider various policy alternatives, their costs, benefits, and uncertainties, so that governments might take action on them.

The Genesis Strategy itself consists of a series of programs combined under the terms of a Global Survival Compromise, a name chosen to expose the seriousness of the food problems. Schneider emphasizes that these measures "must begin immediately and be rapidly implemented." The great dangers, in Schneider's estimation, are "a widening gulf in the material living standards between rich and poor [nations]" and the growth of world population beyond the earth's capacity.[34]

In response he wants both to raise food production in poor nations via short-term aid and technological assistance from the rich and, in return, commit the poor to reducing their population growth.[35] Schneider worries that the short-term perspective of immediate economic gain and election to office will lead to "selling food to the rich" rather than using food surpluses to help poor countries increase their standards of living rapidly.[36]

My comments will be brief. The prospects of suffering from famine due to lack of ability to grow food (as differentiated from civil war, gangsterism, and price controls) have been greatly reduced by the rise of democracies capable of growing vast amounts. As this knowledge and capacity spread around the world, the chances of famine by food shortage greatly diminish. The redundancy of food supplies that Schneider desires already exists. He is mistaken in his belief that the existing world food-production system is inadequate. Indeed, it is the institutional structure of nations—their possession of democracy, market capitalism, and science, rather than any special plan—that protects them against disasters.[37]

Similarly, I agree with Schneider (and many others) that diversity in supply of vital resources, such as energy, is desirable. But I differ in observing that the capitalist countries already provide considerable diversity without the need for additional governmental intervention. The only threat I see to variety in energy production lies in the effort to stop nuclear power generation.

Though the claim of disaster from global cooling is now old (perhaps temporarily superseded by global warming), I am struck by the contemporary character of Schneider's prescriptions: act at once, right now, before it is too late. He finds costs only in economic efficiency, not in health. He does not envisage streams of false positives, only the necessity of preventing something awful. With an Institute of Imminent Disasters one might expect many more equivocal threats like global cooling and hence a weakening from unnecessary expenditures.

Most striking is Schneider's remedy—international redistribution of income—because it need have nothing to do with the causes of famine.[38] The difficulties of some poor countries in growing food are long-standing and need no impetus from hypothetical global cooling. Population control, usually by other people, is a favorite measure to this very day, even without the impetus of imminent disaster. The difference is that Schneider couples disasters with redistributive measures. Hot or cold, cooling or freezing, global egalitarian measures are required.

A political activist since teenage years, I am the last person to criticize Schneider for engaging in politics. It is his view of risk that I find wanting. Recall Schneider's response to a reporter's question about scientists possibly crying wolf in the case of global warming: he replied that uncertainty due to lack of scientific knowledge does not imply inaction; nor does action require precise location of the danger. Rather, he asserted, knowledge of the probability of wolves in the forest is sufficient grounds for deciding on preventive action.[39] No, it is not sufficient. The example is misleading in respect to global warming in particular; it is also misleading in general with regard to the thrust of environmental and safety measures. Its best use is to illustrate what is wrong with the precautionary principle.

Let us start with this question: should we protect ourselves against the possibility that there may be wolves in all forests on the grounds that we know there are wolves in a few forests? What is wrong with Schneider's wolf example, to begin with, is that it takes something we know exists—wolf sightings—which we can be confident of predicting within two or three instances, and compares it with a phenomenon—global warming—whose timing and extent are dubious. Suppose, however, that wolves and weather were comparable. That still would not do. Before acting, we would want to know about the following considerations: (1) How much harm is expected to humans versus harm to wolves? (2) Would safety measures, such as closing off forests or poisoning wolves, harm more people or animals than they helped? (3) Would the cost of these measures in unemployment and financial loss hurt more than it helped?

sji

(Loss of revenue might also harm other animals by lessening measures designed to aid them.) (4) Would false alarms diminish public trust as well as reduce alertness to other dangers? There is no free health.

In an interview in *Discover* magazine, Stephen Schneider stated:

> On the one hand, as scientists, we are ethically bound to the scientific method, in effect promising to tell the truth, the whole truth, and nothing but—which means that we must include all the doubts, the caveats, the ifs, ands, and buts. On the other hand we are not just scientists but human beings as well. And like most people we'd like to see the world a better place, which in this context translates into our working to reduce the risk of potentially disastrous climatic change. To do that we need to get some broad-based support, to capture the public's imagination. That, of course, entails getting loads of media coverage. So we have to offer up scary scenarios, make simplified, dramatic statements, and make little mention of any doubts we might have . . . Each of us has to decide what the right balance is between being effective and being honest.[40]

I would not put Schneider down for being forthright, but those who walk this line should understand the consequences. If everyone is trying to counter someone else's comments, it will not be possible to figure out even approximately where the truth, insofar as it is known, lies. Turning to today's news, one reads headlines like "Environmentalists Hope for Scorcher: Aim Is to Avert Governmental Complacency on Greenhouse Effect."[41]

In a preventive society,[42] the people in charge of raising the alarm will rule. Who wants this? Why provide such strong incentives for predicting disasters? Why organize our lives around predictions unlikely to come true?

The Opposite Prescriptions for Environmental and Safety Regulations

The main outlines of the case for modern technology endangering humans can usefully be summarized in a few propositions. I shall call this the environmentalist paradigm. (1) *Possibility should replace probability as a criterion for regulation designed to protect human health and safety.* A corollary is: (1a) *The mode of risk assessment that increases prediction of harm by the largest amount should be adopted in the absence of proof to the contrary.*

The second fundamental proposition is related but not equivalent to the first: (2) *No cause, however weak, is incapable of producing substantial harmful effects.* Put positively, any chemical harmful at high doses is harmful at exceedingly low doses. A corollary of the interest in weak interactions is: (2a) *It is desirable to search for weak effects of weak causes.*

The third general proposition speaks to the spirit of the enterprise: (3) *The purpose of risk regulation is to prevent health detriments, not to secure health benefits.* It follows that improvements in health must not be allowed to offset harms from the same object.

The fourth proposition has broad applicability: (4) *What is not explicitly permitted is forbidden; substances or processes must be demonstrated to be benign before they can be used.* Whereas before the time of extensive environmental and safety regulations action proceeded unless there was good reason to stop it, nowadays this presumption has been reversed. Its corollary is: (4a) *The burden of proof rests on a conclusive demonstration that a substance does not cause cancer or other specified harm.* From this corollary follows the use of animal rodent tests as predictive of cancer in human beings, however inaccurate. In a more general application, a much higher standard of proof is required to show that significant global warming will not occur than that it will.

From my point of view, health and safety will be better secured by exactly the opposite prescriptions. We should be guided by the probability and extent of harm, not by its mere possibility. The search for possibilities is endless and it trivializes the subject. There is bound to be great diversion of resources without reducing substantial sources of harm. Consternation is created but health is not enhanced. The environmentalist paradigm, presented in column one of Table C.1, should be abandoned and the proposed prescriptions in column two followed.

Weak causes are likely to have weak effects. Our search should be for strong causes with palpable effects, like cigarette smoking. They are easier to find, and their effects are much more important to control. Methods should be used that give the most accurate results, like epidemiological studies. No doubt health and safety might be improved, but "might be" is not a rationale for a crash program that is likely to turn up only trivia at great expense.

Endless stultification is produced by the principle that what is not explicitly permitted is forbidden. Much has been done in the name of nature and health that otherwise would have been strongly resisted. A network of regulation more extensive and intensive than any previously experienced by Americans has been instituted. The ability of individuals and businesses to use their property and their labor has been curtailed on the grounds that they might do harm. The past necessity of proving harm has been replaced by a reversal of causality: now the individuals and businesses must prove that they will do no harm. My objection to this form of prior restraint, as the reader knows by now, is profound: our liberties are curbed and our health is harmed.

The multiple restraints are obvious, for they are written into laws and regulations. The harm to our health and safety is not, which is why I have written this book. If one prevents enough of life from happening, presumably some of the harm that is part of life is avoided. But the alleged harms are too tiny, as with most chemicals, or too unlikely, as with global warming, or just plain in error, as with the effects of asbestos or acid rain. And there has been no voice for the health lost by growing restraints.

Table C.1   Reversing the prescriptions

Environmentalist paradigm	Proposed prescription
Possibilities If there is a possibility of harm, regulate.	*Probabilities* Regulate if there is probable harm.
Weak causes and effects Consequences: science inadequate because an impossibility theorem (proof of no harm) is required.	*Strong causes and effects* Consequences: science adequate; can often meet standard of probable or improbable harm.
Rodent tests Predict many more harms but invalid; choice of statistical models overdetermines results.	*People tests (epidemiology)* Predict far fewer harms but valid; statistical models either not required, because there is no need to extrapolate from rodents to people, or tests are chosen with knowledge of causal mechanism.
Biased cost-benefit For industry: consider health harms but not benefits; for regulation, consider health benefits but not harms.	*Full cost-benefit* For industry and for regulation: consider health harms and benefits.

The truth value of the environmental-cum-safety issues of our time is exceedingly low. With the exception of CFCs thinning the ozone layer, the charges are false, mostly false, unproven, or negligible. What, in my vision, is left of environmentalism? There is respect for nature, for all life. There are the moral questions of human relationships to all of creation. What is left out? Only the falsehoods.

Notes

Introduction

1. Leo Levenson also influenced my decision to return to the traditional method of setting standards for poisons. In working together on the cranberry scare, we wondered long and hard about the implications of the case's involving an amount of suspect chemical a million times less than the traditional standard.
2. Brendon Swedlow and Aaron Wildavsky, "Is Egalitarianism Really on the Rise?" in Wildavsky, *The Rise of Radical Egalitarianism* (New Brunswick, N.J.: Transaction Publishers, 1991), pp. 63–98.
3. Jesse Malkin and Aaron Wildavsky, "Why the Traditional Distinction between Public and Private Goods Should Be Abandoned," *Journal of Theoretical Politics,* 3 (Oct. 1991): 355–378. Nelson Polsby and Aaron Wildavsky, *Presidential Elections,* 8th ed. (New York: Free Press, 1990).
4. Mary Douglas and Aaron Wildavsky, *Risk and Culture* (Berkeley: University of California Press, 1982). For an empirical test, see Karl Dake and Aaron Wildavsky, "Theories of Risk Perception: Who Fears What and Why?" *Daedalus,* 119 (Fall 1990): 41–60.
5. Aaron Wildavsky, *Searching for Safety* (New Brunswick, N.J.: Transaction Publishers, 1988). For a quantitative model embodying this thesis, see Ralph Keeney, "Mortality Risks Induced by Economic Expenditures," *Risk Analysis,* 10 (1990): 147–159.
6. John D. Graham, Laura C. Green, and Marc J. Roberts, *In Search of Safety* (Cambridge, Mass.: Harvard University Press, 1991). John D. Graham, ed., *Harnessing Science for Environmental Regulation* (New York: Praeger, 1991). Kenneth R. Foster, David E. Bernstein, and Peter W. Huber, eds., *Phantom Risk: Scientific Inference and the Law* (Cambridge, Mass.: MIT Press, 1993).
7. See Aaron Wildavsky, "Wealthier Is Healthier," *Regulation Magazine,* Jan./Feb. 1980, pp. 10–12, 55; and Wildavsky, "But Is It True? The Unasked Question in the Analysis of Risk," a review of Risa I. Palm, *Natural Hazards,* in *Minerva,* 24 (Summer 1991): 231–234.

8. Julian Simon and Aaron Wildavsky, "Species Loss Revisited," *Society,* 30 (Nov./ Dec. 1992): 41–46.
9. Aaron Wildavsky, "Public Policy," in Bernard E. Davis, ed., *The Genetic Revolution* (Baltimore: John Hopkins University Press, 1991), pp. 77–104.
10. Michael Polanyi, "The Republic of Science: Its Political and Economic Theory," in Marjorie Grene, ed., *Knowing and Being. Essays by Michael Polanyi* (Chicago: University of Chicago Press, 1969): 49–72.
11. Harry Sootin, *Isaac Newton* (New York: Messner, 1959).
12. Ian Hacking, *The Emergence of Probability: A Philosophical Study of Early Ideas about Probability, Induction, and Statistics Inference* (London: Cambridge University Press, 1975).

1, I. The Cranberry Scare of 1959

1. T. H. Jukes and C. B. Shaffer, "Antithyroid Effects of Aminotriazole," *Science,* 132 (July 29, 1960): 296.
2. Eugene Feingold, "The Great Cranberry Crisis," in Edwin A. Bock, ed., *Government Regulation of Business; A Casebook* (Englewood Cliffs, N.J.: Prentice-Hall, 1965), pp. 1–31. See also Edward W. Lawless, *Technology and Social Shock* (New Brunswick, N.J.: Rutgers University Press, 1977).
3. Jukes and Shaffer, "Antithyroid Effects," p. 296.
4. Ibid.; A. M. Chesney, T. A. Clawson, and B. Webster, "Endemic Goiter in Rabbits: Incidents and Characteristics," *Bulletin of Johns Hopkins Hospital,* 43 (1928): 261–277; and L. S. Goodman and Alfred Goodman, *Pharmacological Basis of Therapeutics,* 5th ed. (New York: Macmillan, 1975), p. 1420.
5. Jukes and Shaffer, "Antithyroid Effects."
6. Arthur S. Flemming, "Aminotriazole in Cranberries," testimony before the House Committee on Interstate and Foreign Commerce, 86th Congress, vol. 1, Jan. 26, 1960.
7. Clarence Dean, "Cranberry Sales Curbed, U.S. Widens Taint Check: $45 Million Loss Feared," *New York Times,* Nov. 11, 1959, p. 1.
8. Letter from William Dufort (Oregon manager of Ocean Spray) and Jack Dean to U.S. Senator Wayne Morse (Oregon), reprinted in *Congressional Record,* vol. 106, no. 3 (Feb. 25, 1960): 3410–3411.
9. Bess Furman, "Accord Reached on Cranberries: Flemming and Industry Say They Will Disclose Plan at Conference Today," *New York Times,* Nov. 19, 1959, p. 26.
10. Dean, "Cranberry Sales Curbed"; Flemming, "Aminotriazole in Cranberries."
11. Flemming, "Aminotriazole in Cranberries."
12. "Some Cranberry Crop Tainted by a Weed-Killer, U.S. Warns," *New York Times,* Nov. 10, 1959, p. 1.
13. Richard E. Mooney, "Color Additives to Stir New Feud: Debates Similar to Those of Cranberry Tiff Will Be Heard in Congress," *New York Times,* Nov. 30, 1959, p. 39.
14. "Tainted Cranberries Sold in Missouri," *New York Times,* Nov. 26, 1959, p. 46.
15. John R. Fenton, "Cape Cod Fears Cranberry Loss: Fruit Growing Major Factor in Region's Economy—Aid May Be Sought," *New York Times,* Nov. 12, 1959, p. 21.

16. Lawless, *Technology and Social Shock*. Lawless also mentions that about 10 percent of the total cranberry crop of 1959 may have been eventually banned but does not cite a reference for the estimate (p. 63).

17. Greg MacGregor, "Cranberry Men Sight Recovery: Growers Say Sauce Will Be Back with Turkey—Weed Killer Dropped," *New York Times*, Oct. 16, 1960, p. 71.

18. "$8 Million for Cranberry Scare," *New York Times*, Oct. 3, 1962, p. 23.

19. "Weed-Killer Move by U.S. Minimized," *New York Times*, Nov. 13, 1959, p. 46; "Weed-Killer Plant Closed," *New York Times*, Nov. 19, 1959, p. 33.

20. Paul Eck, *The American Cranberry* (New Brunswick, N.J.: Rutgers University Press, 1990).

21. " '60 Cranberry Crop Is Approved by U.S.," *New York Times*, Nov. 18, 1960, p. 26.

22. John R. Fenton, "Cranberries Gain New Dining Role; Can Be a Cocktail," *New York Times*, Nov. 26, 1964, p. 50. For a perspective on Ocean Spray's success in marketing cranberries since the scare, see Harold Thorkilsen, "Lessons of the Great Cranberry Crisis," *Wall Street Journal*, Dec. 21, 1987, p. 22.

23. Flemming, "Aminotriazole in Cranberries," p. 41.

24. Ibid., pp. 52, 56.

25. Ibid., p. 55.

26. "Some of Cranberry Crop Tainted by a Weed-Killer, U.S. Warns," *New York Times*, Nov. 10, 1959, p. 1.

27. "Cranberries," *New York Times*, Nov. 11, 1959, p. 34.

28. "Cranberries and Mr. Flemming," *New York Times*, Nov. 14, 1959, p. 20.

29. "Those Cranberries," *New Republic*, 141 (Nov. 30, 1959): 20.

30. David Rutstein, "Cranberry Ban Approved: Background of Secretary Flemming's Action on Product Discussed," *New York Times*, Nov. 18, 1959, p. 40; Brian Lipton, "Acting on Sales of Tobacco," *New York Times*, Nov. 25, 1959, p. 28.

31. "Cranberries, Anyone?" *Commonweal*, 71 (Nov. 27, 1959): 254–255.

32. *Today Show*, interviews with Dr. C. Boyd Shaffer and FDA Commissioner George P. Larrick, Nov. 10, 1959.

33. International Agency for Research on Cancer, *IARC Monographs on the Evaluation of Carcinogenic Risk of Chemicals to Humans*, 41 (1986): 293–317.

34. Personal communication with Dr. Richard Teske, FDA, Apr. 2, 1991.

1, II. Silent Spring *and Dieldrin*

1. H. Patricia Hynes, *The Recurring Silent Spring* (New York: Pergamon, 1989), p. 2.

2. Ibid., p. 4.

3. E. B. White, in a letter to Rachel Carson sometime in 1958, quoted in Frank Graham, *Since Silent Spring* (Boston: Houghton Mifflin, 1970), p. 19.

4. Ibid., Rachel Carson in a letter to E. B. White dated Feb. 3, 1958.

5. Rachel Carson, *Silent Spring* (Boston: Houghton Mifflin; Cambridge, Mass.: Riverside Press, 1962), p. 8.

6. Ibid., pp. 275, 278, 297.

7. Ibid., p. 272.

8. Ibid., p. 113.

9. Ibid., p. 25.

10. For a summary of the conflict between the need to control pests and the need to protect the environment, from the point of view of the U.S. Department of Agriculture, see George W. Irving, Jr., "Agricultural Pest Control and the Environment," *Science,* 168 (June 19, 1970): 1419–1424.

11. A. A. Belisle et al., "Residues of Organochlorine Pesticides, Polychlorinated Biphenyls, and Mercury as Autopsy Data for Bald Eagles," *Pesticide Monitor,* 6 (Jan. 1969/1970): 133–138.

12. John Robinson, "Organochlorine Insecticides and Bird Populations in Britain," in M. W. Miller and G. G. Berg, eds., *Chemical Fallout* (Springfield, Ill.: Charles C. Thomas, 1969), pp. 157–158.

13. R. E. Geenelly and R. L. Rudd, "Effects of DDT, Taxophene, and Dieldrin on Pheasant Reproduction," *Auk,* 73 (1956): 529–539.

14. V. K. H. Brown, J. Robinson, A. Richardson, and D. E. Stevenson, "The Effects of Aldrin and Dieldrin on Birds," *Food and Cosmetics Toxicology,* 3 (1965): 675–679.

15. World Health Organization, "Aldrin and Dieldrin," in *Environmental Health Criteria,* vol. 91 (Geneva: WHO, 1989), p. 175.

16. Ibid.

17. R. B. Dahlenger and R. L. Linder, "Effects of Dieldrin in Penned Pheasants through the Third Generation," *Wildlife Management,* 38 (1974): 320–330.

18. D. J. Call and B. E. Harrel, "Effects of Dieldrin and PCBs on the Production and Morphology of Japanese Quail Eggs," *Bulletin of Environmental Contamination and Toxicology,* 11 (1974): 70–77.

19. World Health Organization, "Aldrin and Dieldrin," p. 175.

20. Ibid.

21. D. A. Ratcliffe, "Decrease in Eggshell Weight in Certain Birds of Prey," *Nature,* 215 (July 8, 1967): 208–210.

22. P. N. Lehner and A. Egbert, "Dieldrin and Eggshell Thickness in Ducks," *Nature,* 224 (Dec. 20, 1969): 1218–1219.

23. M. A. Haegele and R. K. Tucker, "Effects of 15 Common Environmental Pollutants on Eggshell Thickness in Mallards and Cournix," *Bulletin of Environmental and Contamination Toxicology,* 11 (1974): 98–102.

24. K. J. Davis and O. G. Fitzhugh, "Tumoric Potential of Aldrin and Dieldrin for Mice," *Toxicology of Applied Pharmacology,* 4 (1962): 187–189.

25. Ibid.

26. E. Thorpe and A. I. T. Walker, "The Toxicology of Dieldrin (HEOD), II: Comparative Long Term Oral Toxicity in Mice, with Special Reference to Aldrin and Dieldrin," *Journal of Agriculture, Food, and Chemistry,* 3 (1973): 402–408.

27. O. G. Fitzhugh, A. A. Nelson, and M. L. Quail, "Chronic Oral Toxicity of Aldrin and Dieldrin in Rats and Dogs," *Food and Cosmetics Toxicology,* 2 (1964): 551–561.

28. A. S. Wright et al., "The Effects of Prolonged Ingestion of Dieldrin on the Livers of Male Rhesus Monkeys," *Ecotoxicology and Environmental Safety,* 1, no. 4 (Mar. 1978): 477–502.

29. P. C. Gupta, "Neurotoxicity of Chronic Chlorinated Hydrocarbon Insecticide

Poisoning: A Clinical and Electroencephalographic Study in Man," *Indian Journal of Medical Research,* 63 (1975): 601–606.

30. World Health Organization, "Aldrin and Dieldrin," p. 220.
31. Carson, *Silent Spring,* p. 22.
32. C. G. Hunter and J. Robinson, "Aldrin, Dieldrin, and Man," *Food and Cosmetics Toxicology,* 6 (1968): 253–260.
33. Carson, *Silent Spring,* p. 23.
34. World Health Organization, "Aldrin and Dieldrin," p. 235.
35. C. G. Hunter and J. Robinson, "Pharmacodynamics of Dieldrin (HEOD), I: Ingestion by Human Subjects for 18 Months," *Archives of Environmental Health,* 15 (Nov. 1967): 614–626.
36. Ibid., p. 625.
37. World Health Organization, "Aldrin and Dieldrin," p. 235.
38. Ibid., p. 234.

1, III. *The Saccharin Debate*

1. Richard A. Merrill and Michael R. Taylor, "Saccharin: A Case Study of Government Regulation of Environmental Carcinogens," *Virginia Journal of Natural Resources Law,* 5, no. 1 (1985): 25.
2. D. L. Arnold, D. Krewski, and I. C. Monroe, "Saccharin: A Toxicological and Historical Perspective," *Toxicology,* 7 (1983): 179–181.
3. John P. Wiley, "Phenomena, Comment, and Notes," *Smithsonian,* Dec. 1980, pp. 24–25.
4. Merrill and Taylor, "Saccharin: A Case Study," p. 27.
5. "The Sour Taste of the Saccharin Ban," *Time,* Mar. 28, 1977, p. 109.
6. Subcommittee on Nonnutritive Sweeteners, Committee on Food Protection, Food and Nutrition Board, National Research Council, *Safety of Saccharin and Sodium Saccharin in the Human Diet* (Springfield, Ill.: National Technical Information Service, 1974), p. 4.
7. Ibid.
8. Ibid., pp. 2–4.
9. "Sweet News," *Newsweek,* Aug. 3, 1970, p. 43.
10. Gardner quoted in Merrill and Taylor, "Saccharin: A Case Study," p. 48.
11. "Saccharin: Where Do We Go from Here?" *FDA Consumer,* 12 (1978): 16.
12. "For Saccharin Decision, Revamp Food Laws," *Science News,* 115, no. 9 (Mar. 10, 1979): 150.
13. *Federal Register,* 36, no. 123 (June 25, 1971): 2109–2110.
14. G. T. Bryan, E. Erturk, and O. Yoshida, "Production of Urinary Bladder Carcinomas in Mice by Sodium Saccharin," *Science,* 168 (June 5, 1970): 1238–1240.
15. "Sweet News," p. 43.
16. Lincoln Pierson Brower, "Sodium Cyclamate and Bladder Carcinoma," *Science,* 170 (Oct. 30, 1970): 558.
17. Leo B. Ellwein and Samuel M. Cohen, "The Health Risks of Saccharin Revisited," *Toxicology,* 20 (1990): 312.
18. J. M. Taylor, M. A. Weinberger, and L. Friedman, "Chronic Toxicity and Carci-

nogenicity in Utero-Exposed Rats," *Toxicology and Applied Pharmacology,* 54 (1980): 74.

19. Ibid.
20. Ibid., p. 57.
21. M. O. Tisdel et al., "Long Term Feeding of Saccharin in Rats," in G. E. Inglett, ed., *Symposium: Sweeteners* (Westport, Conn.: Avi Publishing, 1974), p. 147.
22. Subcommittee on Nonnutritive Sweeteners, *Safety of Saccharin,* p. 64.
23. Ibid., p. 63.
24. Arnold, Krewski, and Monroe, "Saccharin," p. 210.
25. Subcommittee on Nonnutritive Sweeteners, *Safety of Saccharin,* pp. 51–52a.
26. Merrill and Taylor, "Saccharin: A Case Study," p. 36.
27. D. L. Arnold et al., "Long-Term Toxicity of Ortho-Toluenesulfonamide and Sodium Saccharin in the Rat," *Toxicology and Applied Pharmacology,* 52 (1980): 114–115.
28. Ibid., p. 148.
29. U.S. Congress, Office of Technology Assessment, *Cancer Testing Technology and Saccharin* (Washington, D.C.: Government Printing Office, 1977), p. 22.
30. Bruce Ames et al., "Ranking Possible Carcinogenic Hazards," *Science,* 236 (Apr. 17, 1987): 271–280.
31. Gardner quoted in Merrill and Taylor, "Saccharin: A Case Study," p. 48.
32. "Reappraising Saccharin—and the FDA," *Time,* Apr. 25, 1977, pp. 75–76.
33. "The Sour Taste of the Saccharin Ban," *Time,* Mar. 28, 1977, pp. 76–77.
34. Anthony Wolff, "Of Rats and Men," *New York Times Magazine,* May 15, 1977, pp. 88–94.
35. Charles W. Wurster, "For the Saccharin Ban," *New York Times,* Mar. 20, 1977, section 4, p. 17.
36. Wiley, "Phenomena, Comment, and Notes," p. 32.
37. "Should Saccharin Be Banned?" *U.S. News and World Report,* Apr. 4, 1977, pp. 59–60.
38. "The Sour Taste of the Saccharin Ban," pp. 76–77.
39. "Second Opinions," *Time,* Aug. 7, 1978, p. 68.
40. "Fight Starts to Beat the Ban on Saccharin," *U.S. News and World Report,* Mar. 28, 1977, p. 50.
41. "A Bitter Reaction to the FDA Ban," *Time,* Mar. 21, 1977, pp. 60–62.
42. "Fight over Proposed Saccharin Ban Will Not Be Settled for Months," *Science,* 196 (Apr. 15, 1977): 276–277.
43. G. R. Howe, J. D. Burch, and A. B. Miller, "Artificial Sweeteners and Human Bladder Cancer," *Lancet,* 2 (1977): 579.
44. Ibid., pp. 579–581.
45. Brendon Swedlow, "Does Dioxin Cause Soft Tissue Sarcoma?" unpublished manuscript, University of California, Berkeley, 1991, pp. 10–11.
46. Merrill and Taylor, "Saccharin: A Case Study," p. 64.
47. Ellwein and Cohen, "Health Risks," p. 311.
48. "A Bitter Reaction to the FDA Ban," p. 60.
49. R. N. Hoover, and P. H. Strasser, "Artificial Sweeteners and Human Bladder Cancer: Preliminary Results," *Lancet,* 1 (1980): 837–840; O. M. Jensen and C. Camby, "Intrauterine Exposure to Saccharin and Risk of Bladder Cancer in

Man," *International Journal of Cancer,* 29 (1982): 507–509; Council on Scientific Affairs, American Medical Association, "Saccharin: Review of Safety Issues," *Journal of the American Medical Association,* 254 (1985): 2622–2624. See also the discussion in Sidney Shindell et al., "Low Calorie Sweeteners: Aspartame, Saccharin, Cyclamate," report by the American Council on Science and Health, 2d ed., revised and updated, August 1986, pp. 23–32. See also Elizabeth M. Whelan and William R. Havender, "Sweet Truth: What Do Scientists Really Know about Saccharin? And What Does It Mean for the Regulators?" *Reason,* 16 (Oct. 1984): 33–38.

50. Clifford Grobstin, "A Scientist's View," in Robert W. Crandall and Lester B. Lave, eds., *The Scientific Basis of Health and Safety Regulation* (Washington, D.C.: Brookings Institution, 1981), pp. 121–122.

2, I. Which Regulations Governing PCB Residues Are Justified?

1. Cecil K. Drinker, Madeline F. Warren, and Granville A. Bennett, "The Problem of Possible Systemic Effects from Certain Chlorinated Hydrocarbons," *Journal of Industrial Hygiene and Toxicology,* 19 (Sep. 1937): 283–311; U.S. Department of Health, Education, and Welfare and National Institute for Occupational Safety and Health, *Criteria Document Recommendations for an Occupational Exposure Standard for Polychlorinated Biphenyls,* publication no. 77-225 (Washington, D.C.: Government Printing Office, 1977), pp. 77–225.

2. "Report of a New Chemical Hazard," *New Scientist,* 32 (1966): 612; Soren Jensen, "The PCB Story," *Ambio,* 1 (Aug. 1972): 123–131.

3. See U.S. Environmental Protection Agency [EPA], *Conference Proceedings, National Conference on Polychlorinated Biphenyls,* report no. EPA-560/6-75-004, Mar. 1976, pp. 4–29; Masanori Kuratsune et al., "Some of the Recent Findings concerning Yusho." Masanori Kuratsune et al., "A Cohort Study on Mortality of 'Yusho' Patients: A Preliminary Report," in R. W. Miller et al., eds., *Unusual Occurrences as Clues to Cancer Etiology* (Tokyo: Japan Scientific Society Press, 1988), pp. 61–66.

4. W. J. Rogan, "Yu-Cheng," in R. D. Kimbrough et al., eds., *Halogenated Biphenyls, Terphenyls, Naphtalenes, Dibenzodioxins, and Related Products,* 2d ed. (Amsterdam: Elsevier Science Publishers, 1989), pp. 401ff.

5. Junya Nagayama, Masanori Kuratsune, and Yoshito Masuda, "Determination of Chlorinated Dibenzofurans in Karechlors and 'Yusho Oil,' " *Bulletin of Environmental Contamination and Toxicology,* 15 (1976): 9–13; Kuratsune, "Some Recent Findings"; S. Safe, ed., *Polychlorinated Biphenyls (PCBs): Mammalian and Environmental Toxicology* (Berlin: Springer-Verlag, 1987), pp. 134–145: Safe, "PCBs and Human Health"; T. Kanshimoto et al., "Role of Polychlorinated Dibenzofuran in Yusho (PCB Poisoning)," *Archives of Environmental Health,* 36 (1981): 321–326.

6. EPA, *Review of PCB Levels in the Environment,* Office of Toxic Substances, report no. EPA-560/7-76-001 (1976); National Research Council [NRC] Committee on the Assessment of Polychlorinated Biphenyls in the Environment, *Polychlorinated Biphenyls* (Washington, D.C.: National Academy of Sciences, 1979).

7. NRC, *Polychlorinated Biphenyls.*

8. C. Walker, "The Occurrence of PCB in the National Fish and Wildlife Monitoring Program," in EPA, *National Conference* (1976), pp. 161–176.

9. Massachusetts Audubon Society, *Criteria Document for PCBs,* prepared for EPA Office of Water Planning and Standards, report no. EPA-440/9-76-021 (1976), pp. 53–59.

10. D. S. Dennis, "Polychlorinated Biphenyls in the Surface Waters and Bottom Sediments of the Major Drainage Basins of the United States," in EPA, *National Conference* (1976), pp. 183–198.

11. NRC, *Polychlorinated Biphenyls,* pp. 14–15. A small percentage of PCBs attached to surface soil does enter the atmosphere through evaporation and suspension of PCB-containing dust. The EPA later concluded that a maximum of about 0.2 percent/day of PCBs in a spill could enter the atmosphere this way (see *National Conference* [1976] p. 17). PCBs on surface soil can also be carried into streams with storm runoff. In the early 1970s some samples of surface runoff water from Michigan landfills reportedly contained a fraction of a part per billion of PCBs; J. L. Hesse, "Polychlorinated Biphenyl Usage and Sources of Loss to the Environment in Michigan," in EPA, *National Conference* (1976), pp. 127–133.

12. D. E. Rosenblam, "Monsanto Plans to Curb Chemical," *New York Times,* July 15, 1970, p. 27.

13. William M. Blair, "U.S. Fears Poison in Some Chickens," *New York Times,* July 24, 1971, p. 30. The concentration of PCBs in the meal was reported in Massachusetts Audubon Society, *Criteria Document,* p. 73.

14. A review of the *New York Times, Washington Post,* and *Readers Guide to Periodical Literature* found no mention of the Yusho episode before 1971.

15. "The Menace of PCB," *Time,* Oct. 11, 1971, pp. 91–92.

16. "The PCB Crisis," *Newsweek,* Oct. 11, 1971, p. 60.

17. In reporting on the Holly Farms incident, *Newsweek* also treated the effects that PCBs apparently had on humans in the Yusho episode of 1968 ("The PCB Crisis," p. 66).

18. Food and Drug Administration [FDA], "Polychlorinated Biphenyls (PCBs). Contamination of Animal Feeds, Foods, and Food-Packaging Materials," *Federal Register,* 38 (July 6, 1973): 18096–18102.

19. Note that the fish standard of 5 ppm is in units different from the TDI of 1 μg/kg/day. Parts per million or billion refer to the concentration of a chemical in food, water, soil, or air. They usually are calculated by weight, so that in a gram of fish with 1 millionth of a gram (1 μg) of PCBs, the concentration is considered to be 1 ppm. A billionth of a gram of PCBs in a gram of fish is 1 ppb. These units are useful for conveying how concentrated a chemical is in its surrounding medium. The units of 1 μg/kg/day for the TDI refer to the amount of a chemical consumed per kilogram of the person's body weight (not the weight of the food) per day. These units take into account that a heavier person can usually eat more of a poison than a lighter one before suffering ill effects.

20. FDA, "Contamination of Animal Feeds," p. 18096.

21. FDA, "Polychlorinated Biphenyls (PCBs) in Fish and Shellfish: Reduction of Tolerances—Final Decision," *Federal Register,* 49 (May 22, 1984): 21514–21520.

22. FDA, "Contamination of Animal Feeds," p. 18097.

23. Ibid.

24. Kanshimoto et al., "Role of Polychlorinated Dibenzofuran."

25. In the United States, regulators have frequently set standards for human consumption of chemical substances at about 100 times less than the highest levels that do not cause observable harm to animals in laboratory experiments. This margin allows for the possibility that the most sensitive people might be up to 100 times more susceptible to the toxic effects of the substance than the experimental animals. The number 100 is admittedly arbitrary. For a discussion of safety factors and alternative methods for setting standards see Daniel Krewski et al., "Determining 'Safe' Levels of Exposure: Safety Factors or Mathematical Models?" *Fundamental and Applied Toxicology,* 4 (1984): 383–394.

26. Richard Severo, "Warning Ignored on Striped Bass: Fish Is Still Being Ordered by Diners Here despite Report of a Toxic Peril," *New York Times,* Aug. 9, 1975, p. 21.

27. "U.S. Health Official Asks Immediate Ban on Hudson Pollutant," *New York Times,* Sep. 9, 1975, p. 33.

28. The two primary replacements are specially refined and modified mineral oils and silicone fluid. See D. Branson, "DOW XF6-41692: An Environmentally Acceptable Capacitor Fluid," in EPA, *National Conference* (1976), pp. 314–316; and R. H. Montgomery, "The Use of Dow Corning Q2-11090 Dielectric Liquid in Power Transference," in EPA, *National Conference* (1976), pp. 312–313.

29. Section 6(e) of the Toxic Substances Control Act of 1976, U.S. Public Law 94-469, Oct. 11, 1976.

30. A. V. Nebeker, "Summary of Recent Information regarding the Effects of PCBs on Freshwater Organisms," in EPA *National Conference* (1976), pp. 284–290. Nebeker reported that 50 percent of fathead minnows (a species chosen for high sensitivity to organic chemicals) died after sixty days in water containing 4.6 ppb of the commercial PCB Aroclor 1254 (54 percent chlorine), well over the 3 ppb maximum level of PCBs found in U.S. waters. A paper published in 1991 reported that minnows suffered no acute toxicity to most PCB types tested at maximum saturation. See T. M. Dillon and W. D. S. Burton, "Acute Toxicity of PCB Congeners to *Daphnia magna* and *Pimephales promelas,*" *Bulletin of Environmental Contamination and Toxicology,* 46 (1991): 208–215.

31. NRC, *Polychlorinated Biphenyls,* p. 166.

32. A review of toxicity studies available in 1976 can be found in Massachusetts Audubon Society, *Criteria Document,* pp. 75–275.

33. R. D. Kimbrough et al., "Induction of Liver Tumors in Sherman Strain Female Rats by Polychlorinated Biphenyl Aroclor 1260," *Journal of the National Cancer Institute,* 55 (1975): 1453–1456.

34. Ibid.

35. Ekkehard Schaeffer et al., "Pathology of Chronic Polychlorinated Biphenyl (PCB) Feeding in Rats," *Toxicology and Applied Pharmacology,* 75 (1984): 278–288.

36. Ibid., p. 287.

37. J. R. Allen et al., "Pathobiological Responses of Primates to Polychlorinated Biphenyl Exposure," in EPA, *National Conference* (1976), pp. 43–49; and D. A. Barsotti et al., "Reproductive Dysfunction in Rhesus Monkeys," *Food and Cosmetics Toxicology,* 14 (1976): 99–103. We calculated that 2.5 ppm in the diet was equivalent to 0.1 mg/kg/day from information presented in the latter article.

38. N. H. Altman et al., "A Spontaneous Outbreak of Polychlorinated Biphenyl (PCB)

Toxicity in Rhesus Monkeys *(Macaca mulatta):* Clinical Observations," *Laboratory Animal Science,* 29 (1979): 661–665.

39. Kuratsune, "Some Recent Findings," p. 16.
40. Ibid., p. 17.
41. Ibid., p. 17. Massachusetts Audubon Society, *Criteria Document,* p. 37.
42. EPA, "Polychlorinated Biphenyls (PCBs) Manufacturing, Processing, Distribution in Commerce, and Use Prohibitions: Use in Electrical Equipment," *Federal Register,* 47 (Aug. 25, 1982): 37342–37355; EPA, "Polychlorinated Biphenyls in Electrical Transformers," *Federal Register,* 50 (July 17, 1985): 29170–29197. The EPA estimated that the 1982 regulations would require $136 million in capacitor removals and the 1985 regulations $635 million in commercial transformer removals and modifications. Judging from the escalation in costs for transformer flushing to six times EPA estimates in some cases (see text), actual costs have probably been much higher.
43. EPA, "Polychlorinated Biphenyls (PCBs) Manufacturing, Processing, Distribution in Commerce, and Use Prohibitions," *Federal Register,* 44 (May 21, 1979): 31514–31568.
44. EPA, "Polychlorinated Biphenyls" (1982).
45. Ibid., pp. 37348 and 37350 for cost estimates.
46. Calculated from EPA, "Polychlorinated Biphenyls Spill Cleanup Policy," *Federal Register,* 52 (Apr. 2, 1987): 10695. The 17-pound figure was derived from the EPA assumption that 9,000 capacitor leaks per year would result in the release of 154,000 pounds of PCBs.
47. EPA, "Polychlorinated Biphenyls" (1982), p. 37348.
48. Calculated from D. Mackay, *Environmental Pathways of Polychlorinated Biphenyls,* cited in ibid., p. 37348.
49. See, e.g., U.S. Public Health Service, Agency for Toxic Substances and Disease Registry, *Toxicological Profile for Selected PCBs (Aroclor -1260, -1254, -1248, -1242, -1232, -1221, and -1016,* ATSDR/TP-88/21 (1989); and R. D. Kimbrough, "Human Health Effects of Polychlorinated Biphenyls (PCBs)," *Annual Review of Pharmacological Toxicology,* 27 (1987): 87–111.
50. EPA, "Polychlorinated Biphenyls" (1982), pp. 37348–37349.
51. Ibid., p. 37346.
52. Personal communication, Dr. Gil Addis, Electric Power Research Institute, July 1991.
53. EPA, *Environmental Protection Agency Support Document/Voluntary Environmental Impact Statement for Polychlorinated Biphenyls (PCBs) Manufacturing, Processing, Distribution in Commerce, and Use Ban Regulation (Section 6(e) of TSCA),* (Washington, D.C.: EPA, Office of Toxic Substances, Apr. 1979), p. 57.
54. Ibid., p. 60.
55. L. G. Hanson, "Environmental Toxicology of Polychlorinated Biphenyls," in Safe, *Polychlorinated Biphenyls,* pp. 19–20.
56. W. M. Leis et al., *Study of Migration of PCBs from Landfills and Dredge Spoil Sites and Related to the Hudson River,* final report submitted by Weston Environmental Consultants-Designers, Westchester, Penn., to New York State Department of Environmental Conservation, Albany, N.Y., 1978. Cited in NRC, *Polychlorinated Biphenyls,* pp. 56–57.

57. EPA, *Support Document/Impact Statement* (1979), pp. 315–316.
58. EPA, "Polychlorinated Biphenyls" (1982), p. 37346.
59. C. J. Queenan et al., "Regulatory Impact Analysis of the Proposed Rule for PCB-Containing Electrical Equipment," EPA Office of Pesticides and Toxic Substances, 1982.
60. R. K. Kump, "Retrofitting or Retrofilling PCB Transformers," in *Conference Record of 1988 Annual Pulp and Paper Industry Technical Conference, June 6–10, 1988, IEEE, Piscatawny, N.J.* catalog no. 88CH2570-0, 1988, pp. 130–134.
61. We used an inflation factor between 1988 and 1982 of 1.16, derived from the "all items" category of the Consumer Price Index, reported in *Economic Indicators,* prepared for the Joint Economics Committee of the U.S. Congress by the Council of Economic Advisors (Washington, D.C.: Government Printing Office, any post-1988 edition).
62. Personal communication, Carl Manger, Baltimore Gas and Electric Company. The accident did not occur at his company.
63. L. B. Clarke, *Acceptable Risk? Making Decisions in a Toxic Environment* (Berkeley: University of California Press, 1989), p. 141.
64. Recent research has bolstered the concept of rating different types of PCBs, PCDFs, and dioxins as to their TCDD equivalency. See Safe et al., "Polychlorinated Biphenyls, Polychlorinated Dibenzo-p-Dioxins, and Polychlorinated Dibenzofurans and Related Compounds: Environmental and Mechanistic Considerations Which Support the Development of Toxic Equivalency Factors," *Critical Reviews of Toxicology,* 21, no. 1 (1990): 51–88.
65. G. Eadon et al., "Chemical Data on Air Samples from the Binghamton State Office Building," New York State Department of Health, Albany, N.Y., 1983; as cited in N. Kim et al., "Draft Revised Risk Assessment, Binghamton State Office Building," Bureau of Toxic Substance Assessment, Division of Health Risk Control, New York State Department of Health, Albany, N.Y., July 22, 1983.
66. Calculated from information presented in N. Kim and John Hawley, "Dioxin/Dibenzofuran Re-entry Guidelines, Binghamton State Office Building," Bureau of Toxic Substance Assessment, Division of Environmental Health Assessment, New York State Department of Health, Albany, N.Y., document 0529P, July 17, 1985.
67. Personal communication with a senior official, New York State Department of Health, July 3, 1991.
68. Center for Environmental Health and Wadsworth Center for Laboratories and Research, "Comparison of PCBs and PCDD/PCDFs in the Air and on Surfaces of the Binghamton State Office Building and Utica State Office Building," New York State Department of Health, Albany, N.Y., June 1989.
69. Personal communication with a senior official, New York State Department of Health, Dec. 17, 1991.
70. EPA, "Polychlorinated Biphenyls in Electrical Transformers," pp. 29170–29199.
71. Personal communication, Dr. Gilbert Addis, Electric Power Research Institute, July 1991.
72. EPA, "Polychlorinated Biphenyls in Electrical Transformers."
73. "U.S. EPA Signs Consent Agreement with University of California," press release from EPA Region 9, Jan. 21, 1992.

74. EPA, "Polychlorinated Biphenyls Spill Cleanup Policy," p. 10688.
75. Ibid., p. 10699.
76. Abt Associates, Inc., for the U.S. EPA Office of Toxic Substances, "Summary of State PCB Management Programs," Cambridge, Mass., Feb. 19, 1991.
77. Calculated from T. E. McKone et al., "Estimating Human Exposure through Multiple Pathways from Air, Water, and Soil," *Regulatory Toxicology and Pharmacology,* 13 (1991): 36–61.
78. J. Mariani, "PCBs to Jack up Price of New Police Station," *Honolulu Star-Bulletin,* Nov. 13, 1990, p. A5. The costs of PCB cleanup in Hawaii are especially high because all containment materials must be transported to the mainland.
79. Personal communication, Joe Karkoski, EPA Region 9, Sep. 1991.
80. P. A. Bertazzi et al., "Cancer Mortality of Capacitor Manufacturing Workers," *American Journal of Industrial Medicine,* 11 (1987): 165–176.
81. D. P. Brown et al., "Mortality and Industrial Hygiene Study of Industrial Workers Exposed to Polychlorinated Biphenyls," *Archives of Environmental Health,* 36 (1981): 120; D. P. Brown, "Mortality of Workers Exposed to Polychlorinated Biphenyls—An Update," *Archives of Environmental Health,* 42 (1987): 333–339.
82. M. A. Hayes, *Carcinogenic and Mutagenic Effects of PCBs,* Environmental Toxin Series, vol. 1 (Berlin: Springer-Verlag, 1987), pp. 81–95.
83. Greta G. Fein et al., "Prenatal Exposure to Polychlorinated Biphenyls: Effects on Birth Size and Gestational Age," *Journal of Pediatrics,* 105 (1984): 315–320.
84. P. R. Taylor et al., "The Relation of Polychlorinated Biphenyls to Birth Weight and Gestational Age in the Offspring of Occupationally Exposed Mothers," *American Journal of Epidemiology,* 129 (1989): 395–406.
85. U.S. Public Health Service, Agency for Toxic Substances and Disease Registry, *Toxicological Profile;* Kimbrough, "Human Health Effects."
86. W. J. Rogan et al., "Neonatal Effects of Transplacental Exposure to PCBs and DDE," *Journal of Pediatrics,* 109 (1986): 335–341.
87. B. C. Gladen et al., "Effects of Perinatal Polychlorinated Biphenyls and Dichlordiphenyl Dichloroethene on Later Development," *Journal of Pediatrics,* 119 (1991): 58–63.
88. U.S. Public Health Service, Agency for Toxic Substances and Disease Registry, *Toxicological Profile.*
89. David Perlman, "PCB-Cancer Controversy," *San Francisco Chronicle,* May 19, 1983, p. 6.
90. Don Wegars, "New Chemical Peril from Highrise Fire," *San Francisco Chronicle,* May 21, 1983, p. 1.

2, II. Is DDT a Chemical of Ill Repute?

We would like to acknowledge our intellectual debt to previous chroniclers of the DDT saga—in particular, Robert Ackerly, Wayland Hayes, George Claus, Karen Bolander, and Edward Lawless.

1. Elizabeth M. Whelan, *Toxic Terror* (Ottawa, Ill.: Jameson Books, 1985), p. 70; George Ware, *Pesticides: Theory and Application* (San Francisco: Freeman, 1983), p. 36.

2. Thomas Dunlap, *DDT: Scientists, Citizens, and Public Policy* (Princeton, N.J.: Princeton University Press, 1981), p. 37.

3. M. M. Ellis et al., "Toxicity of Dichloro-Diphenyl-Trichloroethane (DDT) to Goldfish and Frogs," *Science,* 100 (Nov. 24, 1944): 477.

4. Geoffrey Woodward et al., "Accumulation of DDT in the Body Fat and Its Appearance in the Milk of Dogs," *Science,* 102 (Aug. 17, 1945): 177–178; H. S. Telford and J. E. Guthrie, "Transmission of the Toxicity of DDT through the Milk of White Rats and Goats," *Science,* 102 (Dec. 21, 1945): 647.

5. Kenneth Mellanby, "With Safeguards, DDT Should Still Be Used," *Wall Street Journal,* Sep. 12, 1989, p. A26.

6. See, for example, M. S. Biskind, "DDT Poisoning and the Elusive 'Virus X': A New Cause of Gastro-Enteritis," *American Journal of Digestive Diseases,* 16 (Mar. 1949): 79–84.

7. "DDT Danger Refuted," *Science Digest,* 26 (July 1949): 47.

8. E. G. Hunt and A. I. Bischoff, "Inimical Effects on Wildlife of Periodic DDT Applications in Clear Lake," *California Fish and Game,* 46 (Jan. 1960): 91–109.

9. Edward Lawless, *Technology and Social Shock* (New Brunswick, N.J.: Rutgers University Press, 1977), p. 282.

10. A. W. Breidenbach and J. J. Lichtenberg, "DDT and Dieldrin in the Rivers: A Report of the National Water Quality Network," *Science,* 141 (Sep. 6, 1963): 899–901; P. Antommaria, "Airborne Particulates in Pittsburgh: Association with p,p'-DDT," *Science,* 150 (Dec. 10, 1965): 1476–47.

11. W. J. L. Sladen et al., "DDT Residues in Adelic Penguins and a Crabeater from Antarctica: Ecological Implications," *Nature,* 210 (May 14, 1966): 670–673; J. L. George and E. H. Frear, "Pesticides in the Antarctic," *Journal of Applied Ecology,* 3, suppl. (1966); and T. J. Peterly, "DDT in Antarctic Snow," *Nature,* 224 (Nov. 8, 1969): 620.

12. D. A. Ratcliffe, "Decrease in Eggshell Weight in Certain Birds of Prey," *Nature,* 215 (July 8, 1967): 208–210.

13. See, e.g., J. R. Innes et al., "Bioassay of Pesticides and Industrial Chemicals for Tumorigenicity in Mice: A Preliminary Note," *Journal of the National Cancer Institute,* 42 (June 1969): 1101–1114.

14. Lawless, *Technology and Social Shock,* p. 282. There were exceptions to the ban: disease fighting in the case of epidemics, shipment to countries where malaria was a problem, and use on onions, green peppers, and sweet potatoes in areas that were particularly vulnerable to pests.

15. "Super-Delouser," *Newsweek,* June 12, 1941, p. 96. Other early laudatory articles included: "DDT Can Wipe Out Plagues," *Science News Letter,* 48 (Sep. 8, 1945): 147; "Superior New Insecticide Called DDT," *Science Digest,* 15 (Mar. 1944): 93; and "War against Lice," *Newsweek,* Feb. 28, 1944, p. 102.

16. "DDT Warning," *Time,* Aug. 7, 1944, p. 66; "DDT Dangers," *Time,* Apr. 16, 1945, p. 91; "Careful with DDT," *Time,* Oct. 22, 1945, p. 46; "Fishermen Beware," *Time,* Dec. 10, 1945, p. 88; "Fish Killed by DDT in Mosquito Tests," *New York Times,* Aug. 9, 1945, p. 23; V. B. Wigglesworth, "DDT and the Balance of Nature," *Atlantic,* Dec. 1945, pp. 107–123; John K. Terres, "Dynamite in DDT," *New Republic,* Mar. 25, 1946, pp. 415–416; D. E. Howell, "A Case of DDT Storage in Human Body Fat," *Proceedings of the Oklahoma Academy of*

Science, 29 (1948): 31; E. P. Laug et al., "Occurrence of DDT in Human Fat and Milk," *AMA Archives of Industrial Hygiene and Occupational Medicine,* 3 (Mar. 1951): 235–236; and J. M. Ginsburg and J. P. Reed, "A Survey of DDT Accumulation in Soils in Relation to Different Crops," *Journal of Economic Entomology,* 47 (1954): 467–474.

17. Lawless, *Technology and Social Shock,* p. 277. Thalidomide, a sleep-inducing drug, played a role in producing fetal deformities in the late 1950s and early 1960s.

18. "Beyond the Bug," *Time,* Apr. 18, 1969, p. 25.

19. "The Place of DDT in Operations against Malaria and Other Vector-Borne Disease," *Official Records of the World Health Organization,* no. 190 (Geneva, Apr. 1971), p. 176. Cited by Thomas H. Jukes, "Insecticides in Health, Agriculture, and the Environment," *Naturwissenschaften,* 61 (Jan. 1974): 9.

20. George Claus and Karen Bolander, *Ecological Sanity* (New York: David Mackay, 1977), p. 322.

21. *Newsweek* carried a story saying that DDT was going to be used to control the tussock moth, but no magazine published an after-the-fact piece about DDT's success.

22. E. M. Sweeney, *Hearing Examiner's Recommended Findings, Conclusions, and Orders* (consolidated DDT hearings) (Washington, D.C.: Environmental Protection Agency, Apr. 25, 1972), p. 93.

23. William Ruckelshaus, "Consolidated DDT Hearings: Opinion and Order of the Administrator," *Federal Register,* 37 (July 7, 1972): 13375.

24. This is according to Environmental Protection Agency [EPA], *DDT: A Review of Scientific and Economic Aspects of the Decision to Ban Its Use as a Pesticide* (Washington, D.C.: EPA, July 1975), p. 83.

25. R. Tarjan and T. Kemeny, "Multigeneration Studies on DDT in Mice," *Food and Cosmetics Toxicology,* 7 (May 1969): 215–222.

26. EPA, *DDT* (1975), p. 83. See also Wayland J. Hayes, letter to Washington DDT Hearing Coordinator, in *Selected Statements from State of Washington DDT Hearings Held in Seattle, October 14, 15, 16, 1969, and Other Related Papers,* compiled by Max Sobelman (Torrance, Calif.: DDT Producers of the United States, 1970), p. 69.

27. Hayes, letter to Washington DDT Coordinator, p. 69.

28. Innes et al., "Bioassay of Pesticides," pp. 1101–1114.

29. Cited by Edith Efron, *The Apocalyptics: Cancer and the Big Lie* (New York: Simon and Schuster, 1984), p. 269.

30. EPA, *DDT* (1975), p. 83.

31. C. S. Weil, "Selection of the Valid Number of Sampling Units and a Consideration of Their Combination in Toxicological Studies Involving Reproduction, Teratogenesis, or Carcinogenesis," *Food and Cosmetics Toxicology,* 8 (Apr. 1970): 177–182.

32. L. Tomatis et al., "The Effect of Long-Term Exposure to DDT on CF-1 Mice," *International Journal of Cancer,* 10 (Nov. 1972): 489–506; L. Tomatis et al., "Effect of Long-Term Exposure to 1,1-dichloro-2,2-bis (p-Chlorophenyl) Ethylene, to 1,1-dichloro-2,2-bis (p-Chlorophenyl) Ethane, and to the Two Chemi-

cals Combined on CF-1 Mice," *Journal of the National Cancer Institute,* 52 (1974): 883–891.

33. B. Terracini et al., "The Effects of Long-Term Feeding of DDT to BALB/c Mice," *International Journal of Cancer,* 11 (May 1973): 747–764.

34. E. Thorpe and A. I. T. Walker, "The Toxicology of Dieldrin (HEOD), II: Comparative Long-Term Oral Toxicity Studies in Mice with Dieldrin, DDT, Phenobarbitone, β-BHC and γ-BHC," *Food and Cosmetics Toxicology,* 11 (June 1973): 433–442.

35. See, for example, J. H. Weisburger and Elizabeth K. Weisburger, "Food Additives and Chemical Carcinogens: On the Concept of Zero Tolerance," *Food and Cosmetics Toxicology,* 6 (Aug. 1968): 235–242.

36. Allan B. Okey, "Dimethylbenzanthracene-Induced Mammary Tumors in Rats: Inhibition by DDT," *Life Sciences,* 11 (Sep. 1, 1972): 833.

37. Ruckelshaus, "Consolidated DDT Hearings," p. 13375.

38. L. Tomatis et al., "The Predictive Value of Mouse Liver Tumor Induction in Carcinogenicity Testing—A Literature Survey," *International Journal of Cancer,* 12 (July 1973): 120. Cited by EPA, *DDT* (1975), p. 87.

39. EPA, *DDT* (1975), p. 87.

40. Jukes, "Insecticides," p. 14. See also Michael Fumento, "The Politics of Cancer Testing," *American Spectator,* Aug. 1990, p. 20.

41. "Dietary Pesticides: 99.9 Percent All Natural" and "Nature's Chemicals and Synthetic Chemicals: Comparative Toxicology," *Proceedings of the National Academy of Sciences,* 87 (1990): 7772–7786.

42. Ruckelshaus, "Consolidated DDT Hearings," p. 13371.

43. Max Sobelman, *DDT: A Case Study* (Washington, D.C.: National Research Council), 1975, p. 17.

44. Wayland Jackson Hayes, "Epidemiology of Pesticides," in E. Link and R. J. Whitaker, eds., *Proceedings of the Short Course on the Occupational Health Aspects of Pesticides* (Normal: University of Oklahoma, 1964), pp. 109–130; Wayland Jackson Hayes, "Pesticides and Human Toxicity," *Annals of the New York Academy of Sciences* 160 (1969): 40–54.

45. Mark F. Ortelee, "Study of Men with Prolonged Intensive Occupational Exposure to DDT," *AMA Archives of Industrial Health,* 18 (Nov. 1958): 433–440; cited in Robert Ackerly, "DDT: A Re-evaluation, Part I," *Chemical Times and Trends,* Oct. 1981, p. 48.

46. David Laws et al., "Men with Intensive Occupational Exposure to DDT: A Clinical and Chemical Study," *Archives of Environmental Health,* 15 (Dec. 1967): 774.

47. World Health Organization, "The Place of DDT in Operations against Malaria," p. 176. Cited by Jukes, "Insecticides," p. 9.

48. Hayes, letter to DDT Hearing Coordinator, pp. 69–70. The abstract of the Radomski study was included in an antipesticide handbook with the words "Several Good Reasons Why You Shouldn't Use Hard Poisons in Your Home and Garden" printed on the side. The study was published in J. L. Radomski et al., "Pesticide Concentrations in the Liver, Brain, and Adipose Tissue of Terminal Hospital Patients," *Food and Cosmetics Toxicology,* 6 (1968): 209–220.

49. William D. Ruckelshaus, Brief for the Respondents, U.S. Court of Appeals for the District of Columbia, no. 23813, on Petition for Review of the Order of the Secretary of Agriculture, Aug. 31, 1970. Cited in Thomas Jukes, "DDT Stands Trial Again," *Bioscience,* 22 (Nov. 1972), 672.

50. David H. Garabrant, "DDT and Related Compounds and Risk of Pancreatic Cancer," *Journal of the National Cancer Institute,* 10 (May 20, 1992): 764–771.

51. Mary S. Wolff et al., "Blood Levels of Organochlorine Residues and Risk of Breast Cancer," *Journal of the National Cancer Institute,* 85 (Apr. 21, 1993): 648–652.

52. See, for example, Nancy Sniderman's TV report on KPIX, Channel 5, Apr. 20, 1993.

53. Barry A. Miller, Eric J. Feuer, and Benjamin F. Hankey, "Recent Incidence Trends for Breast Cancer in Women and the Relevance of Early Detection: An Update," *CA—A Cancer Journal for Clinicians,* 43 (Jan./Feb. 1993): 37–38.

54. Paul Ehrlich, "Eco-Catastrophe!" *Ramparts,* 8 (1969): 24; cited in Jukes, "Insecticides," p. 12.

55. Charles F. Wurster, "DDT Reduces Photosynthesis by Marine Phytoplankton," *Science,* 159 (Mar. 29, 1968): 1474–1475.

56. Thomas Jukes, "DDT," *Journal of the American Medical Association,* 229 (July 29, 1974): 572.

57. L. Machta and E. Hughes, "Atmospheric Oxygen in 1967 to 1970," *Science,* 168 (June 26, 1970): 1582–1584; Wallace S. Broecker, "Man's Oxygen Reserves," *Science,* 168 (June 26, 1970): 1537–1538.

58. Roy J. Barker, "Notes on Some Ecological Effects of DDT Sprayed on Elms," *Journal of Wildlife Management,* 22 (July 1958): 270.

59. Ibid., pp. 272–273.

60. Ibid., p. 271.

61. Wayland J. Hayes, *Toxicology of Pesticides* (Baltimore: Williams and Wilkins, 1974), p. 502.

62. Robert Boyle, "Poison Roams Our Coastal Seas," *Sports Illustrated,* 33 (Oct. 26, 1970): 74.

63. Hayes, *Toxicology of Pesticides,* p. 490.

64. See, e.g., Rachel Carson, *Silent Spring Twenty-Fifth Anniversary Edition* (Boston: Houghton Mifflin, 1987), pp. 104–105, 118.

65. Cited in Whelan, *Toxic Terror,* p. 75.

66. *Hearings before the House Committee on Agriculture,* 92d Congress (1971): see vol. 3, p. 686, for Hickey's comments.

67. Roger Tory Peterson, *The Birds* (New York: Life Nature Company, 1963); cited in Whelan, *Toxic Terror,* p. 75.

68. Whelan, *Toxic Terror,* p. 75. Cited by William Hazeltine in *Hearings before the House Committee on Agriculture* (1971), vol. 3, p. 199.

69. Whelan, *Toxic Terror,* p. 75. See also John C. Devlin, "Herring Gull Becoming a Nuisance on Jamaica Bay," *New York Times,* Apr. 15, 1971, p. 104.

70. Frederick C. Schmid, "The Status of the Osprey in Cape May County, New Jersey, between 1939 and 1963," *Chesapeake Science,* 7 (Dec. 1966): 220–223.

71. Joseph Hickey, *A Guide to Bird Watching* (London: Oxford University Press, 1943), p. 177.

72. Byron Porterfield, "Ospreys' Return to L.I. Is Limited," *New York Times,* July

18, 1965, p. 46; Francis X. Clines, "Decline of Osprey Is Reported on L.I.," *New York Times,* Apr. 24, 1967, p. 27.

73. See Joseph Hickey, ed., *Peregrine Falcon Populations: Their Biology and Decline* (Madison: University of Wisconsin Press, 1969), esp. Roger Tory Peterson, "Population Trends of Ospreys in the Northeastern United States," p. 336.

74. William H. Stickel, "Ospreys in the Chesapeake Bay Area," in ibid., p. 167.

75. Robert White-Stevens in *Hearings before the House Committee on Agriculture* (1971), vol. 3, p. 473.

76. There is less than 1 percent probability that this increase was a random occurrence. James C. Bednarz et al., "Migration Counts of Raptors at Hawk Mountain, Pennsylvania, as Indicators of Population Trends, 1934–1986," *Auk,* 197 (Jan. 1990): 101.

77. Ackerly, "DDT: A Re-evaluation, Part I," p. 52.

78. Bednarz et al., "Migration Counts," p. 104.

79. Charles L. Broley, "The Plight of the American Bald Eagle," *Audubon Magazine,* 60 (July/Aug. 1958): 162.

80. Cited by Edwards in *Hearings before the House Committee on Agriculture* (1971), vol. 3, p. 579.

81. *Hungry Horse News* (West Glacier, Mont.), Nov. 21 and Dec. 5, 1969; cited in Jukes, "Insecticides," p. 11.

82. Alexander Sprunt quoted in Hickey, ed., *Peregrine Falcon Populations,* p. 349.

83. Dunlap, *DDT,* p. 131.

84. D. A. Ratcliffe, "The Status of the Peregrine in Great Britain," *Bird Study,* 10 (June 1963): 56–90.

85. Hickey in *Hearings before the House Committee on Agriculture* (1971), vol. 3, p. 686.

86. Joseph J. Hickey, *Auk,* 59 (1942): 176; cited in Jukes, "Insecticides," p. 11.

87. Frank L. Beebe, *The Myth of the Vanishing Peregrine,* rev. ed. Privately distributed from Saanighton, British Columbia (1971).

88. J. Gordon Edwards, "DDT's Effects on Bird Abundance and Reproduction," in Jay H. Lehr, ed., *Rational Readings on Environmental Concerns* (New York: Van Nostrand-Reinhold, 1992), p. 202.

89. John C. Devlin, "DDT in Food Held Fatal to Eagles," *New York Times,* Nov. 11, 1962, p. 49.

90. N. C. Coon et al., "Causes of Bald Eagle Mortality, 1960–1965," *Journal of Wildlife Diseases,* 6 (Jan. 1970): 76; Bernard Mulhern et al., "Organochlorine Residues and Autopsy Data from Bald Eagles: 1966–68," *Pesticides Monitoring Journal,* 4 (Dec. 1970): 144; Eugene Cromartie et al., "Residues of Organochlorine Pesticides and Polychlorinated Biphenyls and Autopsy Data for Bald Eagles, 1971–72," *Pesticides Monitoring Journal,* 9 (June 1975): 13; Richard Prouty et al., "Residues of Organochlorine Pesticides and Polychlorinated Biphenyls and Autopsy Data for Bald Eagles, 1973–74," *Pesticides Monitoring Journal,* 11 (Dec. 1977): 136; T. Earl Kaiser et al., "Organochlorine Pesticide, PCB, and PPB Residues and Necropsy Data for Bald Eagles from 29 States, 1975–77," *Pesticides Monitoring Journal,* 13 (Mar. 1980): 148.

91. Robert Riseborough, "Pesticides and Bird Populations," in Richard F. Johnson, ed., *Current Ornithology,* vol. 3, (New York: Plenum Press, 1986), p. 401. See

C. E. Grue et al., "Assessing Hazards of Organophosphate Pesticides to Wildlife," *Transactions of the North American Wildlife Natural Resource Conference,* 48 (1983): 200–220.

92. Mark Stalmaster, *The Bald Eagle* (New York: Universe Books, 1987), p. 142. See also A. S. Cooke, "Shell Thinning in Avian Eggs by Environmental Pollutants," *Environmental Pollution,* 4 (Feb. 1973): 91.

93. Ratcliffe, "Decrease in Eggshell Weight," pp. 208–210; Derek Ratcliffe, *The Peregrine Falcon* (Vermillion, S.D.: Buteo Books, 1980), pp. 312–336; Dunlap, *DDT,* p. 137.

94. Hayes, *Toxicology of Pesticides,* p. 499.

95. Joseph Hickey and Daniel Anderson, "Chlorinated Hydrocarbons and Eggshell Changes in Raptorial and Fish-Eating Birds," *Science,* 162 (Oct. 11, 1968): 271–273.

96. Daniel Anderson and Joseph Hickey, "Oological Data on Egg and Breeding Characteristics of Brown Pelicans," *Wilson Bulletin,* 82 (Mar. 1970): 14–28; J. O. Keith et al., "Reproductive Failure in Brown Pelicans on the Pacific Coast," *Transactions of the North American Wildlife Conference,* 35 (1970): 56–64; Robert Riseborough et al., "Reproductive Failure of the Brown Pelican on Anacapa Island in 1969," *American Birds,* 25 (Feb. 1971): 8–9.

97. Hayes, *Toxicology of Pesticides,* p. 500.

98. Derek Ratcliffe, "Changes Attributable to Pesticides in Egg Breakage Frequency and Eggshell Thickness in Some British Birds," *Journal of Applied Ecology,* 7 (Apr. 1970): 67–115.

99. Hayes, *Toxicology of Pesticides,* p. 500.

100. Richard W. Fyfe et al., "DDE, Productivity, and Eggshell Thickness Relationship in the Genus *Falco,*" in Thomas J. Cade et al., eds., *Peregrine Falcon Populations* (Boise, Id.: Peregrine Fund, 1988), p. 319.

101. Hickey and Anderson, "Chlorinated Hydrocarbons," pp. 271–273.

102. Tom J. Cade et al., "DDE Residues and Eggshell Changes in Alaskan Falcons and Hawks," *Science,* 172 (May 28, 1971): 955–957; James Enderson and Daniel D. Berger, "Pesticides: Eggshell Thinning and Lowered Production of Young in Prairie Falcons," *Bioscience,* 20 (Mar. 15, 1970): 355–366; N. Fimreite et al., "Mercury Contamination of Canadian Prairie Seed-Eaters and Their Avian Predators," *Canadian Field Naturalist,* 84 (1970): 269–276; Lawrence Blus et al., "Eggshell Thinning in the Brown Pelican: Implication of DDE," *Bioscience,* 21 (Dec. 15, 1971): 1213–1215; Daniel Armstrong et al., "Significance of Chlorinated Hydrocarbon Residues to Breeding Pelicans and Cormorants," *Canadian Field Naturalist,* 83 (1969): 91–112.

103. Bruce Switzer et al., "Shell Thickness, DDE Levels in Eggs, and Reproductive Success in Common Terns *(Sterna hirundo)* in Alberta," *Canadian Journal of Zoology,* 49 (Jan. 1971): 69–73.

104. R. A. Faber and Joseph Hickey, "Eggshell Thinning, Chlorinated Hydrocarbons, and Mercury in Inland Equatic Bird Eggs, 1969 and 1970," *Pesticides Monitoring Journal,* 7 (June 1973): 27–36.

105. James H. Enderson and Daniel D. Berger, "Chlorinated Hydrocarbons, Residues in Peregrines and Their Prey Species from Northern Canada," *Condor,* 70 (Apr. 1968): 153.

106. Tom J. Cade and Richard Fyfe, "The North American Peregrine Survey, 1970," *Canadian Field Naturalist,* 84 (1970): 231–245.
107. Hayes, *Toxicology of Pesticides,* p. 500.
108. Hickey in *Hearings before the House Committee on Agriculture* (1971), vol. 3, pp. 686–688.
109. Robert Riseborough et al., "Organochlorine Pollutants in Peregrines and Merlins Migrating through Wisconsin," *Canadian Field Naturalist,* 84 (1970): 250.
110. For a review of these studies, see Cooke, "Shell Thinning," pp. 94–95.
111. Robert G. Heath et al., "Marked DDE Impairment of Mallard Reproduction in Controlled Studies," *Nature,* 224 (Oct. 4, 1969): 47–48.
112. Richard K. Tucker and H. A. Haegle, "Eggshell Thinning as Influenced by Method of DDT Exposure," *Bulletin of Environmental Contamination and Toxicology,* 5 (May/June 1970): 191–194.
113. S. N. Wienmeyer and R. D. Porter, "DDE Thins Eggshells of Captive American Kestrels," *Nature,* 227 (Aug. 15, 1970): 737–738.
114. Riseborough, "Pesticides and Bird Populations," p. 406.
115. Panel on Mercury, Coordinating Committee for Scientific and Technical Assessments of Environmental Pollution, National Research Council, *An Assessment of Mercury in the Environment* (Washington, D.C.: National Academy of Sciences, 1978), p. 64.
116. Harold Faber, "Once-Imperiled Osprey Makes New York Comeback," *New York Times,* Jan. 18, 1982, p. B1. See also Sarah Lyall, "Ospreys Are Back on L.I., Some in Custom-Built Nests," *New York Times,* Apr. 25, 1991, p. B1.
117. James W. Grier, "Ban of DDT and Subsequent Recovery of Reproduction in Bald Eagles," *Science,* 218 (Dec. 17, 1982): 1232–1234.
118. Mark Wexler, "A Case of Urban Renewal," *National Wildlife,* June/July 1989, p. 12.
119. Riseborough, "Pesticides and Bird Populations," p. 412.
120. But see Edwards, "DDT's Effects on Bird Abundance and Reproduction."
121. John Noble Wilford, "Deaths from DDT Successor Stir Concern," *New York Times,* Aug. 21, 1970, p. 1.
122. W. W. Fletcher, *The Pest War* (New York: Wiley, 1974), p. 49.
123. Ruckelshaus, "Consolidated DDT Hearings," p. 13374.
124. Wilford, "Deaths from DDT Successor," p. 5; Kenneth P. DuBois, "DDT Substitutes Pose More Medical Hazards," *Modern Medicine,* Dec. 13, 1971, p. 46.
125. Claus and Bolander, *Ecological Sanity,* p. 544.
126. Frank Graham, Jr., *The Dragon Hunters* (New York: Truman Talley Books, 1984), p. 70.
127. George Ware, *Pesticides: Theory and Application* (San Francisco: Freeman, 1983), pp. 56–57.
128. Quoted in Graham, *Dragon Hunters,* p. 248.
129. Ibid., p. 250.
130. G. T. Brooks, *Chlorinated Hydrocarbons,* vol. 2, *Biological and Environmental Aspects* (Cleveland: CRC Press, 1974), p. 53.
131. Russell Train, "State of Louisiana Request for Emergency Use of DDT on Cotton," *Federal Register,* 40 (Apr. 8, 1975): 15942.
132. EPA, *DDT* (1975), pp. 155, 156, 158.

133. Ibid., p. 158.
134. Ibid., p. 151.
135. Claus and Bolander, *Ecological Sanity,* p. 543.
136. E. A. Eagan, "Cotton Pest Control Problems from the Ginner's Viewpoint," Proceedings of the Cotton Symposium on Insect and Mite Control Problems and Research, Berkeley, Calif., 1968, pp. 19–21. Cited by John E. Swift, "Unexpected Effects from Substitute Pest Control Methods," Symposium on the Biological Impact on Pesticides in the Environment, Oregon State University, Corvalis (Aug. 1969), p. 5.
137. Rita Gray Beatty, *The DDT Myth: Triumph of the Amateurs* (New York: John Day, 1973), p. 130.
138. Swift, "Unexpected Effects," pp. 2–3.
139. L. H. Foote, "California Beekeeping Status and Trends," an analysis prepared for the California Assembly Advisory Committee on Bee and Pesticide Problems, cited in ibid., p. 3.
140. Cited by Beatty, *The DDT Myth,* p. 142.
141. "Agency Forbids Use of DDT Moth Kill," *Los Angeles Times,* Apr. 21, 1973, p. 11.
142. Train, "State of Louisiana Request," p. 15938; Claus and Bolander, *Ecological Sanity,* p. 322.
143. Jack Mounts, "1974 Douglas-Fir Tussock Moth Control Project," *Journal of Forestry,* 74 (Feb. 1976): 86.
144. Statement of Ralph Sherman in *Hearings before the House Committee on Agriculture* (1971), vol. 3, p. 239.
145. Rodney E. Garrett, "DDT: What Some European Countries Are Doing about It," *American Forests,* Sep. 1971; p. 19.
146. Ackerly, "DDT: A Re-evaluation," p. 50.
147. Graham, *Dragon Hunters,* p. 32.
148. V. G. Dethier, *Man's Plague?* (Princeton, N.J.: Darwin Press, 1976), p. 152.
149. Claus and Bolander, *Ecological Sanity,* p. 318.
150. John H. Perkins, *Insects, Experts, and the Insecticide Crisis* (New York: Plenum Press, 1982), p. 59.
151. Graham, *Dragon Hunters,* p. 296.
152. Perkins, *Insects, Experts,* p. 59.
153. Ackerly, "DDT: A Reevaluation," p. 50; Sobelman, *DDT: A Case Study,* pp. 3–4.
154. Sweeney, *Recommended Findings,* p. 93.
155. Ruckelshaus, "Consolidated DDT Hearings," p. 13373.
156. Sobelman, *DDT: A Case Study,* pp. 3–4. There were also leaps of logic in Ruckelshaus's order. One passage read: "The petitioner-registrants' assertion that there is no evidence of declining aquatic or avian populations, even if actually true, is an attempt at confession and avoidance. It does not refute the basic proposition that DDT causes damage to wildlife species." The conclusions of this argument do not seem to follow from the premises.
157. Sobelman, *DDT: A Case Study,* p. 17.
158. Ruckelshaus, "Consolidated DDT Hearings," p. 13375; emphasis added.
159. Ackerly, "DDT: A Re-evaluation," p. 50.

Doing More Harm Than Good

1. See Aaron Wildavsky, *Searching for Safety* (New Brunswick, N.J.: Transaction Publishers, 1988).
2. See Aaron Wildavsky, "Doing More and Using Less: Utilization of Research as a Result of Regime," in Meinoff Dierkes, Hans Weiler, and Ariane Berthoin Antal, eds., *Comparative Policy Research: Learning from Experience* (Aldershot, Eng.: Gower, 1986), pp. 56–93, for an analysis of why the United States does a lot of research but uses little.

3. Dioxin, Agent Orange, and Times Beach

1. A. L. Young, "The Military Use of Herbicides in Vietnam," in Young and G. M. Reggiani, eds., *Agent Orange and Its Associated Dioxin: Assessment of a Controversy* (New York: Elsevier, 1988), p. 9.
2. Michael Gough, *Dioxin, Agent Orange: The Facts* (New York: Plenum, 1986), p. 158.
3. G. M. Reggiani, "Historical Overview of the Controversy Surrounding Dioxin," in Young and Reggiani, *Agent Orange and Its Associated Dioxin*, pp. 47–48.
4. Gough, *Dioxin, Agent Orange*, p. 161.
5. Ibid., pp. 161–162.
6. Reggiani, "Historical Overview," p. 42.
7. Dow Chemical Company, *Dioxin, Agent Orange, and Human Health*, pamphlet, Apr. 1984, pp. 44–45.
8. Michael Gough, "Human Health Effects: What the Data Indicate," *Science of the Total Environment*, 104 (1991): 129.
9. Michael Gough, "Human Exposures from Dioxin in Soil—A Meeting Report," *Journal of Toxicology and Environmental Health* (1991): 227.
10. Ellen K. Silbergeld and Thomas A. Gasiewicz, "Dioxins and the Ah Receptor," *American Journal of Industrial Medicine*, 16 (1989): 459.
11. Gough, *Dioxin, Agent Orange*, p. 187.
12. Young, "Military Use of Herbicides," p. 9.
13. Dow, *Dioxin, Agent Orange, and Human Health*, p. 43.
14. Shelby L. Stanton, "Area-Scoring Methodology for Estimating Agent Orange Exposure Status of U.S. Army Personnel in the Republic of Vietnam," in Frank DeStefano et al., *Comparison of Serum Levels of 2,3,7,8-Tetrachlorodibenzo-p-Dioxin with Indirect Estimates of Agent Orange Exposure Among Vietnam Veterans*, final report, appendix A (Atlanta: Centers for Disease Control, Sep. 1989), p. 35.
15. Elmo Zumwalt, Jr., and Elmo Zumwalt III with John Pekkanen, *My Father, My Son* (New York: Macmillan, 1986), p. 47.
16. Ibid., p. 162.
17. Ibid., p. 163.
18. Young, "Military Use of Herbicides," p. 10.
19. Ibid., pp. 13–14.
20. Ibid., p. 11.

21. Ibid.
22. Peter Schuck, *Agent Orange on Trial* (Cambridge, Mass.: Harvard University Press, 1986), p. 86.
23. Ibid., pp. 87, 141.
24. Ibid., p. 86.
25. Young, "Military Use of Herbicides," pp. 12, 14.
26. Gough, *Dioxin, Agent Orange,* pp. 52, 49.
27. Ibid., p. 12.
28. Young, "Military Use of Herbicides," p. 13; Gough, *Dioxin, Agent Orange,* p. 52.
29. Young, "Military Use of Herbicides," p. 16.
30. Reggiani, "Historical Overview," p. 37.
31. Joel Primack and Frank von Hippel, *Advice and Dissent: Scientists in the Political Arena* (New York: Basic Books, 1974), p. 154.
32. Ibid., p. 75.
33. Ibid., p. 154; Reggiani, "Historical Overview," p. 37.
34. Primack and von Hippel, *Advice and Dissent,* p. 155.
35. Phillip M. Boffey, *The Brain Bank of America: An Inquiry into the Politics of Science* (New York: McGraw-Hill, 1975), p. 148.
36. Ibid., p. 152.
37. Dow, *Dioxin, Agent Orange, and Human Health,* p. 46.
38. Boffey, *Brain Bank,* p. 152.
39. Ibid., p. 153.
40. Ibid., p. 152.
41. Dael Wolfle, "Vietnam, Herbicides, and AAAS," *AAAS Observer,* supplement to *Science,* Nov. 3, 1989, pp. 11–14; see p. 12.
42. Reggiani, "Historical Overview," p. 38.
43. Primack and von Hippel, *Advice and Dissent,* p. 156.
44. Ibid.
45. Gough, *Dioxin, Agent Orange,* p. 55.
46. Primack and von Hippel, *Advice and Dissent,* p. 156.
47. Ibid., p. 75.
48. Ibid., p. 157.
49. Gough, *Dioxin, Agent Orange,* p. 56.
50. Primack and von Hippel, *Advice and Dissent,* p. 76.
51. Ibid., p. 77. (Matthew S. Meselson's mother, Ann Swedlow, is my grandfather's sister.—B.S.)
52. Ibid., p. 81.
53. Ibid., pp. 78, 77, 157.
54. Gough, *Dioxin, Agent Orange,* p. 56.
55. Primack and von Hippel, *Advice and Dissent,* p. 78.
56. Ibid., p. 79.
57. Ibid., p. 80.
58. Boffey, *Brain Bank,* p. 157.
59. Primack and von Hippel, *Advice and Dissent,* p. 158.
60. Boffey, *Brain Bank,* p. 157.
61. Primack and von Hippel, *Advice and Dissent,* p. 160.
62. Ibid., p. 157.

63. Young, "Military Use of Herbicides," p. 12.
64. Primack and von Hippel, *Advice and Dissent,* p. 82.
65. Boffey, *Brain Bank,* p. 160.
66. Ibid., p. 161.
67. Ibid., p. 162.
68. Ibid.
69. Primack and von Hippel, *Advice and Dissent,* p. 160.
70. Clifford Linedecker with Michael Ryan and Maureen Ryan, *Kerry: Agent Orange and an American Family* (New York: St. Martin's, 1982), p. 124.
71. John Coombs, "The Agent Orange Phenomenon: The Report of the Australian Royal Commission," in Young and Reggiani, *Agent Orange and Its Associated Dioxin,* p. 301.
72. Linedecker, *Kerry,* pp. 125–126.
73. Coombs, "Agent Orange Phenomenon," pp. 301–302.
74. Quoted in Young, "Social Assessment of the Agent Orange Controversy," in Young and Reggiani, *Agent Orange and Its Associated Dioxin,* p. 200.
75. Linedecker, *Kerry,* p. 127.
76. Gough, *Dioxin, Agent Orange,* p. 46.
77. Linedecker, *Kerry,* p. 130; Gough, *Dioxin, Agent Orange,* p. 68.
78. Schuck, *Agent Orange on Trial,* p. 45.
79. Arvin Maskin, "A Legal Assessment of the Controversy," in Young and Reggiani, *Agent Orange and Its Associated Dioxin,* p. 173.
80. Schuck, *Agent Orange on Trial,* p. 45.
81. Ibid., p. 72.
82. Ibid., p. 47.
83. Ibid., p. 105.
84. Ibid., p. 52.
85. Ibid., p. 104.
86. Environmental Protection Agency, "Report of Assessment of a Field Investigation of Six-Year Spontaneous Abortion Rates in Three Oregon Areas in Relation to Forest 2,4,5-T Spray Practices," Feb. 28, 1979, in Gough, *Dioxin, Agent Orange,* pp. 140–143.
87. As quoted in Gough, *Dioxin, Agent Orange,* p. 143.
88. Ibid., p. 144.
89. L. Hardell and A. Sandstroem, "Case-Control Study: Soft-Tissue Sarcomas and Exposure to Phenoxyacetic Acids or Chlorophenols," *British Journal of Cancer,* 39 (1979): 713–714.
90. Reggiani, "Historical Overview," p. 64; Gough, *Dioxin, Agent Orange,* p. 245.
91. Gough, *Dioxin, Agent Orange,* p. 90.
92. Ibid., p. 91.
93. Michael Gough, "The Political Assessment: A Congressional View," in Young and Reggiani, *Agent Orange and Its Associated Dioxin,* p. 185.
94. Gough, *Dioxin, Agent Orange,* p. 52.
95. Gough, "The Political Assessment," pp. 185–186.
96. Gough, *Dioxin, Agent Orange,* pp. 92–93.
97. Ibid., p. 95.
98. Gough, "The Political Assessment," p. 186.

99. Ibid., p. 184.
100. Ibid., p. 187.
101. Gough, *Dioxin, Agent Orange,* p. 95.
102. Schuck, *Agent Orange on Trial,* p. 81.
103. Ibid., p. 99.
104. Ibid., p. 100.
105. Gough, *Dioxin, Agent Orange,* p. 79.
106. Ibid., p. 74.
107. J. D. Erickson, J. Mulinare, P. W. McClain et al., "Vietnam Veterans' Risks for Fathering Babies with Birth Defects," *Journal of the American Medical Association* 252 (Aug. 17, 1984): 903–912.
108. Ibid., p. 907.
109. Gough, "The Political Assessment," p. 184.
110. Gough, *Dioxin, Agent Orange,* pp. 112–113.
111. Erickson et al., "Vietnam Veterans' Risks," p. 912.
112. Schuck, *Agent Orange on Trial,* p. 110.
113. Ibid., p. 181.
114. Ibid., p. 147.
115. Ibid., p. 156. The liability (and market share) of the other defendants were as follows: Diamond Shamrock, 12 percent (5.1 percent market share); Hercules, 10 percent (19.7 percent market); T. H. Agriculture and Nutrition, 6 percent (7.2 percent market); Uniroyal, 5 percent (6.5 percent market); Thompson Chemicals, 2 percent (2.2 percent market share).
116. Ibid., p. 181.
117. Ibid., p. 227.
118. Ibid., p. 229.
119. Ibid., pp. 232–233.
120. Ibid., p. 234.
121. Ibid., p. 248.
122. Maskin, "A Legal Assessment," pp. 178–179.
123. Steven D. Stellman, Jeanne Mager Stellman, and John F. Sommer, Jr., "Health and Reproductive Outcomes among American Legionnaires in Relation to Combat and Herbicide Exposure in Vietnam," *Environmental Research,* 47 (Dec. 1988): at 150.
124. Ibid., p. 159.
125. Frank DeStefano et al., *Comparison of Serum Levels of 2,3,7,8-Tetrachlorodibenzo-p-Dioxin with Indirect Estimates of Agent Orange Exposure among Vietnam Veterans,* final report (Atlanta: Centers for Disease Control, Sep. 1989).
126. James L. Pirkle et al., "Estimates of the Half-Life of 2,3,7,8-Tetrachlorodibenzo-p-Dioxin in Vietnam Veterans of Operation Ranch Hand," *Journal of Toxicology and Environmental Health,* 27 (1989): 165.
127. Ted Weiss, *Oversight Review of CDC's Agent Orange Study,* prepared for the House Subcommittee on Human Resources and Intergovernmental Relations of the Committee on Government Operations, 101st Congress, July 11, 1989, vol. 32.
128. Ibid.
129. William H. Wolfe et al., "Health Status of Air Force Veterans Occupationally

Exposed to Herbicides in Vietnam, I: Physical Health," *Journal of the American Medical Association,* 264 (Oct. 10, 1990): 1829.

130. Joel E. Michalek et al., "Health Status of Air Force Veterans Occupationally Exposed to Herbicides in Vietnam, II: Mortality," *Journal of the American Medical Association,* 264 (Oct. 10, 1990): 1834.

131. Wolfe et al., "Health Status," p. 1829.

132. Edward A. Brann et al., *The Association of Selected Cancers with Service in the U.S. Military in Vietnam,* final report (Atlanta: Centers for Disease Control, Sep. 1990), p. 7.

133. Ibid., p. 27.

134. Ibid., p. 2.

135. Michael Fumento, *Science under Siege: Balancing Technology and the Environment* (New York: William Morrow, 1993), p. 179.

136. Vera Haller, "First Agent Orange Payments in '89," *Santa Rosa Press Democrat,* Jan. 3, 1989.

137. *Santa Rosa Press Democrat,* Apr. 3, 1989.

138. Eric Felten, "The Times Beach Fiasco," *Insight,* Aug. 12, 1991, p. 12.

139. Unless otherwise indicated, quotations and information about dioxin in Missouri are taken from the Sep. 14, 1983, *St. Louis Post-Dispatch* special section, "Dioxin: Quandary for the 80's," reported by William Freivogel, Marjorie Mandel, Jo Mannies, and Lawrence M. O'Rourke.

140. Gough, *Dioxin, Agent Orange,* p. 122.

141. Ibid., p. 124.

142. Ibid., p. 122.

143. Toxic Substances Control Act, Section 4(a)(1)(i), 15, U.S.C.A. sections 2601 to 2671.

144. Ibid., section 6(a).

145. Randolph L. Hill, "An Overview of RCRA: The 'Mind-Numbing' Provisions of the Most Complicated Environmental Statute," *Environmental Law Reporter,* 21 (May 1991): 10254.

146. *United States v. Northeastern Pharmaceutical,* 810 F.2d 726 (8th Cir. 1986).

147. "Doctors Accuse Press over Dioxin Panic," *New Scientist,* 98, no. 1363 (June 30, 1983): 926.

148. "Chemical Risks: Fears, Facts, and the Media," Media Institute study, Washington, D.C., 1985, conducted and written under the direction of Sarah J. Midgley, director of research. Another Media Institute study, by Edward J. Burger, Jr., analyzes the treatment of herbicides, PCBs, and chemotherapy in the print media. Burger, a medical doctor who was director of Georgetown University's Institute for Health Policy Analysis in 1984 and a senior policy analyst to the president's science advisor from 1969 to 1976, concluded of herbicides that "treatment of the 2,4,5-T issue by the media was heavily colored by the military use of defoliants (and by the polarized positions concerning those practices), and by the then fast-rising tide of the environmental movement. Scientific issues were of little real interest, except to the extent that selected scientific reports seemed to support a point of view. Thus, with two exceptions, there was essentially no discussion in the press of any of the scientific background of the several government announcements about 2,4,5-T or of the opinions offered by private parties. The

assertions (birth defects, cancer, and genetic alteration) were simply repeated or reinforced by suggestions of authority." See "Health Risks: The Challenge of Informing the Public," Media Institute Study, Washington, D.C., 1984, p. 26.

149. All material was analyzed twice by two coders. Once for "the volume and placement of stories; the volume devoted to charges, responses, or background information; and sources, i.e., distinct categories of people." The second analysis looked for "the use of speculative anecdotes (unsubstantiated stories about the effects of the chemical at issue) and the use of scientific data and sources." Media Institute, "Chemical Risks," p. 3.

150. Ibid., p. xii.

151. Ibid., p. 40.

152. Ibid., p. 27.

153. Ibid., p. 23.

154. This and the following statement are based on the Media Institute's analysis of media treatment of all three chemical risk episodes.

155. Media Institute, "Chemical Risks," p. xii–xiii.

156. Ibid., p. 24.

157. Ibid., p. 25.

158. Ibid., p. xiii.

159. Paul E. Stehr-Green et al., "An Overview of the Missouri Dioxin Studies," *Archives of Environmental Health,* 43 (Mar./Apr. 1988): 175.

160. Richard E. Hoffman et al., "Health Effects of Long-Term Exposure to 2,3,7,8-Tetrachlorodibenzo-*p*-Dioxin," *Journal of the American Medical Association,* 255 (Apr. 18, 1986): 2031.

161. Ibid., p. 2033. PCT is a disturbance in the body's capacity to break down hemoglobin, resulting in liver damage and a blistering and fragility of skin (Gough, *Dioxin, Agent Orange,* pp. 178–179).

162. Hoffman et al., "Health Effects of Long-term Exposure," p. 2034.

163. Ibid., p. 2035.

164. R. Gregory Evans et al., "A Medical Follow-Up of the Health Effects of Long-Term Exposure to 2,3,7,8-Tetrachlorodibenzo-*p*-Dioxin," *Archives of Environmental Health,* 43 (July/Aug. 1988): 273, 275, 277.

165. Karen B. Webb et al., "Medical Evaluation of Subjects with Known Body Levels of 2,3,7,8-Tetrachlorodibenzo-*p*-Dioxin," *Journal of Toxicology and Environmental Health,* 28 (1989): 190.

166. Even without the results of an important 1991 study by Marilyn Fingerhut and colleagues, most scientists reviewing evidence of dioxin's health effects in the late 1980s and early 1990s concluded that the chemical was not a serious threat. Only one set of reviewers remained equivocal because "no study appears to be identifiably more credible than any other" (S. A. Skene, I. C. Dewhurst, and M. Greenberg, "Polychlorinated Dibenzo-*p*-dioxins and Polychlorinated Dibenzofurans: The Risks to Human Health. A Review," *Human Toxicology,* 8 [1989]: 53). For the rest, dioxin's health effects were clearly limited: "We conclude that the weak-to-moderately strong carcinogenic effect of the chemical manifests itself in only one type of cancer, i.e., non-Hodgkin's lymphoma" (David E. Lilienfeld and Michael A. Gallo, "2,4-D, 2,4,5-T, and 2,3,7,8-TCDD: A Overview," *Epidemiologic Reviews,* 11 [1989]: 51). "The total weight of evidence currently avail-

able does not support the conclusion that any of the phenoxy herbicides present a carcinogenic hazard to man" (Gregory G. Bond, Kenneth M. Bodner, and Ralph R. Cook, "Phenoxy Herbicides and Cancer: Insufficient Epidemiologic Evidence for a Causal Relationship," *Fundamental and Applied Toxicology,* 12 [1989]: 185). "No human illness, other than the skin disease chloracne . . . has been convincingly associated with dioxin" (Gough, "Human Health Effects").

167. Kristine Napier, "Reevaluating Dioxin: The Implications for Science Policy," *Priorities,* Winter 1992, p. 36.
168. Ibid., p. 37.
169. "EPA Calls Dioxin Most Potent Material," *St. Louis Post-Dispatch,* July 24, 1983; "Dioxin Scare Called Mistake," *St. Louis Post-Dispatch,* May 23, 1991; "On 2nd Thought, Toxic Nightmares May Be Unpleasant Dreams," *Chicago Tribune,* Sep. 1, 1991; "Dioxin Joins List of Costly False Alarms," *Los Angeles Times,* Aug. 9, 1991; *New York Times,* Sept. 1, 1991. As cited in Liane Clorfene Casten, "Dioxin Charade Poisons the Press," *Extra!* 5, no. 1. (Jan./Feb. 1992): 12.
170. Reed Irvine, The Dioxin Un-Scare—"Where's the Press?" *Wall Street Journal,* Aug. 6, 1991, p. A14.
171. "U.S. Health Aide Says He Erred on Times Beach," *New York Times,* May 26, 1991, p. A20.
172. Felten, "Times Beach Fiasco," p. 19.
173. Marilyn A. Fingerhut et al., "Cancer Mortality in Workers Exposed to 2,3,7,8-Tetrachlorodibenzo-*p*-Dioxin," *New England Journal of Medicine,* 324 (Jan. 24, 1991): 212.
174. Ibid., p. 214.
175. Leslie Roberts, "Dioxin Risks Revisited," *Science,* 251 (Feb. 8, 1991): 625.
176. Richard Stone, "Dioxin: Still Deadly," *Science,* 260 (Apr. 2, 1993): 31.
177. Karen F. Schmidt, "Dioxin's Other Face: Portrait of an 'Environmental Hormone,' " *Science News,* 141 (Jan. 11, 1992): 24.
178. Ibid.

4. Love Canal

1. Costs are still accumulating. "Containment" of Love Canal is estimated to cost $1.5 million per year. See Andrew Danzo, "The Big Sleazy: Love Canal Ten Years Later," *Washington Monthly,* Sep. 1988, p. 11.
2. Carr quoted in Sam Howe Verhovek, "At Love Canal, Land Rush on a Burial Ground," *New York Times,* July 26, 1990, p. A1; the epigraph quoting Carr is from this article.
3. Ralph Nader and Ronald Brownstein, "Beyond the Love Canal," *Progressive,* May 1980, p. 28.
4. See Michael P. Zweig, "The Federal Connection: A History of U.S. Military Involvement in the Toxic Contamination of Love Canal and the Niagara Frontier Region," *New York State Assembly Task Force on Toxic Substances Report,* vol. 1, Jan. 29, 1981.
5. Donald Baeder, "Love Canal—What Really Happened," *Chemtech,* 10 (Dec. 1980): 740.

6. Statement of Bruce Davis, *Hearings on Hazardous Waste Disposal,* House Sub-committee on Oversight and Investigations of the Committee on Interstate and Foreign Commerce, 96th Congress, Mar. 21, 1979, vol. 22, p. 500.

7. Ibid., p. 501.

8. See Eric Zuesse, "Love Canal: The Truth Seeps Out," *Reason,* 12 (Feb. 1981): 17–31, for a complete account of the dealings between Hooker and the Board of Education.

9. Donald McNeil, "Upstate Waste Site May Endanger Lives," *New York Times,* Aug. 2, 1978, p. B9.

10. See, for example, Lois Gibbs, *Love Canal: My Story* (Albany: State University of New York Press, 1982), p. 4.

11. John Elliot, "Lessons from Love Canal," *Journal of the American Medical Association,* 240, no. 19 (Nov. 3, 1978): 2034.

12. Baeder, "Love Canal—What Really Happened," p. 742.

13. Michael Brown, *Laying Waste: The Poisoning of America* (New York: Pantheon, 1982), pp. 10–11.

14. Zuesse, "The Truth Seeps Out," p. 20.

15. Ibid., p. 27.

16. Ralph Blumenthal, "Bid to Curb Toxic Sites: Doubts Persist as U.S. Acts to Limit Dumping," *New York Times,* June 30, 1980, p. B11 (our emphasis).

17. Andrew Brown, "The Devil's Brew in Love Canal," *Fortune,* 100 (Nov. 19, 1979): 76.

18. "A Formula to Settle Toxic Dump Problems," *Business Week,* Jan. 14, 1980, p. 34.

19. Michael Brown, "A Toxic Ghost Town," *Atlantic Monthly,* July 1989, p. 28.

20. R. D. Kimbrough, "Health Impact of Toxic Wastes: Estimation of Risk," in Vincent T. Covello et al., eds., *The Analysis of Actual versus Perceived Risks* (New York: Plenum Publishing, 1983), p. 260.

21. Sheldon Wolff, "Problems and Prospects in the Utilization of Cytogenetics to Estimate Exposure at Toxic Chemical Waste Dumps," *Environmental Health Perspectives,* 48 (1983): 25.

22. Irwin Molotsky, "Damage to Chromosomes Found in Love Canal Tests," *New York Times,* May 17, 1980, p. A1.

23. Josh Barbanel, "Homeowners at Love Canal Hold 2 Officials until FBI Intervenes," *New York Times,* May 20, 1980, p. B4.

24. Jurgen Schulz-Schaeffer, *Cytogenetics: Plants, Animals, Humans* (New York: Springer-Verlag, 1980), p. 2.

25. W. W. Lowrance, ed., *Assessment of Health Effects at Chemical Disposal Sites: Proceedings of a Symposium Held in New York City* (New York: Rockefeller University, 1981), pp. 8, 9.

26. Gina Bari Kolata, "Chromosome Damage: What It Is, What It Means," *Science,* 208 (June 13, 1980): 1240.

27. Wolff, "Problems and Prospects in the Utilization of Cytogenetics," pp. 25–27.

28. Sidney Green, "Review of Report and Slides on the 'Pilot Cytogenetic Study on the Residents of Love Canal, NY,' " Environmental Protection Agency [EPA] memorandum, May 28, 1980, p. 2.

29. Dante Picciano, "Pilot Cytogenetic Study of the Residents of Love Canal, New York," performed by Biogenics Corporation, Houston, Tex., May 14, 1980, p. 2.

30. Gina Bari Kolata, "Damage to Body's Chromosomes Can Be Caused in Several Ways," *New York Times,* May 18, 1980, p. A37.
31. Gina Bari Kolata, "Love Canal: False Alarm Caused by Botched Study," *Science,* 208 (June 13, 1980): 1240.
32. Sheldon Wolff, "Love Canal Revisited," *Journal of the American Medical Association,* 251 (Mar. 16, 1984): 1464.
33. Picciano, "Pilot Cytogenetic Study," cited in Samuel S. Epstein et al., *Hazardous Waste in America* (San Francisco: Sierra Club Books, 1982), p. 113.
34. Kolata, "False Alarm," p. 1241.
35. Lois Ember, "Uncertain Science Pushes Love Canal Solutions to Political, Legal Arenas," *Chemical and Engineering News,* Aug. 11, 1980, p. 24. See also Adeline Levine, *Love Canal: Science, Politics, and People* (Lexington, Mass.: Lexington Books, 1982), p. 139.
36. Picciano, "Pilot Cytogenetic Study," p. 6.
37. Wolff, "Love Canal Revisited," p. 1464.
38. Ibid.
39. Dante Picciano, "Pilot Cytogenetic Study of the Residents Living near Love Canal, a Hazardous Waste Site," *Mammalian Chromosome Newsletter,* 21 (1980): 86–93.
40. Wolff, "Love Canal Revisited," p. 1464.
41. "More Cruelty at Love Canal," *New York Times,* May 22, 1980, p. A34.
42. Barbara J. Culliton, "Continuing Confusion over Love Canal," *Science,* 209 (Aug. 29, 1980): 1002.
43. Roy Albert quoted in letter to the editor by Stephen J. Gage, *Science,* 209 (Aug. 5, 1980): 752.
44. David Axelrod quoted in Robin Herman, "Carey Criticizes Facets of Study on Love Canal," *New York Times,* May 19, 1980, p. A1.
45. Health and Human Services Panel Report, quoted in Kolata, "False Alarm," p. 1240.
46. Lewis Thomas, "Report of the Governor's Panel to Review Scientific Studies and the Development of Public Policy on Problems Resulting from Hazardous Wastes," Albany, N.Y., Oct. 1980, pp. 6–17.
47. Margery Shaw, "Love Canal Chromosome Study," letter to *Science,* 209 (Aug. 15, 1980): 751–752.
48. Ibid.
49. Michael P. Greenberg, *Public Health and the Environment: The United States Experience* (New York: Guilford Press, 1987), p. 192.
50. Culliton, "Continuing Confusion over Love Canal," p. 1003.
51. Wolff, "Problems and Prospects in the Utilization of Cytogenetics," pp. 25–27; Arthur Bloom, ed., *Guidelines for Studies of Human Populations Exposed to Mutagenic and Reproductive Hazards,* (White Plains, N.Y.: March of Dimes Birth Defects Foundation, 1981), quoted in *Morbidity and Mortality Weekly Report,* 32 (May 27, 1983): 262.
52. Joe Grisham, *Health Aspects of the Disposal of Waste Chemicals* (New York: Pergamon Press, 1986), p. 81.
53. Clark Heath et al., "Cytogenetic Findings in Persons Living Near the Love Canal," *Journal of the American Medical Association,* 251 (Mar. 16, 1984): 1437–1440.

54. Greenberg, *Public Health and the Environment,* p. 192.
55. Heath et al., "Cytogenetic Findings," p. 1438; Wolff, "Love Canal Revisited," p. 1464.
56. Heath et al., "Cytogenetic Findings," pp. 1437–1440.
57. Michael Bender, "Cytogenetic Patterns in Persons Living near Love Canal—New York," *Morbidity and Mortality Weekly Report,* 32 (May 27, 1983): 261–262; Heath et al., "Cytogenetic Findings."
58. Heath et al., "Cytogenetic Findings," p. 1440.
59. Ibid.
60. New York State Department of Heath [NYSDOH], "In the Matter of the Love Canal Chemical Waste Landfill Site Located in the City of Niagara Falls, Niagara County," 1979, Collected Documents, section 28.
61. Levine, *Love Canal,* p. 42.
62. NYSDOH memorandum, submitted to the House Committee on Interstate and Foreign Commerce, serial no. 96-48, Mar. 21, 1979, vol. 22, p. 120.
63. Nicholas J. Vianna et al., "Adverse Pregnancy Outcomes in the Love Canal Area," provisional report, Apr. 1980, New York State Department of Health microfiche no. 80-04453.
64. Testimony of David Axelrod, New York State Health Commissioner, *Hearings on Hazardous Waste,* House Committee on Interstate and Foreign Commerce, Mar. 21, 1979, vol. 22, p. 288.
65. See, e.g., Epstein et al., *Hazardous Waste in America,* pp. 103–104; Lewis Regenstein, *America the Poisoned* (Washington, D.C.: Acropolis Books, 1982), pp. 138–140; and Brown, *Laying Waste,* pp. 46–47.
66. Julian B. Andelman and Dwight H. Underhill, eds., *Health Effects from Hazardous Waste Sites* (Chelsea, Mich.: Lewis Publishers, 1987), p. 12.
67. Levine, *Love Canal,* p. 89.
68. Constance Holden, "Love Canal Residents under Stress," *Science,* 209 (June 13, 1980): 1243.
69. Grisham, *Health Aspects,* p. 139.
70. Vianna et al., "Adverse Pregnancy Outcomes," p. 7–8.
71. Ibid.
72. Ibid., p. 11.
73. Ibid., pp. 10–11.
74. Dorothy Warburton and F. C. Fraser, "Spontaneous Abortion Risks in Man: Data from Reproductive Histories Collected in a Medical Genetics Unit," *American Journal of Human Genetics,* 16, no. 1 (1964): 1–25.
75. Vianna et al., "Adverse Pregnancy Outcomes," pp. 9–10.
76. Ember, "Uncertain Science," p. 24.
77. Vianna et al., "Adverse Pregnancy Outcomes," p. 10.
78. Levine, *Love Canal,* p. 130.
79. Vianna et al., "Adverse Pregnancy Outcomes," pp. 9–10.
80. Grisham, *Health Aspects,* p. 311.
81. Vianna et al., "Adverse Pregnancy Outcomes," p. 10.
82. Clark Heath, "The Effects of Hazardous Waste on Public Health," in Syamal K. Majumdar and E. Willard Miller, eds., *Hazardous and Toxic Wastes: Technology, Management, and Health Effects* (Easton: Pennsylvania Academy of Science, 1984), p. 384.

83. Vianna et al., "Adverse Pregnancy Outcomes," pp. 19–20.
84. Levine, *Love Canal*, p. 112.
85. Grisham, *Health Aspects*, p. 311.
86. Thomas, "Report of the Governor's Panel."
87. Vianna et al., "Adverse Pregnancy Outcomes," p. 6.
88. Levine, *Love Canal*, p. 89.
89. Gibbs, *Love Canal*, p. 66.
90. Ibid., p. 67.
91. Levine, *Love Canal*, p. 91.
92. Gibbs, *Love Canal*, pp. 66–67.
93. Beverly Paigen, "Health Problems in a Community Living near a Hazardous Waste Site Known as Love Canal," paper presented at the American Public Health Association Meeting, Detroit, Oct. 21, 1980, p. 2.
94. Ibid.
95. Zena Stein et al., "Epidemiological Considerations in Assessing Health Effects," in *Assessment of Health Effects*, pp. 137.
96. David P. Rall et al., "Report of Meetings between Scientists from HEW and EPA with Dr. Beverly Paigen and New York State Department of Health Scientists," submitted to the House Committee on Interstate and Foreign Commerce, 96th Congress, May 22, 1980, vol. 28, p. 55.
97. Epstein et al., *Hazardous Waste*, p. 117; Irwin Bross, "Muddying the Waters at Niagara," *New Scientist* (Dec. 11, 1980): 728–729.
98. George Lumb, "Health Effects of Hazardous Wastes," in Syamal K. Majumdar and E. Willard Miller, eds., *Hazardous and Toxic Wastes: Technology, Management, and Health Effects*, (Easton: Pennsylvania Academy of Science, 1984), pp. 391–392.
99. Rall et al., "Report of Meetings between Scientists," p. 56.
100. Thomas, "Report of the Governor's Panel," p. 20.
101. Kathy Trost, "Love Canal: Despite Warning Signs, Some Residents Stay On," *Washington Post*, June 23, 1980, pp. A1, A4.
102. Brown, *Laying Waste*, p. 329.
103. Kathy Trost, "Long-Buried Poison Haunts Neighborhood: Chemicals Blight Dream of Good Life," *Detroit Free Press*, Aug. 7, 1978, p. 16A; Gibbs, *Love Canal*, p. 40.
104. Dwight Janerich et al., "Cancer Incidence in the Love Canal Area," *Science*, 212 (June 19, 1981): 1404.
105. Samuel Epstein, *The Politics of Cancer* (Garden City, N.Y.: Anchor Press, 1979), p. xi.
106. Elizabeth Whelan, *Preventing Cancer* (New York: W.W. Norton, 1980), p. 38.
107. Larry Agran, *The Cancer Connection: And What We Can Do about It* (Boston: Houghton Mifflin, 1977), p. xvi.
108. Nader Quoted in Edith Efron, *The Apocalyptics: Cancer and the Big Lie* (New York: Simon and Schuster, 1984), p. 22.
109. Epstein, *Politics of Cancer*, pp. 5, 11.
110. Ibid., p. 489.
111. Efron, *The Apocalyptics*, p. 69.
112. Janerich et al., "Cancer Incidence in the Love Canal Area," p. 1404. The New York Cancer Registry began collecting reports on cases in 1940, when it initially

recorded date of report as reference data. The investigators used data from 1966 on, when the registry changed reference data to include date of diagnosis (rather than date of report).

113. Ibid., table 1.
114. Ibid., p. 1406, table 2.
115. Ibid., p. 1406.
116. Clark Heath, "The Effects of Hazardous Waste," p. 384.
117. Andelman and Underhill, *Health Effects,* p. 275.
118. Epstein et al., *Hazardous Waste,* p. 275.
119. Ibid., pp. 117–118.
120. Aug. 7, 1978, p. 16A.
121. Brown, *Laying Waste,* pp. 18–19.
122. Brown, "Toxic Ghost Town," p. 28.
123. Josh Barbanel, "For One Love Canal Family, the Road to Despair," *New York Times,* May 19, 1980, p. A1; Trost, "Despite Warning Signs, Some Residents Stay On," p. A1.
124. CBS News and NBC News, from *Television News Index Abstracts,* May 17, 1980.
125. Martin Linsky, *Impact: How the Press Affects Federal Policymaking* (New York: Norton, 1986), p. 80.
126. John Elliot, "Lessons from Love Canal," *Journal of the American Medical Association,* 240 (Nov. 3, 1978): 2033.
127. "The Chemicals around Us," *Newsweek,* Aug. 21, 1978, p. 25.
128. Linsky, *Impact,* p. 79.
129. Nader and Brownstein, "Beyond the Love Canal," p. 28.
130. Zuesse, "The Truth Seeps Out," p. 30.
131. Irwin Molotsky, "Damage to Chromosomes," p. A1.
132. Ibid.
133. Levine, *Love Canal,* p. 4.
134. "The Toxicity Connection," *Time,* Sep. 22, 1980, p. 63.
135. Charles Kaiser, "Hell Holes," *Newsweek,* Aug. 18, 1980, p. 79.
136. Richard Cohen, book review, *New Republic,* May 24, 1980, pp. 36–37.
137. Ellen Williams, book review, *Technology Review,* 85 (Oct. 1982): 85–86.
138. Harvey Molotch, "Ethnography of a Disaster," *Science,* 216 (June 25, 1982): 1401–1402.
139. Martha Folkes and Patricia Miller, "Love Canal: The Social Construction of Disaster," final report for the Federal Emergency Management Agency, Oct. 1982, p. 64.
140. Lois Gibbs in *Hearings on Hazardous Waste,* House Committee on Interstate and Foreign Commerce, March 21, 1979, vol. 22, p. 36.
141. "Awards for Global Environment Crusaders," *Science,* 248 (Apr. 27, 1990): 447.
142. Donald McNeil, "Upstate Waste Site May Endanger Lives," *New York Times,* Aug. 6, 1978, p. A1.
143. Levine, *Love Canal,* p. 18.
144. John LaFalce in *Love Canal: Health Studies and Relocation,* House Committee on Interstate and Foreign Commerce, 96th Congress, May 22, 1980, vol. 28, pp. 4–5.

145. See ibid., pp. 55–57, for a full report.

146. Levine, *Love Canal*, pp. 156–157.

147. Danzo, "The Big Sleazy," p. 14.

148. U.S. Environmental Protection Agency [EPA], *Environmental Monitoring at Love Canal*, report no. EPA-600/4-82-030, May 1982, vol. 1.

149. Ibid., p. iv; "Design of the EPA Monitoring Study," appendix B in *Habitability of the Love Canal Area: An Analysis of the Technical Basis for the Decision on Habitability of the Emergency Declaration Area—A Technical Memorandum*, report no. OTA-TM-M-13 (Washington, D.C.: U.S. Congress, Office of Technology Assessment, June 1983).

150. EPA, *Environmental Monitoring at Love Canal*, p. iv.

151. *Hearings of the House Energy and Commerce Committee*, 97th Congress, Aug. 9, 1982, no. 197, p. 89, table 2, "Power Calculations for Statistical Analysis of the Love Canal Data."

152. Ibid., pp. 60, 51.

153. *DHHS Evaluation of Results of Environmental Chemical Testing by EPA in the Vicinity of Love Canal—Implications for Human Health—Further Considerations concerning Habitability* (Washington, D.C.: Department of Health and Human Services, July 13, 1982).

154. *Habitability of the Love Canal Area*, p. 3.

155. David Axelrod, "Habitability Decision: Report on Habitability, Love Canal Emergency Declaration Area," NYSDOH report, Albany, N.Y., Sep. 27, 1988, p. 5.

156. Ibid., p. 19.

157. Ibid., pp. 30, 29.

158. Verhovek, "Land Rush on a Burial Ground," p. B2.

159. Brown, "Toxic Ghost Town," p. 26.

160. Peter Huber's lively title for his book about junky science in the courtroom illustrates the point beautifully. Every time an allegedly expert witness made a ludicrous statement, controverted by massive evidence, he deflected the critique by claiming that Galileo was similarly criticized. (See Peter W. Huber, *Galileo's Revenge: Junk Science in the Courtroom* (New York: Basic Books, 1991).

5. *Superfund's Abandoned Hazardous Waste Sites*

1. U.S. Office of Technology Assessment [OTA], *Coming Clean: Superfund's Problems Can Be Solved* (Washington, D.C.: OTA, 1989), p. 11. Frank Viviano, "Superfund Costs May Top S & L Bailout," *San Francisco Chronicle*, May 29, 1991, p. A1.

2. See, e.g., Mark Landy, "Passing Superfund," in Mark Landy et al., *The Environmental Protection Agency* (New York: Oxford University Press, 1990), p. 145; Viviano, "Superfund Costs May Top S & L Bailout," p. A4.

3. Frank Viviano, "Toxic Cleanup a Bonanza for the Legal Profession," *San Francisco Chronicle*, May 29, 1991, p. A4. On local governments as responsible parties, see Cleansites, *Main Street Meets Superfund: Local Government Involvement at Superfund Hazardous Waste Sites* (Alexandria, VA.: Cleansites, 1992).

4. On insurers' involvement with site cleanups, see Jan Paul Acton and Lloyd S.

Dixon, *Superfund and Transaction Costs: The Experiences of Insurers and Very Large Industrial Firms* (Santa Monica, Calif.: RAND), 1992.

5. U.S. General Accounting Office [GAO], *EPA Has Not Corrected Long Standing Contract Management Programs* (Washington, D.C.: GAO, 1991), p. 4.

6. U.S. Office of Technology Assessment [OTA], *Coming Clean: Superfund's Problems Can Be Solved* (Washington, D.C.: OTA, 1989). See also Acton and Dixon, *Superfund and Transaction Costs.*

7. President Jimmy Carter, letter to Walter F. Mondale, president of the Senate, June 13, 1979. Addendum to Senate Committee on Environment and Public Works, *Environmental Emergency Response Act,* report no. 96-848 (July 11, 1980), pp. 104–105.

8. Senate Committee on Environment and Public Works, *Environmental Emergency Response Act,* p. 2.

9. Mark Landy, "Cleaning up Superfund," *Public Interest,* 85 (Fall 1986): 58–71.

10. Jack Lewis, "Superfund, RCRA, and UST: The Cleanup Threesome," *EPA Journal,* 17, no. 3 (1991): 7–14.

11. Senator Robert T. Stafford, "Why Superfund Was Needed," *EPA Journal,* 7 (June 1981): 9–10.

12. U.S. Environmental Protection Agency [EPA], *Research Summary: Controlling Hazardous Wastes* (Washington, D.C.: EPA Office of Research and Development, 1980), p. 12.

13. Ibid., p. 13.

14. Ibid., pp. 14–15.

15. Gary M. Marsh and Richard J. Caplan, "Evaluating Health Effects of Exposure at Hazardous Waste Sites: A Review of the State-of-the-Art, with Recommendations for Future Research," in Julian B. Andelman and Dwight W. Underhill, *Health Effects from Hazardous Waste Sites* (Chelsea, Mich.: Lewis Publishers, 1987), p. 6.

16. R. B. Pojasek, "Disposal of Hazardous Chemical Wastes," *Environmental Science and Technology,* 13 (1979): 810–814.

17. Hart Associates (for the EPA), *Assessment of Hazardous Waste Mismanagement Damage Case Histories* (Washington, D.C.: EPA Office of Solid Waste and Emergency Response, 1982).

18. These limits were doubled by Congress in 1986 to twelve months and $2 million.

19. EPA, *Research Summary,* p. 5; GAO, *Cleaning up Hazardous Wastes: A Review of Superfund Reauthorization Issues* (Washington, D.C.: GAO, 1985).

20. "Buried Time Bombs," *Economist,* Aug. 2, 1980, p. 27.

21. "Superfund Contaminated by Partisan Politics," *Congressional Quarterly Weekly Report,* Mar. 17, 1984, p. 615.

22. EPA, *Research Summary;* Hart Associates, *Assessment of Hazardous Wastes.*

23. EPA, *Research Summary.*

24. Stafford, "Why Superfund Was Needed," p. 9.

25. U.S. Surgeon General, *Health Effects from Toxic Pollution* (Washington, D.C.: Department of Health and Human Services, 1980), p. 19.

26. EPA, *Research Summary.* "Lined" ponds have at their bottoms a layer of material that prevents or slows the leaching of waste chemicals through to underlying soil and groundwater.

27. Ibid.
28. Senate Committee on Environmental and Public Works, *Environmental Emergency Response Act,* p. 4; Stafford, "Why Superfund Was Needed," p. 10.
29. See again Senate Committee on Environmental and Public Works, *Environmental Emergency Response Act,* p. 4, and Stafford, "Why Superfund Was Needed," p. 10.
30. EPA, *Research Summary,* p. 1.
31. See, e.g., Philip H. Abelson, "Waste Management," *Science,* 220 (June 3, 1983): 1003.
32. See, e.g., Landy, "Passing Superfund," pp. 132–167.
33. John P. Dwyer, "The Pathology of Symbolic Legislation," *Ecology Law Quarterly,* 17, no. 2 (1990): 233–316.
34. Surgeon General, *Health Effects from Toxic Pollution,* p. iii.
35. Senate Committee on Environmental and Public Works, *Environmental Emergency Response Act,* p. 3.
36. Congressional Research Service, *A Brief Review of Selected Environmental Contamination Incidents with a Potential for Health Effects* (Washington, D.C.: Congressional Research Service, 1980), p. 153.
37. Ibid., p. 157.
38. Ibid., p. 160.
39. Landy, "Cleaning Up Superfund," p. 61.
40. Stafford, "Why Superfund Was Needed," p. 10.
41. Congressional Research Service, *A Brief Review,* p. 160.
42. Joe W. Grisham, ed., *Health Aspects of the Disposal of Waste Chemicals: A Report of the Executive Scientific Panel* (New York: Pergamon Press, 1986), p. v.
43. The sites were: McColl, Calif.; New Bedford, Mass.,: Silresim, Mass.; Woburn, Mass.; McKim, Maine; Montague, Mich.; Perham, Minn.; GEMs Landfill, N.J.; Jackson Township Landfill, N.J.; Krysowaty Farm, N.J.; Pamona Oaks, N.J.; Price Landfill, N.J.; Reich's Farm, N.J.; Rutherford, N.J.; Brookfield Landfill, N.Y.; Hooker Hyde Park, N.Y.; Love Canal, N.Y.; ABM Wade, Penn.; Drake Chemical, Penn.; Stanley Kesser, Penn.; Hardeman County, Tenn. Ibid., pp. 285–293.
44. Grisham, *Health Aspects,* pp. 242–245.
45. Ibid., p. 344.
46. Ibid., p. 266.
47. Ibid., p. 254.
48. Ibid., p. 246.
49. Marsh and Caplan, "Evaluating Health Effects of Exposure," p. 65.
50. Ibid., p. 67; Grisham, ed., *Health Aspects,* p. 256.
51. Grisham, *Health Aspects,* p. 202.
52. Ibid., p. 254; Marsh and Caplan, "Evaluating Health Effects of Exposure," p. 67.
53. Grisham, *Health Aspects,* p. 244.
54. Ibid., p. 256; Marsh and Caplan, "Evaluating Health Effects of Exposure," p. 67; Michael McDowell, *The Identification of Man-Made Environmental Hazards to Health* (Basingstoke: Macmillan, 1987), p. 238.
55. Grisham, *Health Aspects,* pp. 134, 253.

56. Grisham, *Health Aspects,* p. 256; McDowell, *Identification,* p. 239.
57. Grisham, *Health Aspects,* p. 141.
58. Ibid., p. 245; Marsh and Caplan, "Evaluating Health Effects of Exposure," p. 67; McDowell, *Identification,* p. 116.
59. Grisham, *Health Aspects,* p. 245.
60. McDowell, *Identification,* p. 117.
61. Grisham, *Health Aspects,* p. 134.
62. Ibid., p. 344.
63. Ibid., p. 346.
64. Richard J. Hickey, "Scientific Malpractice in Epidemiology," *Chemtech,* May 1989, pp. 269–271.
65. Marvin A. Schneiderman, "Expectation and Limitation of Human Studies and Risk Assessment," in Julian B. Andelman and Dwight W. Underhill, *Health Effects from Hazardous Waste Sites* (Chelsea, Mich.: Lewis Publishers, 1987), p. 160.
66. National Research Council, *Environmental Epidemiology: Public Health and Hazardous Wastes* (Washington, D.C.: National Academy Press, 1991).
67. Grisham, *Health Aspects,* p. 128. McDowell likewise maintains that "although epidemiological evidence in many cases may be a relatively weak tool in identifying whether a specific exposure is a hazard to health, it may be easier, after the accumulation of suitable evidence, to draw conclusions along the lines that if any risk does exist, it must be less than a particular level" (*Identification,* p. 118).
68. EPA, *Fiscal Year 1989 Implementation Report* (Washington, D.C.: EPA Office of Emergency and Remedial Response, 1990), pp. 24–28.
69. OTA, *Coming Clean,* p. 11.
70. EPA, "Hazard Ranking System (HRS) for Uncontrolled Hazardous Substances Releases," *Federal Register,* 53 (Dec. 23, 1988): 51963.
71. Ibid., p. 51965.
72. Ibid., p. 51966.
73. Ibid., pp. 51962–51963; see also GAO, *Cleaning Up Hazardous Wastes;* OTA, *Superfund Strategy* (Washington, D.C.: OTA, 1985).
74. EPA, *Superfund Exposure Assessment Manual* (Washington, D.C.: EPA Office of Emergency and Remedial Response, 1988); EPA, *Superfund Public Health Evaluation Manual* (Washington, D.C.: EPA Office of Emergency and Remedial Response, 1986).
75. EPA, *Public Health Evaluation Manual,* p. 14.
76. EPA, *Exposure Assessment Manual,* p. 98.
77. Ibid., pp. 95, 96. On the general problem of generating solid data at waste sites, see also Gary L. McKown et al., "Effects of Uncertainties of Data Collection on Risk Assessment," in *Management of Uncontrolled Hazardous Waste Sites* (Silver Spring, Md.: Hazardous Materials Control Research Institute, 1984), pp. 283–286.
78. EPA, *Exposure Assessment Manual,* p. 96.
79. E.g., ibid., pp. 2–3.
80. Ibid., p. 97
81. Ibid., p. 3.
82. Ibid., p. 98.

83. Ibid., p. 95.
84. Ibid., pp. 4, 98.
85. Leo Levenson, "EPA Cleanup Efforts Have Wrong Targets," *Sacramento Bee,* June 4, 1991, p. B5.
86. E.g., Daniel E. Koshland, Jr., "Toxic Chemicals and Toxic Laws," *Science,* 253 (Aug. 30, 1991): 949; OTA, *Coming Clean,* p. 3.
87. James Wilson, on Elizabeth Anderson et al., "Risk Assessment Issues Associated with Cleaning Up Inactive Hazardous Waste Sites" in Howard Kunreuther and Rajeev Gowa, eds., *Integrating Insurance and Risk Management for Hazardous Wastes* (Norwell, Mass.: Kluwar Academic Publishers, 1990), p. 37.
88. EPA, *Public Health Evaluation Manual,* pp. 42–60.
89. Ibid., pp. 61–72.
90. Ibid., p. 61.
91. Ibid.
92. Chemicals most commonly found at Superfund sites include organic volatiles such as benzene, chloroform, methylene chloride, and toluene; organic semivolatiles such as bis (2-tehylhexyl) phthalate and phenol; and metals such as aluminum, chromium, cadmium, zinc, and arsenic. William P. Eckel et al., "Distribution and Concentration of Chemicals and Toxic Materials Found at Hazardous Waste Dump Sites," in *Management of Uncontrolled Hazardous Waste Sites,* pp. 250–257.
93. "Low-dose" here denotes concentrations very much lower than those that may cause acute effects; in environmental media they are typically measured in parts per million or parts per billion. E.g., Grisham, *Health Aspects,* p. 3.
94. EPA, *Public Health Evaluation Manual,* p. 91.
95. Ibid., p. 163.
96. Ibid., pp. 137–140.
97. Grisham, *Health Aspects,* pp. 195–230.
98. Ibid., p. 201.
99. On the deep uncertainties involved in extrapolations of this sort, see e.g. ibid., pp. 65–76; Don G. Scroggin, "The Interaction of Science, Policy, and the Law in Agency Use of Risk Assessments for the Regulation of Carcinogens," *Hazardous Waste,* 1, no. 3 (1984): 363–375.
100. Scroggin, "Interaction," p. 367.
101. Michael Gough, "How Much Cancer Can EPA Regulate Away?" *Risk Analysis,* 10, no. 1 (1990): 1–6.
102. EPA, *Public Health Evaluation Manual,* p. 74.
103. EPA profiles of noncarcinogenic toxics common to waste sites are also rooted principally in animal test evidence (ibid., p. 73) but do not employ as conservative an extrapolation technique. Chronic intake levels derived from the exposure assessment are expressed as a percentage of "acceptable intake for chronic exposure" (AIC) values, with the latter typically set at 1/100 of the "highest no observable adverse effect level" (NOAEL) in animals. Unlike carcinogen profiles, there is no presumption of "one-hit" toxicity. However, because stringent carcinogen standards will ultimately drive most cleanup designs, assessment of noncarcinogenic toxic risks will often be immaterial to remedial decisions. For more on animal test extrapolation models, see the next chapter.

104. E.g., OTA, *Coming Clean.*
105. 42 U.S.C. section 9601 (24).
106. EPA, "National Contingency Plan," *Code of Federal Regulations,* vol. 40, chap. 1 (July 1, 1990): 58.
107. Ibid., pp. 61–62.
108. GAO, *Cleaning Up Hazardous Wastes,* pp. 38–39.
109. Ibid., pp. 40–41; OTA, *Superfund Strategy,* p. 112–119.
110. GAO, *Cleaning Up Hazardous Wastes,* pp. 38–39; William N. Hedeman et al., "The Superfund Amendments and Reauthorization Act of 1986: Statutory Provisions and EPA Implementation," *Hazardous Waste and Hazardous Materials,* 4, no. 2 (1987): 193–221.
111. The National Contingency Plan dictates that several remedial options be developed, with corresponding residual carcinogenic risks ranging from 10^{-4} through 10^{-7}, but that the 10^{-6} risk level "shall be used as the point of departure for determining remediation goals." In practice most remedial designs selected will correspond to the 10^{-6} level (see, e.g., EPA, *Fiscal Year 1989 Implementation Report*).
112. Leo Levenson, "Chemical Risks and Public Risk Perceptions," unpublished manuscript, University of California, Berkeley, 1990.
113. David W. Gaylor, "Preliminary Estimates of the Virtually Safe Dose for Tumors Obtained from the Maximum Tolerated Dose," *Regulatory Toxicology and Pharmacology,* 9 (1989): 101–108.
114. OTA, *Coming Clean.*
115. EPA, "Corrective Action for Solid Waste Management Units at Hazardous Waste Management Facilities," *Federal Register,* 55, no. 145 (July 27, 1990): 30798–30853.
116. On treatment technologies, see EPA, *Fiscal Year 1989 Implementation Report.*
117. EPA, "National Contingency Plan," p. 58.
118. Hedeman et al., "The Superfund Amendments."
119. GAO, *Hazardous Waste: EPA's Consideration of Permanent Cleanup Remedies* (Washington, D.C.: GAO, 1986), p. 9.
120. EPA, *Fiscal Year 1989 Implementation Report,* p. 12. Doubts have recently been raised about whether treatment technologies can accomplish what is being asked of them, at least in the case of groundwater contamination. In a provocative article, Curtis Travis and Carolyn Doty of Oak Ridge National Laboratory contend: "No matter how much money the federal government is willing to spend, at present contaminated aquifers cannot be restored to a condition compatible with [EPA] standards." Leading groundwater scientists and modelers have recently argued that even under ideal conditions it would take 100–200 years of pumping and treatment to reduce aquifer contamination by a factor of 100; EPA cleanup goals commonly call for even greater reductions. In less ideal circumstances, such as those involving relatively insoluble substances, prospects are even bleaker. Curtis C. Travis and Carolyn B. Doty, "Can Contaminated Aquifers at Superfund Sites Be Remediated?" *Environmental Science and Technology,* 24, no. 10 (1990): 1465.
121. Sources for this site: University of Tennessee, Hazardous Waste Remediation Project, *The Superfund Process: Site-Level Experience* (Knoxville: Waste Man-

agement Research and Education Institute, 1991); EPA, *Superfund Record of Decision: Old Springfield Landfill, VT* (Boston: EPA Region 1, 1988).

122. Safe Drinking Water Act MCLs are set by the EPA at or near a 1-in-1 million lifetime cancer risk level (for chronic intake) and are rooted principally in animal test data on MCLs; see, e.g., Edward J. Calabrese et al., *Safe Drinking Water Act: Amendments, Regulations, and Standards* (Chelsea, Mich.: Lewis Publishers, 1989).

123. EPA, *Record of Decision: Old Springfield*, responsiveness summary, pp. 38–40.

124. Ibid.

125. Ibid., p. 1.

126. Source for this site: EPA, *Superfund Record of Decision: Reich Farm, NJ.* (New York: EPA Region 2, 1988).

127. Ibid., p. 8.

128. EPA, *Record of Decision: Reich's Farm*, comment and response summary, p. 10.

129. Ibid., p. 8.

130. Ibid., p. 11.

131. Ibid., comment and response summary, letter from Union Carbide.

132. Sources for this site: University of Tennessee, *The Superfund Process;* EPA, *Superfund Record of Decision: Palmerton Zinc* (Philadelphia: EPA Region 3, 1988).

133. University of Tennessee, *The Superfund Process*, p. 5.

134. Ibid., p. 7.

135. EPA, *Record of Decision: Palmerton Zinc*, responsiveness summary.

136. Source for this site: EPA, *Superfund Record of Decision: Selma Pressure Treating Company.* (San Francisco: EPA Region 9, 1988).

137. Ibid., p. 22.

138. Ibid., p. 19.

139. Sources for this site: University of Tennessee, *The Superfund Process;* EPA, *Superfund Record of Decision: MOTCO, TX* (Dallas: EPA Region 6, 1989).

140. Koshland, "Toxic Chemicals and Toxic Laws," p. 949.

141. Thomas P. Grumbly, "Superfund: Candidly Speaking," *EPA Journal*, 17, no. 3 (1991): 21.

142. E.g., ibid.

6, I. Is Asbestos in Schoolrooms Hazardous to Students' Health?

1. Environmental Protection Agency [EPA], *Support Document for Proposed Rule on Friable Asbestos-Containing Materials in School Buildings—Health Effects and Magnitude of Exposure*, (Washington, D.C.: EPA Office of Pesticides and Toxic Substances, 1980), p. 92.

2. EPA, *Support Document for Asbestos Containing Materials in Schools—Economic Impact Analysis of Identification and Notification Proposed Rule, Section 6 Toxic Substances Control Act*, (Washington, D.C.: EPA Office of Pesticides and Toxic Substances, 1980), p. 11.

3. H. Schreier, *Studies in Environmental Science, 37: Asbestos in the Natural Environment* (New York: Elsevier Science Publishing, 1989), p. 12.

4. Malcolm Ross, "A Survey of Asbestos-Related Diseases in Trades and Mining

Occupations and in Factory and Mining Communities as a Means of Predicting Health Risks of Nonoccupational Exposure to Fibrous Minerals," in Benjamin Levadie, ed., *Definitions for Asbestos and Other Health-Related Silicates,* special technical publication no. 834 (Philadelphia: American Society for Testing and Materials, 1984), p. 54.

5. I. J. Selikoff and D. H. Lee, *Asbestos and Disease* (New York: Academic Press, 1978), p. 3.
6. Ross, "Survey of Asbestos-Related Diseases," pp. 12, 54.
7. R. W. Goldberg, *Occupational Diseases* (New York: Columbia University Press, 1931), p. 13.
8. Paul Gross and D. C. Braun, *Toxic and Biomedical Effects of Fibers* (Park Ridge, N.J.: Noyes Publications, 1984), p. 25.
9. Marguerite Villeco, "Technology: Spray Fireproofing Faces Control or Ban as Research Links Asbestos to Cancer," *Architectural Forum,* 133 (1970): 52; Selikoff and Lee, *Asbestos and Disease,* p. 22.
10. W. E. Cooke, "Pulmonary Asbestosis," *British Journal of Industrial Medicine,* 11 (1927): 1024.
11. E. R. A. Merewether and C. V. Price, "Report on the Effects of Asbestos Dust on the Lungs and Dust Suppression in the Asbestos Industry" (London: HM Stationery Office, 1930).
12. Villeco, "Technology," p. 51.
13. Kenneth M. Lynch and W. Atmar Smilth, "Pulmonary Asbestosis, III. Carcinoma of the Lung in Asbesto-Silicosis," *American Journal of Cancer,* 24 (May 1935): 56.
14. Richard Doll, "Mortality from Lung Cancer in Asbestos Workers," *British Journal of Industrial Medicine,* 12 (Apr. 1955): 81–86.
15. Ibid., p. 81.
16. J. C. Wagner, C. A. Sleggs, and P. Marchand, "Diffuse Pleural Mesothelioma and Asbestos Exposure in the North-West Cape Province," *British Journal of Industrial Medicine,* 17 (1960): 260–271.
17. Gross and Braun, *Toxic and Biomedical Effects,* pp. 87, 88.
18. I. J. Selikoff, C. E. Hammond, and J. Churg, "Asbestos Exposure, Smoking, and Neoplasia," *Journal of the American Medical Association,* 204, no. 2 (Apr. 8, 1968): 106–110.
19. Ross, "Survey of Asbestos-Related Diseases," p. 61.
20. I. J. Selikoff and E. C. Hammond, "Editorial: Asbestos and Smoking," *Journal of the American Medical Association,* 242, no. 5 (Aug. 3, 1979): 458–459.
21. Dan Levin, "Asbestos in Schools: Walls and Halls of Trouble," *American School Board Journal,* 165, no. 11 (Nov. 1978): 29–32.
22. Ibid.
23. EPA, *Friable Asbestos-Containing Materials,* p. 12.
24. EPA, *Asbestos-Containing Materials in School Buildings: A Guidance Document,* (Washington, D.C.: EPA Office of Pesticides and Toxic Substances, 1979), pp. 3, 4, 7.
25. S. L. Campbell, "Asbestos in Schools—How Much Hazard?" *American Council on Science and Health,* brochure no. 6 (1985): 4.

26. EPA, *Airborne Asbestos Health Assessment Update*, (Washington, D.C.: EPA Office of Health and Environmental Assessment, 1984), p. 4.
27. U.S. Attorney General, *The Attorney General's Asbestos Liability Report to Congress* (Washington, D.C.: U.S. Government Printing Office, 1981), p. 53.
28. W. J. Nicholsen, A. N. Rohl, and E. F. Ferrand, "Asbestos Air Pollution in New York City," *Proceedings of the Second International Clean Air Congress,* Washington, D.C., December 1970 (New York: Academic Press, 1970), pp. 136–139.
29. *Attorney General's Asbestos Liability Report,* p. 53.
30. Ibid., p. 59.
31. Testimony of Dr. I. J. Selikoff, *Hearings on the Asbestos School Hazard and Control Act,* Senate Subcommittee on Education, Arts, and Human Resources of the Committee on Labor and Human Resources, 96th Congress, Mar. 17, 1980, vol. 2, p. 59.
32. National Cancer Institute and National Institute of Environmental Health Sciences, Draft Summary of "Estimates of the Fraction of Cancer Incidence in the United States Attributable to Occupational Factors," mimeographed document issued as a press release, with the following statement on its first page: "Not for use before 3:30 P.M. EDT, September 11, 1978," pp. 1–2, cited in Edith Efron, *The Apocalyptics: Cancer and the Big Lie* (New York: Simon and Schuster, 1984), p. 437.
33. Efron, *Apocalyptics,* p. 443.
34. Ibid., pp. 437–440.
35. Richard Doll and Richard Peto, "The Causes of Cancer: Quantitative Estimates of Avoidable Risks of Cancer in the United States Today," *Journal of the National Cancer Institute,* 66 (1981): 1191–1308.
36. Rule-making procedures are based on authority contained in section 6 of the 1976 Toxic Substances Control Act, by which the EPA can control chemical substances that present an unreasonable public health risk.
37. EPA, *Economic Impact Analysis of Proposed Identification and Notification Rule on Friable Asbestos-Containing Materials in Schools,* (Washington. D.C.: EPA Office of Pesticides and Toxic Substances, 1980), p. 2.
38. Ibid., p. 1.
39. Ibid., p. 59.
40. EPA, "Guidance Document."
41. EPA, *Friable Asbestos-Containing Materials,* pp. 7, 83.
42. Levin, "Asbestos in Schools: Walls and Halls of Trouble," p. 31–32.
43. In December 1980 three parents filed suit against a Philadelphia school district asking the court to order the district to (1) establish a forty-five-year $20 million trust fund to pay off any future medical claims filed by students who developed health problems due to asbestos exposure in school; (2) pay $50,000 to each student; (3) pay $10,000 to the parents of each child; and (4) pay $10,000,000 in punitive damages to children and parents. *Steigelman v. The School District of Philadelphia,* CA 8-4729, in *Asbestos Litigation Reporter,* Dec. 12, 1980, p. 2651.
44. *Attorney General's Asbestos Liability Report,* p. vii.
45. Gross and Braun, *Toxic and Biomedical Effects,* p. 4.

46. Ibid., p. 35.
47. Ibid., p. 7.
48. Richard Doll and Julian Peto, *Effects on Health of Exposure to Asbestos,* (London: Health and Safety Commission, HM Stationery Office, 1985), p. 14.
49. B. T. Mossman et al., "Asbestos: Scientific Development and Implications for Public Policy," *Science,* 247 (Jan. 19, 1990): 294–295.
50. Ibid., p. 296.
51. J. C. Wagner et al., "The Effects of the Inhalation of Asbestos in Rats," *British Journal of Cancer,* 29 (1974): 252–269.
52. Doll and Peto, *Effects on Health,* p. 14.
53. Malcolm Ross, "Suspect Minerals and the Survival of the U.S. Mining Industry: A Commentary," paper presented at the U.S. Geological Survey and U.S. Bureau of Mines Meeting on Mineral Commodities, Reston, Va., Jan. 21, 1987.
54. Marcel Cossette, "Defining Asbestos Particulates for Monitoring Purposes," in Benjamin Levadie, ed., *Definitions for Asbestos and Other Health- Related Silicates,* special technical publication no. 834 (Philadelphia: American Society for Testing and Minerals, 1984), pp. 11–12.
55. Ibid., p. 12.
56. Health Effects Institute—Asbestos Research, "Asbestos in Public and Commercial Buildings: A Literature Review and Synthesis of Current Knowledge," report by the institute, Cambridge, Mass., 1991, part 4, pp. 43–44.
57. William J. Nicholsen et al., "Asbestos Contamination in United States Schools from the Use of Asbestos Surfacing Materials," in I. J. Selikoff and E. C. Hammond, eds., *Health Hazards of Asbestos Exposure,* published in *Annals of the New York Academy of Sciences,* 330 (1979): 587–596. Mossman et al., "Asbestos: Scientific Developments," pp. 294–300.
58. Health Effects Institute—Asbestos Research, "Asbestos in Buildings," part 4, p. 81.
59. H. Weill and J. M. Hughes, "Asbestos as a Public Health Risk: Disease and Policy," *Annual Review of Public Health,* 7 (1986): 171–192.
60. EPA, *Friable Asbestos-Containing Materials,* p. 55.
61. Doll and Peto, *Effects on Health,* p. 2.
62. E. G. Hammond and I. J. Selikoff, "Relation of Cigarette Smoking to Risk of Death of Asbestos-Associated Disease among Insulation Workers," in J. C. Wagner, ed., *Biological Effects of Asbestos,* IARC Scientific Publication, 8 (1973): 312–316.
63. I. J. Selikoff, E. C. Hammond, and H. Seidman, "Mortality Experience of Insulation Workers in the U.S. and Canada, 1943–1976," *Annals of the New York Academy of Science,* 33 (1979): 91–116.
64. Ross, "Survey of Asbestos-Related Diseases," p. 80.
65. Ibid., p. 82.
66. Ibid.
67. Cited in ibid., pp. 82–84.
68. Selikoff and Lee, *Asbestos and Disease,* pp. 27–28.
69. Martin Rutstein, "Fibrous Minerals, Mining, and Disease," discussion paper presented at the annual meeting of the Geological Society of America, Nov. 1, 1988.
70. Gross and Braun, *Toxic and Biomedical Effects,* p. 16.

71. Cossette, "Defining Asbestos," p. 12.
72. Michael Fumento, "The Asbestos Rip-Off," *American Spectator,* 22 (Oct. 1989): 25.
73. Health Effects Institute–Asbestos Research, "Asbestos in Buildings," part 5, p. 33.
74. Fumento, "Asbestos Rip-Off," p. 25.
75. Health Effects Institute–Asbestos Research, "Asbestos in Buildings," part 5, p. 53.
76. William K. Reilly, "Asbestos, Sound Science, and Public Perceptions: Why We Need a New Approach to Risk," address to the American Enterprise Institute Environmental Policy Conference, Vista Hotel, Washington, D.C., June 12, 1990, p. 1.
77. Ibid., p. 3.
78. Ibid., pp. 4–5.
79. Ibid., p. 3.
80. William Booth, "Risk of One Type of Asbestos Discounted," *Washington Post,* Jan. 19, 1990, p. A8. See also the article in the *New York Times* by William K. Stevens, "Despite Asbestos Risk, Experts See No Cause for 'Fiber Phobia,' " Sep. 5, 1989, p. C4. This article, with the subtitle "Asbestos Danger Varies by Type," gives pictures of different types of asbestos.
81. Lawrence Mosher, "Lone Mineral Expert Tries to Calm Government Down about Asbestos," *National Journal,* 16, no. 41 (Oct. 13, 1984): 1928–1929.
82. Doll and Peto, *Effects on Health.*
83. Weil and Hughes, "Asbestos as a Public Health Risk," p. 35.
84. Cited in ibid.
85. Royal Commission Study, cited in Fumento, "Asbestos Rip-Off," pp. 21–26.
86. J. A. Mazoué, "I Saved My Family from Asbestos Contamination," *Good Housekeeping,* 204 (Apr. 1987): 108.
87. Brooke T. Mossman and J. Bernard Gee, "Asbestos-Related Disease," *New England Journal of Medicine,* 320, no. 26 (June 29, 1989): 1724, 1729. The best article on this subject for the lay reader is Donald N. Dewees, "Does the Danger from Asbestos in Buildings Warrant the Cost of Taking It Out?" *American Scientist,* 75 (May/June 1987): 285–288. He concludes, "Furthermore, removal shifts risks from building occupants and maintenance workers to removal workers. Yet despite the many variations, one conclusion seems reasonably clear: we should resist squandering our resources on crash programs of asbestos removal to reduce already insignificant risks lest we find ourselves unprepared to cope with more acute risks from other hazards or even from the programs themselves" (p. 288).
88. Mossman et al., "Asbestos: Scientific Developments," pp. 294–300.
89. Of course, the controversy is not over. For a representative sample of thrust and counterthrust, see the objections to the article by Mossman and her colleagues and their response in *Science,* 248 (May 18, 1990): 795–802.
90. Mossman et al., "Asbestos: Scientific Developments," p. 299 (emphasis added).
91. Robert C. McNally, letter, *Asbestos Issues,* 3, no. 4 (Apr. 1990): 22.
92. Mossman et al., "Asbestos: Scientific Developments," p. 300.
93. Ibid., p. 299.
94. See Malcolm Ross, U.S. Geological Survey, "Minerals and Health: The Asbestos Problem," in H. Wesley Peirce, ed., *Proceedings of the 21st Forum on the Geology of Industrial Minerals,* special paper no. 4, 1987, pp. 83–89.

95. See Aaron Wildavsky, *Searching for Safety* (New Brunswick, N.J.: Transaction Press, 1988); Ralph Keeney, "Mortality Risks Induced by Economic Expenditures," *Risk Analysis,* 10, no. 1 (1990): 147–159.

96. Michael J. Bennett, *The Asbestos Racket: An Environmental Parable* (Bellevue, Wash.: Free Enterprise Press, 1991), pp. 65–87.

97. Ralph D'Agostino, Jr., and Richard Wilson, "Asbestos: The Hazards, the Risk, and Public Policy," in Kenneth R. Foster, David E. Bernstein, and Peter W. Huber, eds., *Phantom Risk: Scientific Inference and the Law* (Cambridge, Mass.: MIT Press, 1993), p. 200.

98. Ibid., pp. 204–205.

99. Marvin Schneiderman quoted in Tom Reynolds, "Asbestos-Linked Cancer Rates up Less Than Predicted," *Journal of the National Cancer Institute,* 84 (Apr. 15, 1992): 560, 562.

100. "Asbestos Removal, Health Hazards, and the EPA," *Journal of the American Medical Association,* 266, no. 5 (Aug. 7, 1991): 696–697; National Research Council Committee on Nonoccupational Health Risks of Asbestiform Fibers, *Asbestiform Fibers: Nonoccupational Health Risks* (Washington, D.C.: National Academy Press, 1984).

101. Editorial, *New York Times,* Sep. 18, 1989, p. A18. Reynolds, "Asbestos-Linked Cancer Rates up Less Than Predicted," pp. 560–562. Readers who would like to understand the intersection of personal and scientific disagreement on asbestos should consult Richard Stone, "No Meeting of the Minds on Asbestos: The Debate on the Health Hazards of Asbestos Has Become So Polarized That Researchers from One Camp No Longer Go to the Other Camp's Meeting," *Science,* 254 (Nov. 15, 1991): 928–931.

6, II. Does Alar on Apples Cause Cancer in Children?

1. Quoted in Andrea Arnold, *Fear of Food* (Bellevue, Wash.: Free Enterprise Press, 1990), chap. 4.

2. EPA Special Review Technical Support Document for Daminozide, May 1989, Executive Summary.

3. Bradford H. Sewell and Robin M. Whyatt, principal authors, Natural Resources Defense Council [NRDC], *Intolerable Risk: Pesticides in Our Children's Food* (Washington, D.C.: NRDC, Feb. 27, 1989).

4. Senator Lieberman, "Apples and Pesticide," *Congressional Record,* vol. 135, no. 9 (Mar. 15, 1989): S2539.

5. "How a PR Firm Executed the Alar Scare," editorial excerpting memo by NRDC's hired public relations agent, David Fenton, *Wall Street Journal,* Oct. 3, 1989, p. A22.

6. Ibid., p. A22. Robert James Bidinotto, "The Great Apple Scare," *Readers Digest,* Oct. 1990, pp. 53–58; Kenneth Smith, "Alar, One Year Later: A Media Analysis of a Hypothetical Health Risk," American Council on Science and Health, special report, Mar. 1990, p. 5.

7. "How PR Firm Executed," p. A22.

8. "Health Official Rebukes Schools over Apple Bans," *New York Times,* Mar. 16, 1989, p. B10.

9. Bidinotto, "Great Apple Scare," p. 54.
10. "Board Returns Apples to New York Schools," *New York Times,* Mar. 21, 1989, p. B3; "Apples Pass Tests for Return to Los Angeles School Menu," *New York Times,* Mar. 18, 1989, p. 50.
11. Philip Shabecoff, "3 U.S. Agencies to Allay Public's Fears, Declare Apples Safe," *New York Times,* Mar. 17, 1989, p. A16. See also the joint publication by the USDA, FDA, and EPA, "Statement on Alar," Mar. 16, 1989.
12. Timothy Egan, "Apple Growers Bruised and Bitter after Alar Scare," *New York Times,* July 9, 1991, p. A1. Interviewed by Eliot Marshall for a news article in *Science,* 254 (Oct. 4, 1991): 20–22, Boyd Buxton of the USDA estimated the total losses to growers from the apple scare of 1989 at $120 million (box, p. 21). Losses would have been concentrated among small West Coast orchards and East Coast orchards specializing in red MacIntosh apples.
13. Quoted by Marshall in *Science,* 254 (Oct. 4, 1991), p. 21. See also Egan, "Apple Growers Bruised," and "After Scare, Suit by Apple Farmers," *New York Times,* Nov. 29, 1991, p. A22.
14. "Government Will Buy Apples Leftover from Scare on Alar," *New York Times,* July 8, 1989, p. A6.
15. Philip Shabecoff, "Apple Industry Says It Will End Use of Chemical," *New York Times,* May 16, 1989, p. A1.
16. "Bad Apples: Alar—Not Gone, Not Forgotten," *Consumer Reports,* May 1989, pp. 288–292.
17. Philip Shabecoff, "Apple Chemical Being Removed in U.S. Market," *New York Times,* June 3, 1989, p. A1.
18. Senator John Warner, "Introduction of Bills and Joint Resolutions," *Congressional Record,* vol. 135, no. 64 (1989): S5611.
19. Senate Subcommittee on Toxic Substances, Environmental Oversight, and Research and Development of the Committee on Environment and Public Works, *Government Regulation of Pesticides in Food: the Need for Administrative and Regulatory Reform,* 101st Congress, Oct. 1989, Senate Print 101-55, p. 411.
20. Philip Shabecoff, "EPA Proposing Quicker Action against Suspect Farm Chemicals," *New York Times,* July 20, 1989, p. A1.
21. Ibid.
22. See especially the report by Marshall in *Science,* 254 (Oct. 4, 1991): 20–22.
23. Gary M. Meunier, "The U.S. Environmental Protection Agency's Regulation of the Plant Growth Regulator Daminozide" (Ph.D. diss., University of California, Los Angeles, 1986), p. 8.
24. Joseph D. Rosen, "Much Ado about Alar," *Issues in Science and Technology,* 7, no. 1 (Fall 1990): 86.
25. Meunier, "Environmental Protection Agency's Regulation," p. 9.
26. J. Sandlin, private communication.
27. EPA Special Review Document for Daminozide, p. i, Executive Summary.
28. Ibid., p. III-4.
29. Kenneth Smith, "Alar Three Years Later," American Council on Science and Health special report, Feb. 1992.
30. Meunier, "Environmental Protection Agency's Regulation," pp. 10–13. See also Arnold, *Fear of Food,* esp. pp. 44–45.

31. EPA, Special Review Document for Daminozide, p. i.

32. NRDC, *Intolerable Risk*, p. 1.

33. Ibid., chap. 2, pp. 38–39, and appendix 3, "Methodology for Estimating Lifetime Cancer Risk from Preschooler Exposure to Carcinogenic Pesticides in Food."

34. See, however, the EPA's Position Document (PD-1) in *Federal Register*, 49 (July 18, 1984), where at p. 29138, UDMH is described as "a potent animal carcinogen."

35. NRDC, *Intolerable Risk*, pp. 2–3.

36. Ibid., p. 37.

37. Ibid.

38. Mary F. Argus and Cornelia Hoch-Ligeti, "Comparative Study of the Carcinogenic Activity of Nitrosamines," *Journal of the National Cancer Institute*, 27 (1961): 695–709; Margaret G. Kelly et al., "Comparative Carcinogenicity of N-Isopropyl-a-(2-Methylhydrazino)-p-Toluamide.Hcl (Procarbazine Hydrochloride), Its Degradation Products, Other Hydrazines, and Isonicotinic Acid Hydrazide," *Journal of the National Cancer Institute* 42 (1969): 337–344; F. J. C. Roe, G. A. Grant, and D. M. Millican, "Carcinogenicity of Hydrazine and 1,1-Dimethylhydrazine for Mouse Lung," *Nature*, 216, no. 5113 (Oct. 28, 1967): 375–376.

39. Bela Toth, "1,1-Dimethylhydrazine (Unsymmetrical) Carcinogenesis in Mice," *Journal of the National Cancer Institute* 50 (1973): 181–187; Bela Toth, "Induction of Tumors in Mice with the Herbicide Succinic Acid 2,2-Dimethylhydrazide," *Cancer Research*, 37 (1977): 3497–3500.

40. Bela Toth, "The Large Bowel Carcinogenic Effects of Hydrazines and Related Compounds Occurring in Nature and in the Environment," *Cancer*, 40 (1977): 2427–2431.

41. National Cancer Institute, "Bioassay of Daminozide for Possible Carcinogenicity," NCI Carcinogenesis Technical Report Series no. 83, DHEW publication no. NIH 78-1333, 1978.

42. Meunier, "Environmental Protection Agency's Regulation," pp. ix, 5.

43. The legal standard for keeping a pesticide on the market is that it can be used without "any unreasonable risk to man or the environment, taking into account the economic, social, and environmental costs and benefits of the use of the pesticide." Federal Insecticide, Fungicide, and Rodenticide Act [FIFRA], section 2(bb), cited in ibid., p. 5. The phrase "valid laboratory study" comes from an EPA official, Losi Rossi, quoted by Winston Williams in "Polishing the Apple's Image," *New York Times*, May 25, 1986, section 3, p. 4. See also "EPA Won't Ban Use of Chemical on Apples," by Philip Shabecoff, *New York Times*, Jan. 23, 1986, p. A18.

44. Meunier, "Environmental Protection Agency's Regulation," pp. 4–5

45. Ibid., p. 101.

46. The events described are narrated in the 1989 Senate subcommittee report, *Government Regulation of Pesticides in Food: The Need for Administrative and Regulatory Reform* (pp. 25 ff.). The source cited by the Senate report is Congressional Research Service, "Apple Alarm: Public Concern about Pesticide Residues in Fruit and Vegetables," report, Mar. 1, 1989.

47. EPA, PD-1, pp. 29136–29141.

48. Note that at p. 29138 of PD-1 particular concern is expressed for young children who consume large quantities of apple products.

49. Meunier, "Environmental Protection Agency's Regulation," p. 6.

50. EPA Special Review Document for Daminozide, p. ii; Bidinotto, "Great Apple Scare," p. 55. For a more detailed account of the SAP, its purpose, its intended composition, and the kinds of permissible relations between its members and the chemical industry, see Arnold, *Fear of Food,* pp. 49–50.

51. Sheila Jasanoff, "EPA's Regulation of Daminozide: Unscrambling the Messages of Risk," *Science, Technology, and Human Values,* 12, nos. 3 and 4 (Summer/Fall 1987): 118.

52. Ibid.

53. See, e.g., Bidinotto, "Great Apple Scare," pp. 55–56. An especially detailed account of the meeting is given in Jasanoff, "EPA's Regulation of Daminozide." Jasanoff's sources include interviews with EPA staff and an SAP member and the transcript of the meeting. Notice that already in 1987, well before the big apple scare, Jasanoff had identified the case of daminozide as a good example of the potential misunderstandings to arise from the character of the regulatory process—pitting public officials, industry representatives, and environmentalist groups against each other. Taking yet another angle on the SAP meeting, the Senate subcommittee report *Government Regulation of Pesticides in Food: The Need for Administrative and Regulatory Reform* makes much of the fact that at the time of the meeting on daminozide, "Seven out of eight members of the SAP that recommended against the EPA staff proposal to ban Alar in 1985 were paid consultants to the chemical industry or to organizations supported by the industry at the same time that they served on the panel" (see pp. 33–37 and 61–62).

54. FIFRA SAP, "Review of a Set of Scientific Issues being Considered by EPA in Connection with the Special Review of Daminozide" (available upon request from the EPA).

55. *Federal Register,* 51 (Apr. 16, 1986): 12889.

56. Shabecoff, "EPA Won't Ban Use of Chemical," p. A18; Williams, "Polishing the Apple's Image," *New York Times,* p. 4.

57. Quoted in Williams, "Polishing the Apple's Image."

58. Ibid.

59. Uniroyal Chemical Company, Inc., "Two-Year Oncogenicity Study in Mice (Daminozide)," International Research and Development Corporation, Mattawan, Mich., 1988, as summarized in EPA Special Review Document for Daminozide, pp. II-2ff.; Uniroyal Chemical Company, Inc., "Two-Year Oncogenicity Study in Rats (Daminozide)," International Research and Development Corporation, Mattawan, Mich., 1988, as summarized in EPA Special Review Document for Daminozide, p. II-4.

60. EPA Special Review Document for Daminozide, p. II-4.

61. Uniroyal Chemical Company, Inc., "Two-Year Oncogenicity Study in Rats (UDMH)," International Research and Development Corporation Study no. 399-062, 1989; Uniroyal Chemical Company, Inc., "Two-Year Oncogenicity Study in Mice (UDMH)," International Research and Development Corporation Study no. 399-063, 1989, also called "low-dose study"; Raymond A. Cardona, "Discussion of Results of 1,1-dimethylhydrazine (UDMH) Oncogenicity Studies," Jan.

1990, received from Uniroyal by mail. The EPA required that interim results from these studies be submitted in 1988, and its regulatory decision of 1989 was based largely on the interim results. In the rat study, UDMH was given to rats in drinking water at concentrations of 1, 50, and 100 parts per million for two years. In the interim results, there was no evidence that UDMH at any of the doses caused cancer in rats. The final results also showed no conclusive evidence that UDMH caused cancer in rats.

62. EPA Special Review Document for Daminozide, p. II-5.
63. Cardona, "Discussion of Results," p. 1.
64. Uniroyal Chemical Company, Inc., "Two-Year Oncogenicity Study in Mice (UDMH)," International Research and Development Corporation Study no. 399-065, 1990, also called "high-dose study."
65. EPA Special Review Document for Daminozide, p. II-5.
66. Ibid., pp. II-5 and II-6: "Uniroyal believed that the results of a 13-week subchronic study (Cranmer, M. and Frith, C., 1987) supported 20 ppm as the MTD for mice and that elevating the dose would threaten the lives of the animals and the validity of the study. Uniroyal's opinion was based on: (1) microscopic examination of liver, spleen and bone marrow which they believe suggests significant cellular alterations, (2) evaluation of hematological effects from which they suggested significant changes had occurred to critical blood elements resulting in life-threatening anemia, and (3) changes in alkaline phosphatase levels which they considered to be significant and which they correlate with histopathological changes in the liver. The Agency considered this interpretation . . . but was not satisfied that the changes noted in the report were biologically meaningful. The Agency was of the opinion that the effects reported by Uniroyal were not life-threatening in nature."
67. Cardona, "Discussion of Results."
68. EPA Special Review Document for Daminozide, p. II-6.
69. Quoted by Marshall, *Science,* 254 (Oct. 4, 1991): 22.
70. Richard Wilson and E. A. Crouch, "Risk Assessment and Comparisons: An Introduction," *Science,* 236 (Apr. 17, 1987): 267–270.
71. EPA, "Guidelines for Carcinogen Risk Assessment," *Federal Register,* 51 (Sep. 24, 1986): 33994.
72. Ibid., p. 33995.
73. The EPA explains the basis for these classifications in its Special Review Document for Daminozide, p. II-15 and p. II-17. Concerning daminozide, it states, "Although deficiencies in the Toth and NCI studies have been noted, it is significant that the studies showed vascular and lung tumors, as found in the recent Uniroyal studies. These studies support the findings in the Uniroyal studies and the Agency believes the weight-of-the-evidence is sufficient to classify daminozide as a B_2 carcinogen, probable human carcinogen. The Agency also believes the oncogenic response seen in the daminozide studies is likely caused by the presence of UDMH in the test material and/or metabolic conversion to UDMH." We should explain that "weight-of-the-evidence" is a technical expression referring to the EPA's established classification system for tested chemicals. For more details on this term, see "Guidelines," especially section C, on p. 33996. About UDMH: "As with daminozide, although deficiencies have been noted in the earlier

data (NCI, Toth, Haun), the findings of similar tumor types and sites (liver, vascular, and lung) in both this data and the Uniroyal 1-year interim sacrifice data support a B_2 classification (probable human carcinogen) for UDMH." "Sacrifice" here refers to the killing of some animals in the study after one year to investigate the effects of treatment.

74. EPA, "Guidelines," p. 33997.
75. Ibid.
76. EPA Special Review Document for Daminozide, p. II-20.
77. Ibid., pp. II-6, II-39.
78. Ibid., p. II-19.
79. Ibid.
80. EPA, "Guidelines," p. 33998.
81. EPA Special Review Document for Daminozide, p. II-20.
82. R. B. Howe, K. S. Crump, and C. Van Landingham, "A Computer Program to Extrapolate Quantal Animal Toxicity Data to Low Doses," unpublished report, 1986. EPA Special Review Document for Daminozide, pp. II-20, II-21.
83. Robin M. Whyatt, "NRDC on Alar" (letter), *Science*, 254 (Sep. 1, 1989): 910–911.
84. Meunier, "Environmental Protection Agency's Regulation," p. 107. See also NRDC, *Intolerable Risk*, appendix 3, p. 129, table A3-1, where the earlier EPA potency estimate for UDMH is listed as 8.9; a couple of EPA sources are cited here that we have not seen. For our purposes, there is not a significant difference between 8.7 and 8.9.
85. EPA Special Review Document for Daminozide, pp. II-39, II-40.
86. *Federal Register*, 54 (Feb. 10, 1989): 6394.
87. EPA Special Review Document for Daminozide, p. II-43.
88. We cannot find a statement of the individual factors in the NRDC's equation.
89. NRDC, *Intolerable Risk*, pp. 2–3.
90. The assumption of a background cancer risk of 25 percent or 1 out of 4, is commonplace in contemporary discussions of the subject. Our source for the figure is John A. Moore, "Speaking of Data: The Alar Controversy," *EPA Journal*, May/June 1989, pp. 5–9.
91. NRDC, "Fact Sheet on the NRDC Study *Intolerable Risk*" (Washington, D.C.: NRDC, 1989).
92. Ibid.; see also Leslie Roberts, "Alar: The Numbers Game," *Science*, 243 (Mar. 17, 1989): 1430.
93. NRDC, "Fact Sheet"; the source cited here is USDA, Human Nutrition Information Service, *CSFII-Nationwide Food Consumption Survey, Continuing Survey of Food Intakes by Individuals, Women 19–50 and Their Children 1–5 Years, 1 Day*, report no. 85-1, 1985.
94. Wilson and Crouch, "Risk Assessment," table 2. The uncertainty is 10 percent, since the actual estimate covers a range between 2.16×10^{-4} and 2.64×10^{-4}.
95. NRDC, *Intolerable Risk*, p. 3: "[The risk from exposure to UDMH via Alar] is 240 times greater than the cancer risk considered acceptable by EPA [1.0×10^{-6}] following a full lifetime of exposure."
96. Ibid, p. 33.
97. Ronald J. Prokopy et al., Department of Entomology, University of Massachu-

setts, Amherst, "11th Annual March Message to Massachusetts Tree Fruit Growers," 1989, section D, "How Loss of Alar Will Affect IPM Practices"; Stephen M. Wood, New England Fruit Growers' Council on the Environment, West Lebanon, N.H., "Food Safety and Alternative Agriculture: Harmony or Conflict?" paper presented to the American Chemical Society food safety conference, Jan. 22, 1990.

98. NRDC, "Fact Sheet"; "NRDC's Response to Technical Criticisms of the *Intolerable Risk* Report" and "NRDC's Response to Common Criticisms of the *Intolerable Risk* Report and the Public Debate over Pesticides" (Washington, D.C.: NRDC, 1989).

99. Fact Sheet #1 in NRDC, "Response to Common Criticisms," criticism #3.

100. Fact Sheet #3 in NRDC, "Response to Technical Criticisms," criticism #3.

101. Reported in Marshall, *Science,* 254 (Oct. 4, 1991): 20–22.

102. NRDC, "Response to Common Criticisms," criticism #4.

103. Quoted in Marshall, *Science,* 254 (1991): 21.

104. Daniel Koshland, "Credibility in Science and the Press," *Science,* 254 (Nov. 1, 1991): 629.

Regulation without Evidence of Harm

1. Ralph L. Keeney, "Mortality Risks Induced by Economic Expenditures," *Risk Analysis,* 10, no. 2 (1990): 147–159.

7, I. Is Arsenic in Drinking Water Harmful to Our Health?

1. Agency for Toxic Substances and Disease Registry [ATSDR], *Toxicological Profile for Arsenic* (Oak Ridge National Laboratory: U.S. Public Health Service, 1989).

2. William H. Lederer and Robert J. Fensterheim, eds., *Arsenic: Industrial, Biomedical, Environmental Perspectives* (New York: Van Nostrand Reinhold, 1983): Kurt J. Irgolic et al., "Determination of Arsenic and Arsenic Compounds in Water Supplies," pp. 284–285.

3. R. A. Schraufnagel, "Arsenic in Energy Sources: A Future Supply or an Environmental Problem?" in Lederer and Fensterheim, *Arsenic,* pp. 18–19.

4. See the following chapters in Lederer and Fensterheim, *Arsenic:* J. R. Abernathy, "Role of Arsenical Chemicals in Agriculture"; J. C. Alden, "The Continuing Need for Inorganic Arsenical Pesticides"; W. J. Baldwin, "The Use of Arsenic as a Wood Preservative"; and R. K. Willardson, "Arsenic in Electronics."

5. ATSDR, *Toxicological Profile for Arsenic,* pp. 1–6.

6. From this point on "arsenic" will denote inorganic arsenic; it is this form that is of concern to health researchers and officials.

7. Ken Nelson, "Industrial Sources: Introduction," in Lederer and Fensterheim, *Arsenic,* p. 1.

8. ATSDR, *Toxicological Profile for Arsenic,* p. 11.

9. Gerhard Stohrer, "Arsenic: Opportunity for Risk Assessment," *Archives of Toxicology,* 65 (1991): 525–531.

10. W. P. Tseng et al., "Prevalence of Skin Cancer in an Endemic Area of Chronic

Arsenicism in Taiwan," *Journal of the National Cancer Institute,* 40 (1968): 453–463.

11. Mariano Cebrian et al., "Chronic Arsenic Poisoning in the North of Mexico," *Human Toxicology,* 2 (1983): 121–133.

12. Ibid., p. 125.

13. V. U. Fierz, "Katamnestische untersuchungen über die nebenwirkungen der therapie mit anorganischem arsen bei hautkrankheiten," *Dermatolog,* 131 (1965), summarized in Environmental Protection Agency [EPA], *Risk Assessment Forum: Special Report on Ingested Arsenic* (Washington, D.C.: EPA, July 1988), p. 15.

14. William Morton et al., "Skin Cancer and Water Arsenic in Lane County, Oregon," *Cancer,* 37 (1976): 2523–2532.

15. Ibid., p. 2527.

16. J. Malcolm Harrington et al., "A Survey of a Population Exposed to High Concentrations of Arsenic in Well Water in Fairbanks, Alaska," *American Journal of Epidemiology,* 108, no. 5 (1978): 377–385.

17. J. W. Southwick et al., "An Epidemiological Study of Arsenic in Drinking Water in Millard County, Utah," in Lederer and Fensterheim, *Arsenic,* pp. 210–255.

18. Ibid., p. 224.

19. Jane L. Valentine et al., "Health Response by Questionnaire in Arsenic-Exposed Populations," *Journal of Clinical Epidemiology,* 45, no. 5 (1992): 487–494.

20. Ibid., p. 490.

21. Arthur Furst, "A New Look at Arsenic Carcinogenesis," in Lederer and Fensterheim, *Arsenic;* ATSDR, *Toxicological Profile for Arsenic;* Anna M. Fan, "The Carcinogenic Potential of Cadmium, Arsenic, and Selenium and the Associated Public Health and Regulatory Implications," *Journal of Toxicological Sciences,* 15 (1990): 162–175 supplement.

22. Stohrer, "Arsenic," p. 528.

23. E.g., ibid. and EPA, *Risk Assessment Forum,* p. 6.

24. Kenneth Brown et al., "A Dose-Response Analysis of Skin Cancer from Inorganic Arsenic in Drinking Water," *Risk Analysis,* 9, no. 4 (1989): 519–528; EPA, *Risk Assessment Forum,* p. 6.

25. EPA, *Risk Assessment Forum,* p. 6; I. Harding-Barlow, "What Is the Status of Arsenic as a Human Carcinogen?" in Lederer and Fensterheim, *Arsenic,* pp. 206–207.

26. EPA, *Risk Assessment Forum,* p. 13.

27. Harding-Barlow, "What Is the Status of Arsenic."

28. ATSDR, *Toxicological Profile for Arsenic,* p. 11.

29. Morton et al., "Skin Cancer"; Harrington et al., "A Survey"; Southwick, "An Epidemiological Study"; EPA, *Risk Assessment Forum.*

30. EPA, *Risk Assessment Forum,* p. 9.

31. ATSDR, *Toxicological Profile for Arsenic,* p. 6.

32. E.g., ibid., p. 17; Harding-Barlow, "What Is the Status of Arsenic," pp. 206–207.

33. Harrington, "A Survey," p. 384; Harding-Barlow, "What Is the Status of Arsenic," p. 207; ATSDR, *Toxicological Profile for Arsenic,* p. 17.

34. Stohrer, "Arsenic."

35. Ibid, p. 526.

36. Ibid., p. 527.

37. Ibid., pp. 527–528.
38. Ibid., pp. 528–529.
39. EPA, *Risk Assessment Forum,* p. 28.
40. ATSDR, *Toxicological Profile for Arsenic,* p. 17.
41. Ibid., p. 6
42. EPA, *Risk Assessment Forum,* pp. 28, 5.
43. Ibid., p. 28
44. Stohrer, "Arsenic," p. 526.
45. E.g., ibid., pp. 529–530; EPA, *Risk Assessment Forum,* p. 2.
46. ATSDR, *Toxicological Profile for Arsenic,* p. 17.
47. EPA, *Risk Assessment Forum,* p. 9.
48. Ibid., p. 3.
49. Harrington, "A Survey," p. 384.

7, II. Whom Can You Trust? The Nitrite Controversy

1. Karen DeWitt, "U.S. Will Not Seek to Ban Nitrite from Foods as a Cause of Cancer," *New York Times,* Aug. 20, 1980, p. A1.
2. U.S. Department of Agriculture [USDA], "An Analysis of a Ban on Nitrite Use in Curing Bacon," Economics, Statistics, and Cooperatives Service, no. 48, March 1979, p. 2. National Academy of Science [NAS], *The Health Effects of Nitrate, Nitrite, and N-Nitroso Compounds,* part 1 of a two-part study by the Committee on Nitrite and Alternative Curing Agents in Food (Washington, D.C.: National Academy Press, 1981), p. 1-1.
3. USDA, "Analysis of a Ban," p. 2.
4. Cited in NAS, *Health Effects of Nitrate,* p. 2-8.
5. U.S. General Accounting Office [GAO], "Does Nitrite Cause Cancer? Concerns about the Validity of FDA-Sponsored Study Delay Answer," report by the Comptroller General of the United States, Jan. 31, 1980, pp. 5–6.
6. NAS, *Health Effects of Nitrate,* p. 2-9.
7. GAO, "Does Nitrite Cause Cancer?" p. 9.
8. R. C. Shank and P. M. Newberne, "Dose-Response Study of the Carcinogenicity of Dietary Sodium Nitrite and Morpholine in Rats and Hamsters," *Food and Cosmetics Toxicology,* 14, no. 1 (1976): 1.
9. Ibid., p. 4.
10. Ibid., p. 5.
11. GAO, "Does Nitrite Cause Cancer?" p. 11.
12. Paul M. Newberne, "Nitrite Promotes Lymphoma Incidence in Rats," *Science,* 204 (June 8, 1979): 1079–1080.
13. Ibid.
14. GAO, "Does Nitrite Cause Cancer?" pp. 13–14.
15. Donald Kennedy, letter to authors, Nov. 19, 1991.
16. GAO, "Does Nitrite Cause Cancer?" p. 14.
17. Ibid., pp. 18, 19.
18. Ibid., pp. 17–18.
19. Ibid., p. 21.
20. Ibid., p. 47.

21. U.S. Food and Drug Administration [FDA], "Report of the Interagency Working Group on Nitrite Research," Department of Health and Human Services, Aug. 15, 1980, p. 3.
22. GAO, "Does Nitrite Cause Cancer?" p. 21.
23. Quoted in Marian Burros and Victor Cohn, "New Study Calls Nitrite Carcinogen," *Washington Post,* Aug. 12, 1978, p. A1.
24. Quoted in Victor Cohn, "U.S. Planning to Ban Sodium Nitrite from Meat," *Washington Post,* Aug. 18, 1978, p. C12.
25. Ibid.
26. Penny Girard, "Nitrite Ban: One Hazard vs. Another," *Los Angeles Times,* Sep. 24, 1978, p. 1.
27. "Danger of Nitrite in Food?" *U.S. News and World Report,* Oct. 23, 1978, p. 82.
28. Ibid., p. 81.
29. Ibid., p. 82.
30. Girard, "Nitrite Ban," p. 1.
31. Ibid.
32. Robert G. Cassens, *Nitrite-Cured Meat: A Food Safety Issue in Perspective* (Trumbull, Conn.: Food and Nutrition Press, 1990), p. 86.
33. Ibid., p. 78.
34. "Group Seeks Public's Help to Ban Nitrite," *Los Angeles Times,* Feb. 26, 1979, sec. 1, p. 13.
35. Marian Burros, "New Rule Clears Uncured Meat Product Names," *Los Angeles Times,* Aug. 27, 1979, sec. 1, p. 13.
36. "Handle Nitrite-Free Meats with Care, Agriculture Dept. Warns," *Los Angeles Times,* Sep. 26, 1979, sec. 1, p. 8.
37. Kennedy, letter to authors, Nov. 19, 1991.
38. GAO, "Does Nitrite Cause Cancer?" pp. i–v.
39. FDA, "Report of the Interagency Working Group," p. 30.
40. Ibid., p. 10.
41. Ibid., p. 11.
42. Ibid., p. 15.
43. Ibid., p. 29.
44. Paul Jacobs, "No Link between Nitrite in Meat, Cancer, U.S. Claims," *Los Angeles Times,* Aug. 20, 1980, sec 1, p. 26.
45. FDA, "Report of the Interagency Working Group," p. 16.
46. Ibid.
47. Newberne, "Nitrite Promotes Lymphoma," p. 1080.
48. FDA, "Report of the Interagency Working Group," p. 16.
49. Ibid., pp. 18–19, 20–27, 29.
50. DeWitt, "U.S. Will Not Seek to Ban Nitrite," p. A1.
51. Ibid., p. C14.
52. Jacobs, "No Link," p. 26.
53. "No Nitrate Ban," *Time,* Sep. 1, 1980, p. 55.
54. DeWitt, "U.S. Will Not Seek to Ban Nitrite," p. C14.
55. NAS, *Health Effects of Nitrate,* p. v.
56. Cassens, *Nitrite-Cured Meat,* p. 115.
57. NAS, *Alternatives to the Current Use of Nitrite in Foods,* part 2 of a two-part

study by the Committee on Nitrite and Alternative Curing Agents in Food (Washington, D.C.: National Academy Press, 1982).

58. NAS, *Health Effects of Nitrate,* p. 1-11.
59. Ibid., p. 9-3.
60. Ibid., p. 7-15.
61. Ibid., p. 7-36.
62. Ibid., p. 9-59.
63. NAS, *Alternatives,* pp. 8–13.
64. Ibid., pp. 1–5.
65. Ibid., pp. 1–8.
66. Ibid., pp. 9–10.
67. Cassens, *Nitrite-Cured Meat,* p. 115.
68. Burros and Cohn, "New Study Calls Nitrite Carcinogen," p. A1.
69. Nicholas von Hoffman, "Nitrite Ban Aftereffects," *Washington Post,* Sep. 7, 1978, p. D19.
70. "Nitrite Danger," *U.S. News and World Report,* Aug. 28, 1978, p. 52.
71. Jacobs, "No Link."
72. "Nitrites Get a Nod—At Least for Now," *Newsweek,* Sep. 1, 1980, p. 43.
73. Ralph W. Moss, "The Nitrite Fiasco: Look What's Back in Our Hot Dogs," *Nation,* Feb. 7, 1981, p. 141.

8. Do Rodent Studies Predict Cancer in Human Beings?

1. Michael A. Gallo and John Doull, "History and Scope of Toxicology," in M. O. Amadur et al., eds., *Casarett and Doull's Toxicology: The Basic Science of Poisons,* 4th ed. (New York: Pergamon Press, 1991), pp. 3–11.
2. Ibid.
3. One of the earliest references to the 100-fold safety factor approach in the United States is found in A. J. Lehman and O. G. Fitzhugh, "100-Fold Margin of Safety," *Association of Food and Drug Officials of the United States, Quarterly Bulletin,* 18 (1954): 33–35, cited in M. L. Dourson and J. F. Stara, "Regulatory History and Experimental Support of Uncertainty (Safety) Factors," *Regulatory Toxicology and Pharmacology,* 3 (1983): 224–238.
4. Calculated from Leo Levenson, "The Cranberry Scare of 1959," unpublished manuscript, University of California, Berkeley, 1991. See also the first part of Chapter 1.
5. An expanded statement of these arguments can be found in California Health and Welfare Agency, *Carcinogen Identification Policy: A Statement of Science as a Basis of Policy* (Sacramento: California Department of Health Services, July 1982), pp. 7–12.
6. Levenson, "Cranberry Scare."
7. Ibid.
8. Quoted in U.S. Food and Drug Administration [FDA], "Chemical Components in Food-Producing Animals: Criteria and Procedures for Evaluating Assays for Carcinogenic Residues," *Federal Register,* 44 (1979): 17072.
9. FDA, "Sponsored Compounds in Food-Producing Animals: Criteria and Proce-

dures for Evaluating the Safety of Carcinogenic Residues; Animal Drug Safety Policy," *Federal Register,* 52 (Dec. 31, 1987): 49572–49586.

10. FDA, "Chemical Components," p. 17072.

11. J. M. Sontag, "Aspects in Carcinogen Bioassay," in H. H. Hiatt, J. D. Watson, and J. A. Winsten, eds., *Origins of Human Cancer, Book C, Human Risk Assessment* (Cold Spring Harbor, N.Y.: Cold Spring Harbor Laboratory, 1977), pp. 1327–1338.

12. FDA, "Chemical Components," p. 17088.

13. Alvin Weinberg, "Science and Transcience," *Minerva,* 10, no. 2 (1972): 209–222. No epidemiological study could identify a lifetime excess cancer risk as small as 1 in 1 million, because the extra deaths would be impossible to notice against the background incidence of hundreds of thousands of cancer deaths per year. See Daniel Byrd and Lester Lave, "Narrowing the Range: A Framework for Risk Regulators," *Issues in Science and Technology,* Summer 1987, pp. 92–100.

14. See Aaron Wildavsky, *Searching for Safety* (New Brunswick, N.J.: Transaction Publishers, 1988).

15. U.S. Environmental Protection Agency [EPA], "Guidelines for Carcinogenic Risk Assessment," *Federal Register,* 51 (Sep. 24, 1986): 33992–34003.

16. Elizabeth L. Anderson and the Carcinogen Assessment Group of the EPA, "Quantitative Approaches in Use to Assess Cancer Risk," *Risk Analysis,* 3, no. 4 (1983): 280.

17. EPA, "National Oil and Hazardous Substances Pollution Contingency Plan: Final Rule," *Federal Register,* 55, no. 46 (Mar. 8, 1990): 8848.

18. D. W. Gaylor, "Preliminary Estimates of the Virtually Safe Dose for Tumors Obtained from the Maximum Tolerated Dose," *Regulatory Toxicology and Pharmacology,* 9 (1989): 101–108.

19. Leo Levenson, "Chemical Risks and Public Risk Perceptions," unpublished manuscript, University of California, Berkeley, 1990.

20. See K. J. Rothman, "Causation and Causal Inference," in David Schottenfeld and J. F. Fraumeni, Jr., eds., *Cancer Epidemiology and Prevention* (Philadelphia: W. B. Saunders, 1982), pp. 15–29; International Agency for Research on Cancer [IARC], *IARC Monographs on the Evaluation of the Carcinogenic Risk of Chemicals to Humans,* vol. 40 (Lyons, France: IARC, 1986), pp. 1–32; B. MacMahon, "Epidemiological Methods in Cancer Research," *Yale Journal of Biological Medicine,* 37 (1965): 508–522; C. C. Harris, ed., *Biochemical and Molecular Epidemiology of Cancer* (New York: Alan R. Liss, 1986); A. Berlin et al., eds., *Monitoring Human Exposure to Carcinogenic and Mutagenic Agents,* IARC Scientific Publications no. 59 (Lyons, France: IARC, 1984); U.S. Interagency Staff Group on Carcinogens, "Chemical Carcinogens: A Review of the Science and Its Associated Principles," *Environmental Health Perspectives,* 67 (1986): 201–282. A good selection for nonscientists is contained in a pamphlet by the American Council on Science and Health, "Of Mice and Men: The Benefits and Limitations of Animal Cancer Tests," March 1984.

21. Dale Hattis and David Kennedy, "Assessing Risks from Health Hazards: An Imperfect Science," *Technology Review,* 89, no. 4 (May/June 1986): 63.

22. Ibid., p. 64.

23. Douglas Hanahan, "Transgenic Mice as Probes into Complex Systems," *Science,* 246 (Dec. 8, 1989): 1265–1275.

24. Volker Schirrmacher, "Immunobiology and Immunotherapy of Cancer Metastasis: Ten-Year Studies in an Animal Model Resulting in the Design of an Immunotherapy Procedure Now under Clinical Testing," *Interdisciplinary Science Reviews,* 14, no. 3 (1989): 291–303.

25. D. A. Freedman and H. Zeisel, "From Mouse-to-Man: The Quantitative Assessment of Cancer Risks," *Statistical Science,* 3, no. 11 (Feb. 1988): 14.

26. Ibid.

27. Ibid., p. 10.

28. John Higginson, "Changing Concepts in Cancer Prevention: Limitations and Implications for Future Research in Environmental Carcinogenesis," *Cancer Research,* 48 (Mar. 15, 1993): 1381. See also Isaac Berenblum, "Theoretical and Practical Aspects of the Two-Stage Mechanism of Carcinogenesis," in A. Clark Griffin and Charles R. Shaw, eds., *Carcinogens: Identification and Mechanisms of Action* (New York: Raven Press, 1979), pp. 25–36.

29. American Council on Science and Health, "Of Mice and Men," p. 7.

30. Kenneth C. Chu, Cipriano Cueto, Jr., and Jerrold M. Ward, "Factors in the Evaluation of 200 National Cancer Institute Carcinogen Bioassays," *Journal of Toxicology and Environmental Health,* 8, nos. 1 and 2 (July/Aug. 1981): 251–280; J. M. Ward et al., "Quality Assurance for Pathology in Rodent Carcinogenesis Tests," *Journal of Environmental Pathology and Toxicology,* 2 (1978): 371–378.

31. Chu et al., "Factors in the Evaluation," pp. 256–257.

32. See the discussion in Andrew N. Rowan, *Of Mice, Models, and Men: A Critical Evaluation of Animal Research* (Albany: State University of New York Press, 1984). See also L. G. Stevenson, "Science down the Drain," *Bulletin of Historical Medicine,* 29 (1955): 1–26.

33. Chu et al., "Factors in the Evaluation," pp. 252–253.

34. J. Gordon Edwards, "Worried about Pesticides in Food and Water? Here Are the Facts," pamphlet distributed by National Council for Environmental Balance, Inc., Louisville, Ky. (no date, no page nos.).

35. G. E. Paget, ed., *Methods in Toxicology* (Oxford: Blackwell Scientific Publications, 1970).

36. *Human Toxicology,* 3 (1984): 85–92, cited in Robert Sharpe, *The Cruel Deception* (Wellingborough, Northamptonshire, England: Thornson's Publishing, 1988), pp. 94–95.

37. Rowan, *Of Mice, Models, and Men,* p. 207.

38. Ibid., citing S. B. de C. Baker, "The Study of the Toxicity of Potential Drugs in Laboratory Animals," in *The Use of Animals in Toxicological Studies* (Potters Bar, Eng.: Universities Federation for Animal Welfare, 1969), p. 23.

39. Rowan, *Of Mice, Models, and Men,* p. 207, citing G. Zbinden, "A Look at the World from Inside the Toxicologist's Cage," *European Journal of Clinical Pharmacology,* 9 (1976): 333–338.

40. See Sharpe, *Cruel Deception,* pp. 100–101, for a list of authorities with negative verdicts on the animal tests.

41. See the numerous examples in Rowan, *Of Mice, Models, and Men,* pp. 207–208.

42. Jay H. Lehr, "Toxicological Risk Assessment Distortions, Part II: The Dose Makes the Poison," American Ground Water Trust, Dublin, Ohio, May 1990.

43. Rowan, *Of Mice, Models, and Men,* p. 190, citing E. M. Boyd and I. L. Godi, "Acute Oral Toxicity of Distilled Water in Albino Rats," *Industrial Medicine and Surgery,* 36 (1967): 609–613.

44. Rowan, *Of Mice, Models, and Men,* p. 238, citing S. J. Jaffe, "Summary: Pediatrician's View," in R. H. Schwarz and S. J. Yaffe, eds., *Drug and Chemical Risks to the Fetus and Newborn* (New York: Alan R. Liss, 1980), pp. 157–161.

45. Rowan, *Of Mice, Models, and Men.* See also J. G. Wilson, "Reproduction and Teratogenesis: Current Methods and Suggested Improvements," *Journal of the Association of Official Analytical Chemists,* 58 (1975): 657–667.

46. National Public Radio, *Morning Edition,* transcript from Mar. 12, 1992, 10:00 A.M. EDT, pp. 8–10.

47. See, e.g., Jean Marx, "Animal Carcinogen Testing Challenged," *Science,* 250 (Nov. 9, 1990): 743–745.

48. See Allen H. Smith and Dan S. Sharp, "A Standardized Benchmark Approach to the Use of Cancer Epidemiology Data for Risk Assessment," presented at the EPA Symposium on Advances in Health Risk Assessment for Systemic Toxicants and Chemical Mixtures, Cincinnati, Oct. 23–25, 1984, p. 3.

49. "Revision without Revolution in Carcinogen Policy," *Regulation,* 8, no. 4 (July/ Aug. 1984): 5–7.

50. Rowan, *Of Mice, Models, and Men,* pp. 234–235.

51. Anderson et al., "Quantitative Approaches in Use to Assess Cancer Risk," p. 281.

52. Ibid.

53. Ibid., pp. 289–290.

54. See Suresh H. Moolgavkar, "Carcinogenesis Modeling: From Molecular Biology to Epidemiology," *Annual Review of Public Health,* 7 (1986): 151–169.

55. Richard Peto, "Epidemiology, Multistage Models, and Short-Term Mutagenicity Tests," in Hiatt et al., *Origins of Human Cancer,* p. 1404.

56. David A. Freedman and William C. Navidi, "Multistage Models for Carcinogenesis," *Environmental Health Perspectives,* 81 (1989): 172.

57. Moolgavkar, "Carcinogenesis Modeling," and Freedman and Navidi, "Multistage Models."

58. Thomas R. Fears, Robert E. Tarone, and Kenneth C. Chu, "False-Positive and False-Negative Rates for Carcinogenicity Screens," *Cancer Research,* 37 (July 1977): 1941.

59. David Salsburg, "The Lifetime Feeding Study in Mice and Rats: An Examination of Its Validity as a Bioassay for Human Carcinogens," *Fundamental and Applied Toxicology,* 3 (1983): 65.

60. Ibid., p. 63.

61. Ibid., p. 66. See Salsburg's "Decision Rules Used" in "Lifetime Feeding Study," p. 67:

Seven decision rules were used to examine the ability of the lifetime feeding study to distinguish between carcinogenesis and its negation. They were:

(1) A compound will be called "carcinogenic" ("c") or "protecting" from cancer ("p") if there is a single statistical significance ($p \leq 0.05$) indicative of such an effect. This is a

rule which ignores the statistical false positives that might result from multiple testing but it is apparently the rule intended by the authors of the OSHA generic policy on carcinogens (U.S. Dept. of Labor, 1980).

(2) A compound will be called "c" or "p" if there is a statistical significance after protecting from statistical false positives by a Bonferroni adjustment (Feller, 1950). This is a rule which was suggested by Fears et al. (1977).

(3) A compound will be called "c" or "p" if there are significant statistical tests in two species or two sexes. This rule appears to be the one favored by Fears et al. (1977). It is also close to the rule used by Chu et al. (1981) for their tables (3) and (4).

(4) A compound will be called "c" or "p" if there are significant tests in two species or two sexes after significance has been adjusted by the Bonferroni inequality to protect from statistical false positives. This rule is an obvious extension of the discussion in Fears et al. (1977).

(5) A compound will be called "c" or "p" if there is a statistically significant dose response in the same organ in two sexes or two species.

(6) A compound will be called "c" or "p" if there is a statistically significant dose response in the total tumor burden of at least one sex or species. This rule has been proposed to me by Nathan Mantel.

(7) A compound will be called "c" or "p" if there is a statistically significant dose response in total tumor burden of at least two sex/species. This is an obvious extension of rule 6.

To avoid ambiguity the use of the Bonferroni inequality requires a more detailed description of the statistical tests performed on this data. The Armitage test was applied to every type of tumor tabulated in a given report for which there were a total of seven or more animals with that specific type of tumor. This led to a number of statistical tests varying from 5 to 35 on each report. The p-value which was considered significant was then taken as $0.05/N$ where N is the total number of tests performed on that report.

See also David Salsburg, "The Effects of Lifetime Feeding Studies on Patterns of Senile Lesions in Mice and Rats," *Drug and Chemical Toxicology,* 3 (1980): 1–33; and Lorenzo Tomatis et al., "Evaluation of the Carcinogenicity of Chemicals: A Review of the Monograph Program of the International Agency for Research on Cancer (1971 to 1977)," *Cancer Research,* 38 (1978): 877–885.

62. Bruce N. Ames and Lois Swirsky Gold, "Too Many Rodent Carcinogens: Mitogenesis Increases Mutogenesis," *Science,* 249 (Aug. 31, 1990): 970–971; Jon C. Mirsalis and Karen L. Steinmetz, "The Role of Hyperplasia in Liver Carcinogenesis," in Donald E. Stevenson et al., eds., *Mouse Liver Carcinogenesis: Mechanisms and Species Comparisons,* proceedings of the symposium held in Austin, Tex., Nov. 30–Dec. 3, 1988 (New York: Wiley-Liss, 1990), p. 150; Ames and Gold, "Too Many Rodent Carcinogens." See also Ames and Gold, Response to F. P. Perera in Letters, *Science,* 250 (Dec. 21, 1990): 1645–1646; and Ames and Gold, "Endogenous Mutagens and the Causes of Aging and Cancer," *Mutation Research,* 250, nos. 1/2 (1991): 3–16.

63. Samuel M. Cohen and Leon B. Ellwein, "Cell Proliferation in Carcinogenesis," *Science,* 249 (Aug. 31, 1990): 1007.

64. Marx, "Animal Carcinogen Testing," p. 744.

65. Thomas H. Maugh II, "Carcinogen Test Process Challenged," *Los Angeles Times,* Aug. 31, 1990, p. A38.

66. Frederica P. Perera, "Carcinogens and Human Health: Part 1," Letters, *Science,* 250 (Dec. 21, 1990): 1644. See Ames and Gold's response in the same issue, pp. 1645–1646.

67. Vincent James Cogliano et al., "Carcinogens and Human Health: Part III," Letter to the Editor, *Science,* 251 (Feb. 8, 1991): 607.

68. Michael Gough, "Chemical Risk Assessment Is Not Science," *In Chemistry,* 3 (Dec./Jan. 1993): 23–26. *In Chemistry* is a publication of the American Chemical Society for undergraduates.

69. Hattis and Kennedy, "Assessing Risks from Health Hazards," p. 65; Alice S. Whittemore, "Facts and Values in Risk Analysis for Environmental Toxicants," *Risk Analysis,* 3, no. 1 (1983): 23–33.

70. National Academy of Sciences, *Saccharin: Technical Assessment of Risks and Benefits* (Washington, D.C.: National Research Council Committee for Study on Saccharin and Food Safety Policy, 1978), pp. 72 and 61ff.

71. Bruce N. Ames and Lois Swirsky Gold, "Dietary Carcinogens, Environmental Pollution, and Cancer," *Medical Oncology and Tumor Pharmacotherapy,* 7, nos. 2–3 (1990): 69–85; Bruce Ames, "What Are the Major Carcinogens in the Etiology of Human Cancer? Environmental Pollution, Natural Carcinogens, and the Causes of Human Cancer; Six Errors," in V. T. De Vita, Jr., Samuel Hellman, and S. A. Rosenberg, eds., *Important Advances in Oncology* (Philadelphia: J. B. Lippincott, 1989), pp. 237–247.

72. See Bruce Ames, Margie Profit, and Lois Swirsky Gold, "Nature's Chemicals and Synthetic Chemicals: Comparative Toxicology," *Proceedings of the National Academy of Sciences,* 87 (Oct. 1990): 7782–7786.

73. Dennis Coyle and Aaron Wildavsky, "The Battle Within: How the Human Body Defends Itself," in Wildavsky, *Searching for Safety* (New Brunswick, N.J.: Transaction Publishers, 1988), pp. 149–168.

74. Ames and Gold, "Endogenous Mutagens," p. 11; Bruce Ames, "Of Mice and Men: Finding Cancer's Causes," *Reason,* Dec. 1991, p. 18.

75. John C. Bailar and Elaine M. Smith, "Progress against Cancer," *New England Journal of Medicine,* 314, no. 19 (May 8, 1980): 1226.

76. Ibid., p. 1231.

77. Richard Doll, "Health and the Environment in the 1990s," *American Journal of Public Health,* 82, no. 7 (July 1992): 936.

78. See, e.g., Curtis C. Travis, ed., *Biologically Based Methods for Cancer Risk Assessment* (New York: Plenum Press, 1989); and Freedman and Zeisel, "From Mouse-to-Man."

79. See, e.g., Frederica P. Perera, "Perspectives on the Risk Assessment for Nongenotoxic Carcinogens and Tumor Promoters," *Environmental Health Perspectives,* 94 (1991): 231–235.

80. "Delaney Clause philosophy" refers not just to the reasoning used by those involved in passing the original clause but also to the point of view of those advocating the clause's "zero tolerance" during the past three decades.

81. See, e.g., U.S. Occupational and Safety Administration, "Identification, Classification, and Regulation of Potential Occupational Carcinogens, Part VII, Book 2," *Federal Register* 45 (Jan. 22, 1980): 5024: "There is evidence that the first stage in initiation of carcinogenesis may result from the interaction of a molecule of the carcinogen with DNA or other genetic material in the cell . . . Hence it would follow from the fact that cancer develops from a single cell, that cancer may be initiated by the interaction of a single molecule of a carcinogen with the

critical target site in a cell." Quoted in Edith Efron, *The Apocalyptics: Cancer and the Big Lie,* (New York: Simon and Schuster, 1984), p. 224.

82. See, e.g., Thomas H. Corbett, *Cancer and Chemicals* (Chicago: Nelson-Hall, 1977); and Samuel S. Epstein, *The Politics of Cancer* (San Francisco: Sierra Club Books, 1978).

83. See American Cancer Society, *Cancer Facts and Figures 1991* (Atlanta: American Cancer Society, 1991).

84. Bruce N. Ames et al., "Ranking Possible Carcinogenic Hazards," *Science,* 236 (Apr. 17, 1987): 271–280.

85. Lois Swirsky Gold, Thomas H. Slone, Bonnie R. Stern, Neela B. Manley, and Bruce N. Ames, "Rodent Carcinogens: Setting Priorities," *Science,* 258 (Oct.9, 1992): 261–265.

9. The Effects of Acid Rain on the United States

1. National Research Council [NRC], Committee on Monitoring and Assessment of Trends in Acid Deposition, *Acid Deposition: Long Term Trends* (Washington, D.C.: National Academy Press, 1986), p. 1.

2. Ellis B. Cowling, "Acid Precipitation in Historical Perspective," *Environmental Science and Technology,* 16 (Feb. 1982): 111A; R. A. Smith, *Air and Rain: The Beginnings of a Chemical Climatology* (London: Longmans, Green, 1872).

3. Formally, the antilogarithm of the concentration of hydrogen (hydronium) ions. A pH of 7 indicates a concentration of 10^{-7} hydrogen ions per liter.

4. C. E. Junge and R. T. Werby, "The Concentration of Chloride, Sodium, Potassium, Calcium, and Sulfate in Rainwater over the United States," *Journal of Meteorology,* 15 (1958): 417–425.

5. NRC, *Acid Deposition* (1986).

6. Gene E. Likens and Thomas J. Butler, "Recent Acidification of Precipitation in North America," *Atmospheric Environment,* 15 (1981): 1103.

7. Ibid.

8. James N. Galloway et al., "The Composition of Precipitation in Remote Areas of the World," *Journal of Geophysical Research,* 87, no. C11 (Oct. 20, 1982): 8784.

9. U.S. National Acid Precipitation Assessment Program [NAPAP], *Acidic Deposition—State of Science and Technology: Summary Report of the U.S. National Acid Precipitation Assessment Program,* ed. Patricia M. Irving (Washington D.C.: GPO, 1991): report no. 6, Douglas L. Sisterson, "Deposition Monitoring: Methods and Results," pp. 70–71.

10. Ibid., p. 69.

11. R. J. Charlson and H. Rodhe, "Factors Controlling the Acidity of Natural Rainwater," *Nature,* 295 (Feb. 25. 1982): 683–685; the authors indicate on p. 683 that "pH values might range from 4.5 to 5.6 due to the variability of the sulphur cycle alone." See also NRC, *Acid Deposition: Atmospheric Processes in Eastern North America* (Washington D.C.: National Academy Press, 1983), which indicates on p. 13 that the possible range of natural pH grows to 4.9 to 6.5 when one includes possible variations in all other known contributing factors.

12. Charlson and Rodhe, "Factors Controlling the Acidity," p. 685.

13. NRC, *Acid Deposition* (1986), p. 181.

14. See ibid. for a discussion of these trends. Using the available data from two temporary networks of sites some scientists found that rain in the eastern United States
had become much more acidic between 1955 and 1981. Ibid., p. 163, citing Likens
and Butler, "Recent Acidification of Precipitation in North America,"
pp. 1103–1109, and C. V. Cogbill, G. E. Likens and T. A. Butler, "Uncertainties
in Historical Aspects of Acid Precipitation: Getting It Straight," *Atmospheric
Environment,* 18 (1984): 2261–2270. See also Gene E. Likens et al., "Acid Rain,"
Scientific American, 241, no. 4 (1979): 43–51, for discussion of how large the
increases of hydrogen ion concentrations from 1955 to 1979 appeared to be.
Likens and associates attributed this change to increased influence of industrial
emissions on the acid-base chemistry of precipitation. Other scientists, questioning this conclusion, have pointed to possible biases introduced by sampling
and analysis problems that may have caused the early pH data to be skewed
toward higher pH (less acidity) than the rain actually had at the time (NRC, *Acid
Deposition* [1986], p. 163, citing D. A. Hansen and G. M. Hidy, "Review of
Questions regarding Rain Acidity Data," *Atmospheric Environment,* 16 [1982]:
2107–2126), and to climatological influences such as a drought, which may have
led to unusually high pH (low acidity) levels over much of eastern North America
during the mid-1950s and early 1960s (NRC, *Acid Deposition* [1986]). On p. 93
the NRC report introduces the notion of climate having a long-term impact on
precipitation chemistry. The drought most likely led to an increase in the amount
of dust carried up into the air from midwestern soils and dirt roads, which is
known to be a source of calcium, magnesium, and potassium in precipitation
(see, e.g., Eville Gorham, Frank B. Martin, and Jack T. Litzau, "Acid Rain: Ionic
Correlations in the Eastern United States, 1980–1981," *Science,* 225 (July 27,
1984): 408). The calcium and magnesium would have partially neutralized the
acids, as they exist as cations (positive ions) in water and thereby take the place
of some of the hydrogen ions. When the drought is corrected for, it appears that
the pH of the rain without the drought would have been comparable to the rain
of 1980. See NRC, *Acid Deposition* (1986), p. 98 and fig. 3.4 on p. 99; the report
also cites G. J. Stensland and R. G. Semonin, "Another Interpretation of the pH
Trend in the United States," *Bulletin of the American Meteorological Society,* 63
(1982): 1277–1284.

15. James N. Galloway, Gene E. Likens, and Eric S. Edgerton, "Acid Precipitation
in the Northeastern United States: pH and Acidity," *Science,* 194 (Nov. 12, 1976):
723.

16. See, e.g., John S. Eaton, Gene E. Likens, and F. Herbert Bormann, "The Input
of Gaseous and Particulate Sulfur to a Forest Ecosystem," *Tellus,* 30 (1978). See
p. 546 and table 1 on p. 547, for an early record of the dominance of sulfuric
acid.

17. James N. Galloway, Gene E. Likens, and Mark E. Hawley, "Acid Precipitation:
Natural versus Anthropogenic Components," *Science,* 226 (Nov. 16, 1984): 829.

18. NAPAP studies confirm the earlier conclusions. They estimate that the median
sulfur deposition in North America is nine times greater than the median deposi-

tion in remote regions. Their estimate was derived from averaging the six Global Precipitation Composition Project sites. See NAPAP, *Acidic Deposition*, Sisterson, "Disposition Monitoring," pp. 68–69.

19. James N. Galloway and Gene E. Likens, "Acid Precipitation: The Importance of Nitric Acid," *Atmospheric Environment*, 15 (1980) : 409.

20. NAPAP, *Acidic Deposition*, report no. 1, Marylynn Placet, "Emissions Involved in Acidic Deposition Processes," p. 31.

21. Actually the background estimate is even smaller than previous ones that generally ranged around 10 percent with significant uncertainties allowing for a possible maximum somewhat larger. See, e.g., NAPAP, *Annual Report to the President and Congress* (Washington D.C.: GPO, 1986), p. 43; and James N. Galloway and Douglas M. Whelpdale, "An Atmospheric Sulfur Budget for Eastern North America," *Atmospheric Environment*, 14 (1980): 412 (table 4). L. Granat, H. Rodhe, and R. O. Hallberg, "The Global Sulphur Cycle," in B. H. Svenson and R. Söderland, eds., *Nitrogen, Phosphorus, and Sulphur—Global Cycle: SCOPE Report 7*, Ecological Bulletins (Stockholm: Swedish National Research Council, 1976), vol. 22, pp. 104–105, offers a table of early estimates of global sulfur. Estimates of eastern U.S. background emissions from these global estimates are all less than one-quarter of industrial emissions.

22. NAPAP, *Acidic Deposition*, Placet, "Emissions," p. 26, table 1-2.

23. According to Robert M. Garrels, Fred T. MacKenzie, and Cynthia Hunt, *Chemical Cycles and the Global Environment: Assessing Human Influences* (Los Altos, Calif.: William Kaufmann, Inc., 1975), p. 83, the average piece of coal burned in the United States as of the mid-1970s was 2 percent sulfur by weight.

24. NRC, *Acid Deposition: Atmospheric Processes*, p. ix.

25. U.S. Office of Technology Assessment, *Acid Rain and Transported Pollutants: Implications for Public Policy* (New York: UNIPUB, 1985), p. 149.

26. NAPAP, *Acidic Deposition*, Placet, "Emissions," p. 29. Also, according to Placet (pp. 25–26), by 1985 emissions had dropped to 21 Tg of sulfur dioxide.

27. G. R. Carmichael and L. K. Peters, "Eulerian Modeling of the Transport and Chemical Processes Affecting Long-Range Transport of Sulfur Dioxide and Sulfate," in Jerald Schnoor, ed., *Modeling of Total Acid Precipitation Impacts*, Acid Precipitation Series 9 (Boston: Butterworth Publishers, 1984), p. 25.

28. NRC, Committee on the Atmosphere and the Biosphere, *Atmosphere-Biosphere Interactions: Toward a Better Understanding of the Ecological Consequences of Fossil Fuel Combustion* (Washington, D.C.: National Academy Press, 1981), p. 181. R. Jeffrey Smith, "Administration Views on Acid Rain Assailed: A New Report Asserts That Acid Rain Is a Serious Problem in Need of Prompt Regulation," *Science*, 214 (Oct. 2, 1981): 38; "Dropping Acid" (editorial), *Washington Post*, Oct. 16, 1981, p. A26.

29. NRC, *Acid Deposition: Atmospheric Pressures*, p. ix.

30. Stephen E. Schwartz, "Acid Deposition: Unraveling a Regional Phenomenon," *Science*, 243 (Feb. 10, 1989): 753–763.

31. NAPAP, *Acidic Deposition*, report no. 5, Robin L. Dennis, "Evaluation of Regional Acidic Deposition Models," p. 63.

32. NRC, *Acid Deposition: Atmospheric Processes*, p. 57, citing Galloway and Whelpdale, "An Atmospheric Sulfur Budget." See William D. Bischoff, Virginia L.

Peterson and Fred T. MacKenzie, "Geochemical Mass Balance for Sulfur- and Nitrogen-Bearing Acid Components: Eastern United States," in Owen P. Bricker, ed., *Geological Aspects of Acid Deposition*, Acid Precipitation Series 7 (Boston: Butterworth Publishers, 1984), pp. 15–16, where these authors estimate that 55 percent of the sulfur dioxide and 75 percent of the nitrogen oxide emissions leave the United States by wind. These percentages vary from year to year depending on the winds, the cloud cover, the temperature, and the timing and amount of precipitation, but the basic breakdown stays fairly constant at least with current emission rates.

33. NRC, *Acid Deposition: Atmospheric Processes*, p. 10.
34. Ibid., p. 75.
35. Ibid., p. 83.
36. Ibid., p. 69.
37. Ibid., pp. 74–75.
38. Ibid., p. 69. This amount was generally 1–4 percent per hour.
39. Ibid., p. 74.
40. See ibid., p. 75. The continuity equation depends on S_{ij} (the magnitude of the source term at position i from source j) such that a term of the form $q_{ij}(S_{ij})S_{ij}$ appears in the solution for the magnitude of sulfate deposition. This term makes the equation nonlinear, as the dependent variable c_i no longer depends on the first order of the independent variable S_{ij}. This term arises because of the nonlinear nature of many atmospheric processes that provide the theoretical base for the use of the continuity equation.
41. Ibid., p. 6, italics added.
42. NAPAP, *Acidic Deposition*, report no. 8, Akula Venkatram, "Relationships between Atmospheric Emissions and Deposition/Air Quality," p. 86.
43. How this works can perhaps be illustrated with a hypothetical example. Suppose there is only one source of sulfur dioxide in a given region. As the SO_2 travels in the air, some of it is deposited in dry form, some is transformed and deposited in wet form, and the remainder is dispersed by the winds. All three of these processes lead to a decrease in the concentration of SO_2 in the air as a function of distance from the source. Near the source the transformation process is oxidant-limited. There is more SO_2 than there is hydrogen peroxide (and other aqueous-phase oxidants). If the amount of SO_2 emitted is decreased, then the region that is oxidant-limited will shrink. The rate of decrease of the total amount of SO_2 in the air as a function of distance will remain constant, because it is fixed by the fact that the region is oxidant-limited, but the initial concentration of SO_2 will be lower. The line described by the concentration of SO_2 as a function of distance from the source increases when the reaction is no longer impeded by a lack of oxidant. Now in reality the concentration of hydrogen peroxide varies with time and location, so that any source-receptor model would need to contain a model of these variations. The function of SO_2 concentration with distance is also not linear but the thought experiment contains the basic concept that the oxidant-limitation regime will shrink in size when emissions are reduced. Furthermore, it should be noted that the total range of the sulfur has been decreased. At all points beyond the new total range, the decrease in deposition is 100 percent. For most of the region that is not oxidant-limited, the fractional decrease in deposition must

be greater than the fractional decrease in emissions. See Likens et al., "Acid Rain," p. 47, figure 2.

44. NAPAP, *Acidic Deposition,* Venkatram, "Relationship," p. 88. The discrepancy between emissions and deposition decreases may be small even in the United States. According to NAPAP, Sisterson, "Deposition Monitoring," p. 69, 30–60 percent of total deposition is dry and therefore not oxidant-limited.

45. In Noye M. Johnson et al., " 'Acid Rain,' Dissolved Aluminum and Chemical Weathering at the Hubbard Brook Experimental Forest, New Hampshire," *Geochimica et Cosmochimica Acta,* 45 (Sep. 1981): 1435. The authors state that streams of third order or greater are unaffected by acid precipitation. The order of a stream is defined by how it is fed. A first-order stream is fed by no other streams; a second-order stream is fed by first-order streams; a third-order stream is fed by both first and second; and so on up to ninth or so. Each higher order is generally larger than the previous one. According to Johnson, "it is stream order, not stream elevation, which correlates with the state of streamwater chemistry in the Hubbard Brook area as a whole" (p. 1435). Since 1981, if not earlier, scientists have known that rivers and large streams are unaffected by acid rain, as are lakes to which these streams provide the majority of water. By the time the waters have reached such streams they have encountered enough buffering materials to be neutralized.

46. James Kramer and Andre Tessier, "Acidification of Aquatic Systems: A Critique of Chemical Approaches," *Environmental Science and Technology,* 16 (Nov. 1982): 606A.

47. T. J. Sullivan, *Historical Changes in Surface Water Acid-Base Chemistry in Response to Acidic Deposition,* NAPAP State of Science and Technology 11 (Washington, D.C.: NAPAP, 1990), p. 152.

48. J. J. Magnuson et al., "Effects on Aquatic Biology," in U.S. Environmental Protection Agency [EPA], *The Acidic Deposition Phenomenon and Its Effects: Critical Assessment Review Papers,* vol. 2, Rick A. Linthurst, ed. (Springfield, Va.: National Technical Information Service, 1984), pp. 5-150–5-155.

49. Ibid., p. 5-154.

50. Sullivan, *Historical Changes,* p. 35.

51. NAPAP, *Acidic Deposition,* report no. 10, Robert S. Turner, "Watershed and Lake Processes Affecting Surface Water Acid-Base Chemistry," pp. 104–105.

52. Edward C. Krug, "Assessment of the Theory and Hypotheses of the Acidification of Watersheds," *State Water Survey Contract Report 457* (Champaign: Illinois State Water Survey Division, Atmospheric Chemistry Division, 1989), p. 3-2.

53. Edward C. Krug and Charles R. Frink, "Acid Rain on Acid Soil: A New Perspective," *Science,* 221 (Aug. 5, 1983): 522.

54. Swedish Ministry of Agriculture, Environment '82 Committee, *Acidification Today and Tomorrow,* Simon Harper, trans. (Stockholm: Swedish Ministry of Agriculture, 1982).

55. Lars N. Overrein, Hans Martin Seip, and Arne Tollan, *Final Report of the SNSF Project, 1972–1980: Acid Precipitation—Effects on Forest and Fish,* report no. 19 (Oslo: SNSF Project, 1980).

56. Sullivan, *Historical Changes,* p. 31.

57. Krug, "Assessment of the Theory," p. 2-11, citing Haines and Akielaszak, "A

Regional Survey of the Chemistry of Headwater Lakes and Streams in New England: Vulnerability to Acidification," U.S. Fish and Wildlife Service, Office of Biological Services, report no. 80/40,15 (1983).
58. Ibid.
59. T. J. Sullivan et al., "Quantification of Changes in Lakewater Chemistry in Response to Acidic Deposition," *Nature*, 345 (May 3, 1990): 55.
60. Krug, "Assessment of the Theory," pp. 3-6, 3-7, citing M. H. Pfeiffer and P. J. Festa, *Acidity Status of Lakes in the Adirondack Region of New York in Relation to Fish Resources* (Albany: New York Department of Environmental Conservation, 1980).
61. Krug, "Assessment of the Theory," and a subsequent study came up with much the same results: see NRC, *Acid Deposition*, (1986), p. 37, citing J. Colquhoun, W. Kretzer, and M. Pfeiffer, *Acidity Status Update of Lakes and Streams in New York State,* report WM P-83 (Albany: New York Department of Environmental Conservation, 1984).
62. Some scientists argue that, although the pH data from the old studies are useless, we can determine whether or not the acid-neutralizing capacity of lakes has declined using these data. This determination is based on assumptions about the old techniques that may not be warranted. Nevertheless, the rejection of the old data is not universal. "Corrected" old capacity data are still used by some. See Sullivan, *Historical Changes,* pp. 98–99, for a discussion of recent uses of historical acid-neutralizing capacity data.
63. The compatibility problem across time for colorimetry lies in the vagueness of the endpoint of the methyl orange test used in the past. We know that it was probably between 4.04 and 4.19, but there are no records that document the exact number for each of the lakes surveyed. The NRC determined that if scientists were to compare the data from 1980 or 1984 with the data of the past, assuming an endpoint of 4.04 or 4.19, they could have four different changes in chemistry taking place. None of these assumptions is any better than any of the others. We simply do not know what the endpoint was and we never will. Nevertheless, people continue to use these old data and whichever assumption best fits their preconceptions when they comment on the historical chemistry record. The interpretation of the data from far enough back to tell us something—the data from a period that might predate the era of widespread acid rain—hinges on a set of untestable hypotheses, those being the four possible assumption sets for interpreting the changes from the 1930s to the 1980s. If one assumes that the most optimistic assumption set is correct, then there has not been much change in the acidity of surface waters in the Adirondacks due to acid rain. If one takes the most pessimistic assumption set to be correct, there have been widespread and significant changes in the acidity of those waters. See NRC, *Acid Deposition* (1986), p. 37, for the numbers used in this discussion.
64. Sullivan et al., "Quantification of Changes in Lakewater," p. 55.
65. NRC, *Acid Deposition* (1986), p. 171.
66. Sullivan, *Historical Changes,* p. 31. See also Brian F. Cumming et al., "Appendix A: Paleolimnological Variability," in ibid., pp. A1–A6. Paleolimnological studies do not directly measure the acidity of waters but are proxies. Essentially such a study compares the abundance in the past of certain microorganisms that leave

hard shells with their abundance today. Since these microorganisms are pH sensitive, their abundance is a function of pH and alkalinity. Cores of sediment are taken from lake bottoms and analyzed layer by layer. This technique appears to be a good proxy.

67. NRC, *Acid Deposition* (1986), p. 37.
68. Sullivan et al., "Quantification of Changes in Lakewater," pp. 54–55.
69. Ibid., p. 56.
70. Timothy J. Sullivan, "Long-Term Temporal Trends in Surface Water Chemistry," in Donald F. Charles, ed., *Acidic Deposition and Aquatic Ecosystems: Regional Case Studies* (New York: Springer-Verlag, 1991), p. 638, indicates that changes of 13 meq/l are relatively small.
71. Sullivan, *Historical Changes,* p. 151. The standard error on the paleolimnological technique used is 0.28 pH units.
72. Ibid.
73. NAPAP, *Acidic Deposition,* report no. 13, Joan P. Baker, "Biological Effects of Changes in Surface Water Chemistry," p. 121.
74. Ibid., p. 123.
75. Ibid., p. 122.
76. NAPAP, *Interim Assessment: The Causes and Effects of Acidic Deposition,* ed. Charles N. Merrick (Washington, D.C.: GPO, 1987): see John L. Malanchuk and Robert S. Turner, "Effects on Aquatic Systems," pp. 8–74 of volume 4.
77. NAPAP, *Acidic Deposition,* report no. 15, Harvey Olem, "Liming Acidic Surface Waters," p. 131.
78. Ibid., p. 132.
79. Ibid., p. 132, and see Likens et al., "Acid Rain." See also James M. Omernik and Glenn E. Griffith, "Total Alkalinity of Surface Waters: A Map of the Upper Midwest Region of the United States," *Environmental Management,* 10 (1986): 829–839.
80. Omernik and Griffith, "Total Alkalinity," p. 829.
81. Overrein, Seip, and Tollan, *Final Report of the SNSF Project, 1972–1980,* p. 7.
82. Omernik and Griffith, "Total Alkalinity," p. 829.
83. Sullivan, *Historical Changes.*
84. Omernik and Griffith, "Total Alkalinity," p. 834, table 1.
85. Christopher S. Cronan and Carl L. Schofield, "Aluminum Leaching Response to Acid Precipitation: Effects on High-Elevation Watersheds in the Northeast," *Science,* 204 (Apr. 20, 1979): 305.
86. Carl L. Schofield and John R. Trojnar, "Aluminum Toxicity to Brook Trout *(Salvelinus fontinalis)* in Acidified Waters," in Taft Y. Toribara, Morton W. Miller, and Paul E. Morrow, eds., *Polluted Rain* (New York: Plenum Press, 1980), p. 360.
87. Cronan and Schofield, "Aluminum Leaching Response," p. 305.
88. Charles T. Driscoll, Jr. et al., "Effect of Aluminum Speciation on Fish in Dilute Acidified Waters," *Nature,* 284 (Mar. 13, 1980): 163.
89. Ibid.
90. Noye M. Johnson, "Acid Rain: Neutralization with the Hubbard Brook Ecosystem and Regional Implications," *Science,* 204 (May 4, 1979): 497.
91. N. M. Johnson et al., " 'Acid Rain,' Dissolved Aluminum," p. 1432.

92. Sullivan, *Historical Changes,* p. 43.
93. Driscoll et al., "Effect of Aluminum Speciation," p. 162.
94. Sullivan, "Long-Term Temporal Trends," p. 629.
95. Ibid., p. 632.
96. NAPAP, *Acidic Deposition,* report no. 16, Joseph E. Barnard and Allen A. Lucier, "Changes in Forest Health and Productivity in the United States and Canada," pp. 135, 138; and D. Binkley et al., *Acidic Deposition and Forest Soils: Context and Case Studies in the Southeastern United States,* Ecological Studies Analysis and Synthesis 72 (New York: Springer-Verlag, 1989). p. 7.
97. Allan H. Legge and Sager V. Krupa, eds., *Acid Deposition: Sulfur and Nitrogen Oxides,* Alberta Government/Industry Acid Deposition Research Program (Chelsea, Mich.: Lewis Publishers, 1990), p. 39: "There are no demonstrated cases of negative effects of 'acidic rain' on vegetation under ambient conditions."
98. A. H. Johnson et al., "Recent Changes in Patterns of Tree Growth Rate in the New Jersey Pinelands: A Possible Effect of Acid Rain," *Journal of Environmental Quality,* 10 (1981): 427–430, citing T. Wood and F. H. Bormann, "Increases in Foliar Leaching Caused by Acidification with an Artificial Mist," *Ambio,* 4 (1975): 169–173.
99. See, e.g., W. W. McFee, J. M. Kelly and R. H. Beck, "Acid Precipitation Effects on Soil pH and Base Saturation of Exchange Sites," *Water, Air, and Soil Pollution,* 7 (1977): 401–408.
100. David A. Bennett, Robert L. Goble, and Rick A. Linthurst, eds., *The Acidic Deposition Phenomenon and Its Effects: Critical Assessment Document* report no. EPA 600/8-85/001 (Washington, D.C.: GPO, 1985), p. 11.
101. See Legge and Krupa, *Acid Deposition,* pp. 58–60, table 2.34.
102. Charles V. Cogbill, "The Effect of Acid Precipitation on Tree Growth in Eastern North America," *Water, Air, and Soil Pollution,* 8 (May 1977): 90.
103. Ibid., p. 92. Past growth rate can be determined through tree ring studies. A tree has one growth ring for each year, and the thickness of the ring provides a record of how much the tree grew over the year.
104. Ibid., p. 93; see also A. H. Johnson et al., "Recent Changes in Patterns of Tree Growth Rate in the New Jersey Pinelands: A Possible Effect of Acid Rain," *Journal of Environmental Quality,* 10 (1981): 427–430.
105. A. H. Johnson et al., "Recent Changes in Patterns," p. 430.
106. Ibid., p. 429.
107. N. M. Johnson, "Acid Rain," pp. 497–499.
108. See Legge and Krupa, *Acid Deposition,* pp. 58–60, table 2.34.
109. Bennett, Goble, and Linthurst, *The Acidic Deposition Phenomenon,* pp. 3–63.
110. NAPAP, *Annual Report to the President and Congress* (1986), p. 89.
111. Ibid.; J. W. Hornbeck, R. B. Smith, and C. A. Federer, "Growth Decline in Red Spruce and Balsam Fir Relative to Natural Processes," *Water, Air, and Soil Pollution,* 31 (Nov. 1986): 425, stated that "linking atmospheric deposition to regional tree health in the Northeastern United States has proven difficult."
112. NAPAP, *Acidic Deposition,* Barnard and Lucier, "Changes in Forest Health," pp. 135–137.
113. Ibid., p. 135; Arthur H. Johnson et al., "Assessing the Possibility of a Link between Acid Precipitation and Decreased Growth Rates of Trees in the North-

eastern United States," in Rick A. Linthurst, ed., *Direct and Indirect Effects of Acidic Deposition on Vegetation*, Acid Precipitation Series 5 (Boston: Butterworth, 1984), pp. 85–87.

114. NAPAP, *Acidic Deposition*, Barnard and Lucier, "Changes in Forest Health," p. 135.

115. Ibid., p. 136.

116. Ibid.

117. Ibid., pp. 135–136, is the source for this discussion.

118. Ibid., p. 137.

119. Ibid., p. 138.

120. J. Laurence Kulp, "Information Needs—Terrestrial," *Acid Rain: The Relationship between Sources and Receptors*, ed. James C. White (New York: Elsevier, 1988), p. 95.

121. W. de Vries, "Philosophy, Structure, and Application of a Soil Acidification Model for the Netherlands," in Juha Kämäri, ed., *Impact Models to Assess Regional Acidification* (Dordrecht: Kluwer Academic Publishers, 1990), pp. 3–21, p. 7 indicates that 80–90 percent of the soils in the Netherlands are acid sandy soils.

122. W. W. McFee et al., "Effects on Soil Systems," in Bennett, Goble, and Linthurst, *The Acidic Deposition Phenomenon*.

123. Pekka E. Kauppi, Kari Mielikäinen, and Kullervo Kuusela, "Biomass and Carbon Budget of European Forests, 1971–1990," *Science*, 256 (Apr. 3, 1992): 71.

124. Ibid., p. 74.

125. See esp. Overrein, Seip, and Tollan, *Final Report of the SNSF Project, 1972–1980*, p. 140, which states, "Decreases in forest growth due to acid deposits have not been demonstrated."

126. Ibid., p. 141.

127. W. Uhlmann et al., "The Problem of Forest Decline and the Bavarian Forest Toxicology Research Group," in E. D. Schulze, O. L. Lange, and R. Oren, eds., *Forest Decline and Air Pollution: A Study of Spruce (Picea abies) on Acid Soils*, Ecological Studies 77 (New York: Springer-Verlag, 1989), p. 1.

128. Binkley et al., *Acidic Deposition and Forest Soils*, p. 5.

129. Ibid., p. 6.

130. Uhlmann et al., "The Problem of Forest Decline," p. 1.

131. Kauppi, Mielikäinen, and Kuusela, "Biomass and Carbon Budget," p. 72.

132. Bernhard Prinz, "Causes of Forest Damage in Europe: Major Hypotheses and Factors," *Environment* 29, no. 9 (Nov. 1987): 12.

133. Kauppi, Mielikäinen, and Kuusela, "Biomass and Carbon Budget," p. 72.

134. The fact that the foliar damage peaked in the mid-1980s and has since decreased may indicate that the thinning is slowing. The trees that could not compete at all have died, leaving fewer trees to compete. If acid rain had damaged the soils, this "recovery"—the decrease in defoliated trees—would probably not occur because the soils would continue to be incapable of supporting the trees.

135. This increase was not due to increases in cultivated land. "The increase of growing stock and forest growth observed in Europe between 1971 and 1990 is almost entirely from stands that were already in place in 1971" (Kauppi, Mielikäinen, and Kuusela, "Biomass and Carbon Budget," p. 72).

136. Schulze, Lange, and Oren, *Forest Decline and Air Pollution*. We found no reference to growing stock in the entire book.

137. Schulze, Lange, and Oren, *Forest Decline,* p. 453.

138. Ibid., p. 454.

139. Legge and Krupa, *Acid Deposition,* p. 39. Although it can leach nutrients from the foliage, simple acid rain is apparently not dangerous; there is no evidence that it has caused any direct harm to trees through foliar leaching or otherwise. However, dry deposition in concert with wet is much more effective at leaching nutrients from leaves. The most effective process takes place in two stages: first, sulfate particles and aerosols are deposited on the surface of leaves; second, rainfall dissolves the particles. Depending on the extent of dry deposition on the leaves and the amount of water caught by them the pH of the solution formed on the leaves' surfaces could therefore be many times more acidic.

140. A. H. Johnson et al., "Recent Changes in Patterns," p. 427, citing Wood and Bormann, "Increases in Foliar Leaching."

141. J. L. Kulp, "Effects on Forests," in NAPAP, *Interim Assessment,* vol. 4, pp. 7-i–7-59.

142. NAPAP, *Acidic Deposition,* report no. 18, David S. Shriner, "Responses of Vegetation to Atmospheric Deposition and Air Pollution," p. 150, indicates that neither sulfur dioxide nor nitrogen oxides have been responsible for growth reductions of trees in the United States.

143. See J. O. Reuss and D. W. Johnson, "Effects of Soil Processes on the Acidification of Water by Acid Deposition," *Journal of Environmental Quality,* 14 (1985): 26–31. See also Binkley et al., *Acidic Deposition and Forest Soils;* W. W. McFee et al., "Effects on Soil Systems," in Bennett, Goble, and Linthurst, *The Acidic Deposition Phenomenon;* NRC, *Acid Deposition* (1986); and Legge and Krupa, *Acid Deposition,*

144. McFee, Kelly, and Beck, "Acid Precipitation Effects," p. 404.

145. The first false assumption is that there are three not one source of cations: the precipitation that carries the acid, the decaying forest litter, and the weathering of silicates, which may be accelerated by a decrease in pH of the precipitation (ibid., p. 405). The second false assumption, as acknowledged by the authors, is that base cations (nutrients) are the only ions that are exchanged for hydrogen ions. In most soils aluminum (Al^{3+}), which is considered an acid cation, exchanges as well. A. H. Johnson et al. ("Recent Changes in Patterns") indicate that the dissolution of base cations such as Ca^{2+}, Na^+, Mg^{2+}, and K^+ is slow relative to aluminum dissolution. Aluminum is therefore most likely the first line of neutralization rather than base cations in many soils (see N. M. Johnson, "Acid Rain," and the discussion of aluminum as the first buffer in the Hubbard Brook soils). The third false assumption is that the only cation input by precipitation into the system is hydrogen. This is a part of the first assumption and is clearly wrong, as shown by a number of scientists. See, e.g., Edward C. Krug and Charles R. Frink, "Acid Rain on Acid Soil: A New Perspective," *Science,* 221 (Aug. 5, 1983): 520–525.

146. McFee, Kelly, and Beck, "Acid Precipitation Effects," pp. 407–408.

147. D. W. Johnson et al., *Effects of Acid Rain on Forest Nutrient Status,* Oak Ridge National Laboratory, Environmental Sciences Division, publication no. 2498

(Oak Ridge, Tenn.: ORNL, 1985), point out that a negative result foreseen in five forests (p. 4-1) does not show that there is no regional threat. It shows only that those particular forests are not at risk. Each forest will have its own unique response to acid rain.

148. NAPAP, *Acidic Deposition,* Shriner, "Responses of Vegetation," p. 148.

149. Bennett, Goble, and Linthurst, *The Acidic Deposition Phenomenon,* p. 17. There are approximately 12,000 soil series in the United States, each having a unique combination of properties. When we combine this variability with the fluctuations in acid loading that are experienced in any given region and the differences in deposition rate and character from one region to another, we see that the soil of each stand of trees will probably have its own unique response to acid rain. Nevertheless, there are a number of common features that can be described and quantified and a general framework of discussion that can be constructed to help provide a picture of how acid rain has affected forests through soils and may affect them in the future. Until we specifically discuss the less than 0.1 percent of the forest area that is at high elevation, the term "forest soils" refers to "low-elevation forest soils."

150. From McFee et al., "Effects on Soil Systems," p. 2-2, in which they state that over 95 percent of our food comes directly or indirectly from terrestrial plants.

151. D. W. Cowling and L. H. P. Jones, "A Deficiency in Soil Sulfur Supplies for Perennial Ryegrass in England," *Soil Science,* 110 (1970): 353.

152. Legge and Krupa, *Acid Deposition,* p. 35.

153. McFee et al., "Effects on Soil Systems," p. 2-20.

154. Kulp, "Information Needs—Terrestrial," p. 96.

155. Ibid.

156. NAPAP, *Acidic Deposition,* report no. 22, Judith A. Graham, "Direct Health Effects of Air Pollutants Associated with Acidic Precursor Emissions," p. 180.

157. See Patricia M. Irving, "Effects of Acidic Deposition and Associated Gaseous Pollutants on Human Health and Visibility," in NAPAP, *Interim Assessment,* vol. 4, p. 10-6, for information on the actual levels found in the United States. Note that as of 1987 "short-term air quality data have not been adequately summarized to estimate the size of human populations exposed to elevated levels of air pollutants."

158. E. L. Avol, W. S. Linn, and J. D. Hackney, "Acute Respiratory Effects of Ambient Acid Fog Episodes: Final Report" (1986), NIEHS grant number ESO3291-02, cited in NAPAP, *Interim Assessment,* vol. 4, Irving, "Effects," p. 10-31.

159. NAPAP, *Interim Assessment,* vol. 4, Irving, "Effects," p. 10-6, citing H. G. Houghton, "On the Chemical Composition of Fog and Cloud Water," *Journal of Meteorology,* 12 (1955): 355–357. For instance, in coastal New England and Canada, where the rain is highly acidic, the fogs evidently range from pH 3.5 to 7.4.

160. According to NAPAP, *Interim Assessment,* vol. 4, Irving, "Effects," p. 10-7, table 10-2, peak one-hour mean concentrations of SO_2 in the Los Angeles basin are generally below 1,000 mg/m³.

161. Articles were gathered from the years 1974 to 1991. Articles were chosen to be relevant by their listings in the indexes of each newspaper or in *The Reader's Guide to Periodical Literature* for magazines. Because of the prodigious number

of articles appearing in the *New York Times*—literally thousands of articles—limited numbers of articles from this paper were collected. Any article beginning within the first fifteen pages of the front section, or within the first five pages of any subsequent section, were examined. In all, 586 articles were included in this survey, with the following breakdown among publications: *Wall Street Journal,* 95; *Los Angeles Times,* 90; *New York Times,* 198; *Washington Post,* 175; *Time,* 10; *Newsweek,* 9; *U.S. News and World Report,* 9.

162. Philip Shabecoff, "Study Discounts Immediate Peril of Acid Rain," *New York Times,* Sep. 18, 1987, pp. A1, A27.

163. Philip Shabecoff, "Government Acid Rain Report Comes under Sharp Attack," *New York Times,* Sep. 22, 1987, pp. C1, C2.

164. Reuters, "U.S. Is Denounced on Acid Rain Study," *New York Times,* Jan. 9, 1988, p. A7.

165. William K. Stevens, "Researchers Find Acid Rain Imperils Forests over Time," *New York Times,* Dec. 31, 1989, pp. A1, A22.

166. William K. Stevens, "Study of Acid Rain Uncovers a Threat to Far Wider Area," *New York Times,* Jan. 16, 1990, p. C4.

167. William K. Stevens, "Worst Fears on Acid Rain Unrealized," *New York Times,* Feb. 20, 1990, pp. C1, C11.

168. Philip Shabecoff, "Acid Rain Report Unleashes a Torrent of Criticism," *New York Times,* Mar. 20, 1990, p. C4.

169. Delthia Ricks, "Peril in the Sky," *Los Angeles Times,* Aug. 24, 1986, pp. 1–3.

170. Ralph Blumenthal, "Acid Rain in the Adirondacks Disrupts the Chain of Life," *New York Times,* June 8, 1981, pp. B1, B14.

171. Keith Schneider, "Lawmakers Agree on Rules to Reduce Acid Rain Damage," *New York Times,* Oct. 22, 1990, pp. A1, A14.

172. Associated Press, "Pollution Assessed in Ohio River Valley," *New York Times,* Mar. 2, 1981, p. A12.

173. Donald Charles, "Acid Rain not Only to Blame," *Nature,* 362 (Apr. 29, 1993): 794.

174. Dieter Mueller-Dombois, "Natural Diebacks in Forests: Groups of Neighboring Trees May Die in Response to Natural Phenomenon, As Well As to Stress Induced by Human Activity," *Bioscience,* 37 (Sep. 1987): 575–583.

175. L. W. Blank, T. M. Roberts, and R. A. Skeffington, "New Perspectives on Forest Decline . . .," *Nature,* 336 (Nov. 3, 1988): 27–30.

10. CFCs and Ozone Depletion

1. World Meteorological Organization [WMO] and NASA, *Atmospheric Ozone, 1985: Assessment of Our Understanding of the Processes Controlling Its Present Distribution and Change,* WMO Global Ozone Research and Monitoring Project, report no. 16 (Greenbelt, Md.: NASA Laboratory for Atmospheres, 1986), p. 70.

2. Mario J. Molina and F. S. Rowland, "Stratospheric Sink for Chlorofluoromethanes: Chlorine Atom-Catalysed Destruction of Ozone," *Nature,* 249 (June 28, 1974): 810.

3. Ibid., p. 812.

4. H. Patricia Hynes, *EarthRight: What You Can Do in Your Home, Workplace,*

and Community to Save Our Environment (Rocklin, Calif.: Prima Publishing and Communications, 1990), p. 141.

5. J. E. Lovelock, R. J. Maggs, and R. J. Wade, "Halogenated Hydrocarbons in and over the Atlantic," *Nature*, 241 (Jan. 19, 1973): 194–196.

6. Sharon L. Roan, *Ozone Crisis: The 15-Year Evolution of a Sudden Global Emergency* (New York: John Wiley and Sons, 1989), p. xiii.

7. F. S. Rowland and M. J. Molina, "Chlorofluoromethanes in the Environment," *Reviews of Geophysics and Space Physics,* 13 (1975): 1–36.

8. See, e.g., National Research Council [NRC], *Protection against Depletion of Stratospheric Ozone by Chlorofluorocarbons,* specifically, "Solar Ultraviolet Irradiance at the Earth's Surface" (Washington, D.C.: National Academy of Sciences, 1979), p. 300.

9. F. Sherwood Rowland, "Stratospheric Ozone Depletion," *Annual Review of Physical Chemistry,* 42 (1991): 763, where he cites WMO, *Atmospheric Ozone, 1985.*

10. Mary Murray, "Overexposure," *Sciences* (New York Academy of Sciences), Apr./May 1992, p. 13.

11. NRC, Panel on Atmospheric Chemistry, Assembly of Mathematical and Physical Sciences, *Halocarbons: Effects on Stratospheric Ozone,* (Washington, D.C.: National Academy of Sciences, 1976), p. 173, figure 9.2.

12. Citing Melinda Beck and Mary Hager, "More Bad News for the Planet: A Grim Report on Ozone," *Newsweek,* Mar. 28, 1988, p. 63.

13. Murray, "Overexposure," pp. 13–14.

14. Hugh W. Ellsaesser, "A Reassessment of Stratospheric Ozone: Credibility of the Threat," *Climatic Change,* 1 (1978): 259.

15. See, e.g., Guy Brasseur and A. De Rudder, "Agents and Effects of Ozone Trends in the Atmosphere," in Robert C. Worrest and Martyn M. Caldwell, eds., *Stratospheric Ozone Reduction, Solar Ultraviolet Radiation, and Plant Life,* NATO ASI Series G: Ecological Sciences, vol. 8 (Berlin: Springer-Verlag, 1986), p. 1.

16. Peter Warneck, *Chemistry of the Natural Atmosphere,* International Geophysics Series, vol. 41 (San Diego: Academic Press, 1988), p. 127.

17. See, e.g., J. G. Anderson, D. W. Toohey, and W. H. Brune for the explanation of the chemistry in their paper "Free Radicals within the Antarctic Vortex: The Role of CFCs in the Antarctic Ozone Loss," *Science,* 251 (Jan. 4, 1991): 39–46.

18. Molina and Rowland, "Stratospheric Sink," p. 810.

19. Ibid., "two chlorofluoromethanes . . . have been detected throughout the troposphere in amounts . . . roughly corresponding to the integrated world industrial production to date . . . There are no obvious rapid sinks for their removal."

20. See, e.g., WMO, *Atmospheric Ozone, 1985,* p. 70, and Ivar S. A. Isaksen, "The Beginnings of a Problem," in Robin Russell Jones and Tom Wigley, eds., *Ozone Depletion: Health and Environmental Consequences* (New York: John Wiley and Sons, 1989), p. 20. The lifetime for various CFCs in the two sources that are quoted in each text are slightly different, but the basic notion that the lifetime is long has stayed the same. See also Michael J. Prather and Robert T. Watson, "Stratospheric Ozone Depletion and Future Levels of Atmospheric Chlorine and Bromine," *Nature,* 344 (Apr. 19, 1990): 729–734.

21. S. Fred Singer, "Stratospheric Ozone: Science and Policy" in Singer, ed., *Global*

Climate Change (New York: ICUS, 1989), pp. 157–162; see Richard Stolarski and Ralph Cicerone, "Stratospheric Chlorine: A Possible Sink for Ozone," *Canadian Journal of Chemistry,* 52 (1974): 1610–1615; D. A. Johnston, "Volcanic Contribution of Chlorine to the Stratosphere: More Significant to Ozone than Previously Estimated?" *Science,* 209 (July 2, 1980): 491–493. See also R. B. Symonds, W. I. Rose, and M. H. Reed, "Contributions of Cl- and F-Bearing Gases in the Atmosphere by Volcanoes," *Nature,* 334 (Aug. 4, 1988): 415–418, for a discussion of chlorine from volcanoes; H. B. Singh, L. J. Salas, and R. E. Stiles, "Methyl Halides in and over the Eastern Pacific (40N–32S)," *Journal of Geophysical Research,* 88, no. C6 (Apr. 20, 1983): 3684–3690, for oceanic biota; and B. J. Finlayson-Pitts, M. J. Ezell, and J. N. Pitts, Jr., "Formation of Chemically Active Chlorine Compounds by Reactions of Atmospheric NaCl Particles with Gaseous N_2O_5 and $ClONO_2$," *Nature,* 337 (Jan. 19, 1989): 241, for sea salt.

22. Prather and Watson, "Stratospheric Ozone Depletion." Richard A. Kerr, "New Assaults Seen on the Earth's Ozone Shield," *Science,* 255 (Feb. 14, 1992): 797; Kerr says that according to the United Nations the chlorine concentration in the stratosphere is now 3.4 ppbv.
23. Prather and Watson, "Stratospheric Ozone Depletion," p. 729.
24. C. P. Rinsland et al., "Infrared Measurements of HF and HCl Total Column Abundances above Kitt Peak, 1977–1990: Seasonal Cycles, Long-Term Increases, and Comparisons with Model Calculations," *Journal of Geophysical Research,* 96, no. D8 (Aug. 20, 1991): 15538.
25. Ibid.
26. See, e.g., Rogelio A. Maduro and Ralf Schauerhammer, *The Holes in the Ozone Scare: The Scientific Evidence That the Sky Isn't Falling* (Washington, D.C.: Twenty-first Century Science Associates, 1992), pp. 11–38.
27. Symonds, Rose, and Reed, "Contributions of Cl- and F-Bearing Gases," p. 417.
28. Johnston, "Volcanic Contribution to the Stratosphere," p. 492.
29. Rowland and Molina, "Chlorofluoromethanes in the Environment."
30. NRC, *Halocarbons: Effects on Ozone,* p. 160.
31. NRC, Committee on Impacts of Stratospheric Change, *Halocarbons: Environmental Effects of Chlorofluoromethane Release* (Washington, D.C.: National Academy of Sciences, 1976), p. 18.
32. Ibid., p. 19.
33. James G. Anderson, "Free Radicals in the Earth's Atmosphere: Measurement and Interpretation," in *Ozone Depletion, Greenhouse Gases, and Climatic Change* (Washington, D.C.: National Academy Press, 1989), p. 57.
34. "Report of the Directorate of Air, Noise, and Wastes" (London: U.K. Department of the Environment, 1977), part 2, p. 193.
35. Brasseur and De Rudder, "Agents and Effects of Ozone Trends," p. 23, fig. 8.
36. WMO, *Atmospheric Ozone, 1985,* p. 787.
37. Brasseur and De Rudder, "Agents and Effects of Ozone Trends," p. 17.
38. WMO, *Scientific Assessment of Stratospheric Ozone, 1989,* Global Ozone Research and Monitoring Project, report no. 20 (Washington, D.C.: U.S. Government Printing Office, 1991), chap. 3.
39. WMO *Atmospheric Ozone, 1985,* p. 726. The models were run "at Lawrence Livermore National Laboratory (D. Wuebbles), Harvard (M. Prather), Atmo-

spheric and Environment Research, Inc. (D. Sze), E.I. DuPont de Nemours & Company (A. Owens), Institute for Aeronomy (G. Brasseur) and Max Planck Institute (MPIC) in Mainz, West Germany (Bruehl and Crutzen)." Not all of the models were run for every scenario.

40. Ibid.

41. Richard Stolarski et al., "Measured Trends in Stratospheric Ozone," *Science*, 256 (Apr. 17, 1992): 342–349.

42. WMO *Atmospheric Ozone, 1985*, p. 777–779.

43. NRC, *Causes and Effects of Stratospheric Ozone Reduction: An Update* (Washington, D.C.: National Academy Press, 1982), p. 25.

44. Brasseur and De Rudder, "Agents and Effects of Ozone Trends," p. 18; WMO, *Atmospheric Ozone, 1985*, p. 439.

45. WMO, *Atmospheric Ozone, 1985*, p. 439.

46. Brasseur and De Rudder, "Agents and Effects of Ozone Trends," p. 18; WMO, *Atmospheric Ozone, 1985*, p. 428.

47. United Kingdom Stratospheric Ozone Review Group, *Stratospheric Ozone, 1988: Second Report* (London: HM Stationery Office, 1988), p. 37.

48. WMO *Atmospheric Ozone, 1985*, p. 786.

49. WMO, *Stratospheric Ozone, 1989*, p. 358.

50. This conclusion is based on a comparison of figs. 3.2–17 and 3.2–23, in ibid.

51. Richard Elliot Benedick, *Ozone Diplomacy: New Directions in Safeguarding the Planet* (Cambridge, Mass.: Harvard University Press, 1991), p. 110.

52. Rowland, "Stratospheric Ozone Depletion," pp. 731–768.

53. The 1 percent lower bound comes from Gregory Reinsel et al., "Statistical Analysis of Total Ozone and Stratospheric Umkehr Data for Trends and Solar Cycle Relationship," *Journal of Geophysical Research*, 92, no. D2 (Feb. 20, 1987): 2205; the 3 percent estimate comes from Guy Brasseur and P. C. Simon, "Stratospheric Chemical and Thermal Response to Long-Term Variability in Solar UV Irradiance," *Journal of Geophysical Research*, 86, no. C8 (Aug. 20, 1981): 7354.

54. NRC, *Halocarbons: Effects on Ozone*, p. 83.

55. Guy Brasseur et al., "Ozone Reduction in the 1980s: A Model Simulation of Anthropogenic and Solar Perturbations," *Geophysical Research Letters*, 15, no. 12 (1988): 1363; the authors indicate that solar UV-induced ozone variability can be predicted using the variability of the irradiance at 205 nm.

56. J. L. Lean, "Contribution of Ultraviolet Irradiance Variations to Changes in the Sun's Total Irradiance," *Science*, 244 (Apr. 14, 1989): 198.

57. J. L. Lean et al., "A Three-Component Model of the Variability of the Solar Ultraviolet Flux: 145–200 nm," *Journal of Geophysical Research*, 87, no. A12 (Dec. 1, 1982): p. 10307.

58. Gerald M. Keating, "The Response of Ozone to Solar Activity Variations: A Review," *Solar Physics*, 74 (Dec. 1981): 322. This variability is of the same order as the variations predicted by some models of solar irradiance.

59. See for example G. M. Keating et al., "Global Ozone Long-Term Trends from Satellite Measurements and the Response to Solar Activity Variations," *Journal of Geophysical Research*, 86, no. C10 (Oct. 20, 1981): 9873; Lean et al. "Three-Component Model," p. 10307; and Reinsel et al., "Statistical Analysis of Total Ozone," p. 2201. There are not enough observations of ultraviolet irradiance of

high enough precision to say definitively that 10.7 cm is a good proxy. (If there were enough observations to determine that definitively, there would be no need for a proxy, except in analysis of data from before the time when the high precision observations were made.) In fact, there is good reason to believe that the two phenomena may be quite dissimilar, for ultraviolet radiation and 10.7 cm radiation are emitted from different layers of the sun. The processes that produce each radiation may be different. If that is the case, there is no reason to believe that variations in the intensity of ultraviolet radiation will be the same as those of 10.7 cm radiation. There is some evidence, however, that the variabilities of the two bands are coupled; the variations in the 180 nm range observed by the Astronomical Exploration Experiment were in phase with and of similar magnitude as the variations in the 10.7 cm range (Keating et al., "Global Ozone Long-Term Trends," p. 9879). Although this observation is interesting, it does not represent enough evidence to establish an effective proxy. But tests of the proxy did show it was promising.

60. Keating et al., "Global Ozone Long-Term Trends," p. 9873.
61. Ibid.
62. E.g., J. K. Angell and J. Korshover, "Global Ozone Variations: An Update into 1976," *Monthly Weather Review,* 106 (May 1978): 725–737; the authors suggest that a number of investigators came to the conclusion that there was a 1–3 percent decrease in ozone around the earth between 1970, near the time of the solar maximum, and 1975, the approximate time of the solar minimum, which was not explained by other factors.
63. The model of solar activity that is most widely used was developed by Lean and her colleagues and published in 1982. Lean, and others before her, predict that there will be a total variability of solar irradiance of 25 percent at 185 nm decreasing logarithmically to 1 percent at 300 nm. Also see Brasseur and Simon, "Sratospheric Chemical and Thermal Response," p. 7343. Such variation is likely to lead to significant variability in the total column ozone in the atmosphere over the course of the eleven-year cycle. Models of ozone chemistry using this variability predicted a decrease of 1.2 percent (Reinsel et al., "Statistical Analysis of Total Ozone," p. 2205) to 3 percent (Brasseur and Simon, "Stratospheric Chemical and Thermal Response," p. 7354) in global total ozone from solar maximum to minimum with substantially larger effects in the upper than in the lower stratosphere.
64. A 3 percent variation in ozone may be explained by "solar variability ranging from 20% at 1800Å to 4% at 3000Å" (Keating et al., "Global Ozone Long-Term Trends," p, 345). There was nothing in the data that would have ruled out such variability, because the numbers are good only to ±15 percent. See ibid., pp. 345–346; Brasseur and Simon, "Stratospheric Chemical and Thermal Response"; and Barry M. Schlesinger and Richard P. Cebula, "Solar Variation, 1979–1987, Estimated from an Empirical Model for Changes with Time in the Sensitivity of the Solar Backscatter Ultraviolet Instrument," *Journal of Geophysical Research,* 97, no. D9 (June 20, 1992): 10133.
65. Lean, "Contribution of Ultraviolet Irradiance," p. 198: "Both the UV irradiances and the total irradiance vary throughout the 11-year sunspot cycle, reaching their peak values near the maximum of solar activity." The OTP analysis is based on

data from 1965 to 1986, two full solar cycles. The panel then used a "hockey stick" trend analysis in which scientists assume a level null trend for the first four years and infer a trend for the remainder of the interval.

66. "Report by the Directorate of Air, Noise, and Wastes," part 2, pp. 72–73.

67. WMO, *Atmospheric Ozone, 1985,* p. 274.

68. R. Bojkov et al., "A Statistical Trend Analysis of Revised Dobson Total Ozone Data over the Northern Hemisphere," *Journal of Geophysical Research,* 95, no. D7 (June 20, 1990): 9793.

69. Richard S. Stolarski et al., "Total Ozone Trends Deduced from Nimbus 7 TOMS Data," *Geophysical Research Letters* 18, no. 6 (1991): 1015–1018.

70. Bojkov et al., "Statistical Trend Analysis," p. 9806. According to Joseph Scotto et al. ("Biologically Effective Radiation: Surface Measurements in the United States, 1974 to 1985," *Science,* 239 [Feb. 12, 1988]: 762), "The most effective biological wavelength for producing erythema on typical caucasian skin is 297 nm. The biological effectiveness decreases logarithmically within the UVB range; at 330 nm it is less than 0.1% as effective as at 297 nm. The R-B meter integrates the weighted amounts of UV and provides counts in sunburn units . . . Average annual R-B counts for two consecutive 6-year periods (1974 to 1979 and 1980 to 1985) show a negative shift at each station with decreases ranging from 2 to 7% . . . there are no positive trends in annual R-B counts from 1974 to 1985."

71. Scotto et al., "Biologically Effective Radiation," p. 762.

72. WMO, *Stratospheric Ozone, 1989,* p. 266.

73. Stolarski et al., "Measured Trends in Stratospheric Ozone."

74. C. Brühl and P. J. Crutzen, "On the Disproportionate Role of Tropospheric Ozone as a Filter against UV-B Radiation," *Geophysical Research Letters,* 16, no. 7 (1989): 705.

75. Ibid., p. 706.

76. See, e.g., Sasha Madronich, "Implications of Recent Total Atmospheric Ozone Measurements for Biologically Active Ultraviolet Radiation Reaching the Earth's Surface," *Geophysical Research Letters,* 19, no. 1 (1992): 37–40.

77. Paul J. Crutzen, "Ultraviolet on the Increase," *Nature,* 356 (Mar. 12, 1992): 104.

78. Stolarski et al., "Measured Trends in Stratospheric Ozone."

79. Stolarski et al., "Total Ozone Trends," pp. 1015–1018.

80. H. De Backer and D. De Muer, "Intercomparison of Total Ozone Data Measured with Dobson and Brewer Ozone Spectrophotometers at Uccle (Belgium) from January 1984 to March 1991, Including Zenith Sky Observations," *Journal of Geophysical Research,* 96, no. D11 (Nov. 20, 1991): 20711–20719; D. De Muer and H. De Backer, "Revision of 20 Years of Dobson Total Ozone Data at Uccle (Belgium): Fictitious Dobson Total Ozone Trends Induced by Sulfur Dioxide Trends," *Journal of Geophysical Research,* 97, no. D5 (Mar. 1992): 5921–5928.

81. Stolarski et al., "Total Ozone Trends," p. 1015.

82. Stolarski et al., "Measured Trends in Stratospheric Ozone," pp. 346–347.

83. Ibid., p. 348.

84. Ibid., p. 347.

85. De Muer and De Backer, "Revision of 20 years," pp. 5937, 5934.

86. Bojkov et al., "Statistical Trend Analysis," p. 9791.

87. De Muer and De Backer, "Revision of 20 years," p. 5921.

88. Ibid., p. 5922.
89. G. M. B. Dobson, "Note on the Measurement of Ozone in the Atmosphere," *Quarterly Journal of the Royal Meteorological Society,* 89 (1963): 409–411.
90. W. D. Komhyr and R. D. Evans, "Dobson Spectrophotometer Total Ozone Measurement Errors Caused by Interfering Absorbing Species such as SO_2, NO_2, and Photochemically Produced O_3 in Polluted Air," *Geophysical Research Letters,* 7, no. 2 (1980): 157–158.
91. De Muer and De Backer, "Revision of 20 Years," p. 5933.
92. Bojkov et al., "Statistical Trend Analysis," p. 9791.
93. J. C. Farman, B. G. Gardiner, and J. D. Shanklin, "Large Losses of Total Ozone in Antarctica Reveal Seasonal ClO_x/NO_x Interaction," *Nature,* 315 (May 16, 1985): 207.
94. Susan Solomon et al., "On the Depletion of Antarctic Ozone," *Nature,* 321 (June 19, 1986): 755. See also Michael B. McElroy et al., "Reductions of Antarctic Ozone Due to Synergistic Interactions of Chlorine and Bromine," *Nature,* 321 (June 19, 1986): 759, which indicates that the concentration dropped farther in 1985, to among the lowest ever recorded on earth.
95. N. Chubachi et al., "The Preliminary Results of Ozone Observations at Syowa from 1982 to January 1983," *Memoirs of the National Institute of Polar Research* (Tokyo), special issue, 34 (1984): 13–19.
96. James G. Anderson, personal communication (1993).
97. R. S. Stolarski et al., "Nimbus 7 Satellite Measurements of the Springtime Antarctic Zone Decrease," *Nature,* 322 (Aug. 28, 1986): 808.
98. Ibid.
99. See, e.g., Stolarski et al., "Measured Trends in Stratospheric Ozone."
100. Farman, Gardiner, and Shanklin, "Large Losses," p. 208.
101. Ka-Kit Tung et al., "Are Antarctic Ozone Variations a Manifestation of Dynamics or Chemistry," *Nature,* 322 (Aug. 28, 1986): 812.
102. Ibid., p. 811.
103. Ibid., pp. 811–812.
104. See the following articles in *Geophysical Research Letters,* 13, no. 12 (supp., Nov. 1986): Richard Stolarski and Mark Schoeberl, "Further Interpretation of Satellite Measurements of Antarctic Total Ozone," pp. 1210–1212; Mark R. Schoeberl and Arlin J. Krueger, "The Morphology of Antarctic Total Ozone as Seen by TOMS," pp. 1217–1220; Ka-Kit Tung, "On the Relationship between the Thermal Structure of the Stratosphere and the Seasonal Distribution of Ozone," pp. 1308–1311; Guang-Yu Shi et al., "Radiative Heating Due to Stratospheric Aerosols over Antarctica," pp. 1335–1338; Joan E. Rosenfeld and Mark R. Schoeberl, "A Computation of Stratospheric Heating Rates and the Diabatic Circulations for the Antarctic Spring," pp. 1339–1342.
105. Solomon et al., "On the Depletion of Antarctic Ozone," p. 755 (abstract). See also McElroy et al., "Reductions of Antarctic Ozone Due to Synergistic Interactions," p. 759; J. G. Anderson, D. W. Toohe, and W. H. Brune, "Free Radicals within the Antarctic Vortex: The Role of CFCs in the Antarctic Ozone Loss," *Science,* 251 (Jan. 4, 1991): 39.
106. Processes involving both gas and solid or gas and liquid phases.
107. Solomon et al., "On the Depletion of Antarctic Ozone," p. 757; Michael B.

McElroy, Ross J. Salawitch and Steven C. Wofsy, "Antarctic O$_3$: Chemical Mechanisms for the Spring Decrease," *Geophysical Research Letters,* 13, no. 12 (supp., Nov. 1986): 1296. The cooling trend that they suggest may have taken place could have arisen from a number of causes such as the Pacific current El Niño or aerosols injected by the volcano El Chichon (in Mexico) leading to a positive feedback in which the aerosols provide the seed that triggers formation of polar stratospheric clouds, which allow for greater cooling, which forms more crystals. J. Austin et al. claimed there were virtually no PSCs after September 10 at 20 km in the southern polar regions and that therefore heterogeneous chemistry could not have been responsible for the depletions observed. ("Polar Stratospheric Clouds Inferred from Satellite Data," *Geophysical Research Letters,* 13, no. 12 [supp., Nov. 1986]: 1256). However, another report in the same issue claimed to "show that the ozone hole develops and the clouds dissipate in the same place at the same time" (Patrick Hamill, O. B. Toon, and R. P. Turco, "Characteristics of the Polar Stratospheric Clouds during the Formation of the Antarctic Ozone Hole," p. 1288). The authors pointed out that at the temperatures observed (colder than −88°C) NO$_2$ and NO$_3$ would have combined to form N$_2$O$_5$, which would have reacted with water to form nitric acid. Since the temperature was cold enough to freeze nitric acid, the gaseous N$_2$O$_5$ would have been converted into a crystal of nitric acid. Once there was a seed crystal, the clouds would form quite rapidly (pp. 1290–1291). The postulated heterogeneous chemistry could then have taken place in the clouds on the surface of these crystals.

108. Anderson, Toohe, and Brune, "Free Radicals," p. 43.
109. Mark Schoeberl and Dennis Hartmann, "The Dynamics of the Stratospheric Polar Vortex and Its Relation to Springtime Ozone Depletions," *Science,* 251 (Jan. 4, 1991): 48.
110. Anderson, Toohe, and Brune, "Free Radicals," p. 44.
111. Ibid., p. 45. Mechanism from L. T. Molina and M. J. Molina, "Production of Cl$_2$O$_2$ from the Self-Reaction of the ClO Radical," *Journal of Physical Chemistry,* 91 (Jan. 15, 1987): 433.
112. Schoeberl and Hartmann, "Dynamics of the Stratospheric Polar Vortex," p. 51.
113. Stolarski et al., "Measured Trends in Stratospheric Ozone," p. 345.
114. Ibid., p. 344.
115. Prather and Watson, "Stratospheric Ozone Depletion."
116. "Currents," *Environmental Science and Technology,* 26 (July 1992): 1267.
117. Maduro and Schauerhammer, *The Holes in the Ozone Scare,* p. 141.
118. J. G. Anderson et al., "Kinetics of O$_3$ Destruction by ClO and BRO within the Antarctic Vortex—An Analysis Based on In Situ ER-2 Data," *Journal of Geophysical Research,* 94, no. D9 (Aug. 30, 1989): 11488. The rate-limiting step has a quadratic dependence on the concentration of ClO.
119. WMO, *Stratospheric Ozone, 1989,* p. 11.
120. J. F. Gleason et al., "Record Low Global Ozone in 1992," *Science,* 260 (Apr. 23, 1992): 523.
121. Guy Brasseur and Claire Granier, "Mount Pinatubo Aerosols, Chlorofluorocarbons, and Ozone Depletion," *Science,* 257 (Aug. 28, 1992): 1239.
122. Jose M. Rodriguez, Malcolm K. W. Ko, and Nien Dak Sze, "Role of Heterogeneous Conversion of N$_2$O$_5$ on Sulphate Aerosols in Global Ozone Losses," *Nature,* 352 (July 11, 1991): 134.

123. D. W. Fahey et al., "*In Situ* Measurements Constraining the Role of Sulphate Aerosols in Mid-Latitude Ozone Depletion," *Nature,* 363 (June 10, 1993): 509–514.

124. Ibid., p. 514

125. James G. Anderson, personal communication (October 6, 1993).

126. Murray, "Overexposure," pp. 13–14.

127. Beck and Hager, "More Bad News for the Planet," p. 63.

128. Michael Balter, "Europe: As Many Cancers as Cuisines," *Science,* 254 (Nov. 22, 1991): 1114.

129. Jean-Pierre Cesarini, "Effects of Ultraviolet Radiation on the Human Skin: With Emphasis on Skin Cancer," in W. F. Passchier and B. F. M. Bosnjakovic, eds., *Human Exposure to Ultraviolet Radiation: Risks and Regulations* (New York: Elsevier, 1987), pp. 33–44.

130. Balter, "Europe," p. 1114.

131. Singer, "Stratospheric Ozone," p. 161.

132. Cesarini, "Effects," p. 39.

133. Ellsaesser, "Reassessment of Stratospheric Ozone," p. 259. The change in ultraviolet exposure is not linear with respect to latitude or elevation; nevertheless, on average these numbers are a good first-order approximation.

134. R. H. Biggs and P. G. Webb, "Effects of Enhanced Ultraviolet-B Radiation on Yield, and Disease Incidence and Severity for Wheat," in Robert C. Worrest and Martyn M. Caldwell, eds., *Stratospheric Ozone Reduction, Solar Ultraviolet Radiation, and Plant Life,* NATO ASI Series G: Ecological Sciences, vol. 8 (Berlin: Springer-Verlag, 1986), p. 303.

135. M. Tevini and W. Iwanzik, "Effects of UV-B Radiation on Growth and Development of Cucumber Seedlings," in Worrest and Caldwell, *Stratospheric Ozone Reduction* (cited in previous note), p. 271. Iwanzik, "Interaction of UV-A, UV-B, and Visible Radiation on Growth, Composition, and Photosynthetic Activity in Radish Seedlings," in Worrest and Caldwell, *Stratospheric Ozone Reduction,* p. 287.

136. J. Lydon, A. H. Teramura, and E. G. Summers, "Effects of Ultraviolet-B Radiation on the Growth and Productivity of Field Grown Soybean," in Worrest and Caldwell, *Stratospheric Ozone Reduction,* p. 313.

137. A. H. Teramura, "Interaction between UV-B Radiation and Other Stresses in Plants," in Worrest and Caldwell, *Stratospheric Ozone Reduction,* p. 341.

138. All of the provisions we have paraphrased were taken from Appendix B, "Montreal Protocol on Substances That Deplete the Ozone Layer, September 1987," of Benedick's book *Ozone Diplomacy.* The text of this appendix is from the *Montreal Protocol on Substances That Deplete the Ozone Layer, Final Act* (Nairobi: UNEP, 1987).

139. Benedick, *Ozone Diplomacy,* p. 232, (reprinting *Montreal Protocol on Substances That Deplete the Ozone Layer, Final Act*).

140. F. Sherwood Rowland, "Chlorofluorocarbons, Stratospheric Ozone, and the Antarctic 'Ozone Hole,' " in S. Fred Singer, ed., *Global Climate Change* (New York: ICUS, 1989), p. 150.

141. L. E. Manzer, "The CFC-Ozone Issue: Progress on the Development of Alternatives to CFCs," *Science,* 249 (July 6, 1990): 32.

142. Benedick, *Ozone Diplomacy,* p. 244.

143. Beck and Hager, "More Bad News for the Planet," p. 63.
144. Ibid.
145. Ibid.
146. Ibid.
147. Richard Monastersky, "Ozone Layer Shows Record Thinning," *Science News,* 143 (Apr. 24, 1993): 260.
148. J. F. Gleason et al., "Record Low Global Ozone in 1992," *Science,* 260 (Apr. 23, 1993): 523.
149. Gary Taubes, "The Ozone Backlash," *Science,* 260 (June 11, 1993): 1581.
150. Ibid.
151. Ibid.
152. Maduro and Schauerhammer, *The Holes in the Ozone Scare;* Rush Limbaugh, *The Way Things Ought to Be* (New York: Pocket Books, 1992).
153. Taubes, "The Ozone Backlash," p. 1580.
154. Boyce Rensberger, "After 2000, Outlook for the Ozone Layer Looks Good," *Washington Post,* Apr. 15, 1993, p. 1. For a more comprehensive exposition of his views, see "The Oz Solution: Worldwide Action Is Blunting the Threat of Ozone Depletion," *Washington Post,* national weekly ed., Apr. 26–May 2, 1993, pp. 6–7.
155. M. D. Schwarzkopf and V. Ramaswamy, "Radiation Forcing Due to Ozone in the 1980s: Dependence on Altitude of Ozone Change," *Geophysical Research Letters,* 20, no. 2 (Feb. 5, 1993): 208. Do not think that creative minds are no longer present. Atmospheric scientist Hugh W. Ellsaesser wants society to put some of its surplus resources into studying the benefits of UV-B radiation (memorandum from Hugh Ellsaesser, guest scientist, Lawrence Livermore Laboratory, "Arguments against the Montreal Protocol," Nov. 11, 1992). UV-B may have decreased over time because of industrial pollution, which cuts down on the offending substance (*Washington Post,* Apr. 1, 1993).
156. Singer, "Stratospheric Ozone," p. 162.

11. *Who's on First? A Global Warming Scorecard*

1. Robert M. White, "The Great Climate Debate," *Scientific American,* 263, no. 1 (July 1990): 36–43 (White is president of the National Academy of Engineering).
2. Robert C. Balling, Jr., *The Heated Debate: Greenhouse Predictions versus Climate Reality* (San Francisco: Pacific Research Institute for Public Policy, 1992); see specifically Aaron Wildavsky, "Global Warming as a Means of Achieving an Egalitarian Society: An Introduction," pp. xv–xxxvi.
3. Richard F. Sanford, "Global Warming: All Smoke and No Heat," *Intellectual Activist,* 6, no. 2 (Mar. 1991): 9. See also U. Siegenthaler and H. Oeschger, "Biospheric CO_2 Emissions during the Past 200 Years Reconstructed by Deconvolution of Ice Core Data," *Tellus,* vol. 39B (1987): 140–154; Committee on Science, Engineering, and Public Policy, *Policy Implications of Greenhouse Warming* (Washington, D.C.: National Academy Press, 1991), pp. 10–12. Early ice core studies varied, giving preindustrial backgrounds of 260 to 290 parts per million: Patrick J. Michaels and David E. Stooksbury, "Global Warming: A Reduced Threat?" *Bulletin of the American Meteorological Society,* 73, no. 10 (Oct. 1992):

1563–1577; A. Neftel, H. Oeschger, and B. Stauffer, "Evidence from Polar Ice Cores for the Increase in Atmospheric CO_2 in the Past Two Centuries," *Nature,* 315 (May 2, 1985): 45–47; D. Raynaud and J. M. Barnola, "An Antarctic Ice Core Reveals Atmospheric CO_2 Variations over the Past Few Centuries," *Nature,* 315 (May 23, 1985): 309–311. Research on the Vostok ice core comes in around 270 parts per million.

4. The seemingly innocuous phrase "carbon dioxide equivalent," used by nearly everyone, reflects a victory for those who want to restrain man-made (anthropogenic) emissions. After all, concentration of gases could be calculated in the form of equivalence in water vapor, which is much greater in volume than CO_2. But there is no one to blame for water vapor. By reiterating the term "CO_2 equivalent," the suggestion is driven home that human activities are driving the earth's heat up to dangerous levels.

5. Jack Schreibman, "Heat Rise Will Harm California: Flooded S.F., Oakland Foreseen," *Santa Rosa Press Democrat,* Nov. 22, 1988.

6. *Rachel's Hazardous Waste News,* no. 300 (Aug. 26, 1992): 1.

7. J. E. Hansen et al., "Climate Impact of Increasing Atmospheric Carbon Dioxide," *Science,* 213 (Aug. 28, 1981): 957–966.

8. Richard Lindzen, "Global Warming: The Origin and Nature of Alleged Scientific Consensus," *Regulation,* 15, no. 2 (Spring 1992).

9. R. Pool, "Struggling to Do Science for Society," *Science,* 248 (May 11, 1990): 672.

10. John Noble Wilford, "His Bold Statement Transforms the Debate on Greenhouse Effect (James E. Hanson)," *New York Times,* Aug. 23, 1988, p. B10.

11. Richard Monastersky, "Buying Time in the War on Global Warming," *Science News,* 139, no. 12 (Mar. 23, 1991): 183.

12. James E. Hansen and Andrew A. Lacis, "Sun and Dust versus Greenhouse Gases: An Assessment of Their Relative Roles in Global Climate Change," *Nature,* 346 (Aug. 23, 1990): 713.

13. Ibid., p. 718.

14. *Hearings on the Greenhouse Effect and Global Climate Change,* Senate Committee on Energy and Natural Resources, 100th Congress (Washington, D.C.: U.S. Government Printing Office, 1988), pp. 53, 56.

15. Hansen and Lacis, "Sun and Dust versus Greenhouse Gases," p. 717.

16. Wilford, "Bold Statement Transforms the Debate."

17. Stephen H. Schneider with Lynne E. Meisrow, *The Genesis Strategy: Climate and Global Survival* (New York: Plenum Press, 1976), pp. xi, 299.

18. Michaels and Stooksbury, "Global Warming: A Reduced Threat."

19. Balling, *Heated Debate,* pp. 3–43.

20. J. T. Houghton, G. J. Jenkins, and J. J. Ephraums, eds., *Climate Change: The IPCC Scientific Assessment* (Cambridge, Eng.: Cambridge University Press, 1990).

21. C. Holden, "Global Warming Petition," *Science,* 247 (Feb. 23, 1990): 919.

22. Arthur Gottschalk, "U.N. Conference Near on Global Warming," *Sunday Patriot News* (Harrisburg, Penn.), Mar. 8, 1992.

23. S. Fred Singer, "No Scientific Consensus on Greenhouse Warming," *Wall Street Journal,* Sep. 23, 1991, p. A16.

24. J. E. Hansen et al., "Climate Impact," p. 957.
25. Hugh Ellsaesser, scientific opponent of the doomsday scenario (as he terms it) regarding global warming, nevertheless states, "I feel confident in saying that the mean global temperature has been warming since roughly 1750 and since 1881 it has warmed about 0.5 degrees Centigrade" (Ellsaesser, "Global Warming and the Doomsday Phenomenon," paper prepared for CEO Conference on Global Warming, Minneapolis, Oct. 1, 1991, p. 1; see also Ellsaesser, M. C. MacCracken, J. J. Walton, and S. L. Grotch, "Global Climatic Trends as Revealed by the Recorded Data," *Reviews of Geophysics,* 24, no. 4 [1986]: 745–792). But Ellsaesser adds that these observations do "not match the model deduced 'fingerprint' for greenhouse warming" ("Doomsday Phenomenon," p. 3).
26. See Richard A. Kerr, "New Greenhouse Report Puts Down Dissenters," *Science,* 249, (Aug. 3, 1990): 481.
27. T. R. Karl, H. F. Diaz, and George Kukla, "Urbanization: Its Detection and Effect in the United States Climatic Record," *Journal of Climate,* 1 (1988): 1099–1123; Helmut E. Landsberg, *The Urban Climate* (London: Academic Press, 1981); George Kukla, J. Gavin, and T. R. Karl, "Urban Warming," *Journal of Climate and Applied Meteorology,* 25 (Sep. 1986): 1265–1270.
28. Kirby Hanson, George A. Maul, and Thomas R. Karl, "Are Atmospheric 'Greenhouse' Effects Apparent in the Climatic Record of the Contiguous United States (1895–1987)?" *Geophysical Research Letters,* 16, no. 1 (Jan. 1989): 49–52.
29. Andrew Solow, "Is There a Global Warming Problem?" in Rudiger Dornbusch and James M. Poterba, eds., *Global Warming: Economic Policy Responses* (Cambridge, Mass.: MIT Press, 1991), pp. 7–31.
30. George C. Marshall Institute, *Global Warming Update: Recent Scientific Findings* (Washington, D.C.: George C. Marshall Institute, 1992), p. 7. See also Hanson, Maul, and Karl, "Are Atmospheric 'Greenhouse' Effects Apparent," pp. 49–52.
31. William A. Nierenberg, "Exaggerated Global Warming Scenarios Impede Urgent Climatology Research," *Scientist,* Feb. 5, 1990, p. 18. Nierenberg considers the Marshall Institute worthy of criticism only because it earlier accepted uncritically the view that there had been a 0.5° raise in temperature, which has since been challenged. See also Science and Environmental Policy Project, "Issues Update: Greenhouse Warming" (newsletter), Arlington, Va., Mar. 24, 1992.
32. Richard A. Kerr, "Hansen vs. the World on the Greenhouse Threat," *Science,* 244 (June 2, 1989): 1042.
33. George Kukla and Thomas R. Karl, "Recent Rise of the Nighttime Temperatures in the Northern Hemisphere," in *DOE Research Summary,* Oak Ridge National Laboratory, Environmental Sciences Division, publication no. 14 (Oak Ridge, Tenn: ORNL, Carbon Dioxide Information Analysis Center, 1992), pp. 2–3.
34. Balling, *Heated Debate,* pp. 105–118.
35. Thomas Karl et al., "Global Warming—Evidence for Asymmetric Diurnal Temperature Change," *Geophysical Research Letters,* 18, no. 12 (Dec. 1991): 2253–2256.
36. Joseph Verrengla, "Arctic Analysis Cool to Global Warming Claims," *Rocky Mountain News,* Apr. 20, 1992, p. 34.
37. "No Evidence of Global Warming, Say Scientists," Associated Press, PD3-30-90. See also S. Fred Singer, "Global Environmental Policies: Drastic Actions Based

on Shaky Science Make for Bad Policies," *In Depth: A Journal for Valves and Public Policy,* Spring 1992, pp. 30–31.

38. Stephen H. Schneider and Norman J. Rosenberg, "The Greenhouse Effect: Its Causes, Possible Impacts, and Associated Uncertainties," in Rosenberg et al., eds., *Greenhouse Warming: Abatement and Adaptation* (Washington, D.C.: Resources for the Future, 1989), p. 9.

39. Wayne A. Woodward and Henry L. Gray, "Testing for Trends in the Global Temperature Data," U.S. Department of Energy Research Summary, no. 16, Apr. 1992, p. 1.

40. Ibid., p. 2; J. Hansen and S. Lebedeff, "Global Trends of Measured Surface Air Temperature," *Journal of Geophysical Research,* 92, no. D11 (Nov. 20, 1987): 13,345–13,372.

41. Woodward and Gray, "Testing for Trends," pp. 2–3.

42. A. H. Gordon, "Global Warming as a Manifestation of a Random Walk," *Journal of Climate,* 4, no. 6 (June 1991): 589, 596.

43. Wallace S. Broecker, "Global Warming on Trial," *Natural History,* Apr. 1992, p. 6.

44. Thomas R. Karl, Richard B. Heim, Jr., and Robert G. Gray, "The Greenhouse Effect in Central North America: If Not Now, When," *Science,* 251 (Mar. 1, 1991): 1058–1061.

45. Kerr, "Hansen vs. the World," p. 1043.

46. Lindzen, "Global Warming: The Origin and Nature of Alleged Scientific Consensus."

47. Sherwood B. Idso, *Carbon Dioxide and Global Change: Earth in Transition* (Tempe, Ariz.: Institute for Biospheric Research Press, 1989), p. 11, quoting M. C. MacCracken and F. M. Luther, "Executive Summary," in MacCracken and Luther, eds., *Projecting the Climatic Effects of Increasing Carbon Dioxide* (Washington, D.C.: U.S. Department of Energy, 1985), pp. xvii–xxv.

48. Richard A. Kerr, "A New Way to Forecast Next Season's Climate," *Science,* 244 (Apr. 7, 1989): 30.

49. G. L. Stephens, "Aspects of Cloud-Climate Feedback," Proceedings of the Symposium on the Role of Clouds in Atmospheric Chemistry in Global Climate, American Meteorology Society, Boston, Mass., 1988, pp. 30–34.

50. R. A. Reck, "Confidence in Climate Models Including Those with Suspended Particles," in Reck and J. R. Hummel, eds., *AIP Conference Proceedings,* no. 82, *Interpretation of Climate and Photochemical Models, Ozone, and Temperature Measurements* (New York: American Institute of Physics, 1982), pp. 23–33; quoted in Idso, *Carbon Dioxide and Global Change,* p. 60.

51. Lindzen, "Global Warming: The Origin and Nature of Alleged Scientific Consensus," p. 22.

52. Sanford, "All Smoke and No Heat," p. 13.

53. M. E. Schlesinger and J. F. B. Mitchell, "Climate Model Simulations of the Equilibrium Climatic Response to Increased Carbon Dioxide," *Review of Geophysics,* 25 (1987): 760–798; quoted in Idso, *Carbon Dioxide and Global Change,* p. 17.

54. See Balling, *Heated Debate,* pp. 36–37.

55. Solow, "Problem?" p. 15.

56. Ellsaesser, "Global Warming and the Doomsday Phenomenon," p. 3, and references cited there. See also Sanford, "All Smoke and No Heat," pp. 12–13; and Richard Lindzen, "Some Coolness concerning Global Warming," *Bulletin of the American Meteorological Society,* 71, no. 3 (Mar. 1990): 288–299.

57. Warren T. Brookes, "The Global Warming Panic," *Forbes,* 144, no. 3 (Dec. 25, 1989): 100–101.

58. See Solow, "Problem?" p. 14, Balling, *Heated Debate,* p. 30.

59. John Carey, "Is the World Heating Up? Well, Just Listen," *Business Week,* 3198 (Feb. 4, 1991): 54–55.

60. Michaels and Stooksbury, "Global Warming: A Reduced Threat?" p. 1566; J. E. Hansen et al., "Climate Sensitivity: Analysis of Feedback Mechanisms," in Hansen and T. Takahashi, eds., *Geophysical Monograph Series, 29: Climate Processes and Climate Sensitivity* (Washington, D.C.: American Geophysical Union, 1984), p. 130–163; M. E. Schlesinger, "Climate Model Simulation of CO_2-Induced Climatic Change," *Advanced Geophysics,* 26 (1984): 141–235; J. F. B. Mitchell, "The Seasonal Response of a General Circulation Model to Changes in CO_2 and Sea Temperatures," *Quarterly Journal of the Royal Meteorological Society,* 109 (Jan. 1983): 113–152; Warren M. Washington and Gerald A. Meehl, "General Circulation Model Experiments on the Climatic Effects Due to a Doubling and Quadrupling of Carbon Dioxide Concentration," *Journal of Geophysical Research,* 88, no. C11 (1983): 6600–6610; Intergovernmental Panel on Climate Change [IPCC], *Scientific Assessment of Climate Change, Summary and Report, World Meteorological Organization,* United Nations Environmental Program (Cambridge, Eng.: Cambridge University Press, 1990).

61. Marshall Institute, *Update,* p. 6.

62. Balling, *Heated Debate,* pp. 87–88. See also Roy W. Spencer and John R. Christy, "Precise Monitoring of Global Temperature Trends from Satellites," in Robert Jastrow, William Nierenberg, and Frederick Seitz, eds., *Scientific Perspectives on the Greenhouse Problem* (Ottawa, Ill.: Jameson Books, 1990), p. 101.

63. Balling, *Heated Debate,* pp. 100–101.

64. J. E. Hansen et al., "Global Climate Changes as Forecast by Goddard Institute for Space Studies Three-Dimensional Model," *Journal of Geophysical Research,* 93, no. D8 (Aug. 20, 1988): 9341–9364; L. O. Mearns, R. W. Katz, and S. H. Schneider, "Extreme High Temperature Events: Changes in Their Probabilities with Changes in Mean Temperature," *Journal of Climate and Applied Meteorology,* 23 (1984): 1601–1613.

65. Quoted in Balling, *Heated Debate,* pp. 104–105; R. C. Balling, Jr., and S. B. Idso, "Effects of Greenhouse Warming on Maximum Summer Temperatures," *Agricultural and Forest Meteorology,* 53, nos. 1–2 (Nov. 1990): 143–147.

66. Richard A. Kerr, "Global Temperature Hits Record Again," *Science,* 251 (Jan. 18, 1991): 274.

67. Quoted in Idso, *Carbon Dioxide and Global Change,* p. 124.

68. Hugh W. Ellsaesser, personal communication, May 7, 1992.

69. Solow, "Problem?" p. 22.

70. B. Bolin et al., "The Global Biogeochemical Carbon Cycle," in Bolin et al., eds., *The Global Carbon Cycle* (New York: John Wiley and Sons, 1979), pp. 1–53.

71. H. Sakai et al., "Venting of CO_2-Rich Fluid and Hydrate Formation in Mid-Okinawa through Backarc Basin," *Science,* 248 (June 1, 1990): 1093–1096.

72. Robert A. Berner and Antonio C. Lasaga, "Modeling the Geochemical Carbon Cycle," *Scientific American,* 260, no. 3 (Mar. 1989): 74–81.

73. Ibid., p. 81.

74. Ibid.

75. Balling, *Heated Debate,* pp. 19–23.

76. J. M. Barnola et al., "Vostok Ice Core Provides 160,000 Year Record of Atmospheric CO_2," *Nature,* 329 (Oct. 1, 1987): 408. See also Neftel, Oeschger, and Stauffer, "Evidence from Polar Ice Cores."

77. C. Genthon et al., "Vostok Ice Core: Climatic Response to CO_2 and Orbital Forcing Changes over the Last Climatic Cycle," *Nature,* 329 (Oct. 1, 1987): 414–418; L. Tangley, "Preparing for Climate Change," *Bioscience,* 38 (1988): 14–18.

78. Idso, *Carbon Cycle and Global Change.*

79. W. D. Sellers, *Physical Climatology* (Chicago: University of Chicago Press, 1965).

80. Freeman J. Dyson, "Carbon Dioxide in the Atmosphere and the Biosphere," Radcliffe Lecture given at Green College, Oxford, Oct. 11, 1990. The substance of this claim was published by Dyson as, "Can We Control the Carbon Dioxide in the Atmosphere?" *Energy,* 2 (1977): 287–291.

81. P. P. Tans et al., "Observational Constraints on the Global Atmospheric CO_2 Budget," *Science,* 247 (Mar. 23, 1990): 1431–1438.

82. See the discussion in Richard Monastersky, "Carbon Dioxide: Where Does It All Go?" *Science News,* 136, no. 9 (Aug. 26, 1989): 132.

83. Tans et al., "Observational Constraints," p. 1438.

84. U. Hampicke, "Net Transfer of Carbon between the Land Biota and the Atmosphere, Induced by Man," in Bert Bolin et al., eds., *The Global Carbon Cycle* (New York: John Wiley and Sons, 1979), p. 219.

85. Richard A. Kerr, "Global Change—Fugitive Carbon Dioxide: It's Not Hiding in the Ocean," *Science,* 256 (Apr. 3, 1992): 35.

86. IPCC, *Scientific Assessment of Climate Change,* p. 7.

87. Chauncey Starr (Electric Power Research Institute, Palo Alto, Calif.) and Scott P. Smith, "Atmospheric CO_2 Residence Time," unpublished manuscript, July 8, 1992, p. 1.

88. Hansen and Lacis, "Sun and Dust versus Greenhouse Gases."

89. M. Heimann, "Modeling the Global Carbon Cycle," paper presented at the First Demetra Meeting on Climate Variability and Global Change, Chianciano Therme, Italy, Oct. 28–Nov. 3, 1991.

90. Daniel A. Lashof, "The Dynamic Greenhouse: Feedback Processes that May Influence Future Concentrations of Atmospheric Trace Gases and Climatic Change," *Climatic Change,* 14, no. 3 (June 1989): 220.

91. "Climate Experts Often Call Clouds the Wild Card in the Game of Global Exchange." Richard Monastersky, "Cloudy Concerns: Will Clouds Prevent or Promote a Drastic Global Warming?" *Science News,* 136, no. 7 (Aug. 12, 1989): 106.

92. Richard Monastersky, "Warmer Clouds Could Keep the Earth Cooler," *Science News,* 136 (Sep. 23, 1989): 196.

93. Robert Jastrow, William Nierenberg, and Frederich Seitz, "Clouds and the Greenhouse Problem," in Jastrow, Nierenberg, and Seitz, eds., *Scientific Perspectives on the Greenhouse Problem* (Ottawa, Ill.: Jameson Books, 1990), pp. 17–18.

94. Andrew C. Revkin, "Clouds in the Greenhouse," *Discover*, 10, no. 6 (June 1989): 24

95. Cited in Richard Lindzen, "Absence of a Scientific Basis: (Global Warming Debate)" *Research and Exploration*, 9, no. 2 (Spring 1993): 14–15; V. Ramanathan et al., "Climate and the Earth's Radiation Budget," *Physics Today*, 42, no. 5 (May 1989).

96. Balling, *Heated Debate*; Revkin, "Clouds in the Greenhouse," p. 24; Idso, *Carbon Dioxide and Global Change*, pp. 18–19; A. Slingo and J. M. Slingo, "The Response of a General Circulation Model to Cloud Long Wave Radiative Forcing, I: Introduction and Initial Experiments," *Quarterly Journal of the Royal Meteorological Society*, 114, no. 482, (July 1988): 1027–1062.

97. Monastersky, "Cloudy Concerns," pp. 106–107, 110. See also the article by John Horgan, "Pinning Down Clouds: Scientists Ponder the Role of Clouds in Climatic Change" *Scientific American*, 260, no. 5 (May 1989): 24.

98. William A. Nierenberg, "Atmospheric CO_2: Causes, Effects, and Options," *Chemical Engineering Progress*, 85 (Aug. 1989): 27–36; Richard Monastersky, "Looking for Mr. Greenhouse: Can Scientists Say Whether Humans Have Changed the Climate through the Greenhouse Effect?" *Science News*, 135, no. 14 (Apr. 8, 1989): p. 221; Richard Monastersky, "Global Change: The Scientific Challenge," *Science News*, 135, no. 15 (Apr. 15, 1989): 232–235.

99. Balling, *Heated Debate*; Revkin, "Clouds in the Greenhouse," p. 24.

100. Hansen et al., "Climate Sensitivity," pp. 130–163; W. M. Washington and G. A. Meehl, "Seasonal Cycle Experiment on the Climate Sensitivity Due to a Doubling of CO_2 with an Atmospheric General Circulation Model Coupled to a Simple Mixed-Layer Ocean Model," *Journal of Geophysical Research*, 89, no. D6 (Nov. 20, 1984): 9475–9503; R. T. Wetherald and S. Manabe, "An Investigation of Cloud Cover Change in Response to Thermal Forcing," *Climatic Change*, 8, no. 1 (Feb. 1986): 5–23; C. A. Wilson and J. F. B. Mitchell, "A Doubled CO_2 Climate Sensitivity Experiment with a Global Climate Model Including a Simple Ocean," *Journal of Geophysical Research*, 92, no. D11 (Nov. 20, 1987): 13315–13343; R. T. Wetherald and S. Manabe, "Cloud Feedback Processes in General Circulation Model," *Journal of Atmospheric Science*, 45 (Apr. 15, 1988): 1397–1415.

101. V. Ramanathan and W. Collins, "Thermodynamic Regulation of Ocean Warming by Cirrus Clouds Deduced from Observations of the 1987 El Niño," *Nature*, 351 (May 2, 1991): 27–32.

102. Rong Fu et al., "Cirrus-Cloud Thermostat for Tropical Sea Surface Temperatures Tested Using Satellite Data," *Nature*, 358 (July 30, 1992): 394–397; Graeme Stephens and Tony Slingo, "An Air-Conditioned Greenhouse," *Nature*, 358 (July 30, 1992): 369–370.

103. V. Ramanathan, Bruce R. Barkstrom, and Edwin F. Harrison, "Climate and the Earth's Radiation Budget," *Physics Today*, 42, no. 5 (May 1989): 22.

104. Ibid.; Revkin, "Clouds in the Greenhouse," p. 24.

105. Lashof, "Dynamic Greenhouse," p. 220.

106. V. Ramanathan et al., "Cloud-Radiative Forcing and Climate: Results from the Earth Radiation Budget Experiment," *Science*, 243 (Jan. 6, 1989): 57–63. See also Anne Henderson-Sellers, "Cloud Changes in a Warmer Europe," *Climatic Change*, 8, no. 1 (1986): 25–52.

107. Sanford, "All Smoke and No Heat"; Ramanathan, Barkstrom, and Harrison, "Climate and the Earth's Radiation."
108. Monastersky, "Cloudy Concerns," p. 106.
109. Michaels and Stooksbury, "Global Warming: A Reduced Threat?" pp. 20–21.
110. Balling, *Heated Debate*, p. 44
111. Nierenberg, "Atmospheric CO_2," p. 31.
112. Solow, "A Problem?" p. 14.
113. Paul Wallich, "Murky Water; Just What Role Do Oceans Play in Absorbing Greenhouse Gases?" *Scientific American*, 262, no. 5 (May 1990): 25–26.
114. T. Appenzeller, "Fire and Ice under the Deep-Sea Floor," *Science*, 252 (June 28, 1991): 1790.
115. Anna Maria Gillis, "Why Can't We Balance the Globe's Carbon Budget?—The Calculation Requires a Better Understanding of the Oceans," *Bioscience*, 41, no. 7 (July/Aug. 1991): 442–443.
116. P. D. Quay, B. Tilbrook, and C. S. Wong, "Oceanic Uptake of Fossil Fuel CO_2: Carbon-13 Evidence," *Science*, 256 (Apr. 3, 1992): 74.
117. Wallace S. Broecker and Jeffrey S. Severinghaus, "Global Change—Diminishing Oxygen," *Nature*, 355 (Aug. 27, 1992): 710. See also Tans et al., "Observational Constraints."
118. Broecker and Severinghaus, "Global Change," p. 711.
119. Marshall Institute, *Update*, pp. 12–13.
120. Solow, "Problem?" pp. 7–31.
121. Wallace S. Broecker, "Global Warming on Trial," *Natural History*, Apr. 1992, p. 12.
122. Robert Buderi, "Could a Hotter Sun Be the Culprit in Global Warming?" *Business Week*, 3239 (Nov. 11, 1991): 163–167.
123. E. Friis-Christensen and K. Lassen, "Length of the Solar Cycle: An Indicator of Solar Activity Closely Associated with Climate," *Science*, 254 (Nov. 1, 1991): 698–700; graph showing correlation reprinted in R. A. Kerr, "Could the Sun Be Warming the Climate?" *Science*, 254 (Nov. 1, 1991): 652.
124. Kerr, "Could the Sun Be Warming the Climate?" p. 652.
125. Ibid.
126. Friis-Christensen and Lassen, "Solar Cycle."
127. P. M. Kelly and T. M. L. Wigley, "Solar Cycle Length, Greenhouse Forcing and Global Climate," *Nature*, 360 (Nov. 26, 1992): 328–330.
128. M. E. Schlesinger and N. Ramankutty, "Implications for Global Warming of Intercycle Solar Irradiance Variations," *Nature*, 360 (Nov. 26, 1992): 330–333.
129. See M. E. Raymo and W. F. Ruddiman, "Tectonic Forcing of Late Cenozoic Climate," *Nature*, 359 (Sep. 10, 1992): 117–122.
130. Lashof, "Dynamic Greenhouse," p. 213.
131. P. J. Crutzen, "Atmospheric Chemistry: Methane's Sinks and Sources," *Nature*, 350 (Apr. 4, 1991): 380–381.
132. Robert James Bidinotto, "What Is the Truth about Global Warming?" *Readers Digest*, 136, no. 814 (Feb. 1990): 94.
133. Balling, *Heated Debate*, pp. 26–27; D. R. Blake and F. S. Rowland, "Continuing Worldwide Increase in Tropospheric Methane," *Science*, 239 (Mar. 4, 1988): 1129–1131.

134. Richard A. Kerr, "Greenhouse Skeptic out in the Cold," *Science*, 246 (Dec. 1, 1989): 1119.
135. "How Dinosaurs May Have Helped Make Earth Warmer," Associated Press, *San Francisco Chronicle*, Oct. 23, 1991, p. A3.
136. L. P. Steele et al., "Slowing Down of the Global Accumulation of Atmospheric Methane during the 1980s," *Nature*, 358 (July 23, 1992): 313–316.
137. Balling, *Heated Debate*, p. 29; R. T. Watson et al., "Greenhouse Gases and Aerosols," in J. T. Houghton, G. J. Jenkins, and J. J. Ephraums, eds., *Climate Change: The IPCC Scientific Assessment* (Cambridge, Eng.: Cambridge University Press, 1990), pp. 1–40. CF_4 and C_2F_6 are produced during the electrolysis of aluminum. To give an idea of magnitudes, CF_4 has been estimated to result in 1.7 percent of all man-made atmospheric gas emissions (Dean Abrahamson, "Aluminum and Global Warming," *Nature*, 356 [Apr. 9, 1992]: 484).
138. V. Ramaswamy, M. D. Schwarzkopf, and K. P. Shine, "Radiative Forcing of Climate from Halocarbon-Induced Global Stratospheric Ozone Loss," *Nature*, 355 (Feb. 27, 1992): 810.
139. Lindzen, "Scientific Basis," p. 15.
140. Idso, *Carbon Dioxide and Global Change*, pp. 38–39.
141. Hansen and Lacis, "Sun and Dust versus Greenhouse Gases," p. 717.
142. D. Rind et al., "Positive Water Vapour Feedback from Climate Models Confirmed by Satellite Data," *Nature*, 349 (Feb. 7, 1991): 500–503.
143. Lindzen, "Scientific Basis," pp. 10–11; A. Arking, "The Radiative Effects of Clouds and Their Impact on Climate," *Bulletin of the American Meteorological Society*, 72, no. 6 (June 1991): 795–813.
144. Lindzen, Scientific Basis,'" pp. 13–14.
145. Richard S. Lindzen and De-Zheng Sun, "Upper Level Water Vapor Inferred from the Mountain Snowline Record," unpublished manuscript, Center for Meteorology and Physical Oceanography, MIT, June 22, 1992.
146. Idso, *Carbon Dioxide and Global Change*, p. 29. See also S. B. Idso, "CO_2 and Climate: Where Is the Water Vapor Feedback?" *Archive of the Meteorological and Geophysical Bioclimatological Society*, Ser. B, 31 (1982): 325–329.
147. Hansen and Lacis, "Sun and Dust versus Greenhouse Gases," pp. 715, 718.
148. R. J. Charlson et al., "Oceanic Phytoplankton, Atmospheric Sulfur, Cloud Albedo and Climate," *Nature*, 326 (Apr. 16, 1987): 655–661; S. E. Schwartz, "Are Global Cloud Cover and Climate Controlled by Marine Phytoplankton?" *Nature*, 336 (Dec. 1, 1988): 441–445; T. M. L. Wigley, "Possible Climate Change Due to SO_2-Derived Cloud Condensation Nuclei," *Nature*, 339 (June 1, 1989): 365–367; B. Albrecht, "Aerosols, Cloud Microphysics, and Fractional Cloudiness," *Science*, 245 (Sep. 15, 1989): 1227–1230.
149. Balling, *Heated Debate*, pp. 123–124.
150. Charlson's work cited in Richard Monastersky, "Haze Clouds the Greenhouse (Sulfur Pollution Slows Global Warming)," *Science News*, 141, no. 15 (Apr. 11, 1992): 232.
151. Balling, *Heated Debate*, p. 67.
152. Monastersky, "Haze Clouds Greenhouse," p. 233; Charles Knox, "Sulfur-Climate Link Called Insignificant," *Science News*, 134, no. 24 (Dec. 10, 1988): 375.
153. New York Times News Service, "The Earth Is Taking Cover from the Sun," *Los Angeles Herald Examiner*, July 5, 1987.

154. Monastersky, "Haze Clouds Greenhouse," p. 233.

155. Jos Lelieveld and Jost Heintzenberg, "Sulfate Cooling Effect on Climate through In-Cloud Oxidation of Anthropogenic SO_2," *Science,* 258 (Oct. 2, 1992): 117–120.

156. Richard A. Kerr, "Hot Nights in the Greenhouse," *Science,* 255 (Feb. 7, 1992): 683.

157. Marshall Institute, *Update,* p. 12.

158. Joyce E. Penner, Robert E. Dickinson, and Christine A. O'Neill, "Effects of Aerosol from Biomass Burning on the Global Radiation Budget," *Science,* 256 (June 5, 1992): 1432.

159. Robert Balling, Jr., and Sherwood B. Idso, "Decreasing Diurnal Temperature Range: CO_2 Greenhouse or SO_2 Energy Balance Effect?" *Atmospheric Research,* 26, no. 5 (Sep. 1991): 455–459.

160. R. A. Bryson, "Late Quarternary Volcanic Modulation of Milankovitch Climate Forcing," *Theoretical and Applied Climatology,* 39, no. 3 (1989): 115–125.

161. See Balling, *Heated Debate,* pp. 65–66, and the numerous references pro and con cited there.

162. William Booth, "Global Warming, Meet Mt. Pinatubo," *Santa Rosa Press Democrat,* Aug. 14, 1991.

163. Richard A. Kerr, "Huge Eruption May Cool the Globe," *Science,* 252 (June 2, 1991): 1780.

164. Lee Dye, "Volcano's Fallout Tests Theories on Atmosphere," *Los Angeles Times,* Dec. 14, 1991, p. A18.

165. Idso, *Carbon Dioxide and Global Change,* p. 38.

166. Fred B. Wood, "Global Alpine Glacier Trends, 1960s to 1980s," *Arctic and Alpine Research,* 20, no. 4 (1988): 404.

167. William J. Campbell and Per Gloersen, quoted in Richard Monastersky, "Shrinking Ice May Mean Warmer Earth," *Science News,* 134, no. 15 (Oct. 8, 1988): 230.

168. Alister Doyle, "In Norway, Glaciers Are Growing Bigger," *Los Angeles Times,* Oct. 28, 1990, p. A29.

169. Richard Monastersky, "Predictions Drop for Future Sea-Level Rise," *Science News,* 136, no. 25 (Dec. 16, 1989): 397.

170. Marshall Institute, *Update,* pp. 21–24.

171. Monastersky, "Predictions Drop," p. 397.

172. See Lori Oliwenstein, "Cold Comfort (Global Warming and the Ice Age)," *Discover,* 13, no. 8 (Aug. 5, 1992): 18–20.

173. Sherwood Idso, "A Role for Soil Microbes in Moderating the Carbon Dioxide Greenhouse Effect," *Soil Science,* 149, no. 3 (Mar. 1990): 179–180. See Idso, *Carbon Dioxide and Global Change,* pp. 21–22, for numerous citations. An excellent discussion for the uninitiated is Robert Kunzig, "Invisible Garden: Were It Not for Tiny Plants Floating in the Sea, We Would Be Smothered by a Thick Cloud of Carbon Dioxide," *Discover,* 11, no. 4 (Apr. 1990): 66–72.

174. Idso, "A Role for Soil Microbes."

175. Hansen et al., "Climate Sensitivity," pp. 130–163. See also Lashof, "Dynamic Greenhouse," p. 227. Building on the finding that single-celled phytoplankton in the Antarctic suffer from an iron deficiency that leaves them unable to use many of the rich nutrients that surround them, oceanographers at the Moss Landing,

California, Marine Laboratories suggested fertilizing the Antarctic with iron as an effort to increase the photosynthesis of phytoplankton that would both emit sulfur and draw in carbon dioxide from the atmosphere. Needless to say, huge doubts have been expressed extending from beliefs that the science is wrong to fears of disturbing the region's ecology. See "Ironing Away a Greenhouse Wrinkle: (Fertilizing the Antarctic Ocean with Iron to Slow Global Warming)," *Science News,* 39, no. 4 (Jan. 26, 1991): 63.

176. Sanford, "All Smoke and No Heat," p. 16. "CO_2 enrichment can directly increase plant production by an improved water use efficiency" (J. Goudian and G. L. Ajtay, "The Possible Effects of Increased CO_2 on Photosynthesis," in B. Bolin et al., eds., *The Global Carbon Cycle* (New York: John Wiley and Sons, 1979), p. 237.

177. Quoted in Sanford, "All Smoke and No Heat," p. 17.

178. Ibid.

179. Lowell Ponte, *The Cooling* (Englewood Cliffs, N.J.: Prentice-Hall, 1976), quoted in ibid.

180. W. F. Ruddiman and J. E. Kutzbach, "Plateau Uplift and Climatic Change," *Scientific American,* 264, no. 3 (Mar. 1991): 66–75. See also Balling, *Heated Debate,* pp. 4–6.

181. John Gribbin, *Future Weather and the Greenhouse Effect* (New York: Delacorte Press, 1982), pp. 9–15.

182. D. Rind et al., "Potential Evapotranspiration and the Likelihood of Future Drought," *Journal of Geophysical Research,* 95, no. D7 (June 20, 1990): 9983–10,004. See also G. J. McCabe, Jr., et al., "Effects of Climatic Change on the Thornthwaite Moisture Index," *Water Resources Bulletin,* 26, no. 4 (Aug. 1990): 633–643; W. W. Kellogg and Z. C. Zhao, "Sensitivity of Soil Moisture to Doubling of Carbon Dioxide in Climate Model Experiments, Part I: North America," *Journal of Climate,* 1, no. 4 (Apr. 1988): 348–366; S. Manabe and R. T. Wetherald, "Reduction in Summer Soil Wetness Induced by an Increase in Atmospheric Carbon Dioxide," *Science,* 232 (May 2, 1986): 626–628; Manabe and Wetherald, "Large-Scale Changes of Soil Wetness Induced by an Increase in Atmospheric Carbon Dioxide," *Journal of the Atmospheric Sciences,* 44 (Apr. 15, 1987): 1211–1235.

183. Balling, *Heated Debate,* pp. 107–109.

184. Ibid., p. 118.

185. Ibid., pp. 110–111.

186. Hanson, Maul, and Karl, "Are Atmospheric 'Greenhouse' Effects Apparent."

187. Idso, *Carbon Dioxide and Global Change,* p. 10.

188. Balling, *Heated Debate,* p. 78; Idso, *Carbon Dioxide and Global Change,* pp. 69–70; T. R. Karl et al., "Decreasing Diurnal Temperature Range in the United States and Canada from 1941 through 1980," *Journal of Climate and Applied Meteorology,* 23 (Nov. 1984): 1489–1504; T. R. Karl et al., "Relationship between Decreased Temperature Range and Precipitation Trends in the United States and Canada, 1941–1980," *Journal of Climate and Applied Meteorology,* 25 (Dec. 1986): pp. 1878–1886.

189. Sanford, "All Smoke and No Heat," p. 16.

190. Boyce Rensberger, "Silver Lining in Global Warming's Cloud: New Look at Data

Shows Balmier Nights, Winters, and Springs," *International Herald Tribune*, June 2, 1993, p. 3.

191. Richard Monastersky, "Arctic Shows No Signs of Greenhouse Warmth," *Science News*, 143 (Jan. 30, 1993): 70.

192. Tim Hilchey, "A Forest Absorbs More Carbon Dioxide Than Was Predicted," *New York Times*, June 8, 1993, p. C4; S. C. Wofsy et al., "Net Exchange of CO_2 in a Mid-Latitude Forest," *Science*, 260 (May 28, 1993): 1314–1317.

193. John R. Luoma, "Sharp Decline Found in Arctic Air Pollution," *New York Times*, June 1, 1993, p. B7. For the original study, see Barry Bodhaine and Ellsworth G. Dutton, "A Long-Term Decrease in Arctic Haze at Barrow, Alaska," *Geophysical Research Letters*, 20, no. 10 (May 21, 1993): 947–950.

194. S. Robert Lichter and Linda Lichter (Center for Media and Public Affairs), "The Great Greenhouse Debate," *Media Monitor*, 6, no. 10 (Dec. 1992): 6–8.

195. Ibid., p. 6.

196. Roger A. Pielke, "The 'Warming Mania': It's Absurd to Claim We Can Predict Future Climate," *Rocky Mountain News*, Sep. 27, 1992, p. 105.

12. Reporting Environmental Science

Note from Brendon Swedlow: Aaron Wildavsky died before I could complete all the work of writing this chapter. I take full responsibility for its final form and content, even though I had a great deal of help researching it. Laura Evans and Steven Kim did massive amounts of content analysis of newspapers and news-weeklies, which unfortunately could not be included here, although some of Laura's work with Aaron is found in Chapter 9. Jacob Salas and Sonia Montejano combed the indexes of the Vanderbilt University Television News Archives in search of references to dioxins and toxic waste, and David Schleicher read drafts of this chapter. I am also grateful to Professor Arnold Meltsner and Administrative Assistant Doris Patton, who provided the necessary guidance and assistance. Finally, I am indebted to the Bradley Foundation for their financial support of this research.

1. "One judgment . . . overwhelmingly agreed on was that information on environmental risk is scanty even in articles selected as the best examples of environmental risk reporting," Peter M. Sandman, David B. Sachsman, Michael R. Greenberg, and Michael Gochfeld write in summarizing the results of one of their studies: *Environmental Risk and the Press: An Exploratory Assessment* (New Brunswick, N.J.: Transaction Press, 1987), p. 40. This panel had been asked to evaluate an archive of 248 articles, compiled by inviting the editors of 26 daily New Jersey newspapers each to submit their 5 best articles from 1984 dealing with environmental risk (ibid., p. 6). Twenty-one editors did so. The other 5 asked the researchers to make the selection from the newspapers' files, and they chose the year's longest articles, assuming these would be the best (ibid., p. 40).

2. Ibid.

3. Ibid., p. 45.

4. Ibid., p. 41.

5. Ibid., p. 45.

6. Ibid.

7. Michael R. Greenberg et al., "Network Television News Coverage of Environmental Risk," *Environment*, 31, no. 2, (Mar. 1989): 40.

8. Ibid., p. 20.

9. American Opinion Research, Inc., *The Press and the Environment—How Journalists Evaluate Environmental Reporting*, conducted for the Foundation for American Communications (Los Angeles: Foundation for American Communications, 1993). Investigators randomly selected newspapers with a circulation of 10,000 or more and local stations with the largest markets. They also interviewed journalists at the three major networks and CNN, at the three major newsmagazines (*Time, Newsweek,* and *U.S. News and World Report*), and at the wire services (Associated Press, Gannett News Service, Knight-Ridder News Service, United Press International, and Reuters Information Services). At each outlet investigators tried to interview reporters who "primarily or exclusively" cover environmental stories and to interview supervisors responsible for story selection and assignments to these reporters. In choosing newspapers, investigators looked at circulation and separately sampled small (10,000–24,999), medium (25,000–49,999), large (50,000–99,999), and very large papers (100,000 plus). Interviews conducted at each of these were then weighted to reflect the circulation, so that the results reflect more of the opinions of journalists at larger papers than of those at smaller ones.

10. As cited in "Is Cancer News a Health Hazard?" *Media Monitor*, 7, no. 8, (Nov./Dec. 1993): 3–4. These results are based on a content analysis of all stories regarding environmental (not inherited) causes of cancer that appeared from 1972 through 1992 on the ABC, CBS, and NBC evening newscasts, and in *Time, Newsweek,* and *U.S. News and World Report,* as well as all prominent articles (those appearing on the front page of any section) from the *New York Times, Washington Post,* and *Wall Street Journal*. The sample thus consisted of 794 television news stories, 54 newsmagazine articles, and 299 newspaper pieces.

11. S. Robert Lichter and Stanley Rothman, "Scientific Opinion vs. Media Coverage of Environmental Cancer: A Report on Research in Progress" (Washington, D.C.: Center for Media and Public Affairs, Center for Science, Technology, and Media; Storrs, Conn.: University of Connecticut, Roper Center for Public Opinion Research; Northampton, Mass.: Smith College, Center for the Study of Social Change, 1993).

12. Ninety-two percent of those interviewed said they were currently involved in research on the causes or prevention of cancer, and a majority had published 50 or more articles on these topics in peer-reviewed journals.

13. Lichter and Rothman analyzed the same three 794 television stories, 54 newsmagazine pieces, and 299 newspaper articles described in n. 10.

14. John C. Burnham, "Of Science and Superstition: The Media and Biopolitics," *Gannett Center Journal*, 4, no. 3, (Summer 1990): 34.

15. Teya Ryan, "Network Earth: Advocacy, Journalism, and the Environment," *Gannett Center Journal*, 4, no. 3, (Summer 1990): 68.

16. Albert Gore, Jr., "Steering by the Stars," *Gannett Center Journal*, 4, no. 3, (Summer 1990): 131.

17. Dorothy Nelkin, "The Culture of Science Journalism," *Society*, 24, no. 6, (Sep./Oct. 1987): 20.

18. Phillip J. Tichenor, "Teaching and the 'Journalism of Uncertainty,' " *Journal of Environmental Education*, 10, no. 3, (Spring 1979): 6.

19. Ryan, "Network Earth," p. 63.

20. Ibid., pp. 66 and 68.

21. As quoted in Amal Kumar Naj, "Greens and Greenbacks," *Gannett Center Journal*, 4, no. 3, (Summer 1990): 86.

22. Ryan, "Network Earth," pp. 69–70.

23. As quoted in Naj, "Greens and Greenbacks," p. 86.

24. Ryan, "Network Earth," p. 70.

25. As quoted in Naj, "Greens and Greenbacks," p. 86.

26. Gore, "Steering by the Stars," p. 131.

27. Ryan, "Network Earth," p. 64.

28. Tichenor, "Teaching and the 'Journalism of Uncertainty,' " p. 7.

29. As quoted in Jim Detjen, "The Traditionalist's Tools," *Gannett Center Journal*, 4, no. 3 (Summer 1990): 83.

30. Naj, "Greens and Greenbacks," p. 90.

31. Environmentalist law professor Christopher D. Stone describes how he experienced a change midway through writing a book about how to solve the world's environmental problems, including global warming: "When I started writing, I was swayed by a barrage of news stories to believe that we were veering towards a global-warming crisis that would batter our shores with mighty typhoons, parch the world's grainlands, and melt the ice caps. For most of humankind, this is bad news. But for an academic author, it offered some possibility of cutting my losses. Books about the end of the world (or even history, ideology, nature, or just the booming stockmarket, for that matter) I am told sell. I thought I would join the growing chorus, or commerce, emphasizing what international law might contribute to stave off doom. While I pressed ahead to write the doleful text, I dispatched a student researcher to cull the scientific literature to trail me with the supportive footnotes. (That is how it is done.) In this spirit, having recited in a draft the popular menace that the polar ice caps were ready to melt on us and so on, I waited for the authoritative backing to materialize in memos. I waited in vain. The deeper into the better authorities we fished, the vaguer and more qualified the projections we landed. One view (now, I believe, the prevailing one) turned out to be that global warming, rather than melting the ice caps and thereby raising the sea level, would more likely thicken them, thus to some extent counteracting higher seas caused by anticipated thermal expansion of the waters." Christopher D. Stone, "Bringing Environmentalists Back Down to Earth," *USC Trojan Family*, Winter 1994, pp. 47–48. Stone changed his views and rewrote his book as a critique of many of the central claims made by environmentalists. See *The Gnat Is Older Than Man: Global Environment and Human Agenda* (Princeton: Princeton University Press, 1993).

32. Stephen Klaidman, *Health in the Headlines* (Oxford: Oxford University Press, 1991), p. 115. By comparison, of 1982 reporting Klaidman writes, "Press coverage at the time reflected considerable scientific uncertainty, and its paucity reflected the facts, as far as the press was concerned, that the threat posed by the greenhouse effect was still distant and ill defined, if indeed it was correct to conclude there was any significant threat at all. The news stories also implicitly

reflected the unwillingness of many scientists to say on the record that they now thought the evidence for greenhouse warming sufficient to warrant an immediate policy response."

33. Ibid., p. 112.
34. Quoted in ibid., p. 105.
35. Lichter's analysis was commissioned by the Center for Science, Technology, and Media, in Washington; these data are reproduced from the center's Feb. 13, 1992, press release and supporting documents. Another content analysis of global warming reporting in 1987 and 1988 pinpoints 1988 as the year that coverage both increased dramatically and switched from evaluating the science to considering policy options for halting warming. In 1987 there were 71 stories mentioning the greenhouse effect in the *New York Times, Los Angeles Times, Washington Post,* and *Christian Science Monitor.* In 1988 the *New York Times* and *Los Angeles Times* each published more than that, for a total of 268 articles from all 4 papers. See "Science as Symbol: The Media Chills the Greenhouse Effect," in Lee Wilkins and Philip Patterson, eds., *Risky Business* (New York: Greenwood, 1991), pp. 167–169.
36. Center for Media and Public Affairs, "The Great Greenhouse Debate," *Media Monitor,* 6, no. 10, (Dec. 1992): 2.
37. Center for Media and Public Affairs, "Greenhouse Debate," p. 3.
38. Ibid.
39. The Gallup interviews were conducted by phone Oct. 14–25, 1991; the survey was commissioned by the Center for Science, Technology, and Media.
40. Citizens interested in a related analysis might look at Eleanor Singer, "A Question of Accuracy: How Journalists and Scientists Report Research on Hazards," *Journal of Communication,* 40, no. 4 (1990): 102–115. This study compares reporting of a number of scientific studies to the studies themselves to see what kind of errors journalists introduce in translating them for the public.
41. It is apparently important to distinguish here between sources that provide ideas, information, and data and those that actually appear in reports. The study by Greenberg et al. of 564 environmental risk network news stories found that industry sources were cited and appeared on the air more frequently than experts, who in turn appeared and were cited more than environmental advocates. Government officials, however, were the sources most frequently cited and shown on the air, particularly in stories where only one source was used. One way these results might be reconciled with the American Opinion Research study findings is if in the earlier Greenberg et al. study, ideas, information, and data did in fact also come from these sources in proportion to the amount they were cited or appeared on the air. This would mean that there had been a change in the mix of sources relied on from the period 1984–1986 to 1993. A more likely possibility is that environmentalists consistently play a large role in providing ideas, information, and data, but wherever a government official can be found who will take a similar line, he or she is cited or used on the air rather than the environmentalist source. Meanwhile, industry officials make their frequent appearances as the indicted or guilty parties, ineffectively defending themselves (because of editing) or, where possible, accepting blame and promising to do better next time. This last is all speculation, a set of hypotheses that require further content analysis of

the same and other reports for verification. Greenberg et al., "Network Television News Coverage," p. 40, table 2.

42. The exact numbers here were 36 percent versus 16 percent (American Opinion Research, *The Press and the Environment,* p. 25). Also, when media supervisors were asked to recommend changes in environmental reporting to improve its quality, they suggested more objective reporting (13 percent) second only to more specific training for reporters (30 percent) (ibid., p. 66). The cultural nature of journalists' biases and why they involve distrust of centralized authority and dislike of business are explored more fully in Aaron Wildavsky's review of four scholarly studies of the media, "The Media's American Egalitarians," in Aaron Wildavsky, *The Rise of Radical Egalitarianism* (Washington, D.C.: American University Press, 1990), pp. 115–132.

13. Citizenship in Science

1. See Aaron Wildavsky, "Public Policy," in Bernard D. Davis, ed., *The Genetic Revolution* (Baltimore: Johns Hopkins University Press, 1991), pp. 77–104.
2. See Aaron Wildavsky, "Global Warming as a Means of Achieving an Egalitarian Society: An Introduction," in Robert Balling, *The Heated Debate* (San Francisco: Pacific Research Institute, 1992), pp. xv–xxxvi.
3. This might be read in conjunction with a chapter entitled "Citizens as Analysts" in my earlier book: see Aaron Wildavsky, *Speaking Truth to Power,* 2d ed. (New Brunswick, N.J.: Transaction Publishers, 1986), pp. 252–279.
4. Richard S. Cornfeld and Stuart F. Schlossman, "Immunologic Laboratory Tests: A Critique of the Alcolac Decision," *Toxic Law Reporter,* 4, no. 14 (Sep. 6, 1989): 381–390; reprinted in Kenneth R. Foster, David Bernstein, and Peter W. Huber, *Phantom Risk* (Cambridge, Mass.: MIT Press for the Manhattan Institute, 1993), pp. 401–424.
5. Ibid., p. 418.
6. Phil Brown, "Popular Epidemiology and Toxic Waste Contamination: Lay and Professional Ways of Knowing," *Journal of Health and Social Behavior,* 33 (Sep. 1991): 269.
7. Ibid., p. 270
8. Ibid., p. 273.
9. Ibid.
10. Ibid., p. 277.
11. Stephanie C. Warner and Timothy E. Aldrich, "The Status of Cancer Cluster Investigations Undertaken in State Health Departments," *American Journal of Public Health,* 78, no. 3 (Mar. 1988): 306–307.
12. Lawrence Garfinkel, "Cancer Clusters," in *Cancer Statistics, 1987: Cancer Clusters* (New York: American Cancer Society, 1987), p. 24.
13. Martyn T. Smith, "Quantitative Aspects of the 'Woburn Case,' " *Science,* 235 (Jan. 9, 1987): 144–145. See also Elliot Marshall, "Woburn Case May Spark Explosion of Lawsuits," *Science* 234 (Oct. 24, 1986): 418–420.
14. This topic is covered, and the epigraph quoted, in Persi Diaconis and Frederick Mosteller, "Methods for Studying Coincidence," *Journal of the American Statis-*

tical Association, 84 (1989): 853–861; quoted in Carol Mock and Herbert F. Weisberg, "Political Innumeracy: Encounters with Coincidence, Improbability, and Chance," *American Journal of Political Science,* 236, no. 4 (Nov. 1992): 1023.

15. Mock and Weisberg, "Political Innumeracy," p. 1023.

16. Ibid., p. 1027.

17. Ibid., p. 1028.

18. Ibid., p. 1044.

19. M. G. Marmot, "Epidemiology and the Art of the Soluble," *Lancet,* 8486 (Apr. 19, 1986): 897.

20. Ibid., p. 898.

21. Ibid., p. 899.

22. Gina Kolata, "Dark Side of Low Cholesterol," *Oakland Tribune,* Aug. 11, 1992, p. A1.

23. Gotto quoted in ibid., p. A12. See also James Le Fanu, "A Healthy Diet—Fact or Fashion?" in *Health, Lifestyle, and Environment: Countering the Panic* (Whitton, Middlesex: Social Affairs Unit, 1991), pp. 87–103.

24. Rick Hester, "The Environmental Cancer Debate," *Science for the People,* 21, no. 1 (Jan./Feb. 1989): 9–10.

25. Ibid., p. 10.

26. Ibid., p. 11.

27. Michael Castleman, "Who Dunnit," *Sierra,* 77, no. 5 (Sep./Oct. 1992): 24, 26.

28. Ibid., p. 26.

29. Below are some essays worth reading on the subject: D. J. Fiorino, "Citizen Participation and Environmental Risk: A Survey of Institutional Mechanisms," *Science, Technology, and Human Values,* 15 (1990): 226–243. F. Fischer, *Technology and the Politics of Expertise* (Newbury Park, Calif.: Sage, 1990). E. Gelhorn, "Public Participation in Administrative Procedures," *Yale Law Journal,* 81 (1972): 359–404. H. R. Holman and D. B. Dutton, "A Case for Public Participation in Science Policy Formation and Practice," *Southern California Law Review,* 51 (1978): 1505–1534. J. G. Kennedy, "Saving American Democracy: The Lessons of Three-Mile Island," *Technology Review,* 82 (1980): 65–75. S. Krimsky, "Beyond Technology: New Routes for Citizen Involvement in Social Risk Assessment," in J. C. Peterson, ed., *Citizen Participation in Science Policy* (Amherst: University of Massachusetts Press, 1984), pp. 43–61. F. N. Laird, "The Decline of Deference: The Political Context of Risk Communication," *Risk Analysis,* 9 (1989): 543–550. F. N. Laird, "Technocracy Revisited: Knowledge, Power, and the Crisis of Energy Decision Making," *Industrial Crisis Quarterly,* 4 (1990): 49–61. S. Langton, "Citizen Participation in America: Current Reflections on the State of the Art" in Langton, ed., *Citizen Participation in America: Essays on the State of the Art* (Lexington, Mass.: Lexington Books, 1978), pp. 1–12. D. MacCrae, Jr., "Science and the Formation of Policy in a Democracy," in T. J. Kuehn and A. L. Porter, eds., *Science, Technology, and National Policy* (Ithaca, NY: Cornell University Press, 1981, pp. 496–514. L. W. Milbrath, "Citizen Surveys as Citizen Participation Mechanisms," *Journal of Applied Behavioral Science,* 17 (1981): 478–496. D. Nelkin, "Scientific Knowledge, Public Policy and Democracy," *Knowledge, Creation, Diffusion, Utilization,* 1, no. 1 (Sep. 1979): 106–

122. D. Nelkin, "The Political Impact of Technical Expertise" in J. C. Petersen, ed., *Citizen Participation in Science Policy* (Amherst: University of Massachusetts Press, 1984), pp. 18–39. C. Pateman, "A Contribution to the Political Theory of Organizational Democracy," in G. D. Garson and M. P. Smith, eds., *Organizational Democracy: Participation and Self-Management* (Beverly Hills: Sage, 1976), pp. 9–30. F. A. Rossini and A. L. Porter, "Public Participation and Professionalism in Impact Assessment" in J. C. Petersen, ed., *Citizen Participation in Science Policy* (Amherst: University of Massachusetts Press, 1984), pp. 62–74. A. Schutz, "The Well-Informed Citizen: An Essay on the Social Distribution of Knowledge," *Social Research,* 13, no. 4 (Dec. 1946): 463–472. R. E. Sclove, "Decision Making in a Democracy," *Bulletin of the Atomic Scientists,* 38 (1982): 44–49. R. E. Sclove, "Energy Policy and Democratic Theory" in D. Zinberg, *Uncertain Power* (New York: Pergamon, 1983), pp. 37–68. D. F. Thompson, *The Democratic Citizen: Social Science and Democratic Theory in the Twentieth Century* (Cambridge: Cambridge University Press, 1970).

30. Robert A. Dahl, "The Problem of Civil Competence," *Journal of Democracy,* 3, no. 4 (Oct. 1992): 48.

31. Ibid., pp. 50–51.

32. Ibid., p. 56.

14. *Detecting Errors in Environmental and Safety Studies*

1. Kenneth M. Towe, "Stratospheric Ozone Trends," letter to *Science,* 257 (Aug. 7, 1992): 727.

2. Richard Stolarski, letter to *Science,* 257 (Aug. 7, 1992): 727–728.

3. Society of the Plastics Industry, "Vinyl Chloride and Birth Defects," in *Vinyl Chloride—Position Papers* (New York: Society of the Plastics Industry, 1979), pp. 14–15.

4. National Council on Radiation Protection and Measurements, Commentary no. 6, *Radon Exposure of the U.S. Population—Status of the Problem* (Bethesda, Md.: National Council on Radiation Protection, Mar. 15, 1991), p. 8.

5. According to David J. Brenner, *Radon Risk and Remedy* (New York: Freeman, 1989), p. 89; 70 percent of the miners in one study in Colorado smoked, a rate twice the national average.

6. G. R. Howe, J. D. Burch, and A. B. Miller, "Artificial Sweeteners and Human Bladder Cancer," *Lancet,* 2 (Mar. 12, 1977): 579.

7. J. Taylor, M. Weinberger, and L. Friedman, "Chronic Toxicity and Carcinogenicity in the In Utero Exposed Rat," *Toxicology and Applied Pharmacology,* 54 (1980): 57–75. One study was performed by the Food and Drug Administration and the other by the Wisconsin Alumni Research Foundation.

8. J. R. Allen et al., "Pathobiological Responses of Primates to Polychlorinated Biphenyl Exposure," in Environmental Protection Agency [EPA], *Conference Proceedings, National Conference on Polychlorinated Biphenyls,* report no. EPA-560-6-75-004, Mar. 1976, pp. 43–49; and D. A. Barsotti et al., "Reproductive Dysfunction in Rhesus Monkeys," *Food and Cosmetic Toxicology,* 14 (1976): 99–103.

9. Marilyn A. Fingerhut et al., "Cancer Mortality in Workers Exposed to 2,3,7,8-

Tetrachlorodibenzo-*p*-Dioxin," *New England Journal of Medicine,* 324, no. 4 (Jan. 24, 1991): 212–218.

10. National Research Council [NRC], *Health Risks of Radon and Other Internally Deposited Alpha-Emitters—BEIR IV* (Washington, D.C.: National Academy Press, 1988), p. 52.

11. U.S. General Accounting Office [GAO], "Does Nitrite Cause Cancer? Concerns about the Validity of FDA-Sponsored Study Delay Answer," report by the comptroller general of the United States (Washington, D.C.: GAO, 1980), pp. 17–18.

12. This assumption can be seen in many scientific papers; see, e.g., Gene E. Likens and Thomas J. Butler, "Recent Acidification of Precipitation in North America," *Atmospheric Environment,* 15 (1981): 1103.

13. U.S. National Acid Precipitation Assessment Program [NAPAP], *Acidic Deposition—State of Science and Technology: Summary Report of the U.S. National Acid Precipitation Assessment Program,* ed. Patricia M. Irving (Washington, D.C.: GPO, 1991), report no. 6, Douglas L. Sisterson, "Deposition Monitoring: Methods and Results," pp. 70–71. See R. J. Charlson and H. Rodhe, "Factors Controlling the Acidity of Natural Rainwater," *Nature,* 295 (Feb. 25, 1982): 683–685, for a model of natural variability with pH from 4.5 to 5.6.

14. W. D. Kerns et al., "Carcinogenicity of Formaldehyde in Rats and Mice after Long-Term Inhalation Exposure," *Cancer Research,* 43 (1983): 4382–4392. This study, sponsored by the Chemical Industry Institute of Toxicology, is cited on p. 40 of John D. Graham, Laura C. Green, and Marc J. Roberts, *In Search of Safety: Chemicals and Cancer Risk* (Cambridge: Harvard University Press, 1988). Their source for the incidence numbers is James E. Gibson, ed., *Formaldehyde Toxicity* (Washington, D.C.: Hemisphere Publishing, 1982), p. 297.

15. Lloyd Tatyryn, *Formaldehyde on Trial* (Toronto: James Lorimer, 1983), pp. 117–121.

16. Joe Grisham, ed., *Health Aspects of the Disposal of Waste Chemicals: A Report of the Executive Scientific Panel* (New York: Pergamon Press, 1986), pp. 93–94.

17. William W. Lowrance, ed., *Assessment of Health Effects at Chemical Disposal Sites: Proceedings of a Symposium Held in New York City on June 1–2, 1981, by the Life Sciences and Public Policy Program of the Rockefeller University* (New York: Rockefeller University; Los Altos, Calif.: W. Kaufmann, 1981), p. 9.

18. Beverly Paigen, "Health Problems in a Community Living Near a Hazardous Waste Site Known as Love Canal," paper presented at the American Public Health Association Meeting, Detroit, Oct. 21, 1980.

19. Clark Heath et al., "Cytogenic Findings in Persons Living Near the Love Canal," *Journal of the American Medical Association,* 251 (Mar. 16, 1984): 1437–1440.

20. Paul Brodeur, "Calamity on Meadow Street," *New Yorker,* 66, no. 21 (July 9, 1990): 38–72.

21. "Updates," *Transmission/Distribution Health and Safety Report,* 9, no. 1 (Jan. 31, 1991): 9.

22. T. J. Sullivan et al., "Quantification of Changes in Lakewater Chemistry in Response to Acidic Deposition," *Nature,* 345 (May 3, 1990): 55.

23. William Ruckelshaus, "Consolidated DDT Hearings: Opinion and Order of the Administrator," *Federal Register,* 37 (July 7, 1972): 13375. Wayland J. Hayes,

letter to Washington DDT Hearing Coordinator, in *Selected Statements from State of Washington DDT Hearings Held in Seattle, October 14, 15, 16, 1969, and Other Related Papers* compiled by Max Sobelman (Torrance, Calif.: DDT Producers of the United States, 1970), p. 69.

24. D. L. Arnold, D. Krewski, and I. C. Monroe, "Saccharin: A Toxicological and Historical Perspective," *Toxicology,* 7 (1983): 210.

25. Pekka E. Kauppi, Kari Mielikäinen, Kullervo Kuusela, "Biomass and Carbon Budget of European Forests, 1971–1990," *Science,* 256 (Apr. 3, 1992): 70–74.

26. W. Uhlmann et al., "Introduction: The Problem of Forest Decline and the Bavarian Forest Toxicology Research Group," in E. D. Schulze, O. L. Lange, and R. Oren, eds., *Forest Decline and Air Pollution: A Study of Spruce (Picea abies) on Acid Soils,* Ecological Studies 77 (New York: Springer-Verlag, 1989), p. 1.

27. Kauppi, Mielikäinen, and Kuusela, "Biomass and Carbon Budget of European Forests," p. 72.

28. Edward C. Krug, "Assessment of the Theory and Hypotheses of the Acidification of Watersheds," *State Water Survey Contract Report 457* (Champaign: Illinois State Water Survey Division, Atmospheric Chemistry Division, 1989), pp. 2–11, citing Terry A. Haines and John Akielaszek, "A Regional Survey of the Chemistry of Headwater Lakes and Streams in New England: Vulnerability to Acidification," report no. 15 (Washington, D.C.: Fish and Wildlife Service, U.S. Department of the Interior, 1983).

29. Ibid. A subsequent study came up with much the same result. See NRC, Committee on Monitoring and Assessment of Trends in Acid Deposition, *Acid Deposition: Long-Term Trends* (Washington, D.C.: National Academy Press, 1986), p. 37, citing J. Colquhoun, W. Kretzer, and M. Pfeiffer, *Acidity Status Update of Lakes and Streams in New York State,* report no. WMP-83 (Albany: New York Department of Environmental Conservations, 1984).

30. Sullivan et al., "Quantification of Changes in Lakewater," p. 55.

31. See Mary Douglas and Aaron Wildavsky, *Risk and Culture* (Berkeley: University of California Press, 1982).

32. David E. Lilienfield and Michael A. Gallo, "2,4-D, 2,4,5-T, and 2,3,7,8-TCDD: An Overview," *Epidemiological Reviews,* 11 (1989): 36–37.

33. John Bailer, *Cell Biology and Toxicology* (Princeton, N.J.: Princeton Scientific Publishing, 1989).

34. EPA, *Superfund Public Health Evaluation Manual* (Washington, D.C.: EPA, 1986), pp. 42–60.

35. EPA, "Characterization of Municipal Solid Waste in the U.S., 1990 Update," *Solid-Waste Energy Response,* report no. EPA-530-sw-90-42, June 1990.

36. William Rathje, "Rubbish! An Archaeologist Who Excavates Landfills Believes That Our Thinking about Garbage Has Been Distorted by Powerful Myths," *Atlantic Monthly,* 264, no. 6 (Dec. 1989): 101; William Rathje and Cullen Murphy, *Rubbish! The Archaeology of Garbage* (New York: HarperCollins, 1992).

37. Michel Gough, "Human Health Effects: What the Data Indicate," *Science of the Total Environment,* 104 (May 1, 1991): 129.

38. Ibid., p. 131.

39. Electric Power Research Institute [EPRI], *EPRI Commentary on Initial Results from the USC Study of Childhood Leukemia and Exposure to Electric and Magnetic Fields* (Palo Alto, Calif.: EPRI, Feb.7, 1991).

40. Gary M. Marsh and Richard J. Caplan, "Evaluating Health Effects of Exposure at Hazardous Waste Sites: A Review of the State-of-the-Art, with Recommendations for Future Research," in Julian B. Andelman and Dwight W. Underhill, eds., *Health Effects from Hazardous Waste Sites* (Chelsea, Mich.: Lewis Publishers, 1987), p. 67.

41. Grisham, *Health Aspects of the Disposal of Waste Chemicals,* p. 244.

42. Howe, Burch, and Miller, "Artificial Sweeteners and Human Bladder Cancer," p. 579.

43. Beverly Paigen, "Health Problems in a Community," p. 2.

44. Lois Gibbs, *Love Canal: My Story* (Albany: State University of New York Press, 1982), p. 67.

45. EPA, *Environmental Protection Agency Support Document/Voluntary Environmental Impact Statement for Polychlorinated Biphenyls (PCBs) Manufacturing, Processing, Distribution in Commerce, and Use Ban Regulation (Section 6(e) of TSCA),* (Washington, D.C.: EPA, Office of Toxic Substances, Apr. 1979).

46. James N. Galloway, Gene E. Likens, and Mark E. Hawley, "Acid Precipitation: Natural versus Anthropogenic Components," *Science,* 226 (Nov. 16, 1984): 829.

47. See Gibbs, *Love Canal.*

48. Arthur S. Fleming, "Aminotriazole in Cranberries," testimony before the House Committee on Interstate and Foreign Commerce, 86th Congress, Jan. 26, 1960, vol. 2.

49. Charles F. Wurster, "For the Saccharin Ban," *New York Times,* Mar. 20, 1977, p. IV-17.

50. Society of the Plastics Industry, "Vinyl Chloride and Birth Defects," pp. 4–6.

51. See, e.g., Andrew J. Rogers, "Eagles, Affluence, and Pesticides," *Mosquito News,* 32 (June 1972): 151–157. See also Joseph Hickey, *A Guide to Bird Watching* (London: Oxford University Press, 1943), p. 349.

52. James W. Grier, "Ban of DDT and Subsequent Recovery of Reproduction in Bald Eagles," *Science,* 218 (Dec. 17, 1982): 1232–1234. The peregrine falcon also has shown signs of recovery: see Mark Wexler, "A Case of Urban Renewal," *National Wildlife,* June/July 1989, p. 12; and Robert Riseborough, "Pesticides and Bird Populations," *Current Ornithology,* vol. 3, ed. Richard F. Johnson (New York: Plenum Press, 1986), p. 412.

53. Alexander Sprunt, quoted in Joseph Hickey, ed., *Peregrine Falcon Populations* (Madison: University of Wisconsin Press, 1969), p. 349.

54. *Hungry Horse News,* West Glacier, Montana, Nov. 21, 1969, and Dec. 5, 1969. See also J. Gordon Edwards's testimony in *Hearings before the House Committee on Agriculture,* 92d Congress (1971), vol. 3, p. 579, where he cites a 1964 report from the Bureau of Sports Fisheries and Wildlife.

55. H. L. Needleman et al., "Deficits in Psychologic and Classroom Performance of Children with Elevated Dentine Lead Levels," *New England Journal of Medicine,* 300 (Mar. 29, 1979): 694.

56. H. L. Needleman et al., "The Relationship between Prenatal Exposure to Lead and Congenital Anomalies," *Journal of the American Medical Association,* 251

(June 8, 1984): 2956–2959, cited in Paul Mushak, "Prenatal and Postnatal Effect of Low-Level Lead Exposure: Integrated Summary of a Report to the U.S. Congress on Childhood Lead Poisoning," *Environmental Research,* 50 (1989): 11–36.

57. T. H. Jukes and C. B. Shaffer, "Antithyroid Effects of Aminotriazole," *Science,* 132 (July 29, 1960): 296.

58. International Agency for Research on Cancer [IARC], *IARC Monographs on the Evaluation of Carcinogenic Risk of Chemicals to Humans,* vol. 41 (Lyons, France: IARC, 1986), pp. 293–317.

59. Radomski et al., "Pesticide Concentrations in the Liver, Brain, and Adipose Tissue of Terminal Hospital Patients," *Food and Cosmetics Toxicology,* 6 (1968): 209–220.

60. See, e.g., David Laws et al., "Men with Intensive Occupational Exposure to DDT: A Clinical and Chemical Study," *Archives of Environmental Health,* 15 (Dec. 1967): 774; P. A. Neal et al., "Toxicity and Potential Dangers of Aerosols, Mists, and Dusting Powders Containing DDT," *Public Health Report,* 177 (1944): 1–32; Wayland Jackson Hayes, "Epidemiology of Pesticides," in E. Link and R. J. Whitaker, eds., *Proceedings of the Short Course on Occupational Health Aspects of Pesticides* (Norman: University of Oklahoma, 1964), pp. 109–130; Wayland Jackson Hayes, "Pesticides and Human Toxicity," *Annals of the New York Academy of Sciences,* 160 (1969): 40–54; Robert Ackerly, "DDT: A Re-Evaluation, Part I," *Chemical Times and Trends,* Oct. 1981, p. 48; Harold M. Schmeck, Jr., "Study Finds No Link between Cancer Risk and DDT Exposure," *New York Times,* Feb. 14, 1989, p. B7, where Schmeck cites a decade-long study of 919 people in which there was no statistically significant link found between the amount of DDT in a person's body and the risk of death by cancer.

61. See Aaron Wildavsky, *Searching for Safety* (New Brunswick, N.J.: Transaction Press, 1988); Ralph Keeney, "Mortality Risks Induced by Economic Expenditures," *Risk Analysis,* 10, no. 1 (1990): 147–159.

Conclusion

1. Charles E. Lindblom, "The Science of Muddling Through," *Public Administration Review,* 19 (Spring 1959): 79–88.

2. For example, Herbert Simon, "A Behavioral Model of Rational Choice," *Quarterly Journal of Economics,* 69 (Feb. 1955): 99–118; "Rationality As Process and As Product of Thought," *American Economic Review,* 68 (May 1978): 1–16.

3. John D. Graham, Laura C. Greene, and Marc J. Roberts, *In Search of Safety: Chemicals and Cancer Risk* (Cambridge, Mass.: Harvard University Press, 1988), p. 199.

4. They bolster their position in Lois Swirsky Gold, Bruce N. Ames et al., "Rodent Carcinogens: Setting Priorities," *Science,* 258 (Oct. 9, 1992): 261–265.

5. Matthew Wald, "As Science Gauges Perils in Life, to Learn More is to Know Less," *New York Times,* Aug. 19, 1991, p. A1.

6. Aaron Wildavsky, *Searching for Safety* (New Brunswick, N.J.: Transaction Press, 1988).

7. Ralph Keeney, "Mortality Risks Induced by Economic Expenditures," *Risk Analysis,* 10, no. 1 (1990): 147–159.

8. Ibid.

9. Gene E. Likens and Thomas J. Butler, "Recent Acidification of Precipitation in North America," *Atmospheric Environment,* 15 (1981): 1103.

10. United States National Acid Precipitation Assessment Program [NAPAP], *Acidic Deposition—State of Science and Technology: Summary Report of the U.S. National Acid Precipitation Program,* ed. Patricia M. Irving (Washington, D.C.: GPO, 1991), report no. 6, Douglas L. Sisterson, "Deposition Monitoring: Methods and Results," pp. 70–71.

11. See Chapter 9 for a discussion of increased sulfur deposition. See also Colin Hocking, Jacqueline Barber, and Jan Coonrod, *Acid Rain: Teacher's Guide* (Berkeley: University of California, Lawrence Hall of Science, 1990).

12. A. H. Johnson et al., "Recent Changes in Patterns of Tree Growth Rate in the New Jersey Pinelands: A Possible Effect of Acid Rain," *Journal of Environmental Quality,* 10 (1981): 427–430. See, for example, W. W. McFee, J. M. Kelly, and R. H. Beck, "Acid Precipitation Effects on Soil pH and Base Saturation of Exchange Sites," *Water, Air, and Soil Pollution,* 7 (1977): 401–408.

13. Joseph E. Barnard and Allen A. Lucier, "Change in Forest Health and Productivity in the United States and Canada," NAPAP, *Acidic Deposition,* p. 136.

14. See Table 10.1.

15. Richard Stolarski et al., "Measured Trends in Stratospheric Ozone," *Science,* 256 (1992): 342–349.

16. Hugh W. Ellsaesser, "The Politicizing of Climate Science," *21st Century,* May/June 1989, pp. 7–8.

17. Ibid.

18. Milton Russell, "Cleanup of Old Wastes: Is Superfund Broke?" *Risk Analysis,* 11, no. 1 (1991): 71–74.

19. Ibid., p. 72.

20. Ibid., pp. 72–73.

21. Ibid., p. 73.

22. Edward Groth III and Peter M. Sandman, "Alar and Moral Outrage; or, Why Consumers Are So Upset about a Little Bit of a Chemical in Apples," unpublished op-ed, April 1989. See also Peter M. Sandman, *Responding to Community Outrage: Strategies for Effective Risk Communication* (Fairfax, Va.: American Industrial Hygiene Association, 1993).

23. Mary Douglas and Aaron Wildavsky, *Risk and Culture* (Berkeley: University of California Press, 1982).

24. Lois Swirsky Gold et al., "Rodent Carcinogens: Setting Priorities," *Science,* 259 (Oct. 9, 1992): 261–165.

25. See Aaron Wildavsky, "The Social Construction of Distinctions: Risk, Rape, Public Goods, and Altruism," in Michael Hechter, Lynn Nadel, and Richard E. Michod, eds., *The Origin of Values* (New York: Aldine de Gruyter, 1993), pp. 47–61.

26. No mention is made of the high price of survival. Joseph made serfs of Egyptian peasants and moved them off their land. And the pharaoh whom Joseph strengthened enslaved his people later. See Aaron Wildavsky, *Assimilation versus Separa-*

tion: The Joseph Stories and the Politics of Religion in Biblical Israel (New Brunswick, N.J.: Transaction Publishers, 1993).

27. Stephen H. Schneider, *The Genesis Strategy* (New York: Plenum Press, 1977), p. xiii.
28. Ibid., p. 96.
29. Ibid., pp. 297–298.
30. Ibid., p. 95.
31. Ibid., p. 300.
32. Ibid., p. xv.
33. Ibid., p. 95.
34. Ibid.
35. Ibid., pp. 55–58.
36. Ibid., p. 300.
37. For a deep understanding of famine, see Amaryta Sen, *Poverty and Famines* (Oxford: Oxford University Press, 1981) and (with Jean Dreze) *Hunger and Public Action* (Oxford: Oxford University Press, 1989).
38. See Sen, *Poverty and Famines.*
39. Schneider, *Genesis Strategy,* p. xi.
40. Jonathan Schell, "Our Fragile Earth: Only Now Are We Beginning to Fathom the Gravity of Our Environmental Dilemma," *Discover,* 10, no. 10 (Oct. 1989): 47.
41. William Booth, *Washington Post,* June 21, 1989, p. A1.
42. I once called this a prophylactic society. See Sanford Weiner and Aaron Wildavsky, "The Prophylactic Presidency," *Public Interest,* 52 (Summer 1978): 3–19.

Index

chlorine oxide, 308, 309, 327–328
chlorobenzene, in groundwater, 175
chlorofluorocarbons, 304–339; costs versus
   benefits of, 204, 304, 336; ban, 304, 305,
   313, 316, 333, 335, 336; media coverage,
   305, 306, 322, 333; human health effects,
   305–306, 307, 331–332; and ozone hole,
   329–330, 437; regulation of, 333–338; and
   global warming, 364–365; lifetime in atmosphere, 520n19, 520n20
chloroform, 173
chlorophenols, 96
cholesterol, 405–406
Christy, John, 347
chromium, 179, 180–181
chromosome studies, 130; of Love Canal residents, 129–135; predictive value of, 134
chrysotile (white asbestos): properties and
   sources, 186, 197; health risks of, 191–
   192, 193, 194, 195
Chu, Kenneth, 263
Chubachi, N., 323
cigarette smoking: and cancer risk, 188, 191,
   193, 267, 379, 380; exposure to nitrites
   from, 244; control, 412–413
citizen: competent, 1–2, 6–7, 408–409; press
   distortion of statements of, 116–117; necessity of following good procedures, 151–
   152; choice of risks incurred, 219; methods
   for gaining scientific information, 395–409;
   how to read scientific literature, 396–397;
   "forget it" or "avoid it" responses, 398–
   399; activism, 400–402; querying experts,
   400–402, 407–408; role of skepticism of,
   425–426
Citizens' Clearinghouse for Hazardous
   Wastes, 148
Claus, George, 76, 460
Clean Air Act, 8, 158, 184, 277
cleanup operations: costs, 45, 50–51, 114,
   123, 133, 153–154, 172, 174, 177, 179,
   181, 183, 188, 190, 222, 253, 433, 434,
   458; of PCB-contaminated soil, 45, 51, 52;
   standards for, 45, 49, 50; most expensive,
   49–50; after office building fires, 49–51; of
   hazardous wastes, 114; at Love Canal, 133;
   costs versus benefits, 184, 433; futility of,
   435–436; limited water cleanup technology,
   486n120. *See also* Superfund program
Clean Water Act, 127, 158, 168, 181
cloud feedback, 350, 358–360, 364, 365–
   366, 369, 525n107

coal burning, 276, 277
cobweb method, of literature search, 397
coconut scale, 76
coffee, 266, 441
Cogliano, Vincent, 264
Cohen, Richard, 146
Cohen, Samuel, 264
cohort, defined, 193
cohort study, 143, 160, 193
coincidence, 404–405
Commission on Pesticides and Their Relation
   to Environmental Health, 89
Committee for Environmental Information,
   91
Committee on Nitrite and Alternative Curing
   Agents in Food, 243–244
Commonweal, cranberry scare coverage, 17–
   18
Community Nutrition Institute, 241, 242
Comprehensive Environmental Response,
   Compensation, and Liability Act, 114,
   154–155
comprehensiveness, 429–430
Conestoga Rovers, 128
confounding factors, 161–162, 253
Congress, 97, 205; beliefs about public health
   hazards, 157, 158–159
Congressional Research Service report, 159
Conklin, H. J., 74
Consumer Reports, 203
containment on-site cleanup procedure, 171
contaminant release analysis, 165
contaminant transport analysis, 165–166
context-dropping, fallacy of, 351
continuity equation, 279–280, 511n40
controls: importance of, 151, 240, 411–415,
   417; in epidemiological studies, 161–162;
   and test design, 240; in animal studies,
   256, 263; for age, 412; for cigarette
   smoking, 412–413; for gender, 413–414;
   for preexisting conditions, 414–415
convection, and global temperature trends,
   352
Cooke, W. E., 187
Coon, Julius M.
Cornfield, Richard S., 399, 400
Costle, Douglas, 95–96
costs versus benefits: disregard of, 16; of
   cleanup operations, 158, 184, 433; of regulation, 158, 428–429, 430, 441, 486n120;
   of asbestos removal, 188, 190, 197–198; of
   CFCs ban, 204, 304, 336; of nitrites ban,

2,4,5-trichlorophenoxyacetic acid (2,4,5-T):
annual usage, 82; dioxin contamination,
85, 89–90; teratogenicity, 86, 88, 90; use
restrictions, 89; media coverage, 91,
473n148; ban, 95–96; and miscarriages,
95–96. *See also* Agent Orange
TSCA. *See* Toxic Substances Control Act
Tschirley, Fred S., 88, 89
Tseng, W. P., arsenic study, 225–226, 230–
231, 232–234, 246
tuning, 351
tussock moth infestation, 58, 74–75
20/20, 94
typhus fever, 56

UDMH. *See* 1,1-dimethylhydrazine
ultraviolet radiation: human health effects,
306–307, 319–320, 380, 382, 524n70; and
melanoma, 307, 331–332; eleven-year cycle
of, 316, 317–318, 523n65; erythemal
weighting of, 319; effects on plants and animals, 332–333; increases in, 437; proxy
for, 522n59
uncertainty: in risk assessment and site analysis, 166, 215, 219; and decision making,
429. *See also* safety factor
Underhill, Dwight, 161
UNECE. *See* United Nations Economic Commission for Europe
Unger, Manya, 203
Union Carbide Corporation, 175–177
Union of Concerned Scientists, 345
Uniroyal Chemical Company, 99, 202, 205,
207, 210, 212–214, 217, 472n115,
496nn61,66,73
United Nations: herbicide study, 88; Alar
study, 206, 221
United Nations Economic Commission for
Europe, forest growth studies, 294–295
Universities Associated for Research and Edu-
cation in Pathology, 160, 243
urethane, 239, 242
U.S. Catholic Conference, 196
U.S. Department of Agriculture (USDA): reim-
bursement to cranberry farmers, 11; and
DDT, 57; and Alar, 204, 205; and nitrites,
237, 241–242; pest control vs. environmental protection, 452n10
U.S. Department of Defense: Vietnam defolia-
tion program, 87, 88; suspension of use of
2,4,5-T, 90; and Love Canal, 148
U.S. Department of Health, Education, and

Welfare, 11; and Love Canal, 148; and
asbestos, 195; and cranberry scare, 249
U.S. Department of Health and Human Services, Love Canal study, 149–150
U.S. Department of Justice, nitrites-related
actions, 241
U.S. Fish and Wildlife Service, 178
U.S. Forest Service, 74, 75
U.S. News and World Report, 33; nitrites coverage, 241, 245
utility companies, against PCB regulations,
45, 47

VA. *See* Veterans Administration
Valentine, Jane L., arsenic study, 229
valid laboratory study, 494n43
"valley of the drums," 155
vegetation, role in global warming, 369–370
ventilation, 38
Veterans Administration, and Agent Orange,
93, 94, 97
Vianna, Nicholas, 135–139, 148
Vickers, Carol, 117
video display terminals, 404, 407
Vietnam, herbicide use in, 84–86
Vietnam veterans: dioxin-affected, 81; cancers
of, 92–98, 105–106; CDC epidemiological
studies, 98, 100, 103–106; and birth
defects, 100–101; dioxin levels in serum of,
103–104. *See also* Operation Ranch Hand
Village Voice, Love Canal coverage, 144
vinyl chloride, 168, 173–174, 182, 412, 423
virus X, 56
volcanic eruptions, and ozone depletion,
309–310, 330
Voluntary Asbestos Survey Report, 190
Vostok ice core, 355–356

Wagner, J. C., 187
Wald, Matthew, 432
Walker, Knox, 73
Wall Street Journal, dioxin coverage, 123
Warburton, Dorothy, 138
WARF. *See* Wisconsin Alumni Research Foun-
dation
Warner, John, 205
Warner, Stephanie, 402–403
Washington, Warren M., 352
Washington Post: PCBs coverage, 54; defolia-
tion program coverage, 91; Times Beach
coverage, 116; Love Canal coverage, 144;
nitrites coverage, 241, 245; acid rain cov-